Helicobacter pylori

Molecular Genetics and Cellular Biology

Edited by Yoshio Yamaoka

Department of Medicine–Gastroenterology
Baylor College of Medicine and Michael E. DeBakey Veterans Affairs Medical Center
Houston, TX
USA

Caister Academic Press

Caister Academic Press
Norfolk, UK

www.caister.com

British Library Cataloguing-in-Publication Data
A catalogue record for this book is available from the British Library

ISBN: 978-1-904455-31-8

Printed and bound in Great Britain

Contents

Contributors

Richard A. Alm
Infection Discovery
AstraZeneca R & D Boston
Waltham, MA
USA

Richard.Alm@astrazeneca.com

Clara Belzer
Department of Gastroenterology and Hepatology
Erasmus MC-University Medical Center
Rotterdam
The Netherlands

belzer@microbes.nl

Steven R. Blanke
University of Illinois
Urbana, IL
USA

sblanke@life.uiuc.edu

Shyh-Horng Chiou
Institute of Biochemical Science
National Taiwan University
Taipei
Taiwan

shchiou@ntn.edu.tw

Lu-Ping Chow
Graduate institute of Biochemistry and Molecular Biology
College of Medicine
National Taiwan University
Taipei
Taiwan

luping@ha.mc.ntu.edu.tw

Timothy L. Cover
Vanderbilt University School of Medicine and
Veterans Affairs Tennessee Valley Healthcare System
Nashville, TN
USA

timothy.L.cover@vanderbilt.edu

Wolfgang Fischer
Max von Pettenkofer-Institut für Hygiene und
Medizinische Mikrobiologie
Ludwig-Maximilians-Universität
Pettenkoferstr 9A
D-80336 Munich
Germany

fischer@mvp.uni-muenchen.de

Rainer Haas
Max von Pettenkofer-Institut für Hygiene und
Medizinische Mikrobiologie
Ludwig-Maximilians-Universität
Pettenkoferstr 9A
D-80336 Munich
Germany

haas@mvp.uni-muenchen.de

Olivier Humbert
Division of Human Biology and Molecular and Cellular Biology
Fred Hutchinson Cancer Research Center
Seattle, WA
USA

olive@u.washington.edu

Luisa F. Jimenez-Soto
Max von Pettenkofer-Institut für Hygiene und
Medizinische Mikrobiologie
Ludwig-Maximilians-Universität
Munich
Germany

Jimenez@m3401.mpk.med.uni-muenchen.de

Christine Josenhans
Institute for Medical Microbiology
Medizinische Hochschule Hannover
Hannover
Germany

Josenhans.Christine@MH-Hannover.de

Arno Karnholz
OXOID Deutschland GmbH
Thermo Fisher Scientific
Munich
Germany

arno.karnholz@thermofisher.com

Jaw-Town Lin
Division of Gastroenterology
Department of Internal Medicine
National Taiwan University Hospital
Taipei
Taiwan

jawtown@ha.mc.ntu.edu.tw

Bodo Linz
Department of Molecular Biology
Max-Plank Institute for Infection Biology
Berlin
Germany

linz@mpiib-berlin.mpg.de

Anthony P. Moran
Department of Microbiology
School of Natural Sciences
National University of Ireland – Galway
Galway
Ireland
and
Deparment of Microbiology
Griffith University
Queensland
Australia

anthony.moran@nuigalway.ie

Teruko Nakazawa
Yamaguchi University School of Medicine
Ube Yamaguchi
Japan

nakazawa@yamaguchi-u.ac.jp

Delia M. Pinto-Santini
Division of Human Biology and Molecular and
Cellular Biology
Fred Hutchinson Cancer Research Center
Seattle, WA
USA

dmps@u.washington.edu

Melanie Rust
Institute for Medical Microbiology
Medizinische Hochschule Hannover
Hannover
Germany

Rust.Melanie@mh-hannover.de

George Sachs
Departments of Physiology and Medicine
David Geffen School of Medicine at UCLA
Los Angeles, CA
USA

gsachs@ucla.edu

Nina Salama
Division of Human Biology and Molecular and
Cellular Biology
Fred Hutchinson Cancer Research Center
Seattle, WA
USA

nsalama@fhcrc.org

Stephan C. Schuster
Center for Comparative Genomics and
Bioinformatics,
Center for Infectious Disease Dynamics
Penn State University
University Park, PA
USA

scs@bx.psu.edu

Tobias Schweinitzer
Institute for Medical Microbiology
Medizinische Hochschule Hannover
Hannover
Germany

Schweinitzer.Tobias@mh-hannover.de

David R. Scott
Department of Physiology
David Geffen School of Medicine at UCLA
Los Angeles, CA
USA

dscott@ucla.edu

Jeroen Stoof
Department of Gastroenterology and Hepatology
Erasmus MC-University Medical Center
Rotterdam
The Netherlands

j.stoof@erasmusmc.nl

Hiroaki Takeuchi
Department of Clinical Laboratory Medicine
Kochi University School of Medicine
Kochi
Japan

htake@kochi-u.ac.jp

M. Stephen Trent
Department of Biochemistry and Molecular Biology
Medical College of Georgia
Augusta, GA
USA

strent@mcg.edu

Arnoud H.M. van Vliet
Institute of Food Research
Norwich
United Kingdom

arnoud.vanvliet@bbsrc.ac.uk

Yi Wen
Department of Physiology
David Geffen School of Medicine at UCLA
Los Angeles, CA
USA

ywen@ucla.edu

Nicola E. Wittekindt
Center for Comparative Genomics and Bioinformatics
Center for Infectious Disease Dynamics
Penn State University
University Park, PA
USA

new12@psu.edu

Ming-Shiang Wu
Division of Gastroenterology
Department of Internal Medicine
National Taiwan University Hospital
Taipei
Taiwan

mingshiang@ntu.edu.tw

Yoshio Yamaoka
Department of Medicine–Gastroenterology
Baylor College of Medicine and Michael E. DeBakey Veterans Affairs Medical Center
Houston, TX
USA

yyamaoka@bcm.tmc.edu

Other Books of Interest

Clostridia: Molecular Biology in the Post-genomic Era	2009
Plant Pathogenic Bacteria: Genomics and Molecular Biology	2009
Microbial Production of Biopolymers and Polymer Precursors	2009
Plasmids: Current Research and Future Trends	2008
Vibrio cholerae: Genomics and Molecular Biology	2008
Pathogenic Fungi: Insights in Molecular Biology	2008
Corynebacteria: Genomics and Molecular Biology	2008
Leishmania: After The Genome	2008
Archaea: New Models for Prokaryotic Biology	2008
RNA and the Regulation of Gene Expression	2008
Legionella Molecular Microbiology	2008
Molecular Oral Microbiology	2008
Epigenetics	2008
Animal Viruses: Molecular Biology	2008
Segmented Double-Stranded RNA Viruses	2008
Acinetobacter Molecular Microbiology	2008
Pseudomonas: Genomics and Molecular Biology	2008
Microbial Biodegradation: Genomics and Molecular Biology	2008
The Cyanobacteria: Molecular Biology, Genomics and Evolution	2008
Coronaviruses: Molecular and Cellular Biology	2007
Real-Time PCR in Microbiology: From Diagnosis to Characterisation	2007
Bacteriophage: Genetics and Molecular Biology	2007
Candida: Comparative and Functional Genomics	2007
Bacillus: Cellular and Molecular Biology	2007
AIDS Vaccine Development: Challenges and Opportunities	2007
Alpha Herpesviruses: Molecular and Cellular Biology	2007
Pathogenic *Treponema*: Molecular and Cellular Biology	2007
PCR Troubleshooting: The Essential Guide	2006
Influenza Virology: Current Topics	2006
Microbial Subversion of Immunity: Current Topics	2006
Cytomegaloviruses: Molecular Biology and Immunology	2006
Papillomavirus Research: From Natural History To Vaccines and Beyond	2006
Epstein Barr Virus	2005
HIV Chemotherapy: A Critical Review	2005

Preface

Helicobacter pylori is an important pathogen that infects up to 50% of the human population. As the leading cause of peptic ulcers and gastric cancer worldwide, the organism has been the subject of intensive research to unravel the mysteries of its genetics and cellular biology. In fact, the number of publications in this field has risen dramatically in recent years, making it extremely difficult for even the most diligent reader to stay abreast of progress. In this book, internationally recognized scientists summarize the current state-of-the-art knowledge about molecular genetics and cellular biology of *H. pylori*.

Numerous books on *H. pylori* have already been published. However, many are now out of date. For example, three complete *H. pylori* genomes have now been published. Following the full sequencing and annotation of the genomes of *H. pylori* 26695 and J99 in 1997 and 1999, respectively, it was more than another 7 years before the genome of a third *H. pylori* (strain HPAG1) was published, and there are no books about *H. pylori* taking all three *H. pylori* strains into account. I was therefore very keen to produce a book that presents the topics in the context of comparison of the three genomes, and this is one of the main reasons that I agreed to edit this new book focused on molecular aspects of *Helicobacter* when invited to do so by Annette Griffin of Horizon Scientific Press. I am very grateful to her for her support and encouragement throughout the course of the project.

The International Workshop on Gastroduodenal Pathology and *Helicobacter* is a very important meeting for discussion of the important topics in *Helicobacter* research, and my plan for the contents of this book was drawn up at this meeting in Wroclaw, Poland, in 2006. As a result, the topics I selected include lipopolysaccharides, outer membrane proteins, motility and chemotaxis, vacuolating toxin, type IV secretion systems, acid adaptation, metal metabolism, cell division, molecular mechanisms of host adaptation, genomotyping and proteonomics. Although the title of this book is *Helicobacter pylori*, it also focuses on the genetic aspects of other *Helicobacter* species (e.g. detailed analyses of *Helicobacter acinonychis*). I contacted very active leading international *Helicobacter* researchers in each field, and, to my delight, all the authors I approached provided a great contribution. I am very grateful to all the authors for their efforts. Although the deadlines I requested were somewhat tight, essentially all the researchers kept to the schedules, and this book will have been published very quickly after receipt of the final chapter. Thus, many recent publications in 2007 and even papers in press are included.

I believe that this book distils the most important cutting-edge findings in the field to produce a timely and comprehensive review. Providing a useful introduction to the subject for new researchers as well as an invaluable reference for experienced researchers, this book is essential reading for all researchers working with *Helicobacter* and related organisms. I hope this book will be the 'bible' of *Helicobacter* research.

Yoshio Yamaoka, Houston, Texas
March 2008

Helicobacter: Overview

1

Yoshio Yamaoka

Abstract

After the modern discovery of *Helicobacter* approximately 25 years ago by Robin Warren and Barry J. Marshall, many scientists developed an interest in the unique characteristics of *Helicobacter*, especially *H. pylori*, a bacterium proven to be the cause of gastritis and peptic ulcer and the progression to gastric cancer. *H. pylori* has very unique characteristics, such as microaerophily and nitrogen metabolism; thus, *H. pylori* has been the subject of intensive research to unravel the mysteries of its genetics and cellular biology. In fact, the number of publications in this field has risen dramatically in recent years making it extremely difficult for even the most diligent reader to stay abreast of progress. This book distils the most important cutting-edge findings in the field to produce a timely and comprehensive review. In this chapter, we described a brief background of the field including the discovery of *Helicobacter* and the relationship between *H. pylori* and gastroduodenal diseases. We then briefly summarize each chapter.

Introduction

Helicobacter are Gram-negative spiral-shaped bacteria, initially considered to be members of the *Campylobacter* genus (*Campylobacter pyloridis* [*pylori*]: see below); however, they were grouped into a separate genus in 1989 (Goodwin *et al.*, 1989). Some species have been found living in the lining of the upper gastrointestinal tract, as well as the liver of mammals and some birds. The most widely known *Helicobacter* species is *H. pylori*.

After the modern discovery of *Helicobacter* approximately 25 years ago, many scientists with special expertise from diverse fields developed an interest in the unique characteristics of *Helicobacter*. As of July 2007, the PubMed database contained more than 26,000 entries for '*Helicobacter*', '*Campylobacter pyloridis*' or '*Campylobacter pylori*' (Fig. 1.1). This number is more than twice the number of entries for '*Shigella*' and '*Yersinia*' (approximately 12,000 for both). Since 2000, there have been more articles published on '*Helicobacter*' than on '*Salmonella*'

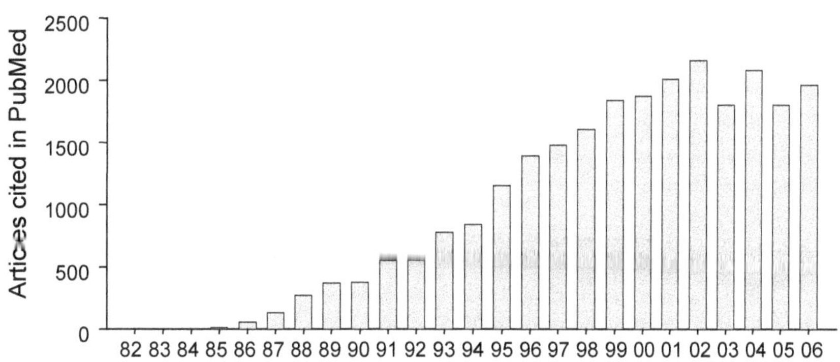

Figure 1.1 The number of articles related to *Helicobacter* in PubMed. The PubMed database was searched for '*Helicobacter*', '*Campylobacter pyloridis*' or '*Campylobacter pylori*'.

(approximately 14,600 vs. 13,400). In other words, an average of 2000 articles has been published each year on *Helicobacter* since 2000. The quantity of research in the *Helicobacter* field, especially the molecular genetics of *Helicobacter*, has increased continuously over the past decade (e.g. 20 articles related to *Helicobacter* have been published in the top journal *Nature*, with 19 published in 1997 or later; 26 of 33 articles related to *Helicobacter* published in *Science* also have been published since 1997).

In this overview, we begin with a brief background of the field including the discovery of *Helicobacter* and the relationship between *H. pylori* and gastroduodenal diseases. We then briefly summarize each chapter.

Discovery of *Helicobacter*

Although the modern discovery of intragastric bacteria occurred approximately 25 years ago, the initial reports about 'spiral-shaped' gastric bacteria were published more than 100 years ago. The first well-known report of gastric *Helicobacter* was in 1893 (Bizzozero, 1893). Bizzozero reported 'spirochaetes' inhabiting the gastric glands of dogs, which now would be classified as *Helicobacter canis*, *Helicobacter felis* and/or *Helicobacter heilmannii*. In the twentieth century, anatomists and pathologists gradually noted spiral organisms in the human gastric mucosa; however, in 1954 Palmer denied the existence of microorganisms in the human stomach by examining more than 1000 gastric biopsies taken with a blind suction biopsy instrument (Palmer, 1954). It was concluded that microorganisms could not survive in the acidic gastric environment. Subsequently, the interest of investigators (especially anatomists and pathologists) in the spiral bacterium gradually disappeared. However, investigators studying peptic ulcer disease generally were unaware of the colonization of the stomach by these spiral bacteria, and few considered that gastroduodenal diseases could result from bacterial infection.

The bacterium was rediscovered in 1979 by Australian pathologist Robin Warren, who collaborated with Barry J. Marshall beginning in 1981; they isolated the organisms from human gastric mucosal specimens and were the first to accomplish *in vitro* culture of the organism (Warren and Marshall, 1983). They subsequently reported that most peptic ulcers and gastritis were caused by infection with this bacterium rather than stress or spicy food as had been assumed before (Marshall and Warren, 1984). The medical community was slow to recognize the role of this bacterium in gastroduodenal diseases; however, Marshall drank a suspension of the bacterium, developed acute gastritis, the bacteria were recovered from his gastric mucosa, and was able to treat himself using a fourteen day dual therapy with bismuth salts and metronidazole, satisfying Koch's postulates ((Marshall *et al.*, 1985). In recognition of their very important discovery, Marshall and Warren were awarded the 2005 Nobel Prize for Medicine and Physiology.

When a spiral bacterium was cultured by Marshall, it was classified as *Campylobacter pyloridis*, which shortly after was changed to *C. pylori* to conform to the Latin grammar. In 1989, after DNA sequencing of 16S rRNA and other data (e.g. flagellum morphology, fatty acid content) showed that the bacterium did not belong in the *Campylobacter* genus, it was transferred to a new genus, *Helicobacter* (Goodwin *et al.*, 1989). Since the last decade, the investigation of the genus *Helicobacter* has expanded tremendously and new species are regularly added (Whary and Fox, 2004).

Helicobacter pylori and gastroduodenal diseases

H. pylori causes a serious chronic transmissible infectious disease that damages gastric structure and function and is recognized as the causative agent in gastric atrophy, peptic ulcer disease, gastric adenocarcinoma and primary gastric B-cell lymphoma. The life-time risk for an *H. pylori*-infected individual to develop peptic ulcer disease has been estimated at one in six and 10–25% of those with peptic ulcers are expected to develop complications such as gastrointestinal haemorrhage or perforation. In 1994, the National Institutes of Health published an opinion stating that peptic ulcer patients with *H. pylori* infection require treatment with antimicrobial agents in addition to antisecretory drugs either on first presentation with the illness or upon recurrence (http://consensus.nih.gov/1994/1994HelicobacterPyloriUlcer094html.htm).

Gastric cancer is the second leading cause of cancer worldwide. It has been estimated that *H. pylori* infection increases the risk of gastric cancer approximately 10-fold, and the World Health Organization's (WHO) International Agency for Research into Cancer (IARC) has classified *H. pylori* as a 'class I carcinogen' i.e. in the same category as cigarette smoking is to cancer of the lung and respiratory tract. The lifetime risk of developing gastric cancer is related to the prevalence of *H. pylori*-associated atrophic gastritis and in high risk countries (such as Changle, China) is as high as 19.6% versus as low as 1.5% in the United Kingdom (Parkin, 2006). These differences reflect differences in *H. pylori* prevalence, in strain virulence, and in host and particularly, environmental factors. Koch's postulates also have been fulfilled for gastric cancer in the animal model (Mongolian gerbil) (Franco *et al.*, 2005; Watanabe *et al.*, 1998).

Physiology and molecular biology

H. pylori has very unique characteristics, such as microaerophily and nitrogen metabolism; therefore, the physiology and molecular biology of *H. pylori* is of wide interest for researchers in many fields in addition to microbiology, but also for molecular biology as well as clinical gastroenterology. Specifically, the cell envelope components are the most important since they are the first points of contact between bacterial factors and the host. In this book, we focus on the important cell envelope components (i.e. Chapter 2, Lipopolysaccharides, by Moran and Trent; Chapter 3, Outer Membrane Proteins, by Yamaoka and Alm; Chapter 4, Flagella, by Rust *et al.*). Many surface proteins of pathogenic bacteria are thought to have roles in adhesion, colonization and the immune response. Some outer membrane proteins also perform transport functions essential for metabolism. In particular, the adherence of *H. pylori* to the gastric mucosa plays an important role in initial colonization and long-term persistence in the human gastric mucosa. *Helicobacter* has specific characteristics with respect to the physiology and molecular biology of these components, and these chapters will review the current knowledge of cell envelope components of *Helicobacter*. The β-barrel domain of the vacuolating cytotoxin (VacA) and type IV secretion components and also are categorized into cell envelope components; and these are believed to be the major virulence factors of *H. pylori*. Therefore, we focus on VacA (Chapter 5 by Blanke and Cover) and the type IV secretion systems (Chapter 6 by Fischer *et al.*). In the type IV secretion system in *H. pylori*, the *cag* pathogenicity island is well known and has been the subject of many reviews. However, it is important to note that three type IV secretion systems have been identified in *H. pylori*; Cag, the ComB, and Tfs3 systems. Therefore, the review in this book summarizes recent progress in our understanding of the structure, function and regulation of the different type IV secretion systems in *H. pylori*.

H. pylori colonizes the highly acidic environment of the gastric mucosa. Specifically, *H. pylori* is able to flourish in its acidic gastric niche by buffering its periplasm to near neutrality using the mechanism of acid acclimation. This unique acid acclimation, in contrast to acid resistance or tolerance expressed by other neutralophiles, is dependent on urease activity; therefore, it is believed that urease activity plays a central role in survival of *H. pylori* in the stomach; acid acclimation also is necessary for gastric colonization, although it may not be sufficient. The longevity of gastric colonization suggests that *H. pylori* must be able to cope with the different stresses encountered in the gastric mucosa; not only the acidic environment but also the restriction of essential metals like iron and nickel. Sacks *et al.* (Chapter 7) and Stoof *et al.* (Chapter 8) review the current knowledge on the role of acid adaptation and metal metabolism in the physiology of *H. pylori*, respectively. They also discuss the contribution of the regulatory mechanisms involved in controlling acid acclimation and metal metabolism in the adaptation of *H. pylori* to its gastric niche.

Finally, chromosome replication and cell division are essential steps for bacterial proliferation. The complete genome sequences of three strains of *H. pylori* revealed that homologues of most genes involved in replication and segregation are present, suggesting that the basic mechanisms are common. However, some orthologues for cell division are missing in *H. pylori* possibly because of its unique niche in gastric mucus. Analysis of the replication machinery of *H. pylori* has begun only recently; in Chapter 9, Nakazawa and Takeuchi provide the latest information of the mechanisms for chromosome replication and cell division in *H. pylori*.

Helicobacter genomes

After *Helicobacter* was recognized as a human pathogen that infects more than 50% of human population, it became a logical target for genome analysis. Currently, three complete *H. pylori* genomes (Alm *et al.*, 1999; Oh *et al.*, 2006; Tomb *et al.*, 1997), and two complete genomes of species *Helicobacter hepaticus* and *Helicobacter acinonychis* have been published (Suerbaum *et al.*, 2003; Eppinger *et al.*, 2006). In Chapter 10, Schuster *et al.* present the updated review of genomic analyses. The molecular mechanisms that drive the genomic evolution of *Helicobacter* in its host have enabled the enormous plasticity of the bacterium's genome. Importantly, a six-step model for host adaptation that also includes *Helicobacter*'s ability for horizontal gene transfer in combination with its natural competence for DNA uptake is proposed.

After the full sequencing and annotation of the genomes of *H. pylori* 26695 and J99 in 1997 and 1999, respectively, it took more than seven years for publication of a third *H. pylori* genome (strain HPAG1); therefore, there are no previous books about *H. pylori* taking the three *H. pylori* strains into account. With this book, we are able to present the topics in the context of comparison between the three genomes. For example, comparison of the outer membrane proteins of three genomes showed that *H. pylori* 26695 and HPAG1 proteins are the most closely related versus *H. pylori* J99 and HPAG1 and as *H. pylori* J99 and 26695 (see Chapter 3). It is unknown whether this is because both HPAG1 and 26695 came from gastritis patients and J99 from a duodenal ulcer patient; however, these results indicate that comparative genomic analyses should provide new insights into the relationship between *H. pylori* infection and gastroduodenal diseases. Furthermore, post-genomic experimental techniques (i.e. microarray and proteomics) gradually are being integrated in *Helicobacter* research. In this book, Humbert *et al.* (Chapter 11) and Wu *et al.* (Chapter 12) discuss current state-of-the-art review on microarray and proteomics, respectively.

Summary

Approximately 25 years have passed since the modern discovery of *Helicobacter*, and a tremendous number of studies have been performed on *Helicobacter*. The quantity of research on the molecular genetics of *H. pylori*, a bacterium proven to be the cause of gastritis and peptic ulcer and the progression to gastric cancer, has increased continuously over the past decade. In this book, internationally recognized scientists who have made the important discoveries in the field summarize the current state-of-the-art knowledge about the molecular genetics of *Helicobacter*.

Finally, we recommend checking website for European *Helicobacter* Study Group (www.helicobacter.org) which contains many useful information as well as useful links related to *Helicobacter* research (e.g. www.infek.lu.se/bakt/english/helicobacter: *Helicobacter* species at a glance).

References

Alm, R.A., Ling, L.S., Moir, D.T., King, B.L., Brown, E.D., Doig, P.C., Smith, D.R., Noonan, B., Guild, B.C., deJonge, B.L., Carmel, G., Tummino, P.J., Caruso, A., Uria-Nickelsen, M., Mills, D.M., Ives, C., Gibson, R., Merberg, D., Mills, S.D., Jiang, Q., Taylor, D.E., Vovis, G.F., and Trust, T.J. (1999). Genomic-sequence comparison of two unrelated isolates of the human gastric pathogen *Helicobacter pylori*. Nature *397*, 176–180.

Bizzozero, G. (1893). Ueber die schlauchformigen Dursen des Magendarmkanals und die Beziehungen ihres Epithels zu dem Oberflachenepithel der Schleimhaut. Arch. Mikr. Anat. *42*, 82–152.

Eppinger, M., Baar, C., Linz, B., Raddatz, G., Lanz, C., Keller, H., Morelli, G., Gressmann, H., Achtman, M., and Schuster, S.C. (2006). Who ate whom? Adaptive *Helicobacter* genomic changes that accompanied a host jump from early humans to large felines. PLoS Genet. 2, e120.

Franco, A.T., Israel, D.A., Washington, M.K., Krishna, U., Fox, J.G., Rogers, A.B., Neish, A.S., Collier-Hyams, L., Perez-Perez, G.I., Hatakeyama, M., Whitehead, R., Gaus, K., O'Brien, D.P., Romero-Gallo, J., and Peek, R.M., Jr. (2005). Activation of beta-catenin by carcinogenic *Helicobacter pylori*. Proc. Natl. Acad. Sci. USA *102*, 10646–10651.

Goodwin, C.S., Armstrong, J.A., Chilvers, T., Peters, M., Collins, D., Sly, L., McConnell, W., and Harper, W.S. (1989). Transfer of *Campylobacter pylori* and *Campylobacter mustelae* to *Helicobacter* gen. nov. as

Helicobacter pylori comb.nov. and *Helicobacter* mustelae comb.nov., respectively. Int. J. Syst. Bacteriol. *4*, 397–405.

Marshall, B.J., Armstrong, J.A., McGechie, D.B., and Glancy, R.J. (1985). Attempt to fulfil Koch's postulates for pyloric *Campylobacter*. Med. J. Aust. *142*, 436–439.

Marshall, B.J. and Warren, J.R. (1984). Unidentified curved bacilli in the stomach of patients with gastritis and peptic ulceration. Lancet *1*, 1311–1315.

Oh, J.D., Kling-Backhed, H., Giannakis, M., Xu, J., Fulton, R.S., Fulton, L.A., Cordum, H.S., Wang, C., Elliott, G., Edwards, J., Mardis, E.R., Engstrand, L.G., and Gordon, J.I. (2006). The complete genome sequence of a chronic atrophic gastritis *Helicobacter pylori* strain: evolution during disease progression. Proc. Natl. Acad. Sci. USA *103*, 9999–10004.

Palmer, E.D. (1954). Investigation of the gastric mucosa spirochetes of the human. Gastroenterology *27*, 218–220.

Parkin, D.M. (2006). The global health burden of infection-associated cancers in the year 2002. Int. J. Cancer *118*, 3030–3044.

Suerbaum, S., Josenhans, C., Sterzenbach, T., Drescher, B., Brandt, P., Bell, M., Droge, M., Fartmann, B., Fischer, H.P., Ge, Z., Horster, A., Holland, R., Klein, K., Konig, J., Macko, L., Mendz, G.L., Nyakatura, G., Schauer, D.B., Shen, Z., Weber, J., Frosch, M., and Fox, J.G. (2003). The complete genome sequence of the carcinogenic bacterium *Helicobacter hepaticus*. Proc. Natl. Acad. Sci. USA *100*, 7901–7906.

Tomb, J.F., White, O., Kerlavage, A.R., Clayton, R.A., Sutton, G.G., Fleischmann, R.D., Ketchum, K.A., Klenk, H.P., Gill, S., Dougherty, B.A., Nelson, K., Quackenbush, J., Zhou, L., Kirkness, E.F., Peterson, S., Loftus, B., Richardson, D., Dodson, R., Khalak, H.G., Glodek, A., McKenney, K., Fitzegerald, L.M., Lee, N., Adams, M.D., Hickey, E.K., Berg, D.E., Gocayne, J.D., Utterback, T.R., Peterson, J.D., Kelley, J.M., Cotton, M.D., Weidman, J.M., Fujii, C., Bowman, C., Watthey, L., Wallin, E., Hayes, W.S., Borodovsky, M., Karp, P.D., Smith, H.O., Fraser, C.M., and Venter, J.C. (1997). The complete genome sequence of the gastric pathogen *Helicobacter pylori*. Nature *388*, 539–547.

Warren, J.R. and Marshall, B.J. (1983). Unidentified curved bacilli on gastric epithelium in active gastritis. Lancet *1*, 1273–1275.

Watanabe, T., Tada, M., Nagai, H., Sasaki, S., and Nakao, M. (1998). *Helicobacter pylori* infection induces gastric cancer in Mongolian gerbils. Gastroenterology *115*, 642–648.

Whary, M.T. and Fox, J.G. (2004). Natural and experimental *Helicobacter* infections. Comp. Med. *54*, 128–158.

Helicobacter pylori Lipopolysaccharides and Lewis Antigens

2

Anthony P. Moran and M. Stephen Trent

Abstract

The outer membrane of *Helicobacter pylori*, like other Gram-negative bacteria, contains lipopolysaccharides (LPS), which are important for the structure of the bacterial cell envelope and the interaction of the bacterium with its environment. They are a family of phosphorylated lipoglycans, composed of a lipid moiety, termed lipid A, a core oligosaccharide and an O-chain polysaccharide. The present chapter reviews the main biological attributes of *H. pylori* LPS and the understanding that has been gained into the molecular genetics, structure and contributing properties of this class of molecule to *H. pylori* pathogenesis. Compared with LPSs of other bacteria, LPS of *H. pylori* has low immunoactivities, the molecular basis for which is the underphosphorylation and unusual acylation pattern of the lipid A component that interacts with immune receptors. Important insights have been gained into the biosynthesis and modification pathways that remodel *H. pylori* lipid A and hence aid persistence of the bacterium in the gastric mucosa. Notwithstanding this, the LPS core exhibits structural properties (i.e. laminin binding and pepsinogen induction) the molecular variation of which may influence the virulence of *H. pylori* strains. Furthermore, the O-chains of most strains, though not all isolates (i.e. those associated with asymptomatic infection), mimic Lewis and related blood group antigens. The nature of these antigens in *H. pylori*, their genetic determination and regulation of expression, particularly of the fucosyltransferases required for synthesis, as well as the biological consequences of this mimicry have provided a better understanding of the molecular pathogenesis of *H. pylori*.

Introduction

Lipopolysaccharide (LPS) molecules are important pathogenic and virulence factors of *Helicobacter pylori* with properties contributing to the severity and chronicity of infection (Moran, 2007). These molecules are a family of phosphorylated lipoglycans found in the outer membrane of Gram-negative bacteria, generally possessing potent immunomodulating and immunostimulating properties (Rietschel *et al.*, 1994; Raetz and Whitfield, 2002). Owing to their ability to adversely affect a broad range of immunological and physiological properties of the host, these molecules have been considered toxic and, since they are present in the bacterial cell wall, they have been termed endotoxins to distinguish them from secreted protein exotoxins (Rietschel *et al.*, 1990; Caroff and Karibian, 2003). Nonetheless, these molecules can be released by multiplying or disintegrating bacteria, as well as by blebbing of outer membrane vesicles from the bacterial cell surface (Hynes *et al.*, 2005). Moreover, LPSs play an essential role in the integrity of the Gram-negative bacterial envelope (Nikaido, 2003), and produce a functional macromolecular matrix for *H. pylori* interaction with the higher organism during infection (Muotiala *et al.*, 1992; Moran, 1995a, 1996b). In general, LPSs strongly activate B-lymphocytes, granulocytes and mononuclear cells and harbour binding sites for immune receptors, antibodies and non-immunoglobulin serum factors, and thus are involved in host immune recognition (Rietschel *et al.* 1990, 1994; Moran, 1995b; Raetz and Whitfield, 2002; Caroff and Karibian, 2003). On the other hand, LPS structural variation may influence the immune response, thereby contributing to the virulence of bacterial strains. As surface antigens (O-antigens) of Gram-negative bacteria, including *H. pylori*, these molecules are capable of inducing an antibody response and whose structural variation can be used as the basis for serotyping strains (Mills *et al.*, 1992; Simoons-Smit *et al.*, 1996). In addition, this structural variation can be detected using non-immunoglobulin factors such as lectins to allow strain differentiation in a lectin typing scheme (Hynes *et al.*, 1999, 2002).

Like members of the Enterobacteriaceae, exemplified by *Escherichia coli*, *H. pylori* produces high-molecular-weight (smooth-form) LPS composed of an outermost saccharide moiety, divided into the O-chain and core oligosaccharide (OS) regions, covalently linked to a lipid moiety, termed lipid A, that anchors the molecule in the outer leaflet of the outer membrane (Rietschel *et al.* 1990; Moran *et al.*, 1992; Moran, 1995b; Caroff and Karibian, 2003). Each of these domains has differing structural and biological properties: the O-chain contributes to the antigenicity and serospecificity of the molecule; the core OS influences permeation properties of the outer membrane, modulates the immune response to lipid A and can influence virulence; whereas the lipid A moiety interacts with immune receptors and endows the LPS molecule with a range of immunological and potential endotoxic properties (Rietschel *et al.* 1990,1994; Moran, 1995b; Yethon *et al.*, 1998, 2000; White *et al.*, 1999; Raetz and Whitfield, 2002; Caroff and Karibian, 2003; Ram *et al.*, 2003). Importantly, the properties of *H. pylori* LPS and its domains, which collectively contribute to both chronicity and disease development (Moran, 2007), conform with the proposal of *H. pylori* as a model for investigating and understanding the dynamics of bacterial persistence and parasitism in chronic infections (Blaser and Kirschner, 1999). The present chapter reviews the main biological attributes of *H. pylori* LPS and the novel insights that have been gained into the molecular genetics, structure and contributing properties of this class of molecule to *H. pylori* pathogenesis.

General structural and biological properties of *H. pylori* LPS

Despite *H. pylori* LPSs conforming to the conserved LPS architectural model by possessing the three domains of the O-chain, core OS and lipid A moiety, these molecules also exhibit novel properties with respect to their expression, fine structure and biological characteristics.

LPS expression

In contrast to the heterogeneous O-chain length seen in enterobacterial LPS preparations, clinical isolates of *H. pylori* produce smooth-form LPS with O-chains of relatively constant chain length (Moran *et al.*, 1992; Moran, 1995a) whose length has been deduced to be determined by an enzymatic molecular ruler mechanism (Nilsson *et al.* 2006). Numerous subcultures on conventional solid media can induce production of low-molecular-weight, rough-form LPS lacking O-chains in many strains (Moran, 1995a), which has proven a useful manipulation when studying the structure and properties of the LPS core. Although strains bearing rough-form LPS can revert to smooth-LPS production when grown in liquid media (Moran, 1995a; Walsh and Moran, 1997), the molecular and environmental factors influencing smooth/rough variation of this LPS have not been determined. Nevertheless, evidence indicates that LPS structure is dependent on the pH of the environment. Acid conditions induce expression of a *wbcJ* homologue (HP0045/JHP0039) (McGowan *et al.*, 1998), which is predicted to encode a protein responsible for the conversion of guanosine diphosphate (GDP)-mannose to GDP-fucose, required for O-chain biosynthesis (Doig *et al.* 1999). The gene was induced under acid shock conditions and at pH 5, and a *wbcJ* mutant did not survive an acid shock of pH 3.5, and was impaired in its survival at pH 4 (McGowan *et al.*, 1998). Thus, expression of LPS appears to be involved in the ability of *H. pylori* to respond to or withstand acid shock. Further emphasizing this response to acid, detailed chemical studies have shown structural differences between the O-chains of *H. pylori* grown in liquid media at pH 7 and pH 5, whereas no differences were observed in the core OS and lipid A regions (Moran *et al.*, 2002c). Other changes in cellular glycolipid composition have been reported when *H. pylori* is grown on solid media at pH 5.5 compared with pH 7, namely increased amounts of lysophosphatidyl ethanolamine and phosphatidyl serine (Bukholm *et al.*, 1997), but these changes can be induced independently of acidic conditions (Moran *et al.*, 2002c) and may simply reflect growth on solid media rather than the acidic environment (Moran *et al.*, 2001b).

Bioactivities and lipid A

Studies on the bioactivities of *H. pylori* LPS, have revealed significantly lower endotoxic and immunological activities when compared with enterobacterial LPS as the gold standard. The pyrogenicity and mitogenicity is 1000-fold lower, lethal toxicity in mice 500-fold lower, and induction of various cytokines and chemokines 1000-fold lower than enterobacterial LPS (Muotiala *et al.*, 1992; Pece *et al.*, 1995). As reviewed extensively (Moran 1999, 2001a; Odenbreit *et al.*, 2006), priming of neutrophils to release toxic oxygen radicals, cytokine and chemokine induction, E-selectin expression on endothelial cells, induction of prostaglandin E_2 and nitric oxide, Natural Killer-cell activity and abolition of suppressor T-cell activity are all significantly lower. Similarly, cyclooxygenase-2 induction in gastric cells by *H. pylori* LPS is negligible compared with that induced by *E. coli* (Smith *et al.*, 2003b). Early studies that tested chemically modified *H. pylori* LPS-derived components in immunological assays indicated that the molecular basis for these low immuno-activities reside in the lipid. A moiety and is modulated by the saccharide component of *H. pylori* LPS, particularly the core. For example, the phosphorylation pattern in lipid A influences cytokine production and induction of procoagulant activity from mononuclear leucocytes, as well as *Limulus* amoebocyte activity, but the core OS modulates these effects (Pece *et al.*, 1995; Semeraro *et al.*, 1996). On the other hand, the lipid A acylation pattern is attributed responsibility for the lack of abolition of suppressor T-cell activity by *H. pylori* LPS (Baker *et al.*, 1994) and for reduced priming of neutrophils to release toxic oxygen radicals (Nielsen *et al.*, 1994).

Consistent with the hypothesis that the structure of lipid A endows *H. pylori* LPS with low immunoactivities (Muotiala *et al.*, 1992), detailed structural studies have found underphosphorylation, underacylation and substitution by long-chain fatty acids in this lipid A (Moran *et al.*, 1992; Moran, 1995a; Moran *et al.*, 1997; Suda *et al.*, 2001), which based on established structure–bioactivity relationships of lipid A molecules (Rietschel *et al.*, 1990, 1994), are likely to translate into reduced immunological activities. Moreover, these attributes also influence the supramolecular shape of *H. pylori* lipid A (Schromm *et al.*, 2000) and other biophysical properties of lipid A that have been correlated previously with lower bioactivity (Seydel *et al.*, 2000). These include a higher phase transition temperature and a lower inclination angle of the lipid A diglucosamine backbone to the membrane plane than encountered with enterobacterial lipid A (Moran *et al.*, 2005c) which would affect interaction with immune receptors.

Core OS properties

The core regions of *H. pylori* strains exhibit an unusual conformation compared with the core OS of enterobacterial LPSs as determined by structural and serological analyses; epitopes are present that are common with other bacterial species (likely to be those present in the inner core) and others that are specific to *H. pylori* (Walsh and Moran, 1996). Whether such epitopes could be exploited in immunodiagnostics remains an open question. As exemplified by the core OS of *H. pylori* NCTC 11637 (Fig. 2.1), a branching occurs from the D-*glycero*-D-*manno*-heptose (DD-heptose) of the core, i.e. the third heptose residue in the inner core, through a second such residue to which the O-chain is attached (Aspinall *et al.*, 1996). However, differences in substitution of the intervening DD-heptose between the core and O-chain occur, whereby heptan and glycan units are present (Monteiro, 2001) (Fig. 2.1), producing what has been termed the third domain (Aspinall *et al.*, 1995; Moran and Aspinall, 1998).

The first LPS biosynthesis gene to be identified and cloned from *H. pylori* was an *rfaC* (also termed *waaC*) homologue (HP0279/JHP0264) encoding heptosyltransferase I that attaches L-*glycero*-D-*manno*-heptose (LD-heptose) to 3-deoxy-D-*manno*-oct-2-ulosonic acid (Kdo) in the inner core (Goldberg *et al.*, 1996). Attempts

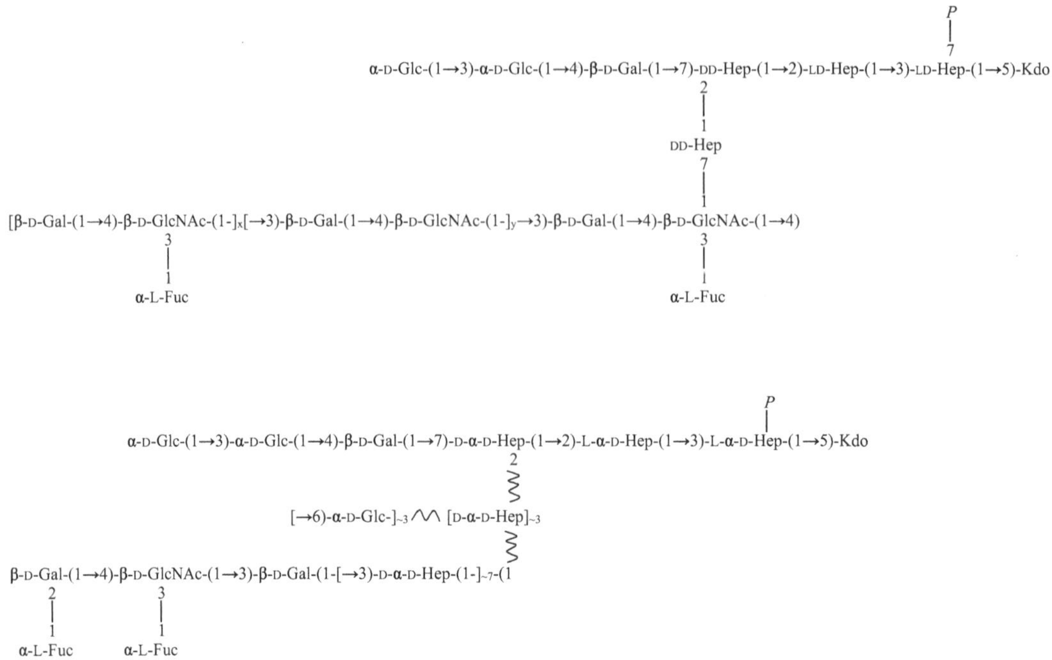

Figure 2.1 Proposed structure of the saccharide components of LPS of *H. pylori* NCTC 11637 (top) and MO19 (bottom) (Aspinall *et al.*, 1994, 1996; Aspinall and Monteiro, 1996). Abbreviations: Fuc, fucose; Gal, galactose; Glc, glucose; GlcNAc, *N*-acetylglucosamine; DD-Hep, D-*glycero*-D-*manno*-heptose; LD-Hep, L-*glycero*-D-*manno*-heptose; Kdo, 3-deoxy-D-*manno*-oct-2-ulosonic acid. Except for DD-Hep and LD-Hep, and fucose, which is in the L-configuration, all other sugars possess the D-configuration.

to produce knockout mutants of the gene have proven unsuccessful (Moran, 2001b), and excluding technical difficulties, indicates that such mutations are lethal, thus emphasizing the importance of the LPS inner core structure for membrane integrity and viability of *H. pylori*.

Comparative genomic analysis of strains 26695 and J99 (Marais *et al.*, 1999; Doig *et al.*, 1999) has identified orthologues of genes potentially involved in core OS synthesis: *rfaF* or *waaF* (HP1191/JHP1116) encoding a heptosyltransferase II, *rfaE* or *waaE* (HP0858/JHP0792) encoding a ADP-DD-heptose synthase, *rfaD* or *waaD* (HP0859/JHP0793) encoding a ADP-LD-heptose-6-epimerase, and which along with HP0279/JHP0264 (*rfaC*-like) are considered to be involved in synthesis of the heptose inner core; *kdsA* (HP0003/JHP0003) encoding a Kdo-8-phosphate synthase, *kdsB* (HP0230/JHP0215) encoding a Kdo cytidyltransferase, and *rpe* (HP1386/JHP1439) encoding a D-ribulose-5-phosphate-3-epimerase are considered relevant for synthesis of the Kdo region; and three homologues of *rfaJ* (HP0159/JHP0147, HP0208/JHP0194, HP1416/JHP1311) which were confusingly deduced to encode α(1,2)-glucosyltransferase activity but more recently assigned other additional activities (Moran *et al.*, 2004), *pgi* (HP1166/JHP1093) encoding glucose-6-phosphate isomerase, and *galU* (HP0646/JHP0591) encoding UDP-glucose pyrophosphorylase have been deduced to be involved in outer core synthesis. Moreover, *galE* (HP0360/JHP1020) mutants (ablating UDP-galactose-4-epimerase activity) of *H. pylori* strains have been shown to result in truncated core OSs because of the role of D-galactose in the structure of the outer core (Moran, 1999; Edwards *et al.*, 2000; Moran *et al.*, 2000; Perepelov *et al.*, 2001).

Similar to the lipid A domain (see below), the inner core region of *H. pylori* LPS is subject to structural modification. The inner heptose sugars of a number of bacterial pathogens are phosphorylated by specific kinases and/or decorated with phosphorylethanolamine (*P*EtN) units, and these modifications of the heptose sugars have been shown to play a role in outer membrane stability, complement resistance and virulence of Gram-negative bacteria (White *et al.*, 1999;

Yethon *et al.*, 1998, 2000; Ram *et al.*, 2003). The first heptose of the *H. pylori* core OS can be substituted with a phosphate or *P*EtN group at the 7-position (Aspinall *et al.*, 1996, 1997; Monteiro *et al.*, 2000a; Moran *et al.*, 2002c; Knirel *et al.*, 1999). In *E. coli* and *Salmonella*, phosphorylation of the inner core heptose sugars is catalysed by two specific kinases termed WaaP and WaaY (Yethon *et al.*, 1998). Addition of these negative charges on the inner core allows for bridging of neighbouring LPS molecules by divalent cations, such as Ca^{2+} and Mg^{2+}, and results in a less fluid and more stable outer membrane (Nikaido, 2003). Interestingly, there are no homologues of WaaP or WaaY in *H. pylori* 26695 or J99, suggesting that this bacterium utilizes a different mechanism for heptose phosphorylation. The origin of *P*EtN residue in the *H. pylori* inner core is unknown. In *Neisserid meningitidis*, deletion of *lpt-3* resulting in the loss of *P*EtN addition to the distal heptose of neisserial LPS (Mackinnon *et al.*, 2002). The *H. pylori* protein encoded by HP1417 (JHP1312) shares homology with Lpt-3, that encoded by HP0022/JHP0020 (see below), and other proteins known to support *P*EtN modification of LPS (Tran *et al.*, 2004). Just upstream of HP1417, is the structural gene HP1416 (JHP1311) encoding a glucosyltransferase that also is involved in the biosynthesis of the core OS of *H. pylori* LPS (Moran *et al.*, 2004). Given that HP1417 is transcribed with other genes encoding LPS biosynthetic enzymes, and the homology of its encoded protein to other known *P*EtN transferases, it is likely that HP1417-enoded protein serves as the heptosyl *P*EtN transferase in *H. pylori*.

The core OS contributes to the pathogenesis and virulence of *H. pylori* strains. For example, in conjunction with a 25 kDa protein adhesin (Valkonen *et al.*, 1997), that is produced *in vivo* (Moran *et al.*, 2005a), *H. pylori* LPS can bind laminin which is an important extracellular matrix glycoprotein found in the basement membrane (Valkonen *et al.*, 1994). Moreover, the LPS-laminin binding may inhibit recognition of the glycoprotein by the laminin receptor (67 kDa integrin) on gastric epithelial cells (Slomiany *et al.*, 1991). This interaction and also that of *H. pylori* with other extracellular matrix proteins (Moran *et al.*, 1993) may play an important role in loss of gastric mucosal integrity (Lee and Moran, 1994; Moran, 1996b), and along with other soluble factors of *H. pylori*, contribute to development of gastric leakiness (Terrés *et al.*, 1998) and potentially trigger cellular apoptosis (Piotrowski *et al.*, 1997). Also, *H. pylori* isolates, particularly those isolated from duodenal ulcer patients, stimulate pepsinogen secretion (Young *et al.*, 1992; Moran *et al.*, 1998), a precursor of the enzyme pepsin, which is considered an aggressive mucolytic factor in the development of duodenal ulcer disease. The contribution of *H. pylori* LPS to these two phenomena and the identity of the structures within the core oligosaccharide (OS) of LPS required for these interactions have been identified (Moran and Aspinall, 1998; Moran *et al.* 1998; Aspinall *et al.*, 1999; Moran, 2001b). Of note, the serological response against the core OS is more developed in duodenal ulcer patients that gastritis patients, reflecting the availability of core structures for the above interactions (Pece *et al.*, 1997).

Other characteristics

Further important contributions of LPS to *H. pylori* pathogenesis concern molecular mimicry within the O-chains. Since the first structural description of Lewis x (Le^x) expression in *H. pylori* O-chains (Aspinall *et al.*, 1994), extensive investigations on Lewis (Le) and related blood group antigen-mimicking O-chains in *H. pylori* LPS have been undertaken (see reviews, Moran, 2001a,b; Monteiro, 2001). Various consequences and implications of this mimicry have been hypothesized, including molecular camouflage, bacterial adaptation to the gastric mucosa, a role in adhesion phenomena, an influence on the inflammatory response and contribution to gastric autoimmunity (Moran 1999, 2001a,b). Supportive evidence of varying strength for these individual concepts has been accumulated (and will be reviewed more extensively below) and continue to be systematically investigated, but despite progress, the complete functions and consequences of this mimicry remain incompletely understood.

Additionally, *H. pylori* LPS has been implicated in a variety of other putative mechanisms of gastric damage that are less well characterized and that require independent verification or whose structure–bioactivity relationships have not been established to date. These include effects on mucus synthesis, glycosylation, sulfation and epithelial cell interaction, leptin induction, acid secretion, and cellular apoptosis (Slomiany *et al.*, 1992; Piotrowski *et al.*, 1993, 1997; Kidd *et al.*, 1997; Brzozowski *et al.*, 2000; Padol *et al.*, 2001; Slomiany and Slomiany, 2005, 2006; Durkin *et al.*, 2006). In contrast, intensive biological and structural studies have been performed on the lipid A and O-chain components of *H. pylori* LPSs, which are central to understanding the pathogenesis and chronicity of *H. pylori*-related disease, and these will form the focus for the remainder of this review.

Lipid A

Typically, the lipid A backbone of Gram-negative bacteria is a β(1→6′)-linked disaccharide of *gluco*-configured sugars, D-glucosamine or 2,3-dimaino-2,3-dideoxy-D-glucose, that is *bis*-phosphorylated and multiply acylated with fatty acid chains (Rietschel *et al.*, 1990; Moran *et al.*, 1991a,b; Moran, 1995b; Raetz and Whitfield, 2002; Caroff and Karibian, 2003). Though key features of this LPS moiety are conserved across bacterial species, there is considerable diversity in the phosphorylation and acylation pattern of lipid A among pathogenic bacteria (Moran, 1995b; Rietschel *et al.*, 1990, 1994; Trent *et al.*, 2006).

H. pylori lipid A fine structure

Two independent research groups using two different strains examined the fine structure of lipid A from *H. pylori* rough-form LPS, one study found a tetra-acyl lipid A (Moran *et al.*, 1997), whereas another reported a triacyl lipid A with an identical phosphorylation pattern to the tetra-acyl form, but lacking acylation at position-3 of the lipid A backbone (Suda *et al.*, 1997). A reinvestigation by the latter group of the same strain found the tetra-acyl form of *H. pylori* lipid A (Suda *et al.*, 2001) and, thus potentially, the triacyl lipid A may have arisen as a degradation product of the former during lipid A isolation (Moran, 2001b). Hence, the major molecular species in *H. pylori* lipid A is composed of a β-(1→6′)-linked D-glucosamine disaccharide backbone acylated by (R)-3-hydroxyoctadecanoic acid [18:0(3-OH)] and (R)-3-hydroxyhexadecanoic acid [16:0(3-OH)] at positions 2 and 3 and with (R)-3-(octadecanoyloxy)octadecanoic acid [3-(18:0-O)-18:0] at the 2′-position and carries phosphate or PEtN groups at position-1 (Moran *et al.*, 1997) (Fig. 2.2). Thus, compared with the lipid A of *E. coli*, the predominant lipid A molecular species of *H. pylori* lacks the usual 4′-phosphate group, as well as the 3′-ester-linked fatty acyl chains, thus containing four rather than six fatty acids with longer chain length (16–18 carbons versus 12–14 carbons), and is derivatized with a PEtN residue at the C-1 position of the proximal glucosamine (Moran *et al.*, 1997; Moran, 1998) (Fig. 2.2).

Nonetheless, in addition to this predominant monophosphorylated tetra-acyl lipid A, *H. pylori* smooth-form LPS also contains as a minor constituent a bisphosphorylated hexa-acyl lipid which is distinguished from tetra-acyl lipid A by carrying 3-(12:0-O)-16:0 or 3-(14:0-O)-16:0 at position-3′ and an extra phosphate group at position-4′ (Moran *et al.*, 1997) (Fig. 2.2). Since this bacterium can synthesize a bisphosphorylated hexa-acyl lipid A, it follows that *H. pylori* expresses enzymes (e.g. a 4′-phoshatase and 3′-acyloxyacyl deacylase) capable of remodelling, and thus modifying to a tetra-acyl form, its lipid A domain after completion of the conserved hexa-acyl lipid A biosynthetic pathway (Tran *et al.*, 2005; Reynolds *et al.*, 2006).

Biosynthesis of the Kdo-lipid A domain

Significant progress has been made towards understanding the molecular mechanisms utilized by *H. pylori* to synthesize and remodel the lipid A domain of its LPS. Nine enzymatic steps are necessary for the biosynthesis of the Kdo-lipid A domain of *E. coli* LPS (reviewed in Raetz and Whitfield, 2002; Tran *et al.*, 2006). The constitutive lipid A pathway is highly conserved, occurs on the cytoplasmic surface of the inner membrane, and generally a single copy of each *lpx* gene can be found in the genome of virtually all Gram-negative bacteria. Table 2.1 compares the lipid A biosynthetic enzymes of *E. coli* (Lpx proteins) to the putative orthologues found in *H. pylori* 26695. On the whole, bioinformatic analysis of the 26695 and J99 (Tomb *et al.*, 1997; Alm *et al.*, 1999) genomes indicates that the

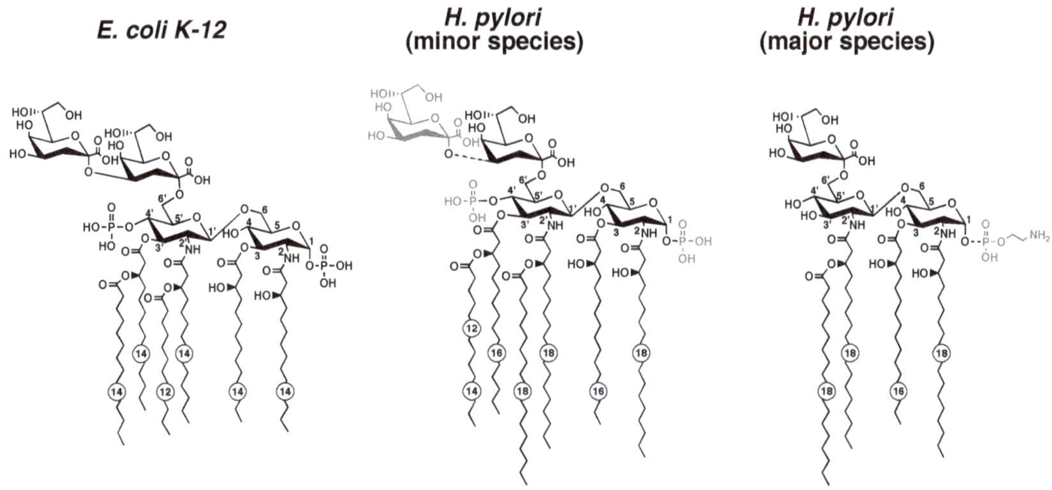

Figure 2.2 Comparison of the chemical structures of the Kdo-lipid A domain of *E. coli* K-12 and the major and minor structures of *H. pylori*. The covalent modifications of lipid A are indicated with dashed bonds, and lengths of the acyl chains are designated with enclosed circles. The major lipid A species of *E. coli* is a hexa-acylated disaccharide of glucosamine that is substituted at the 1- and 4′- positions with phosphate and glycosylated at the 6′ position with two Kdo moieties (Rietschel *et al.*, 1990, 1994; Moran *et al.*, 1995b; Raetz and Whitfield, 2002). The minor lipid A species of *H. pylori* is similar to that seen in *E. coli*, but has slightly longer acyl chains (Moran *et al.*, 1997). The major lipid A species of *H. pylori* is a tetra-acylated structure that is lacking the 4′-phosphate group and is substituted at the C-1 position with a PEtN residue (Moran *et al.*, 1997; Suda *et al.*, 2001). Only a single Kdo sugar is attached to the major lipid A species found in *H. pylori*.

Table 2.1 Comparison of *E. coli* and *H. pylori* biosynthetic enzymes required for synthesis of the Kdo-lipid A domain of LPS

E. coli Lpx protein	Enzymatic function	*H. pylori* orthologue*	Approximate E-values**
LpxA	UDP-*N*-acetylglucosamine O-acyltransferase.	HP1375	4×10^{-48}
LpxC	UDP-(3-*O*-acyl)-N-acetylglucosamine deacetylase	HP1052	7×10^{-65}
LpxD	UDP-3-*O*-acyl-glucosamine N-acyltransferase	HP0196	2×10^{-33}
LpxH	UDP-2,3-diacylglucosamine pyrophosphatase	HP0394	5×10^{-4}
LpxB	Lipid A disaccharide synthase	HP0867	2×10^{-28}
LpxK	Lipid A 4′-kinase	HP0328	2×10^{-3}
WaaA (KdtA)	Kdo transferase	HP0957	2×10^{-27}
LpxL	Transfers secondary acyl chain to 2′-position	HP0280	5×10^{-6}
LpxM	Transfers secondary acyl chain to 3′-position	Not present	

*Designations as per the *H. pylori* 26695 genome.
**Possible orthologues in the NCBI database as of March 1, 2007 were identified with the PSI-Blast algorithm using predicted sequences of the *E. coli* Lpx enzymes.

enzymatic machinery for lipid A biosynthesis is conserved in *H. pylori*.

Differences in the acyl chain length of *H. pylori* and *E. coli* lipid A can be attributed to substrate preferences of the primary and secondary acyltransferases. The primary linked hydroxylated acyl chains of *H. pylori* lipid A are either 16 carbons [16:0(3-OH)] or 18 carbons [18:0(3-OH)] in length (Moran *et al.*, 1997; Suda *et al.*, 1997) (Fig. 2.2). Therefore, the early acyltransferases of the *H. pylori* lipid A pathway (Fig. 2.3), LpxA and LpxD, utilize substrates bearing longer fatty acyl chains when compared to enzymes expressed by most Gram-negative bacteria. Most enterobacterial lipid A structures are characterized by primary-linked acyl chains of 14 carbons [14:0(3-OH)] in length (Moran, 1995b; Rietschel *et al.*, 1990, 1994; Raetz and Whitfield, 2002; Trent *et al.*, 2006). The major lipid A molecular species of *H. pylori* contains a single secondary acyl chain that is 18 carbons in length (18:0) located at the 2′-position (Moran *et al.*, 1997; Suda *et al.*, 1997) which, again, is longer than the secondary acyl chains found in *E. coli* lipid A.

The minor lipid A molecular species (Fig. 2.2) bears two secondary acyl chains and is substituted at the 3′-position with either a lauroyl (12:0) or myristoyl (14:0) group. Interestingly, the *H. pylori* genome contains a single secondary acyltransferase encoded by HP0280 (Tomb *et al.*, 1997; Alm *et al.*, 1999). Thus, either the HP0280-encoded protein is a bifunctional acyltransferase or a second unknown acyltransferase functions to acylate *H. pylori* lipid A. Biochemical characterization of this protein using an *in vitro* assay system has confirmed it to be a monofunctional stearoyl (18:0) acyltransferase (C. M. Stead

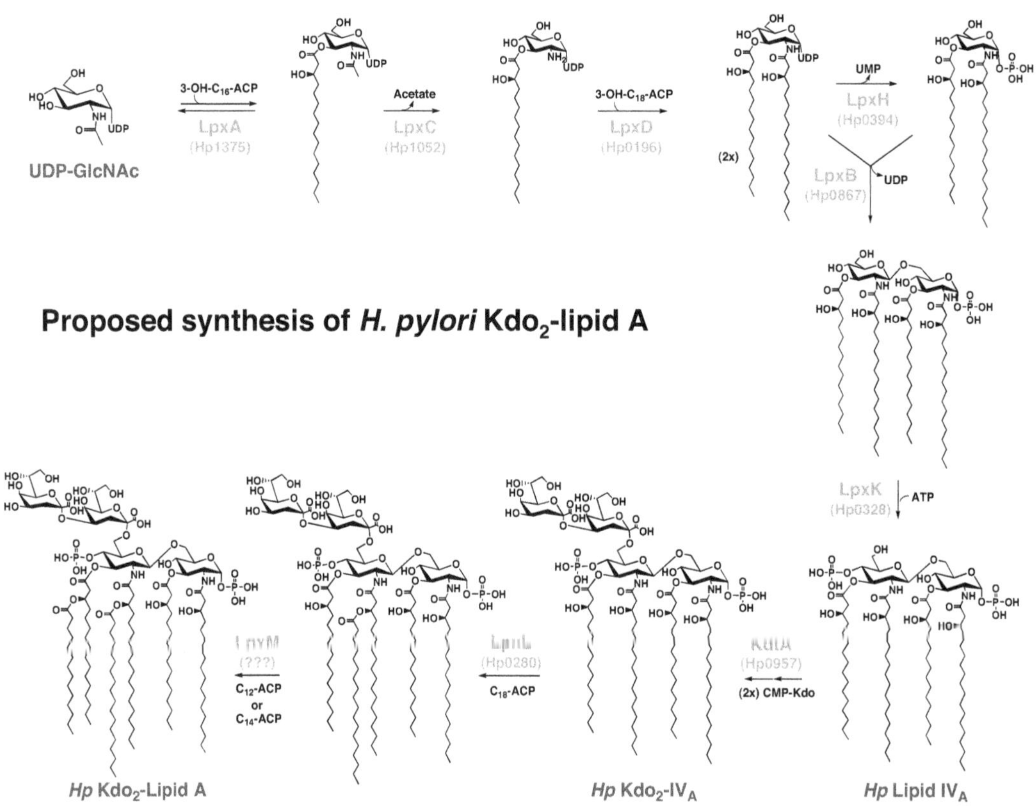

Figure 2.3 Proposed synthesis of *H. pylori* Kdo_2-lipid A. The structures shown are for the constitutive biosynthetic pathway of the Kdo-lipid A domain of *H. pylori* LPS. Putative orthologues of the previously identified *E. coli* Lpx enzymes (Raetz and Whitfield, 2002) are shown in parentheses with the exception of *E. coli* LpxM. Although *H. pylori* Kdo_2-lipid A contains two secondary acyl chains, only a single secondary acyltransferase homologue can be found in the *H. pylori* genome. The major intermediates of the pathway are indicated. Acyl-ACP (acyl carrier protein) is predicted to serve as the acyl donor for the various acyltransferases.

and M. S. Trent, unpublished). The enzyme utilizes acyl-ACP (acyl carrier protein) as its donor, requires the Kdo sugars for activity, and transfers the fatty acyl chain to the 2′-position of *H. pylori* lipid A. Therefore, this acyltransferase functions similarly to *E. coli* LpxL (Clementz *et al.*, 1996) except for acyl chain preference. Upon assaying of *H. pylori* membranes with key lipid A precursors, a second acyltransferase can be detected that utilizes C12 and C14 acyl-ACPs (C. M. Stead and M. S. Trent, unpublished), similar to *E. coli* LpxM (Clementz *et al.*, 1996). The gene encoding the 3′-secondary acyl transferase of *H. pylori* remains, to date, unknown. The predicted biosynthetic pathway of *H. pylori* lipid A is shown in Fig. 2.3.

Of note, only distant orthologues of *E. coli* LpxH (E-value, 10^{-4}) and LpxK (E-value, 10^{-3}) are present in *H. pylori* (Table 2.1). In *E. coli*, *lpxH* and *lpxK* are essential genes encoding for the UDP-2,3-diacylglucosamine pyrophosphatase (Babinski *et al.*, 2002) and the lipid A 4′-kinase (Garrett *et al.*, 1997), respectively (Fig. 2.3). Why certain *lpx* genes are absent or less conserved in *H. pylori* is unclear, but clearly the bacterium has the necessary machinery to synthesize a full lipid A, albeit as a minor molecular species in the outer membrane.

Synthesis of the Kdo-lipid A domain of LPS is followed by the attachment of the core OS and transport of the core-lipid A across the inner membrane (Raetz and Whitfield, 2002). For this process, MsbA, a conserved ABC transporter that is also found in *H. pylori*, serves as the lipid A flippase (Zhou *et al.*, 1998; Doerrler, 2006). The O-chain of *H. pylori* is then attached on the periplasmic side of the inner membrane to form whole LPS. Additional steps are required for translocation of LPS to the outer leaflet of the outer membrane, but are not yet clearly defined (Doerrler, 2006). Following transport across the inner membrane, a number of bacterial species modify the Kdo-lipid A domain of their LPS. Generally the modification systems are found on the periplasmic side of the inner membrane or at the extracellular surface. This allows for separation of the constitutive pathway (Fig. 2.3) from the modification systems (Fig. 2.4) (Trent *et al.*, 2006). Given the structural differences between the major and minor species of *Helicobacter* lipid A (Fig. 2.2), it is reasonable to assume that the bacterium remodels its lipid A following the constitutive pathway.

Modification of the Kdo-lipid A domain

Lipid modification systems of other bacteria have been shown to be important for resistance to cationic antimicrobial peptides (CAMPs) produced by the innate immune system and to polymyxin (Trent *et al.*, 2006). Polymyxin is a cyclic lipopeptide that binds to the phosphate groups of lipid A and kills Gram-negative bacteria in a manner similar to that of CAMPs (Vaara and Vaara, 1981, 1983). During infection of the gastric mucosa, *H. pylori* encounters antimicrobial peptides produced by its human host. For example, expression of the human cathelicidin LL-37 (Hase *et al.*, 2003) and human β-defensin 2 (Wada *et al.*, 1999; Hamanaka *et al.*, 2001) is upregulated in the gastric mucosa of *H. pylori*-infected patients. Furthermore, *H. pylori* itself produces a ceropin-like peptide, Hp(2–20), to which it is resistant, that is derived from the amino-terminal part of its ribosomal protein L1 (Putsep *et al.*, 1999). Ceropin has been shown to bind to *bis*-phosphorylated lipid A (De Lucca, 1995). Presumably, in order to persist in the human gastric mucosa *H. pylori* must be able to resist the action of CAMPs. Some Gram-negative pathogens, such as *Salmonella typhimurium* and *Pseudomonas aeruginosa*, mask the negative charge of their phosphate groups with positively charged amine-containing substituents, such as 4-amino-4-deoxy-L-arabionse (L-Ara4N), to promote CAMP resistance (Guo *et al.*, 1997; Ernst *et al.*, 1999; Trent *et al.*, 2001). On the other hand, in *N. meningitidis* modification with PEtN has been shown to be critical for CAMP resistance (Cox *et al.*, 2003; Tzeng *et al.*, 2005).

H. pylori lipid A contains a PEtN group directly linked to the C-1 hydroxyl group of the proximal glucosamine (Moran *et al.*, 1997; Suda *et al.*, 1997) (Fig. 2.2). Interestingly, PEtN units found in other Gram-negative bacteria, such as *E. coli* (Zhou *et al.*, 2000), *S. typhimurium* (Zhou *et al.*, 2001), and *N. meningitidis* (Kulshin *et al.*, 1992), are attached to the lipid A phosphate group to form a pyrophosphate linkage. The work of Tran *et al.* (2004, 2006) showed that the PEtN moiety in *H. pylori* lipid A arises from the

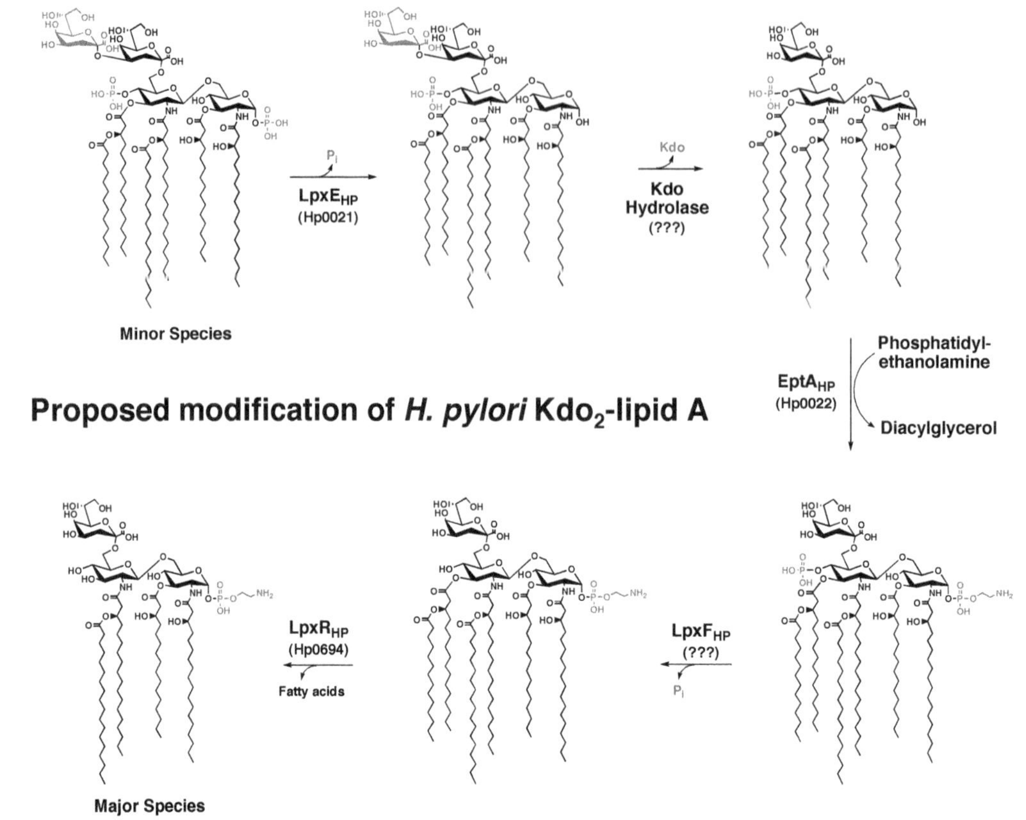

Figure 2.4 Proposed modification of *H. pylori* Kdo$_2$-lipid A. Following the constitutive biosynthetic pathway, *H. pylori* Kdo$_2$-lipid A is modified by several enzymes. First, LpxE catalyses the removal of the 1-phosphate group from *H. pylori* lipid A on the periplasmic side of the inner membrane (Tran *et al.*, 2004). Other modifications include the removal of the outer Kdo sugar (Stead *et al.*, 2005) and the addition of *P*EtN residue to the 1-position catalysed by EptA (Tran *et al.*, 2004), both of which require removal of the 1-phosphate group. Additional enzymatic activities that are considered necessary for the production of the major lipid A molecular species of *H. pylori* include a 4′-phosphatase and a 3′-*O*-deacylase (see text for details). Modification of the Kdo-lipid A domain is thought to aid in evasion of the innate immune response by promoting resistance to cationic antimicrobial peptides and by reducing the immunostimulatory properties of *H. pylori* LPS (Moran, 1999; Tran *et al.*, 2005).

removal of the phosphate group at position-1 of lipid A followed by the addition of *P*EtN to form a glycosidic phosphodiester linkage (Fig. 2.4). Furthermore, this modification of the 1-position of *H. pylori* lipid A occurs on the periplasmic side of the inner membrane, and is thus dependent upon MsbA transport. Removal of the 1-phosphate group and the subsequent addition of *P*EtN is catalysed by the proteins encoded by HP0021 and HP0022, respectively (Tran *et al.*, 2004) (Fig. 2.4 and Table 2.2) . Both proteins contain multiple transmembrane domains and soluble periplasmic regions that contain amino acid sequence predicted to encode active site residues. The genes encoding these enzymes appear to form a typical prokaryotic operon given they are unidirectionally transcribed and separated by only 10 bp of DNA sequence. The HP0021-encoded protein, also known as LpxE, has distant sequence similarity to *E. coli* PgpB, a phosphatidylglycerolphosphatase (Icho and Raetz, 1983; Icho, 1988), but is highly selective for the 1-phosphate group of lipid A. Likewise, a lipid A phosphatase has also been identified in the human pathogen *Francisella novicida* (Wang *et al.*, 2004). HP0022, an orthologue of *eptA*, is homologous to the lipid A *P*EtN transferases identified in other Gram-negative pathogens including *E. coli*, *S. typhimurium*, and *N. meningitidis* (Tran *et al.*, 2004). Again, it is important to note that *H. pylori* EptA transfers the *P*EtN unit after removal of the phosphate group (Tran *et al.*,

Table 2.2 Description of enzymes that modify the inner core OS and lipid A domains of *H. pylori* LPS

Enzyme	Modification	Open reading frame*	Protein localization	Possible role in pathogenesis
EptA	*P*EtN addition to lipid A disaccharide (C-1 position)	HP0022	Inner membrane	CAMP resistance
LpxE	Lipid A 1-phosphatase	HP0021	Inner membrane	CAMP resistance; modulation of TLR4 activation
LpxF	Lipid A 4′-phosphatase	Unknown	Probable inner membrane	CAMP resistance; modulation of TLR4 activation
Kdo hydrolase	Removal of outer Kdo sugar from inner core OS	Unknown	Probable inner membrane	Unknown
LpxR	Removal of 3′-*O*-acyloxylacyl group	HP0694	Outer membrane	Modulation of TLR4 activation
Inner core *P*EtN transferase	Addition of *P*EtN residue to the inner core region	HP1417	Inner membrane	CAMP resistance, complement resistance

*Designations as per the *H. pylori* 26695 genome.

2004, 2006), whereas the EptA of other bacteria modifies the lipid A phosphate group directly (Trent and Raetz, 2002; Lee *et al.*, 2004).

The highly conserved lipid A kinase, LpxK, phosphorylates the 4′-hydroxyl group of the disaccharide backbone (Garrett *et al.*, 1997) (Fig. 2.3). In some organisms, including *H. pylori*, the 4′-phosphate group is removed following the constitutive biosynthetic pathway. An enzymatic activity catalysing the removal of the 4′-phosphate group of lipid A can be easily detected in membranes from *H. pylori* (M. S. Trent *et al.*, unpublished). Although a 4′-phosphatase has been identified in *F. novicida*, which is referred to as LpxF (Wang *et al.*, 2006) (Table 2.2), there is no homologue in the genomes of *H. pylori* J99 or 26695. In *Francisella*, LpxF is localized to the inner membrane and removes the 4′-phosphate group of the *Francisella* lipid A after transport across the inner membrane. Therefore, dephosphorylation of lipid A at the 1- and 4′-positions occurs within the periplasmic region of the bacterial cell. To date, attempts to clone the structural gene encoding *Helicobacter* LpxF have been unsuccessful.

As lipid A phosphate groups are necessary for binding of cationic peptides (Trent *et al.*, 2006), the modified lipid A structure of *H. pylori* should promote resistance to CAMP during host infection. Wild-type strains of *H. pylori* are inherently resistant to polymyxin, displaying a minimal inhibitory concentration (MIC) of > 250 μg/ml. Disruption of *lpxE* (HP0021/JHP0019) in *H. pylori* J99 results in the production of a lipid A bearing a single phosphate group at the 1-position since lack of LpxE function results in loss of *P*EtN transfer as *H. pylori* EptA (HP0022-encoded) requires prior dephosphorylation for activity (Fig. 2.4) (Tran *et al.*, 2006). Exposure of a single unmodified phosphate group within the *H. pylori* outer membrane results in a 25-fold increase in polymyxin sensitivity (MIC of 10 μg/ml). Unexpectedly, it has been observed that disruption of *lpxE* in *H. pylori* 26695 results in the synthesis of lipid A with two phosphate groups and six acyl chains (Tran *et al.*, 2006). The structure is identical to the minor lipid A species previously reported (Fig. 2.2), thereby confirming the findings of Moran *et al.* (1997). Interestingly, the 26695 *lpxE* mutant displayed complete sensitivity to polymyxin with an MIC of 0.2 μg/ml (Tran *et al.*, 2006). These data suggest that removal of the 4′-phosphate group of *H. pylori* lipid A is also critical for CAMP resistance. Why deletion of *lpxE* in *H. pylori* 26695 results in inhibition of additional lipid A modification systems in *H. pylori* is unknown and is currently under investigation. However, phenotypic changes displayed by strain J99 are likely to be more representative of *H. pylori* strains as loss of *lpxE* in fresh clinical isolates of *H. pylori* results in phenotypes identical to that shown by *H. pylori* J99/ΔlpxE (Tran *et al.*, 2006).

The core and O-antigen carbohydrate domains of LPS are linked to the lipid A moiety through Kdo (Rietschel, 1990; Raetz and Whitfield, 2002). Transfer of Kdo residues to the disaccharide backbone of lipid A is catalysed by the glycosyltransferase WaaA (formerly KdtA). In *E. coli* and *Salmonella*, WaaA has been shown to be a bifunctional enzyme transferring two sugars to the tetra-acylated lipid A precursor known as lipid IV_A (Belunis *et al.*, 1995; Clementz and Raetz, 1991) (Fig. 2.3). Nevertheless, not all Gram-negative bacteria contain two Kdo sugars. The WaaA of *Bordetella pertussis* (Isobe *et al.*, 1999) and *Haemophilus influenzae* (Brabetz *et al.*, 2000) is monofunctional, and these organisms contain a single Kdo that is further modified with a phosphate group. A specific Kdo kinase known as KdkA catalyses addition of the phosphate group (White *et al.*, 1999). The inner core region of *H. pylori* LPS contains a single Kdo sugar (Aspinall *et al.*, 1994, 1995, 1996, 1997, 1999; Knirel *et al.*, 1999; Monteiro *et al.*, 2000a; Moran *et al.*, 2002c), thus suggesting that *H. pylori* WaaA is a monofunctional transferase.

Interestingly, Stead and co-workers (2005) demonstrated that *H. pylori* WaaA transfers two Kdo sugars to the disaccharide backbone of lipid A. Although these data were in direct contrast to the reported structure of the inner core region of *H. pylori* LPS (Aspinall *et al.*,1996; Knirel *et al.*, 1999; Monteiro *et al.*, 2000a; Moran *et al.*, 2002c), Stead *et al.* (2005) also reported the discovery of a novel enzyme that removes the outer Kdo sugar of *H. pylori* LPS. This Kdo-trimming enzyme, referred to as a Kdo hydrolase, was predicted to function following lipid A MsbA-dependent transport across the inner membrane since removal of the 1-phosphate group greatly increases the specific activity of the enzyme. Of course, this begs the obvious question as to why the bacterium transfers two Kdo sugars but then removes the second. It is quite possible that two Kdo sugars or phosphorylated Kdo is necessary for proper transport of core-lipid A across the inner membrane by *H. pylori* MsbA. Alternatively, secondary acylation of *H. pylori* lipid A precursors may require the presence of two Kdo sugars. After assembly and transport of the *H. pylori* core-lipid A, removal of the outer Kdo residue may benefit the organism by further reducing the net negative charge of the bacterial surface, thereby promoting resistance to antimicrobial peptides. Perhaps in conjunction with other modifications of the *H. pylori* Kdo_2-lipid A structure, removal of the outer Kdo sugar leads to reduction in the agonistic activity of *Helicobacter* LPS, contributing to persistence of the organism. To date, the structural gene encoding the Kdo-trimming enzyme has not been identified, but is required to determine the role of this modification in *H. pylori* pathogenesis.

The final modification of the *H. pylori* lipid A domain is the removal of the 3′-acyloxyacyl residue resulting in the production of a tetra-acylated lipid A. Hydrolysis of the 3′-linked fatty acids is catalysed by the outer membrane protein encoded by HP0694 (JHP0634), also known as LpxR. Inactivation of HP0694 by insertion of a selectable marker results in the production of a hexa-acylated lipid A structure, but does not inhibit modification of the phosphate groups (C. M. Stead and M. S. Trent, unpublished). LpxR was first identified in *S. typhimurium* by recognizing that its membranes were capable of deacylating lipid A precursors at the 3′-position if the divalent cation Ca^{2+} was present (Reynolds *et al.*, 2006). Although LpxR resides in the outer membrane of *Salmonella* under normal growth conditions, the enzyme remains dormant. On the other hand, the LpxR of *H. pylori* is constitutively active since *H. pylori* lipid A is not substituted at the 3′-position (Fig. 2.2). What factors promote enzymatic activation of *H. pylori* LpxR or the function of the deacylase remain unknown to date.

Implications of lipid A structure for LPS–immune receptor interactions and infection chronicity

Consistent with the low immunological activity of *H. pylori* LPS, its interaction with a variety of immune recognition molecules and receptors has consistently been reported to be low. The binding of *H. pylori* LPS to serum-associated LPS-binding protein (LBP), which acts as a catalytic protein to present complexed LPS to CD14 on the monocyte/macrophage cell surface, is poorer and exhibits slower binding kinetics than *E. coli* LPS, but also shows poorer binding to CD14 (Cunningham *et al.*, 1996). As LBP binds LPS through its lipid A component, these findings are in agreement with the proportionately lower ability of *H. pylori* LPS to activate monocytes, reflecting the unusual phosphorylation and acylation pattern in the lipid A (Moran *et al.*, 1997; Moran, 1998).

Additionally, the interactions of *H. pylori* and its LPS with so-called pattern recognition receptors for detection and response to microbial ligands, particularly the Toll-like receptor (TLR) family that are present on both epithelial cells and macrophages (Akira *et al.*, 2006), have been investigated. The response to *H. pylori* bacteria was found to be mediated predominantly not by TLR4 but rather by TLR2 (Smith *et al.*, 2003a; Mandell *et al.*, 2004). This contrasts with most Gram-negative bacteria that preferentially activate TLR4, notably by their potent LPS ligands, although an exception is the chronically infecting *Porphyromonas gingivalis* whose LPS has been reported to be a TLR2 ligand (Hirschfeld *et al.*, 2001). Using TLR-transfected cell lines, Smith *et al.* (2003a) deduced that *H. pylori* LPS was a TLR2 agonist, whose response was enhanced by CD14, and that this explained the predominant TLR2-mediated response of *H. pylori* bacteria. Moreover, they concluded the response was likely based on similarities in *H. pylori* lipid A structure to that of *P. gingivalis*. However, in contrast, Mandell *et al.* (2004) using highly pure LPS from *H. pylori* clinical isolates, assayed to be free of lipopeptides and other contaminants, reported TLR4- but not TLR2-mediated cytokine production by TLR-transfected cell lines and macrophages from knockout mice. Likewise, another study demonstrated that *H. pylori* acted via TLR4 to stimulate reactive oxygen production from guinea pig gastric pig cells (Kawahara *et al.*, 2001c). The discrepancies between these findings has been attributed to the presence of TLR2-activating contaminants in LPS test preparations (Mandell *et al.*, 2004; Moran, 2007), which have also contributed to inconsistencies in findings when investigating other ligand-immune receptor interactions (Inohara *et al.*, 2003). Also, differences in the LPS dosages (ng versus μg) used in the studies appear likely contributing factors (Smith *et al.*, 2006; Moran, 2007) since μg quantities more closely approximate the *in vivo* levels (Mandell *et al.*, 2004). On the other hand, variation in TLR4-mediated immune responsiveness to LPS from different *H. pylori* strains, particularly clinical isolates, has been noted (Kawahara *et al.*, 2001b; Mandell *et al.*, 2004; Lepper *et al.*, 2005) and likely reflects variations in the extent of acylation and/or phosphorylation of the lipid A moiety. Despite the proposed capacity of *H. pylori* to activate via TLR4 (Kawahara *et al.*, 2001c; Mandell *et al.*, 2004), the response of intact *H. pylori* bacteria is predominantly dependent on TLR2 (Smith *et al.*, 2003a; Mandell *et al.*, 2004) which has been taken to reflect a much lesser potency of the LPS-TLR4 interaction (Mandell *et al.*, 2004), consistent with the weaker activity of *H. pylori* LPS as a cytokine inducer (Moran 1999, 2001a).

Although LPS is apparently not the major determinant of the response to *H. pylori* bacteria, the LPS-TLR4 interaction may, nevertheless, play a role in disease pathogenesis. Initial studies suggested that *H. pylori* LPS does not stimulate human gastric cell lines, particularly MKN5 cells (Aihara *et al.*, 1997; Maeda *et al.*, 2001), but later investigations showed these cells lack MD-2 expression which is a critical cofactor for TLR4-mediated signalling (Smith *et al.*, 2003a). Besides, examination of biopsy specimens demonstrated the essential role of MD-2 in TLR4-dependent signalling during *H. pylori*-associated infection (Ishihara *et al.*, 2004). Notwithstanding a number of attempts, there has been failure to detect TLR2 expression in the stomach (Ortega-Cava *et al.* 2003; Mandell *et al.*, 2004), whereas TLR4 has been detected on gastric pit cells (Kawahara *et al.* 2001a; Ishihara *et al.*, 2004). Thus, in the absence of TLR2 in the gastric environment, the interaction of *H. pylori* LPS with TLR4 may be critical to the early detection of *H. pylori* by the innate immune system and may influence colonization (Sakagami *et al.*, 1997; Kawahara *et al.*, 2001a,b,c; Panthel *et al.*, 2003).

Accordingly, upon initial infection of the gastric mucosa, *H. pylori* may be weakly recognized by interaction of its LPS with TLR4. As seen with other chronic bacterial infections or colonizing commensal bacteria, the induction of low immunological responsiveness, may aid the prolongation of *H. pylori* infection and aid chronicity (Lee and Moran, 1994). It has been hypothesized that *H. pylori* LPS, and its lipid A component in particular, have evolved their present structure, on the one hand, to fulfil their role in producing a functional macromolecular matrix for bacterial interaction with its environment, whereas on the other, reducing the immune response to these necessary components of the bacterial outer membrane (Muotiala *et al.*, 1992; Moran, 1995a, 1996b). Thus, *H. pylori* by expressing an LPS with very weak TLR4 agonist activity and colonizing a TLR2-deficient environment may escape detection and elimination by the immune response initially. Consistent with this evasion of immune detection, though functional TLR5 is expressed in the adult stomach (Schmausser *et al.*, 2004), and TLR5 is considered to bind and respond to bacterial flagellins, *H. pylori* flagellins FlaA and FlaB induce a very low activation of TLR5-mediated responses (Lee *et al.*, 2003).

With progression of the immune response in long-term infection, a substantial inflammatory cytokine response to *H. pylori* may develop only after the infiltration of TLR-2-expressing granulocytes and monocytes into the infected gastric mucosa. *H. pylori* nuclear factor-κB (NF-κB) activation in macrophages has been suggested to involve TLR4 and CD14 (Maeda *et al.*, 2001), but *H. pylori* can activate mononuclear cells by an LPS-independent and TLR4-independent mechanism (Mai *et al.*, 1991; Bäckhed *et al.*, 2003), and hence non-LPS component(s) of the bacterium are the major cytokine-activating molecule(s). In general, since TLR2 recognizes bacterial lipoproteins, lipopeptides and lipoteichoic acids, the numerous putative lipoproteins encoded in the *H. pylori* genome (Doig *et al.*, 1999) represent candidate TLR2 ligands (Moran, 2007). An *H. pylori* heat shock protein (Hsp60) has been implicated in activation of TLR2 (Takenaka *et al.*, 2004), but this protein has also been reported to induce interleukin-6 production by macrophages via a TLR2-independent mechanism (Gobert *et al.*, 2004). Additionally, recognition of non-invasive *H. pylori* by epithelial cells has been shown to be mediated by Nod1, an intracellular pattern recognition molecule with specificity for peptidoglycan-derived muropeptides, which are delivered intracellularly by the type IV secretion system-encoding *cag* pathogenicity island (Viala *et al.*, 2004).

Furthermore, as a consequence of enzymatic degradation of LPS by human phagocytes, some LPS and/or lipid A partially modified structures can be excreted by exocytosis. Such compounds, retaining some immunological activity, could play a role as subliminal, low-grade, persistent stimuli involved in *H. pylori* pathogenesis (Jirillo *et al.*, 1999). In accord with this, *H. pylori* LPS can induce nitric oxide synthase in an *in vivo* animal model, thereby contributing to intestinal damage (Lamarque *et al.*, 2000; Kiss *et al.*, 2001), and also influence gastric motor function (Quintana *et al.*, 2005). Although controversial, chronic *H. pylori* infection has been implicated as a contributing or risk factor in the development of certain extragastric diseases, e.g. coronary heart disease and athersoclerosis. Whether during long-term chronic infection of the gut mucosa by *H. pylori*, low-grade LPS/lipid A stimuli contribute to such extragastric infection sequelae is a tantalizing question (Moran, 1999; Grebowska *et al.* 2006). The structural prerequisites for low-grade expression of tissue factor (leading to fibrin formation) and plasminogen activator inhibitor type 2 (leading to persistence of the formed fibrin) by monocytes are present within *H. pylori* LPS (Semeraro *et al.*, 1996). Moreover, *H. pylori* LPS has been shown to be capable of inducing a microvascular inflammatory response in cardiac, renal, hepatic and pulmonary tissues mediated by inducible nitric oxide synthase (Whittle *et al.*, 2001) and to produce a vasomotor effect on isolated rabbit aorta (Hynes *et al.*, 2003).

O-chain

The expression of Le and related blood group antigens in the O-chains of *H. pylori* (see reviews, Moran, 2001a,b; Monteiro, 2001) has led to intensive investigations of the biosynthesis of these human antigens by *H. pylori* and the implications of their expression for pathogenesis and virulence of the bacterium.

Le antigens in humans

Le antigens are biochemically related to the ABH blood groups that are formed by sequential addition of fucose, galactose and *N*-acetylgalactosamine to backbone carbohydrate chains of both lipids and proteins. These antigens are constituted from two types of backbone structures, containing galactose (Gal) and *N*-acetylglucosamine (GlcNAc) sugars, so-called type 1 chains [composed of Gal-β(1,3)-GlcNAc] giving rise to Lewis^a (Le^a), sialyl-Le^a and Lewis^b (Le^b), and type 2 chains [composed of Gal-β(1,4)-GlcNAc, also termed *N*-acetyllactosamine (LacNAc)] giving rise to Lewis^x (Le^x), sialyl-Le^x, and Lewis^y (Le^y) (Fig. 2.5). In addition to the genes controlling the synthesis of these structures, another gene locus controls the expression of the antigens in secretions and on various cell types (for review of genetic determination and biosynthesis, see Green, 1989). The terms secretor and non-secretor indicate the capacity of an individual to secrete such substances; secretors produce ABH, Le^b and Le^y, whereas non-secretors produce Le^a and Le^x in their saliva (Sakamoto *et al.*, 1984). In the human stomach, Le^a and Le^b are mainly expressed on the surface and foveolar epithelia, whereas Le^x and Le^y are predominantly expressed by the mucous, chief and parietal cells of the gastric glands (Kobayashi *et al.*, 1993), though the physical separation of Le antigens along the gastric pits is not always obvious (Murata *et al.*, 1992). More specifically, in non-secretors the surface and foveolar epithelia express Le^a, whereas these epithelia express Le^b and some Le^y in secretors; but irrespective of secretor status of the individual, the glandular epithelium lacks type 1 antigens (Le^a and Le^b) and expresses type 2 antigens (Le^x and Le^y), though in non-secretors Le^y expression is weaker in the glands (Sakamoto *et al.*, 1989). Thus, a variety of Le antigens are expressed within the ecological niches of *H. pylori*.

Le antigens in *H. pylori* LPS

Typically, the O-chains of *H. pylori* have a poly-(*N*-acetyl-β-lactosamine) chain decorated with multiple lateral α-L-fucose residues forming internal Le^x determinants with terminal Le^x or Le^y determinants (Figs. 2.1 and 2.5) or, in some strains with additional, glucose or galactose residues (Aspinall *et al.*, 1994, 1995, 1996, 1997, 1999; Aspinall and Monteiro, 1996; Knirel *et al.*, 1999; Moran *et al.*, 2002c; Khamri *et al.*, 2005). Furthermore, Le^a, Le^b, sialyl-Le^x, and H-1 determinants have been structurally described in other strains (Monteiro *et al.*, 1998, 2000a,b), as well as the related blood groups A and B (Moran *et al.*, 1999; Monteiro, 2001).

Expression of Le^x or Le^y determinants is a common property of (80–90%) *H. pylori* strains that have been screened serologically (Simoons-Smit *et al.*, 1996; Wirth *et al.*, 1996; Marshall *et al.*, 1999; Heneghan *et al.*, 2000). Nevertheless, serological analysis can under-estimate Le^x and Le^y expression in a population of strains as some strains that are non-typeable with anti-Le antibodies have been shown to express these antigens when examined in structural studies (Knirel *et al.*, 1999). Importantly, false-negative results are dependent upon the serological assay format used when testing strains (Hynes and Moran, 2000). Although it had been suggested that Asian isolates express predominantly type 1 (Le^a, Le^b, H-1) antigens compared with Western strains (predominantly expressing type 2 Le^x or

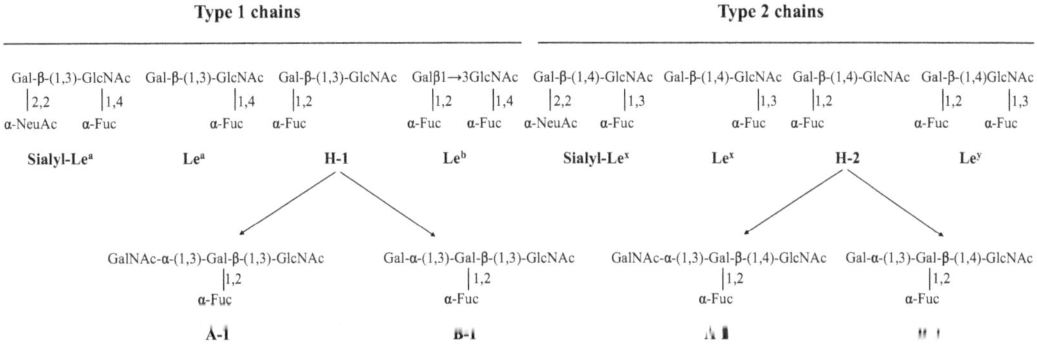

Figure 2.5 Structural relationship between Le and A, B and H blood group antigens. Abbreviations: Fuc, fucose; Gal, galactose; GalNAc, *N*-acetylgalactosamine; Glc, glucose; GlcNAc, *N*-acetylglucosamine; NeuAc, *N*-acetylneuraminic (sialic acid). Except for fucose, which is in the L-configuration , all sugars possess the D-configuration.

Le^y determinants) (Monteiro *et al.*, 2000b), the former have been detected in European isolates along with Le^x and Le^y expression (Heneghan *et al.*, 2000). Thus, a mosaicism of Le antigen and blood group expression can occur in the same O-chain, and thereby, along with some variability in the core OS, gives rise to antigenic diversity that can be detected by antibody and lectin probing (Simoons-Smit *et al.*, 1996; Hynes *et al.*, 1999, 2002).

Nonetheless, there are non-typeable strains that show no reaction with anti-Le^x, -Le^y, -Le^a and -Le^b antibodies (Simoons-Smit *et al.*, 1996) emphasizing that not all strains express Le antigens (Moran 2001a,b). This is particularly true for those strains associated with asymptomatic infection (Kocharova *et al.*, 2000; Senchenkova *et al.*, 2001). Detailed structural analysis has shown that some of these strains possess O-chains composed of L-rhamnose and 3-C-methyl-D-mannose, a sugar not previously found in nature (Kocharova *et al.*, 2000). Other strains have been shown to produce heptoglycans of 2- and 3-linked DD-heptose units (Senchenkova *et al.*, 2001), which has been confirmed in a subsequent independent investigation (Monteiro *et al.*, 2001). The absence of Le antigen mimicry in asymptomatic strains, and animal studies in which a genetically modified *H. pylori* strain lacking Le antigen expression failed to induce gastritis compared to the parental strain (Eaton *et al.*, 2004), support the view that the Le antigen expressing O-chain contributes directly to disease development.

Biosynthesis of Le antigen-mimicking O-chains

For the synthesis of the LacNAc backbone of type 2 antigens such as Le^x and Le^y (Fig. 2.5), the β(1,4)-galactosyltransferase gene (HP0826/JHP0765) of *H. pylori* has been identified, cloned and shown to encode an enzyme that efficiently produces LacNAc units (Endo *et al.*, 2000; Logan *et al.*, 2000). Moreover, as expected, knockouts in this gene produce a truncated LPS lacking an O-chain (Logan *et al.*, 2000). Similarly, *H. pylori* knockout mutants in *galE* result in truncation of LPS (Moran, 1999; Edwards *et al.*, 2000; Moran *et al.*, 2000; Perepelov *et al.*, 2001). Since *H. pylori* lacks galactokinase and cannot utilize exogenous galactose, UDP-galactose-4-epimerase (encoded by *galE*) is a prerequisite for biosynthesis of UDP-galactose that is required in O-chain synthesis. Also, the gene encoding the β(1,3)-glucosaminyltransferase (HP1105/JHP1032) that is required for type 2 chain synthesis has been identified (Logan *et al.*, 2005). Although of low homology, an orthologue of *neuA* (HP0326/JHP309) encoding CMP-*N*-acetylneuraminic acid synthetase was suggested to be involved in sialyl-Le^x expression by *H. pylori* (Berg *et al.*, 1997) and, supporting this, sialyl-Le^x has been confirmed to occur in the O-chain of a strain of *H. pylori* (Monteiro *et al.*, 2000a).

Unlike many enterobacterial species (e.g. *E. coli* and *S. typhimurium*), the genes for LPS synthesis are not clustered but are scattered throughout the *H. pylori* genome, with the exception of those (*rfbM*, *rfbD* and *wbcJ* orthologues) encoding proteins involved in GDP-fucose synthesis (Tomb *et al.*, 1997; Alm *et al.*, 1999; Doig *et al.*, 1999). The latter acts as the donor of fucose in the fucosyltransferase reactions required for Le antigen synthesis, and is itself synthesized in two main steps by a dehydratase and then an epimerase–reductase. Genomic analysis of strains 26695 and J99 has identified the orthologues of *rfbM* (HP0043/JHP0037) predicted to encode a GDP-D-mannose pyrophosphorylase, *rfbD* (HP0044/JHP0038) a GDP-D-mannose dehydratase, and *wbcJ* (HP0045/JHP0039) an epimerase-reductase responsible for the final conversion of GDP-mannose to GDP-fucose (Berg et al ., 1997; Doig *et al.*, 1999). Insertional mutations in the *rfbM* and *wbcJ* genes have confirmed their role in fucosylation of the O-chains (McGowan *et al.*, 1998; Edwards, *et al.*, 2000). However, due to the potential biological importance of fucosylation and Le antigen mimicry in the *H. pylori* O-chains, attention has focused on the fucosyltransferases (FucTs) of *H. pylori*.

FucTs

These enzymes catalyse the reactions in which a fucose residue is transferred from the donor activated sugar (GDP-fucose) to the acceptor molecule. FucTs belong to the glycosyltransferase superfamily which is defined in the category of carbohydrate-active enzymes (CAZY) (www.CAZY.org/fam/acc_GT.html). Based on the site of Fuc addition, FucTs are classified as α(1,2)-, α(1,3/4)-, α(1,6)-, and O-FucTs, and are expressed in prokaryotes and eukaryotes (Ma *et al.*, 2006b). Mammalian α(1,2)- and α(1,3/4)-FucTs are involved in the last steps of synthesis of ABH and Le-related structures, and likewise, *H. pylori* strains have α(1,2)- and α(1,3/4)-FucTs that are involved in Le antigen synthesis in LPS (Ge *et al.*, 1997; Martin *et al.*, 1997; Wang *et al.*, 1999a,b; Rasko *et al.*, 2000b; Moran 2001b). The α(1,2)-FucTs transfer Fuc to the terminal Gal residue of type 1 [Gal-β(1,3)-GlcNAc] and type 2 [Gal-β(1,4)-GlcNAc] chains in an α(1,2)-linkage, and α(1,3/4)-FucTs add Fuc to the subterminal GlcNAc moiety of type 2 or 1 chains in α(1,3)- and α(1,4)-linkages, respectively (Fig. 2.5). As in humans, the biosynthesis of Le^x in *H. pylori* is strictly ordered since Chan *et al.* (1995) showed, using acceptor molecules for fucosylation and analysis of the enzyme reaction products, that addition of β(1,4)-linked Gal is followed by that of α(1,3)-linked Fuc. Of note, *H. pylori* O-chains can contain incompletely fucosylated LacNAc units and evidence suggests that the *H. pylori* O-chain is synthesized by sequential addition of monosaccharides (Aspinall *et al.*, 1994, 1995, 1996; Aspinall and Monteiro, 1996; Berg *et al.*, 1997), which has been described in only a few bacterial species (Senchenkova *et al.*, 1996, 1997), rather than by the polymerization of repeating units to form the O-chain as used in the majority of Gram-negative bacteria (Rietschel *et al.*, 1990). In the case of difucosylated Le antigens (i.e. Le^b and Le^y), synthesis can occur via two pathways: terminal fucosylation involving α(1,2)-linkage and then subterminal fucosylation by α(1,3)- or α(1,4)-linkage, or conversely, subterminal followed by terminal fucosylation. Mammalian cells predominantly use the former pathway (Henry *et al.*, 1995) whereas the latter mechanism of synthesis is primarily used by *H. pylori* (Rasko *et al.*, 2000a; Wang *et al.*, 2000).

H. pylori α*(1,3/4)-FucTs*

The *H. pylori* genome contains two paralogous genes, termed *futA* and *futB* (HP0379/JHP1002 and HP0651/JHP0596) (Tomb *et al.*, 1997; Alm *et al.*, 1999; Doig *et al.*, 1999). Such genes encode enzymes with either α(1,3)- and/or α(1,4)-FucT activities, which thereby use type 2 and/or type 1 acceptor molecules, respectively (Ge *et al.*, 1997; Martin *et al.*, 1997; Rasko *et al.*, 2000a,b). The α(1,3/4)-FucTs of *H. pylori* NCTC 11637 and NCTC 11639 (*futB* and *futA*, respectively) have exclusive α(1,3)-activity (Ge *et al.*, 1997; Martin *et al.*, 1997) like human FucT IV. Those of *H. pylori* UA948 (*futA*) and UA1111 (*futA* and *futB*) possesses both α(1,3)- and/or α(1,4)-FucT activities (Rasko *et al.*, 2000a,b), and like human FucT V, *H. pylori* UA948 FutA has preference for the type 2 acceptor molecule (Rasko *et al.*, 2000b). *H. pylori* DSM6709 has been reported to contain primarily α(1,4)-FucT activity and only limited α(1,3)-activity (Rabbani *et al.*, 2005), like human FucT III. Thus, the activity and specificity of these enzymes may vary between the two paralogues in one strain, as well as between the orthologues in different strains.

DNA motifs near the 5′-end of these genes at two distinct polynucleotide repeats (polyA–polyC) have been deduced to indicate regulation through slipped-strand repair (Alm *et al.*, 1999; Moran, 2001b); these polynucleotide tracts have a higher frequency of addition or deletion of one or more base pairs during DNA replication than the normal mutation frequency, leading to an on/off status of the gene (Wang *et al.*, 2000). Thus, the slipped-strand mispairing in intragenic poly-C tract regions results in alternate reading frames and an on/off switch influencing FucT expression that gives rise to *H. pylori* subclones with different Le glycosylation patterns (Appelmelk *et al.*, 1999; Wirth *et al.*, 1999; Nilsson *et al.*, 2006). Experimentally, mutation in the polyA or polyC tract leading to the on/off status of the gene has been demonstrated in *futA* of *H. pylori* J99 and NCTC 11639 (Wang *et al.*, 2000) and in *futB* of *H. pylori* NCTC 11637 (Appelmelk *et al.*, 1999). The occurrence of these Le-expressing subclones, not only has been demonstrated in laboratory strains (Appelmelk *et al.*, 1999), but also in clinical isolates from different regions of the human stomach and over the course of the infection (Wirth *et al.*, 1999; Nilsson *et al.*, 2006). However, the length of polynucleotide tracts does not always correlate with Le^x or Le^y expression in *H. pylori* clinical isolates (Ryan *et al.*, 2000), indicating that post-translational events and the availability of activated sugar intermediates, in addition to an active enzyme, are important for the expression of the Le phenotype.

Comparing different *H. pylori* strains, the α(1,3/4)-FucTs show >70% sequence identity, with a highly conserved internal catalytic domain, but divergent N- and C-termini (Rasko *et al.*, 2000b). Alignment of the *H. pylori* α(1,3/4)-FucTs with the corresponding mammalian enzyme has demonstrated significant homology only in a very short region within the catalytic domain where the two α(1,3/4)-FucT motifs are localized (Ge *et al.*, 1997; Martin *et al.*, 1997). Despite this, these two enzyme families appear to share functional similarities, with a conserved mechanistic and structural basis for fucose transfer (Ma *et al.*, 2006b). Kinetic studies have shown that *H. pylori* α(1,3/4)-FucTs are comparable to their mammalian counterparts (Rabbani *et al.*, 2005; Ma *et al.*, 2006a). Similar to human α(1,3/4)-FucTs, the hydroxyl group of carbon-6 of Gal in type 1 and 2 acceptor molecules is also essential for the equivalent *H. pylori* enzymatic activity (Ma *et al.*, 2006a). *H. pylori* α(1,3/4)-FucTs are able to use both the sialylated type 1 and 2 acceptors, like the human enzymes (Rabbani *et al.*, 2005; Ma *et al.*, 2006a). Also, like human α(1,3/4)-FucTs, those of *H. pylori* catalyse fucose transfer following a sequential mechanism whereby binding of donor molecule occurs first, followed by that of the acceptor (Ma *et al.*, 2006a).

H. pylori α(1,3/4)-FucTs lack the N-terminal cytosolic tail and transmembrane domain of the mammalian counterpart enzymes that are Type II membrane proteins (Kleene and Berger, 1993), but instead have a heptad-repeat region at the C-terminus that is absent from mammalian enzymes (Ge *et al.*, 1997; Martin *et al.*, 1997; Rasko *et al.*, 2000b). This heptad repeat region contains a leucine-zipper motif potentially mediating dimer formation (Ge *et al.*, 1997; Martin *et al.*, 1997), and the ability of *H. pylori* α(1,3/4)-FucTs to form dimers has been confirmed experimentally (Ma *et al.*, 2006a; Lin *et al.*, 2006). In addition, thermal denaturation studies have confirmed that the heptad repeats facilitate protein folding, and hence helps to maintain a stable protein structure that is aligned with the dimer formation (Lin *et al.*, 2006). The C-terminal heptad repeat region consists of a variable number of DD/NLRV/INY tandem repeats. Nilsson *et al.* (2006) found that the heptad-repeat number of FutA and FutB in *H. pylori* isolates correlated with the size of the O-chain being fucosylated. The data supports a molecular ruler mechanism for how *H. pylori* varies its fucosylation pattern within the O-chain, where one heptad-repeat in the enzyme corresponds to one LacNAc unit in the O-chain. Thus, a model has been proposed in which FutA and FutB form heterodimers or homodimers in which the number of heptad-repeats controls the active fucosylation site from a fixed point and thus determines the size of the O-chain that is fucosylated (Nilsson *et al.*, 2006). Moreover, the number of heptad repeats in FutA and FutB appears to affect subsequent α(1,2)-fucosylation by FutC, probably reflecting availability of terminal Le^x substrates on the O-chains.

It has been deduced that the amphipathic helices of the *H. pylori* α(1,3/4)-FucTs may act as membrane anchors with the hydrophobic face embedded in the membrane and positive charges interacting with phospholipid headgroups (Ma *et al.*, 2003, 2006b). Using domain swapping and site-directed mutagenesis, the C-terminal hypervariable region of *H. pylori* α(1,3/4)-FucTs, which is immediately upstream of the heptad-repeat region, and particularly an aromatic residue (Tyr), was shown to be responsible for type 1 acceptor recognition (Ma *et al.*, 2006a). Overall, mammalian and *H. pylori* FucTs share similar domain architecture but with opposite topology (see review, Ma *et al.*, 2006b). Thus, the membrane anchor region (N-terminus of mammalian but C-terminus of *H. pylori* FucTs) can be truncated without significant loss of enzymatic activity, but removal of a small number of amino acid residues from the opposite terminus (C-terminus in mammalian and N terminus in *H. pylori*) of FucTs can almost completely abolish enzymatic activity (Ge *et al.*, 1997; Lin *et al.*, 2006; Ma *et al.*, 2006a).

H. pylori α(1,2)-FucTs

Initially, no putative gene for α(1,2)-FucTs was found in the genome sequence of *H. pylori* 26695 (Tomb *et al.*, 1997; Berg *et al.*, 1997; Marais *et al.*, 1999). A truncated gene (HP0094) with a C14 tract was found and *in silico* insertion of a C-G pair yielded a full-length protein with strong homology to other α(1,2)-FucTs (Berg *et al.*, 1997). Further analysis of the *H. pylori* 26695 genome showed that it contained two truncated reading frames (HP0093 and HP0094), compared to that of *H. pylori* J99 (JHP0086), and the gene was assigned as *futC* (Tomb *et al.*, 1997; Berg *et al.*, 1997; Alm *et al.*, 1999; Doig *et al.*, 1999). The α(1,2)-FucTs from *H. pylori* strains share very high sequence identity with each other (about 95%) but very low identity (up to 22%) compared with mammalian enzymes (Oriol *et al.*, 1999). *H. pylori* α(1,2)-FucTs lack the N-terminal cytosolic tail and the transmembrane domain compared with mammalian enzymes, and in one *H. pylori* strain α(1,2)-FucT has been demonstrated to be located in the cytoplasm as a soluble protein (Wang *et al.*, 1999a).

Similar to the *futA* and *futB* genes, *futC* also contains the polyA-polyC tract near the 5′-end, and during DNA replication, the addition or deletion of one or more base pairs within these tracts occurs leading to on/off status of the gene (Appelmelk *et al.* 1998b, 1999, 2000a; Wang *et al.*, 1999a, 1999b, 2000). Along with these short repeat sequences, imperfect TAA- or GAA- or AAA-repeats occur at the mid-region of *futC*, immediately downstream of the poly-C tract (Wang *et al.*, 1999b). In addition to these signature sequences, an internal Shine–Dalgarno-like context, a heptamer (AAAAAAG) and the downstream potential stem-loop structure are present also (Wang *et al.*, 1999b; Ma *et al.*, 2006b). The hypermutable sequence provides a possibility of frequent shifting into and out of coding frame by a polymerase slippage mechanism. During the translation process, ribosome slippage can occur to the −1 reading frame within the heptamer at a frequency of 50%, thereby encoding a full-length α(1,2)-FucT rather than two truncated products. Such a mechanism has been confirmed to occur in *H. pylori* 26695 (Wang *et al.*, 1999b). Furthermore, *futC* in some *H. pylori* strains do not possess a valid promoter region resulting in the inability to express the α(1,2)-FucT (Wang *et al.*, 1999b; Ma *et al.*, 2006b). In certain strains, insertions in the hypermutable region generate a frameshift mutation which cannot be compensated for by a translation frameshift cassette (Wang *et al.*, 1999b). Hence, the on/off status of the *H. pylori futC* can be controlled at both the translational and transcriptional levels. Overall, the α(1,2)-FucT activity has been found to be much lower generally than that of the α(1,3/4)-FucTs in the same *H. pylori* strain (Wang *et al.*, 1999a,b; 2000). Also, as discussed above, the number of heptad-repeats in FutA and FutB appears to affect subsequent α(1,2)-fucosylation by FutC (Nilsson *et al.*, 2006).

The frequency of phase variation resulting in on/off switching of Le antigen expression by *H. pylori* is about 0.2–0.5% (Appelmelk *et al.*, 1998b, 1999, 2000a). To predict the Le glycosylation pattern in a *H. pylori* O-chain it is necessary to consider the on/off status of the *futA*, *futB*, and *futC* genes, the number of heptad repeats, the activity and expression levels of the functional enzymes, the preferences of the expressed enzyme for distinctive acceptor molecules, and the availability of activated sugar intermediates.

Roles of LPS-expressed Le antigens in *H. pylori* pathogenesis

Biologically, *H. pylori* Le expression has been implicated in evading the immune response upon initial infection and in influencing bacterial colonization and adhesion (Moran, 1999, 2001a,b), but with the progress of chronic infection, has been suggested to contribute to gastric autoimmunity that leads to gastric atrophy, a precursor state of gastric cancer (Appelmelk *et al.*, 1996; Heneghan *et al.*, 2001; Broutet *et al.*, 2002; Hynes *et al.*, 2005). Although the latter has been considered controversial (Appelmelk *et al.*, 1998a; Moran, 1999; Moran, 2001a), more recent evidence as reviewed elsewhere (Moran, 2008), supports a contributing role of Le mimicry-induced autoantibodies, as well as peptide-directed autoreactive T-cells (D'Elios *et al.*, 2004), in the development of gastric atrophy.

Based on the known expression of Le antigens in the gastric mucosa, it was proposed that bacterial molecular mimicry and its adaptation could provide an escape for *H. pylori* from the host humoral response by preventing the formation of antibodies shared by the host and bacterium (Wirth *et al.*, 1997; Appelmelk *et al.*, 2000b). As detailed previously (Wirth *et al.*, 1997; Moran, 1999; Appelmelk *et al.*, 2000b; Moran *et al.*, 2002b; Wirth *et al.*, 2006) some human population and primate model studies supported this concept, whereas others have not. Importantly, it has been demonstrated that strains expressing Le^x and strains expressing Le^y can be isolated from the same host (Wirth *et al.*, 1999), and extensive diversity of Le^x and Le^y in O-chains can occur over time and in different regions of the human stomach (Nilsson *et al.*, 2006), thereby apparently contradicting the hypothesis of bacterial Le antigen adaptation to that of gastric Le expression-related secretor status of the host.

Notwithstanding this, the diversity of Le expression by *H. pylori* isolates in one host may reflect an ability of the bacterium to adapt to differing microniches and environmental conditions within the human stomach (Moran *et al.*, 2002c; Keenan *et al.*, 2004). Nevertheless, O-chain and Le^x expression have been reported of importance for *in vivo* colonization in mice (Moran, 1999; Moran *et al.*, 2000; Appelmelk *et al.*, 2000b). Polymeric Le^x expression in the *H. pylori* O-chains has been shown to mediate, at least in part, *H. pylori* adhesion to the human antral gastric mucosa (Edwards *et al.*, 2000) and whose gastric receptor has been identified as the galactoside-binding lectin, galectin-3 (Fowler *et al.*, 2006). However, *H. pylori* adhesion is mediated by a number of adhesins (Moran 1995a, 1996a; Moran and Wadström, 1998), with polymeric Le^x playing a distinct role (Mahdavi *et al.*, 2003). Consistent with an adhesion role, strains with a high expression of Le^x cause a higher colonization density in patients than in those with weaker expression (Heneghan *et al.*, 2000). By assisting bacterial adhesion and interaction with the gastric mucosa, bacterial Le^x expression could enhance delivery of secreted products into the gastric mucosa promoting chemotaxis and leucocyte infiltration (Rieder *et al.*, 1997; Moran, 1999). On/off switching of fucosyltransferase activities, and hence phase variation in Le expression (Appelmelk *et al.*, 1999), would allow detachment of *H. pylori* cells from gastric epithelial cells into the mucus layer, which would act as a reservoir for continued bacterial infection. Because of differences in pH within mucus and on the gastric cell surface (pH 2 on the luminal side of the gastric mucus layer to almost pH 7 on the cell surface), the observed influence of pH on Le expression by *H. pylori* (Moran *et al.*, 2002c) would be in accord with this detachment mechanism.

Counterbalancing the gastric adhesion role of the O-chains, surfactant protein D (SP-D), which is a C-type lectin involved in antibody-independent pathogen recognition and clearance (Moran *et al.*, 2005b), binds *H. pylori* LPS resulting in bacterial immobilization and aggregation (Murray *et al.*, 2002; Khamri *et al.*, 2005). Levels of expression of SP-D are significantly increased in *H. pylori*-associated antral gastritis compared to normal gastric mucosa, and SP-D expression co-localizes with *H. pylori* organisms (Murray *et al.*, 2002). Also, experiments in $SP\text{-}D^{-/-}$ mice revealed that *Helicobacter* colonization was more common in the absence of SP-D (Moran *et al.*, 2005b), underscoring an important influence of SP-D binding on the establishment of infection. To evade this important mechanism of innate immune recognition of *H. pylori*, escape variants can arise within the bacterial cell population with modifications in O-chain glycosylation decreasing their interaction with SP-D (Khamri *et al.*, 2005). Likewise, strain-to-strain variability in Le expression was shown to modulate the interaction of *H. pylori* LPS with the cellular innate immune receptor DC-SIGN (another C-type lectin) on gastric dendritic cells which contributes to changed T-cell polarization after innate immune activation (Bergman *et al.*, 2004).

Concluding remarks and future directions

H. pylori has a well-established ability to evade and even subvert innate and adaptive immune responses during long-term infection and, as has been discussed, certain properties of *H. pylori* LPS, particularly those associated with the O-chain and lipid A components, contribute to these. Similar to *H. pylori*, many colonizers of the gut, including commensals and pathogens, have developed mechanisms to vary LPS structure and thus may subvert recognition by innate immune receptors (Miller *et al.*, 2005). This underlines the importance of LPS as a surface structure and immune recognition ligand.

Comparative examination of the two genome sequences of *H. pylori* 26695 and J99 (Tomb *et al.*, 1997; Alm *et al.*, 1999) have revealed important and useful information on LPS biosynthesis, and the conservancy of many LPS biosynthesis genes. The latter has also been confirmed with the more recent availability of the genome sequence of a chronic atrophic gastritis isolate (Oh *et al.*, 2006). Nonetheless, comparative genomic analysis of a wider variety of strains, particularly those from different pathologies, could prove a useful approach to gain information on LPS structures contributing to interstrain differences in virulence properties.

Although the complement of *H. pylori* genes required for lipid A synthesis are known, and the biosynthesis and modification pathways have been described (Tran *et al.*, 2004, 2005, 2006; Stead *et al.*, 2005; Reynolds *et al.*, 2006), further studies are required to determine whether lipid A modifications are somehow regulated during colonization in the gastric mucosa. Likewise, similar intensive studies are required with respect to the other regions of LPS, namely the core OS and O-chain. Some insights have already been gained on the influence of acidic pH on fucosylation of the O-chain (McGowan *et al.*, 1998; Moran *et al.*, 2002c), but how these relate to or occur during *in vivo* colonization needs testing. Importantly, modulation of LPS structure could influence the interaction with TLR4 and affect the interaction of the bacterium with other immune recognition molecules, and thus influence colonization and its disease outcome. The interaction of *H. pylori* LPS with SP-D and the influence of phase variation in FucTs on *H. pylori*–SP-D interaction serves as an obvious paradigm (Khamri *et al.*, 2005). Despite extensive investigations on the molecular mechanisms of regulation of the FucTs that are required for Le antigen synthesis, the environmental factors regulating the expression of these enzymes and their functioning require more detailed examination. With availability of the crystal structures of *H. pylori* FucTs (Sun *et al.*, 2007) further insights can be gained into the molecular recognition and functioning of these enzymes. Comparatively, molecular and genetic interstrain variation of the core OS have received less attention than those of the O-chain (Moran 1999, 2001a,b; Monteiro, 2001), despite the potential contribution of this LPS region to virulence properties.

Finally, despite being at an early stage of investigation, the comparative examination of LPS from the different members of the genus *Helicobacter* (Ferris *et al.*, 2001; Moran *et al.*, 2002a; Hynes *et al.*, 2004) which contains representative bacterial species that induce acute or chronic infections, as well as those that colonize the gastric or hepato-intestinal environments, promises to yield further insights into the role of LPS in chronic infection in various compartments of the gut. Importantly, the availability of the genome sequence of *Helicobacter hepaticus* (Suerbaum *et al.*, 2003) will aid in these investigations.

Acknowledgements

Investigations in the authors' laboratories are supported by the Irish Research Council for Science Engineering and Technology, Millennium Fund, Higher Education Authority PRTL-3 programme of the National Development Plan and European Union Marie Curie programme grant MTKD-CT-2005-029774 (to A.P.M.), and by the National Institutes of Health grant AI-064184 (to M.S.T.).

References

Aihara, M., Tsuchimoto, D., Takizawa, H., Azuma, A., Wakebe, H., Ohmoto, Y., Imagawa, K., Kikuchi, M., Mukaida, N., and Matsushima, K. (1997). Mechanisms involved in *Helicobacter pylori*-induced interleukin-8 production by a gastric cell line, MKN45. Infect. Immun. *65*, 3218–3224.

Akira, S., Uematsu, S., and Takeuchi, O. (2006). Pathogen recognition and innate immunity. Cell *124*, 783–801.

Alm, R.A., Ling, L.L.S., Moir, D.T., King, B.L., Brown, E.D., Doig, P.C., Smith, D.R., Noonan, B., Guild, B.C., deJonge, B.L., Carmel, G., Tummino, P.J., Caruso, A., Uria-Nickelsen, M., Mills, D.M., Ives, C., Gibson, R., Merberg, D., Mills, S.D., Jiang, Q., Taylor, D.E., Vovis, G.F., and Trust, T.J. (1999). Genomic-sequence comparison of two unrelated isolates of the human gastric pathogen *Helicobacter pylori*. Nature 397, 176–180.

Appelmelk, B.J., Faller, G., Claeys, D., Kirchner, T., and Vandenbroucke-Grauls, C.M.J.E. (1998a). Bugs on trial: the case of *Helicobacter pylori* and autoimmunity. Immunol. Today *19*, 296–299.

Appelmelk, B.J., Martin, S.L., Monteiro, M.A., Clayton, C.A., McColm, A.A., Zheng, P., Verboom, T., Maaskant, J.J., van den Eijnden, D.H., Hokke, C.H., Perry, M.B., Vandenbroucke-Grauls, C.M.J.E., and Kusters, J.G. (1999). Phase variation in *Helicobacter pylori* lipopolysaccharide due to changes in the lengths of poly(C) tracts in α3-fucosyltransferase genes. Infect. Immun. *67*, 5361–5366.

Appelmelk, B.J., Martino, M.C., Veenhof, E., Monteiro, M.A., Maaskant, J.J., Negrini, R., Lindh, F., Perry, M., Giudice, G.D., and Vandenbroucke-Grauls, C.M.J.E. (2000a). Phase variation in H type I and Lewis a epitopes of *Helicobacter pylori* lipopolysaccharide. Infect. Immun. *68*, 5928–5932.

Appelmelk, B.J., Monteiro, M.A., Martin, S.L., Moran, A.P., and Vandenbroucke-Grauls, C.M.J.E. (2000b). Why *Helicobacter pylori* has Lewis antigens. Trends Microbiol. *8*, 565–570.

Appelmelk, B.J., Shiberu, B., Trinks, C., Tapsi, N., Zheng, P.Y., Verboom, T., Maaskant, J., Hokke, C.H., Schiphorst, W.E.C.M., Blanchard, D., Simoons-Smit, I.M., van den Eijnden, D.H., and Vandenbroucke-Grauls, C.M.J.E. (1998b). Phase variation in *Helicobacter pylori* lipopolysaccharide. Infect. Immun. *66*, 70–76.

Appelmelk, B.J., Simoons-Smit, I., Negrini, R., Moran, A.P., Aspinall, G.O., Forte, J.G., De Vries, T., Quan, H., Verboom, T., Maaskant, J.J., Ghiara, P., Kuipers, E.J., Bloemena, E., Tadema, T.M., Townsend, R.R., Tyagarajan, K., Crothers, J.M., Monteiro, M.A., Savio, A., and De Graaf, J. (1996). Potential role of molecular mimicry between *Helicobacter pylori* lipopolysaccharide and host Lewis blood group antigens in autoimmunity. Infect. Immun. *64*, 2031–2040.

Aspinall, G.O., Mainkar, A.S., and Moran, A.P. (1999). A structural comparison of lipopolysaccharides from two strains of *Helicobacter pylori*, of which one strain (442) does and the other strain (471) does not stimulate pepsinogen secretion. Glycobiology *9*, 1235–1245.

Aspinall, G.O., and Monteiro, M.A. (1996) Lipopolysaccharides of *Helicobacter pylori* strains P466 and MO19: structures of the O antigen and core oligosaccharide regions. Biochemistry *35*, 2498–2504.

Aspinall, G.O., Monteiro, M.A., Moran, A.P., Pang, H., Penner, J.L., and Shaver, R.T. (1995). Lipopolysaccharides from *Helicobacter pylori*. *Prog. Clin. Biol. Res. 392*, 93–101.

Aspinall, G.O., Monteiro, M.A., Pang, H., Walsh, E.J., and Moran, A.P. (1994). O antigen chains in the lipopolysaccharide of *Helicobacter pylori* NCTC 11637. Carbohydr. Lett. *1*, 151–156.

Aspinall, G.O., Monteiro, M.A., Pang, H., Walsh, E.J., and Moran, A.P. (1996). Lipopolysaccharide of the *Helicobacter pylori* type strain NCTC 11637 (ATCC 43504): structure of the O antigen chain and core oligosaccharide regions. Biochemistry *35*, 2489–2497.

Aspinall, G.O., Monteiro, M.A., Shaver, R.T., Kurjanczyk, L.A., and Penner, J.L. (1997). Lipopolysaccharides of *Helicobacter pylori* serogroups O:3 and O:6. Structures of a classs of lipopolysaccharides with reference to the location of oligomeric units of D-*glycero*-α-D-*manno*-heptose residues. Eur. J. Biochem. *248*, 592–601.

Babinski, K.J., Ribeiro, A.A., and Raetz, C.R. (2002). The *Escherichia coli* gene encoding the UDP-2, 3-diacylglucosamine pyrophosphatase of lipid A biosynthesis. J. Biol. Chem. *277*, 25937–25946.

Baker, P.J., Hraba, T., Taylor, C.E., Stashak, P.W., Fauntleroy, M.B., Zähringer, U., Takayam, K., Sievert, T.R., Hronowski, X., Cotter, R.J., and Perez-Perez, G. (1994). Molecular structures that influence the immunomodulatory properties of the lipid A and inner core region oligosaccharides of bacterial lipopolysaccharides. Infect. Immun. *62*, 2257–2269.

Bäckhed, F., Rokbi, B., Torstensson, E., Zhao, Y., Nilsson, C., Seguin, D., Normark, S., Buchan, A.M., and Richter-Dahlfors, A. (2003). Gastric mucosal recognition of *Helicobacter pylori* is independent of Toll-like receptor 4. J. Infect. Dis. *187*, 829–836.

Belunis, C.J., Clementz, T., Carty, S.M., and Raetz, C.R. (1995). Inhibition of lipopolysaccharide biosynthesis and cell growth following inactivation of the *kdtA* gene in *Escherichia coli*. J. Biol. Chem. *270*, 27646–27652.

Berg, D.E., Hoffman, P.S., Appelmelk, B.J., and Kusters, J.G. (1997). The *Helicobacter pylori* genome sequence: genetic factors for long life in the gastric mucosa. Trends Microbiol. *5*, 468–474.

Bergman, M.P., Engering, A., Smits, H.H., van Vliet, S.J., van Bodegraven, A.A., Wirth, H.-P., Kapensberg, M.L., Vandenbroucke-Grauls, C.M.J.E., van Kooyk, Y., and Appelmelk, B.J. (2004). *Helicobacter pylori* modulates the T helper cell 1/T helper cell 2 balance through phase-variable interaction between lipopolysaccharide and DC-SIGN. J. Exp. Med. *200*, 979–990.

Blaser, M.J., and Kirschner, D. (1999). Dynamics of *Helicobacter pylori* colonization in relation to the host response. Proc. Natl. Acad. Sci. USA *96*, 8359–8364.

Brabetz, W., Muller-Loennies, S., and Brade, H. (2000). 3-Deoxy-D-*manno*-oct-2-ulosonic acid (Kdo) transferase (WaaA) and Kdo kinase (KdkA) of *Haemophilus influenzae* are both required to complement a *waaA* knockout mutation of *Escherichia coli*. J. Biol. Chem. *275*, 34954–34962.

Broutet, N., Moran, A., Hynes, S., Sakarovitch, C., and Mégraud, F. (2002). Lewis antigen expression and other pathogenic factors in the presence of atrophic chronic gastritis in a European population. J. Infect. Dis. *185*, 503–512.

Brzozowski, T., Konturek, P.C., Ptak, A., Konturek, S.J., and Moran, A. (2000). Implication of leptin and gastrin in the adaptation of gastric mucosa to *H. pylori*-lipopolysaccharide (HP-LPS). Gut *47(Suppl. I)*, A40.

Bukholm, G., Tannaes, T., Nedenskov, P., Esbensen, Y., Grav, H.J., Hovig, T., Ariansen, S., and Guldvog, I. (1997). Colony variation of *Helicobacter pylori*: pathogen potential is correlated to cell wall lipid composition. Scand. J. Gastroenterol. *32*, 445–454.

Caroff, M., and Karibian, D. (2003). Structure of bacterial lipopolysaccharides. Carbohydr. Res. *338*, 2431–2447.

Chan, N.W.C., Stangier, K., Sherburne, R., Taylor, D.E., Zhang, Y., Dovichi, N.J., and Palcic, M.M. (1995). The biosynthesis of Lewis X in *Helicobacter pylori*. Glycobiology *5*, 683–688.

Clementz, T., Bednarski, J.J., and Raetz, C.R. (1996). Function of the *htrB* high temperature requirement gene of *Escherichia coli* in the acylation of lipid A: HtrB catalyzed incorporation of laurate. J. Biol. Chem. *271*, 12095–12102.

Clementz, T., and Raetz, C.R. (1991). A gene coding for 3-deoxy-D-*manno*-octulosonic-acid transferase in *Escherichia coli*. Identification, mapping, cloning, and sequencing. J. Biol. Chem. *266*, 9687–9696.

Cox, A.D., Wright, J.C., Li, J., Hood, D.W., Moxon, E.R., and Richards, J.C. (2003). Phosphorylation of the lipid A region of meningococcal lipopolysaccharide: identification of a family of transferases that add phosphoethanolamine to lipopolysaccharide. J. Bacteriol. *185*, 3270–3277.

Cunningham, M.D., Seachord, C., Ratcliffe, K., Bainbridge, B., Aruffo, A., and Darveau, R.P. (1996). *Helicobacter pylori* and *Porphyromonas gingivalis* lipopolysaccharides are poorly transferred to recombinant soluble CD14. Infect. Immun. *64*, 3601–3608.

D'Elios, M.M., Appelmelk, B.J., Amedei, A., Bergman, M.P., and del Prete, G. (2004). Gastric autoimmunity: the role of *Helicobacter pylori* and molecualr mimicry. Trends Mol. Med. *10*, 316–323.

De Lucca, A.J., Jacks, T.J., and Brogden, K.A. (1995). Binding between lipopolysaccharide and ceropin A. Mol. Cell Biochem. *151*, 141–148.

Doerrler, W.T. (2006). Lipid trafficking to the outer membrane of Gram-negative bacteria. Mol. Microbiol. *60*, 542–552.

Doig, P., de Jonge, B.I., Alm, R.A., Brown, E.D., Uria-Nickelsen, M., Noonan, B., Mills, S.D., Tummino, P., Carmel, G., Guild, B.C., Moir, D.T., Vovis, G.F., and Trust, T.J. (1999). *Helicobacter pylori* physiology predicted from genomic comparison of two strains. Microbiol. Mol. Biol. Rev. *63*, 675–707.

Durkin, E., Moran, A.P., and Hanson, P.J. (2006). Apoptosis induction in gastric mucous cells *in vitro*: lesser potency of *Helicobacter pylori* than *Escherichia coli* lipopolysaccharide, but positive interaction with ibuprofen. J. Endotoxin Res. *12*, 47–56.

Eaton, K.A., Logan, S.M., Baker, P.E., Peterson, R.A., Monteiro, M.A., and Altman, E. (2004). *Helicobacter pylori* with a truncated lipopolysaccharide O-chain fails to induce gastritis in SCID mice injected with splenocytes from wild-type C57BL/6J mice. Infect. Immun. *72*, 3925–3931.

Edwards, N.J., Monteiro, M.A., Faller, G., Walsh, E.J., Moran, A.P., Roberts, I.S., and High, N.J. (2000). Lewis X structures in the O antigen side-chain promote adhesion of *Helicobacter pylori* to the gastric epithelium. Mol. Microbiol. *35*, 1530–1539.

Endo, T., Koizumi, S., Tabata, K., and Ozaki, A. (2000). Cloning and expression of β-1,4-galactosyltransferase gene from *Helicobacter pylori*. Glycobiology *10*, 809–813.

Ernst, R.K., Yi, E.C., Guo, L., Lim, K.B., Burns, J.L., Hackett, M., and Miller, S.I. (1999). Specific lipopolysaccharide found in cystic fibrosis airway *Pseudomonas aeruginosa*. Science *286*, 1561–1565.

Ferris, J.A., Perepelov, A.A., Senchenkova, S.N., Kocharova, N.A., Knirel, Y.A., Jansson, P.-E., Wadström, T., Andersen, L.P., and Moran, A.P. (2001). Comparative characterisation of the lipopolysaccharides of gastric and intestinal *Helicobacter* spp. Gut 49 (Suppl II), A51–52.

Fowler, M., Thomas, R.J., Atherton, J., Roberts, I.S., and High, N.J. (2006). Galectin-3 binds to *Helicobacter pylori* O-antigen: it is upregulated and rapidly secreted by gastric epithelial cells in response to *H. pylori* adhesion. Cell. Microbiol. *8*, 44–54.

Garrett, T.A., Kadrmas, J.L., and Raetz, C.R. (1997). Identification of the gene encoding the *Escherichia coli* lipid A 4'-kinase. Facile phosphorylation of endotoxin analogs with recombinant LpxK. J. Biol. Chem. *272*, 21855–21864.

Ge, Z., Chan, N.W.C., Palcic, M.M., and Taylor, D.E. (1997). Cloning and heterologous expression of an α1,3-fucosyltransferase gene from the gastric pathogen *Helicobacter pylori*. J. Biol. Chem. 272, 21357–21363.

Gobert, A.P., Bambou, J.-C., Werts, C., Balloy, V., Chignard, M., Moran, A.P., and Ferrero, R.L. (2004). *Helicobacter pylori* heat shock protein 60 mediates interleukin-6 production by macrophages via a Toll-like receptor (TLR)-2-, TLR-4-, and myeloid differentiation factor 88-independent mechanism. J. Biol. Chem. *279*, 245–250.

Goldberg, J.B., Wright, A., Ramakrishna, J., Moran, A., Kersulyte, D., and Berg, D.E. (1996). Cloning of the lipopolysaccharide (LPS)-core gene, *rfaC*, from *Helicobacter pylori*. Gut 39 *(Suppl. 2)*, A69.

Grebowska, A., Rechcinski, T., Bak-Romaniszyn, L., Czkwianianc, E., Moran, A., Druszczynska, M., Kowalewicz-Kulbat, M., Owczarek, A., Dziuba, M., Krzeminska-Pakula, M., Planeta-Malecka, I., Rudnicka, W., and Chmiela, M. (2006). Potenital role of LPS in the outcome of *Heliocbcater pylori* related disease. Pol. J. Microbiol. 55, 25–30.

Green, C. (1989). The ABO, Lewis and related blood group antigens: a review of structure and biosynthesis. FEMS Microbiol. Immunol. 47, 321–330.

Guo, L., Lim, K.B., Gunn, J.S., Bainbridge, B., Darveau, R.P., Hackett, M., and Miller, S.I. (1997). Regulation of lipid A modifications by *Salmonella typhimurium* virulence genes *phoP-phoQ*. Science *276*, 250–253.

Hamanaka, Y., Nakashima, M., Wada, A., Ito, M., Kurazono, H., Hojo, H., Nakahara, Y., Kohno, S., Hirayama, T., and Sekine, I. (2001). Expression of human β-defensin 2 (hBD-2). in *Helicobacter pylori* induced gastritis: antibacterial effect of hBD-2 against *Helicobacter pylori*. Gut 49, 481–487.

Hase, K., Murakami, M., Iimura, M., Cole, S.P., Horibe, Y., Ohtake, T., Obonyo, M., Gallo, R.L., Eckmann, L., and Kagnoff, M.F. (2003). Expression of LL-37 by human gastric epithelial cells as a potential host defense mechanism against *Helicobacter pylori*. Gastroenterology *125*, 1613–1625.

Heneghan, M.A., McCarthy, C.F., Janulaityte, D., and Moran, A.P. (2001). Relationship of anti-Lewis x and anti-Lewis y antibodies in serum samples from gastric cancer and chronic gastritis patients to *Helicobacter pylori*-mediated autoimmunity. Infect. Immun. *69*, 4774–4781.

Heneghan, M.A., McCarthy, C.F., and Moran, A.P. (2000). Relationship of blood group determinants on *Helicobacter pylori* lipopolysaccharide with host Lewis phenotype and inflammatory response. Infect. Immun. *68*, 937–941.

Henry, S., Oriol, R., and Samuelsson, B. (1995). Lewis histo-blood group system and associated secretory phenotypes. Vox Sang. *69*, 166–182.

Hirschfeld, M., Weiss, J.J., Toshchakov, V., Salkowski, C.A., Cody, M.J., Ward, D.C., Qureshi, N., Michalek, S.M., and Vogel, S.N. (2001). Signalling by Toll-like receptor 2 and 4 agonists results in differential gene expression in murine macrophages. Infect. Immun. *69*, 1477–1482.

Hynes, S.O., Broutet, N., Eurohepygast Study Group, Wadström, T., Mikelsaar, M., O'Toole, P.W., Telford, J., Engstrand, L., Kamiya, S., Mentis, A.F., and Moran, A.P. (2002). Phenotypic variation of *Helicobacter pylori* isolates from geographically distinct regions detected by lectin typing. J. Clin. Microbiol. *40*, 227–232.

Hynes, S.O., Ferris, J.A., Szponar, B., Wadström, T., Fox, J.G., O'Rourke, J., Larsson, L., Yaquian, E., Ljungh, Å., Clyne, M., Andersen, L.P., and Moran, A.P. (2004). Comparative chemical and biological characterization of the lipopolysaccharides of gastric and enterohepatic helicobacters. *Helicobacter 9*, 313–323.

Hynes, S.O., Hirmo, S., Wadström, T., and Moran, A.P. (1999). Differentiation of *Helicobacter pylori* isolates based on lectin binding of cell extracts in an agglutination assay. J. Clin. Microbiol. *37*, 1994–1998.

Hynes, S.O., Keenan, J.I., Ferris, J.A., Annuk, H., and Moran, A.P. (2005). Lewis epitopes on outer membrane vesicles of relevance to *Helicobacter pylori* pathogenesis. *Helicobacter 10*, 146–156.

Hynes, S.O., and Moran , A.P. (2000). Comparison of three serological methods for detection of Lewis antigens on the surface of *Helicobacter pylori*. FEMS Microbiol. Lett. *190*, 67–72.

Hynes, S.O., Moran, A.P., O'Brien, T., and Katusic, Z.S. (2003). Vasomotor effect of *Helicobacter pylori* lipopolysaccharide in isolated rabbit aorta. *Helicobacter 8*, 356–357.

Icho, T. (1988). Membrane-bound phosphatases in *Escherichia* coli: sequence of the *pgpB* gene and dual subcellular localization of the *pgpB* product. J. Bacteriol. *170*, 5117–5124.

Icho, T., and Raetz, C.R. (1983). Multiple genes for membrane-bound phosphatases in *Escherichia coli* and their action on phospholipid precursors. J. Bacteriol. *153*, 722–730.

Inohara, N., Ogura, Y., Fontalba, A., Gutierrez, O., Pons, F., Crespo, J., Fukase, K., Inamura, S., Kusumoto, S., Hashimoto, M., Foster, S.J., Moran, A.P., Fernandez-Luna, J.L., and Nunez, G. (2003). Host recognition of bacterial muramyl dipeptide mediated through NOD2. J. Biol. Chem. *278*, 5509–5512.

Ishihara, S., Rumi, M.A., Kadowaki, Y., Ortega-Cava, C.F., Yuki, T., Yoshino, N., Miyaoka, Y., Kazumori, H., Ishimura, N., Amano, Y., and Kinoshita, Y. (2004). Essential role of MD-2 in TLR4-dependent signaling during *H. pylori*-associated gastritis. J. Immunol. *173*, 1406–1416.

Isobe, T., White, K.A., Allen, A.G., Peacock, M., Raetz, C.R., and Maskell, D J. (1999). *Bordetella pertussis waaA* encodes a monofunctional 2-keto-3-deoxy-D-*manno*-octulosonic acid transferase that can complement an *Escherichia coli waaA* mutation. J. Bacteriol. *181*, 2648–2651.

Jirillo, E., Pece, S., Pellegrino, N.M., Di Leo, A., Russo, F., Linsalata, M., Messa, C., Amati, L., Caradonna, L., Moran, A.P., and Caccavo, D. (1999). The role of *Helicobacter pylori* LPS in the pathogenesis of *H. pylori*-related gastropathy. J. Endotoxin Res. *5*, 222–226.

Kawahara, T., Kuwano, Y., Teshima-Kondo, S., Kawai, T., Nikawa, T., Kishi, K., Rokutan, K. (2001a). Toll-like receptor 4 regulates gastric pit cell responses to *Helicobacter pylori* infection. J. Med. Investig. *48*, 190–197.

Kawahara, T., Kuwano, Y., Teshima-Kondo, S., Sugiyama, T., Kawai, T., Nikawa, T., Kishi, K., Rokutan, K. (2001b). *Helicobacter pylori* lipopolysaccharide from type I, but not type II, strains stimulates apoptosis of cultured gastric mucosal cells. J. Med. Invest. *48*, 167–174.

Kawahara, T., Teshima, S., Oka, A., Sugiyama, T., Kishi, K., and Rokutan, K. (2001c). Type I *Helicobacter pylori* lipopolysaccharide stimulates Toll-like receptor 4 and activates mitogen oxidase 1 in gastric pit cells. Infect. Immun. *69*, 4382–4389.

Keenan, J.I., Ismail, S., McGovern, J.J., and Moran, A.P. (2004). Iron availability influences lipopolysaccharide expression on *Helicobacter pylori* bacteria and outer membrane vesicles. *Helicobacter 9*, 500.

Khamri, W., Moran, A.P., Worku, M.L., Karim, Q.N., Walker, M.M., Annuk, H., Ferris, J.A., Appelmelk, B.J., Eggleton, P., Reid, K.B., and Thursz, M.R. (2005). Variations in *Helicobacter pylori* lipopolysaccharide to evade the innate immune component surfactant protein D. Infect. Immun. *73*, 7677–7686.

Kidd, M., Miu, K., Tang, L.H., Perez-Perez, G.I., Blaser, M.J., Sandor, A., and Modlin, I.M. (1997). *Helicobacter pylori* lipopolysaccharide stimulates histamine release and DNA synthesis in rat enterochromaffin-like cells. Gastroenterology *113*, 1110–1117.

Kiss, J., Lamarque, D., Moran, A.P., Pozsár, J., Morschl, É., László, F., and Whittle, B.J.R. (2001). *Helicobacter pylori* lipopolysaccharide-provoked injury to rat gastroduodenal microvasculature involves inducible nitric oxide synthase. Eur. J. Pharmacol. *420*, 175–179.

Kleene, R., and Berger, E.G. (1993). The molecular and cell biology of glycosyltransferases. Biochim. Biophys. Acta *1154*, 283–325.

Knirel, Y.A., Kocharova, N.A., Hynes, S.O., Widmalm, G., Andersen, L.P., Jansson, P.-E., and Moran, A.P. (1999). Structural studies on lipopolysaccharides of serologically non-typable strains of *Helicobacter pylori*, AF1 and 007, expressing Lewis antigenic determinants. Eur. J. Biochem. *266*, 123–131.

Kobayashi, K., Sakamoto, J., Kito, T., Yamamura, Y., Koshikawa, T., Fujita, M., Watanabe, T., Nkazato, H. (1993). Lewis blood group-related antigen expression in normal gastric epithelium, intestinal metaplasia, gastric adenoma, and gastric carcinoma. Am. J. Gastroenterol. *88*, 919–924.

Kocharova, N.A., Knirel, Y.A., Widmalm, G., Jansson, P.-E., and Moran, A.P. (2000). Structure of an atypical O-antigen polysaccharide of *Helicobacter pylori* containing a novel monosaccharide 3-C-methyl-D-mannose. Biochemistry *39*, 4755–4760.

Kulshin, V.A., Zahringer, U., Lindner, B., Frasch, C.E., Tsai, C.M., Dmitriev, B.A., and Rietschel, E.T. (1992).

Structural characterization of the lipid A component of pathogenic *Neisseria meningitidis*. J. Bacteriol. *174*, 1793–1800.

Lamarque, D., Moran, A.P., Szepes, Z., Delchier, J.-C., and Whittle, B.J.R. (2000). Cytotoxicity associated with induction of nitric oxide synthase in rat duodenal epithelial cells *in vivo* by lipopolysaccharide of *Helicobacter pylori*: inhibition by superoxide dismutase. Br. J. Pharmacol. *130*, 1531–1538.

Lee, A., and Moran, A.P. (1994). Lipopolysaccharide (LPS)-related damage by *H. pylori*. In: *Helicobacter pylori*: Basic Mechanisms to Clinical Cure. Hunt, R.H., Tytgat, G.N.J. eds, (Kluwer Academic Publishers, The Netherlands), pp. 169–179.

Lee, S.K., Stack, A., Katzowitsch, E., Aizawa, S.I., Suerbaum, S., and Josenhans, C. (2003). *Helicobacter pylori* flagellins have very low intrinsic activity to stimulate human gastric epithelial cells via TLR5. Microbes Infect. 5, 1345–1346.

Lee, H., Hsu, F.F., Turk, J., and Groisman, E.A. (2004). The PmrA-regulated *pmrC* gene mediates phosphoethanolamine modification of lipid A and polymyxin resistance in *Salmonella enterica*. J. Bacteriol. *186*, 4124–4133.

Lepper, P.M., Triantafilou, M., Schumann, C., Schneider, E.M., and Triantafilou, M. (2005). Lipopolysaccharides from *Helicobacter pylori* can act as antagonists for Toll-like receptor 4. Cell. Microbiol. *7*, 519–528.

Lin, S., Yuan, T., Li, J., and Lin, C. (2006). Carboxyl terminus of *Helicobacter pylori* α1, 3-fucosyltransferase determines the structure and stability. Biochemistry *45*, 8108–8116.

Logan, S.M., Altman, E., Mykytczuk, O., Brisson, J.-R., Chandan, V., St. Michael, F., Masson, A., Leclerc, S., Hiratsuka, K., Smirnova, N., Li, J., Wu, Y., and Wakarchuk, W.W. (2005). Novel biosynthetic functions of lipopolysaccharide *rfaJ* homologs from *Helicobacter pylori*. Glycobiology *15*, 721–733.

Logan, S.M., Conlan, J.W., Monteiro, M.A., Wakarchuk, W.W., and Altman, E. (2000). Functional genomics of *Helicobacter pylori*: identification of a β-1,4 galactosyltransferase and generation of mutants with altered lipopolysaccharide. Mol. Microbiol. *35*, 1156–1167.

Ma, B., Audette, G.F., Lin, S., Palcic, M.M., Hazes, B., and Taylor, D.E. (2006a). A single aromatic amino acid at the C-terminus of *Helicobacter pylori* α-(1, 3/1, 4)-fucosyltransferases. J. Biol. Chem. *281*, 6385–6394.

Ma, B., Simala-Grant, J.L., and Taylor, D.E. (2006b). Fucosylation in prokaryotes and eukaryotes. Glycobiology *16*, 158R-184R.

Ma, B., Wang, G., Palcic, M.M., Hazes, B., and Taylor, D.E. (2003). C-terminal amino acids of *Helicobacter pylori* α1, 3/4 fucosyltransferases determine type I and type II transfer. J. Biol. Chem. *278*, 21893–218900.

Mackinnon, F.G., Cox, A.D., Plested, J.S., Tang, C.M., Makepeace, K., Coull, P.A., Wright, J.C., Chalmers, R., Hood, D.W., Richards, J.C., and Moxon, E. R. (2002). Identification of a gene (*lpt-3*) required for the addition of phosphoethanolamine to the lipopolysaccharide inner core of *Neisseria meningitidis* and its role in mediating susceptibility to bactericidal killing and opsonophagocytosis. Mol. Microbiol. *43*, 931–943.

Maeda, S., Akanuma, M., Mitsuno, Y., Hirata, Y., Ogura, K., Yoshida, H., Shiratori, Y., and Omata, M. (2001). Distinct mechanism of *Helicobacter pylori*-mediated NF-κB activation between gastric cancer cells and monocytic cells. J. Biol. Chem. *276*, 44856–44864.

Mahdavi, J., Borén, T., Vandenbroucke-Grauls, C., Appelmelk, B.J. 2003. Limited role of lipopolysaccharide Lewis antigens in adherence of *Helicobacter pylori* to the human gastric epithelium. Infect. Immun. *71*, 2876–2880.

Mai, U.E., Perez-Perez, G.I., Wahl, L.M., Wahl, S.M., Blaser, M.J., and Smith, P.D. (1991). Soluble surface proteins from *Helicobacter pylori* activate monocytes/macrophages by lipopolysaccharide-independent mechanism. J. Clin. Invest. *87*, 894–900.

Mandell, L., Moran, A.P., Cocchiarella, A., Houghton, J.M., Taylor, N., Fox, J.G., Wang, T.C., and Kurt-Jones, E.A. (2004). Intact Gram-negative *Helicobacter pylori, Helicobacter felis,* and *Helicobacter hepaticus* bacteria activate innate immunity via Toll-like receptor 2 not Toll-like receptor 4. Infect. Immun. *72*, 6446–6454.

Marais, A., Mendz, G.L., Hazell, S.L., and Mégraud, F. (1999). Metabolism and genetics of *Helicobacter pylori*: the genome era. Microbiol. Mol. Biol. Rev. *63*, 642–674.

Marshall, D.G., Hynes, S.O., Coleman, D.C., O'Morain, C.A., Smyth, C.J., and Moran, A.P. (1999). Lack of a relationship between Lewis antigen expression and *cag A*, CagA, *vacA* and VacA status of Irish *Helicobacter pylori* isolates. FEMS Immunol. Med. Microbiol. 24, 79–90.

Martin, S.L., Edbrooke, M.R., Hodgman, T.C., van den Eijnden, D.H., and Bird, M.I. (1997). Lewis X biosynthesis in *Helicobacter pylori*. Molecular cloning of an α(1, 3)-fucosyltransferase gene. J. Biol. Chem. *272*, 21349–21356.

McGowan, C.C., Necheva, A., Thompson, S.A., Cover, T.L., and Blaser, M.J. (1998). Acid-induced expression of an LPS-associated gene in *Helicobacter pylori*. Mol. Microbiol. *30*, 19–31.

Miller, S.I., Ernst, R.K., and Bader, M.W. (2005). LPS, TLR4 and infectious disease diversity. Nat. Rev. Microbiol. 3, 36–46.

Mills, S.D., Kurjanczyk, L.A., and Penner, J.L. (1992). Antigenicity of *Helicobacter pylori* lipopolysaccharides. Infect. Immun. *30*, 3175–3180.

Monteiro, M.A. (2001). *Helicobacter pylori*: a wolf in sheep's clothing. The glyconjugate families of *Helicobacter pylori* lipopolysaccharides expressing histo-blood groups: structure, biosynthesis, and role in pathogenesis. Adv. Carbohydr. Chem. Biochem. *57*, 99–158.

Monteiro, M.A., Appelmelk, B.J., Rasko, D.A., Moran, A.P., Hynes, S.O., McLean, L.L., Chan, K.H., St. Michael, F., Logan, S.M., O'Rourke, J., Lee, A., Taylor, D.E., and Perry, M.B. (2000a). Lipopolysaccharide structures of *Helicobacter pylori* genomic strains 26695 and J99, mouse model *H. pylori* Sydney strain, *H. pylori* P466 carrying sialyl Lewis X and *H. pylori* UA915 expressing Lewis B. Classification of *H. pylori* lipopolysaccharides into glycotype families. Eur J. Biochem. *267*, 305–320.

Monteiro, M.A., Chan, K.H.N., Rasko, D.A., Taylor, D.E., Zheng, P.Y., Appelmelk, B.J., Wirth, H.-P., Yang, M., Blaser, M.J., Hynes, S.O., Moran, A.P., and Perry, M.B. (1998). Simultaneous expression of type 1 and type 2 Lewis blood group antigens by *Helicobacter pylori* lipopolysaccharides. Molecular mimicry between *H. pylori* lipopolysaccharides and human gastric epithelial cell surface glycoforms. J. Biol. Chem. *273*, 11533–11543.

Monteiro, M.A., St. Michael, F., Rasko, D., Taylor, D.E., Conlan, J.W., Chan, K.H., Logan, S.M., Appelmelk, B.J., and Perry, M.B. (2001). *Helicobacter pylori* from asymptomatic hosts expressing heptoglycan but lacking Lewis O-chains: Lewis blood-group O-chains may play a role in *Helicobacter pylori* induced pathology. Biochem. Cell Biol. *79*, 449–459.

Monteiro, M.A., Zheng, P.-y., Ho, B., Yokota, S.-i., Amano, K.-i., Pam, Z.-j., Berg, D.E., Chan, K.H., MacLean, L.L., and Perry, M.B. (2000b). Expression of histo-blood group antigens by lipopolysaccharides of *Helicobacter pylori* strains from Asian hosts: the propensity to express type 1 blood-group antigens. Glycobiology *10*, 701–713.

Moran, A.P. (1995a). Cell surface characteristics of *Helicobacter pylori*. FEMS Immunol. Med. Microbiol. *10*, 271–280.

Moran, A.P. (1995b). Structure-bioactivity relationships of bacterial endotoxins. J. Toxicol. – Toxin Rev. *14*, 47–83.

Moran, A.P. (1996a). Pathogenic properties of *Helicobacter pylori*. Scand. J. Gastroenterol. *31* (Suppl. 215), 22–31.

Moran, A.P. (1996b). The role of lipopolysaccharide in *Helicobacter pylori* pathogenesis. Aliment. Pharmacol. Therapeut. *10* (Suppl. 1), 39–50.

Moran, A.P. (1998). The products of *Helicobacter pylori* that induce inflammation. Eur. J. Gastroenterol. Hepatol. *10* (Suppl. 1), S3-S8.

Moran, A.P. (1999). *Helicobacter pylori* lipopolysaccharide-mediated gastric and extragastric pathology. J. Physiol. Pharmacol. *50*, 787–805.

Moran, A.P. (2001a). *Helicobacter pylori* lipopolysaccharides. In *Helicobacter pylori*: Molecular and Cellular Biology, Achtman, M. and Suerbaum, S., eds. (Wymondham, UK: Horizon Scientific Press), pp. 207–226.

Moran, A.P. (2001b). Molecular structure, biosynthesis and pathogenic roles of lipopolysaccharides. In *Helicobacter pylori*: Physiology and Genetics, H.L.T. Mobley, G.L. Mendz and S.L. Hazell, eds. (Washington, D.C.: American Society for Microbiology), pp. 81–95.

Moran, A.P. (2007). Lipopolysaccharide in bacterial chronic infection: insights from *Helicobacter pylori* lipopolysaccharide and lipid A. Int. J. Med. Microbiol. *297*, 307–319.

Moran, A.P. (2008). Relevance of fucosylation and Lewis antigen expression in *Helciobacter pylori*. Carbohydr. Res. (in press).

Moran, A.P., and Aspinall, G.O. (1998). Unique structural and biological features of *Helicobacter pylori* lipopolysaccharides. Prog. Clin. Biol. Res. *397*, 37–49.

Moran, A.P., Broaders, S.A., Rapa, A., and Oderda, G. (2005a). In vivo expression of the 25-kDa laminin-binding protein of *Helicobacter pylori*. FEMS Immunol. Med. Microbiol. *43*, 331–337.

Moran, A.P., Ferris, J.A., Kocharova, N.A., Knirel, Y.A., Widmalm, G., Andersen, L.P., and Jansson, P.E. (2002a). Serological and structural characterization of *Helicobacter bizzozeronii* lipopolysaccharide. Gut *51* (Suppl. II), A1.

Moran, A.P., Ferris, J.A., Perepelov, A.V., Blaser, M.J., Kocharova, N.A., Knirel, Y.A., Wirth, H.P., and Jansson, P.E. (2002b). Structural examination of Lewis expression and adaptation in lipopolysaccharides of *Helicobacter pylori* from experimentally infected rhesus monkeys. Gut *51*(Suppl. II), A15.

Moran, A.P., Helander, I.M., and Kosunen, T.U. (1992). Compositional analysis of *Helicobacter pylori* rough-form lipopolysaccharides. J. Bacteriol. *174*, 1370–1377.

Moran, A.P., Hynes, S.O., and Heneghan, M.A. (1999). Mimicry of blood group antigen A by *Helicobacter mustelae* and *H. pylori*. Gastroenterology *116*, 504–505.

Moran, A.P., Kharmi, W., Walker, M.M., and Thursz, M.R. (2005b). Role of surfactant protein D (SP-D) in innate immunity in the gastric mucosa: evidence of interaction with *Helicobacter pylori* lipopolysaccharide. J. Endotoxin Res. *11*, 357–362.

Moran, A.P., Knirel, Y.A., Senchenkova, S.N., Widmalm, G., Hynes, S.O., and Jansson, P.-E. (2002c). Phenotypic variation in molecular mimicry between *Helicobacter pylori* lipopolysaccharides and human gastric epithelial cell surface glycoforms. Acid-induced phase variation in Lewis^x and Lewis^y expression by *H. pylori* lipopolysaccharides. J. Biol. Chem. *277*, 5785–5795.

Moran, A.P., Kuusela, P., and Kosunen, T.U. (1993). Interaction of *Helicobacter pylori* with extracellular matrix proteins. J. Appl. Bacteriol. *75*, 184–189.

Moran, A.P., Lindner, B., and Walsh, E.J. (1997). Structural characterization of the lipid A component of *Helicobacter pylori* rough- and smooth-form lipopolysaccharides. J. Bacteriol. *179*, 6453–6463.

Moran, A.P., Rietschel, E.T., Kosunen, T.U., and Zähringer, U. (1991a). Chemical characterization of *Campylobacter jejuni* lipopolysaccharides containing *N*-acetylneuraminic acid and 2,3-diamino-2, 3-dideoxy-D-glucose. J. Bacteriol. *173*, 618–626.

Moran, A.P., Schromm, A.B., Andrä, J., Koch, M.H.J., Seydel, U., and Brandenburg, K. (2005c). Influence of endotoxic activity of *Helicobacter pylori* lipopolysaccharide on physicochemical parameters. *Helicobacter* *10*, 489.

Moran, A.P., Shiberu, B., Ferris, J.A., Knirel, Y.A., Senchenkova, S.N., Perepelov, A.V., Jansson, P.E., and Goldberg, J.B. (2004). Role of *Helicobacter pylori rfaJ* genes (HP0159 and HP1416) in lipopolysaccharide synthesis. FEMS Microbiol Lett *241*, 57–65.

Moran, A.P., Sturegård, E., Sjunnesson, H., Wadström, T., and Hynes, S.O. (2000). The relationship between O-chain expression and colonisation ability of *Helicobacter pylori* in a mouse model. FEMS Immunol. Med. Microbiol. *29*, 263–270.

Moran, A.P, and Wadström, T. (1998). Pathogenesis of *Helicobacter pylori*. Curr. Opin. Gastroenterol. *14* (Suppl. 1), S9-S14.

Moran, A.P., Young, G.O., and Lastovica, A.J. (1998). Pepsinogen induction by *Helicobacter pylori* lipopolysaccharides. Gut *43* (Suppl. 2), A15.

Moran, A.P., Zähringer, U., Seydel, U., Scholz, D., Stütz, P., and Rietschel, E.Th. (1991b). Structural analysis of the lipid A component of *Campylobacter jejuni* CCUG 10936 (serotype O:2) lipopolysaccharide. Description of a lipid A containing a hybrid backbone of 2-amino-2-deoxy-D-glucose and 2,3-diamino-2, 3-dideoxy-D-glucose. Eur. J. Biochem. *198*, 459–469.

Muotiala, A., Helander, I.M., Pyhälä, L., Kosunen, T.U., and Moran, A.P. (1992). Low biological activity of *Helicobacter pylori* lipopolysaccharide. Infect. Immun. *60*, 1714–1716.

Murata, K. Egami, H., Shibata, Y., Sakamoto, K., Misumi, A., and Ogawa, M. (1992). Expression of blood group-related antigens, ABH, Lewis^a, Lewis^b, Lewis^x, Lewis^y, CA19–9, and CSLEX1 in early cancer, intestinal metaplasis, and uninvolved mucosa of the stomach. Am. J. Clin. Pathol. *98*, 67–75.

Murray, E., Khamri, W., Walker, M.M., Eggleton, P., Moran A.P., Ferris, J.A., Knapp, S., Karim, Q.N., Worku, M., Strong, P., Reid, K.B.M., and Thursz, M.R. (2002). Expression of surfactant protein D in human gastric mucosa and *Helicobacter pylori* infection. Infect. Immun. *70*, 1481–1487.

Nielsen, H., Birkholz, S., Andersen, L.P., and Moran, A.P. (1994). Neutrophil activation by *Helicobacter pylori* lipopolysaccharides. J. Infect. Dis. *170*, 135–139.

Nikaido, H. (2003). Molecular basis of bacterial outer membrane permeability revisited. Microbiol. Mol. Biol. Rev. *67*, 593–656.

Nilsson, C., Skoglund, A., Moran, A.P., Annuk, A., Engstrand, L., and Normark, S. (2006). An enzymatic ruler modulates Lewis antigen glycosylation of *Helicobacter pylori* LPS during persistent infection. Proc. Natl. Acad. Sci. USA *103*, 2863–2868.

Odenbreit, S., Lindner, S., Gerbert-Vogl, B., Rieder, G., Moran, A.P., and Haas, A.P. (2006). Interleukin-6 induction by *Helicobacter pylori* in human macrophages is dependent on phagocytosis. Helicobacter *11*, 196–207.

Oh, J.D., Kling-Bäkhed, H., Giannakis, M., Xu, J., Fulton, R.S., Fulton, L.A., Cordum, H.S., Wang, C., Elliott, G., Edwards, J., Mardis, E.R., Engstrand, L.G., and Gordon, J.I. (2006). The complete genome sequence of a chronic atrophic gastritis *Helicobacter pylori* strain: evolution during disease progression. Proc. Natl. Acad Sci. USA 103, 9999–10004.

Oriol, R., Mollicone, R., Cailleau, A., Balanzino, L., and Breton, C. (1999). Divergent evolution of fucosyltransferase genes from vertebrates, invertebrates, and bacteria. Glycobiology 9, 323–334.

Ortega-Cava, C.F., Ishihara, S., Rumi, M.A., Kawashima, K., Ishimura, N., Kazumori, H., Udagawa, J., Kadowaki, Y., and Kinoshita, Y. (2003). Strategic compartmentalization of Toll-like receptor 4 in the mouse gut. J. Immunol. *170*, 3977–3985.

Padol, I.T., Moran, A.P., and Hunt, R.H. (2001). Effect of purified lipopolysaccharides from strains of *Helicobacter pylori* and *Helicobacter felis* on acid secretion in mouse gastric glands *in vitro*. Infect. Immun. *69*, 3891–3896.

Panthel, K., Faller, G., and Haas, R. (2003). Colonization of C57BL/6J and BALB/c wild-type and knockout mice with *Helicobacter pylori*: effect of vaccination and implications for innate and acquired immunity. Infect. Immun. *71*, 794–800.

Pece, S., Fumarola, D., Giuliani, G., Jirillo, E., and Moran, A.P. (1995). Activity in the *Limulus* amebocyte lysate assay and induction of tumour necrosis factor-α by diverse *Helicobacter pylori* lipopolysaccharide preparations. J. Endotoxin Res. *2*, 455–462.

Pece, S., Messa, C., Caccavo, D., Giuliani, G., Greco, B., Fumarola, D., Berloco, P., Di Leo, A., Jirillo, E., and Moran, A.P. (1997). Serum antibody response against *Helicobacter pylori* NCTC 11637 smooth- and rough-lipopolysaccharide phenotypes in patients with *H. pylori*-related gastropathy. J. Endotoxin Res. *4*, 383–390.

Perepolov, A.A., Senchenkova, S.N., Kocharova, N.A., Knirel, Y.A., Moran, A.P., Ferris, J.A., and Jansson, P.-E. (2001). Studies on new *Helicobacter pylori* lipopolysaccharides. In Abstracts of the 11th European Carbohydrate Symposium, Rauter, A.P., ed. (Lisbon, Portugal: Faculdade de Ciencias da Universidade de Lisboa), p. 330.

Piotrowski, J., Piotrowski, E., Skrodzka, D., Slominay, A., and Slomiany, B. (1997). Induction of acute gastritis and epithelial apoptosis by *Helicobacter pylori* lipopolysaccharide. Scand. J. Gastroenterol. *32*, 203–211.

Piotrowski, J., Slomiany, A., and Slomiany, B. (1993). Inhibition of gastric mucosal mucin receptor by *Helicobacter pylori* lipopolysaccharide. Biochem. Mol. Biol. Int. *31*, 1051–1058.

Putsep, K., Branden, C.I., Boman, H.G., and Normark, S. (1999). Antibacterial peptide from *H. pylori*. Nature *398*, 671–672.

Quintana, E., Ernandez, C., Moran, A.P., Esplungues, J.V., and Barrachina, M.D. (2005). Transcriptional upregulation of nNOS in the dorsal vagal complex during low endotoxemia. Life Sci. *77*, 1044–1054.

Rabbani, S., Miksa, V., Wipf, B., and Ernst, B. (2005). Molecular cloning and functional expression of a novel *Helicobacter pylori* α1,4 fucosyltransferase. Glycobiology *15*, 1076–1083.

Raetz, C.R., and Whitfield, C. (2002). Lipopolysaccharide endotoxins. Annu. Rev. Biochem. *71*, 635–700.

Ram, S., Cox, A.D., Wright, J.C., Vogel, U., Getzlaff, S., Boden, R. , Li, J., Plested, J.S., Meri, S., Gulati, S., Stein, D.C., Richards, J.C., Moxon, E.R., and Rice, P.A. (2003). Neisserial lipooligosaccharide is a target for complement component C4b. Inner core phosphoethanolamine residues define C4b linkage specificity. J. Biol. Chem. *278*, 50853–50862.

Rasko, D.A., Wang, G., Monteiro, M.A., Palcic, M.M., and Taylor, D.E. (2000a). Synthesis of mono- and di-fucosylated type I Lewis blood group antigens by *Helicobacter pylori*. Eur. J. Biochem. *267*, 6059–6066.

Rasko, D.A., Wang, G., Palcic, M.M., and Taylor, D.E. (2000b). Cloning and characterization of the $\alpha(1, 3/4)$-fucosyltransferase of *Helicobacter pylori*. J. Biol. Chem. *275*, 4988–4994.

Reynolds, C.M., Ribeiro, A.A., McGrath, S.C., Cotter, R.J., Raetz, C.R., and Trent, M.S. (2006). An outer membrnae enzyme encoded by *Salmonella typhimurium lpxR* removes the 3′-acyloxyacyl moiety of lipid A. J. Biol. Chem. *281*, 21974–21987.

Rieder, G., Hatz, R.A., Moran, A.P., Walz, A., Stolte, M., and Enders, G. (1997). Role of adherence in interleukin-8 induction in *Helicobacter pylori*-associated gastritis. Infect. Immun. *65*, 3622–3630.

Rietschel, E.T., Brade, L., Holst, O., Kulschin, V.A., Lindner, B., Moran, A.P., Schade, U.F., Zähringer, U., and Brade, H. (1990). Molecular structure of bacterial endotoxin in relation to bioactivity. In: Cellular and Molecular Aspects of Endotoxin Reactions, Vol. I. Nowotny, A., Spitzer, J.J., Ziegler, E.J. eds, (Elsevier Science Publishers, Amsterdam), pp. 15–32.

Rietschel, E.T., Kirikae, T., Schade, F.U., Mamat, U., Schmidt, G., Lopponow, H., Ulmer, A.J., Zähringer, U., Seydel, U., and Di Padova, F. (1994). Bacterial endotoxin: molecular relationships of structure to activity and function. FASEB J. *8*, 217–225.

Ryan, K.A., Moran, A.P., Hynes, S.O., Smith, T., Hyde, D., O'Morain, C.A., and Maher, M. (2000). Genotyping of *cagA* and *vacA*, Lewis antigen status, and analysis of the poly-(C) tract in the $\alpha(1, 3)$-fucosyltransferase gene of Irish *Helicobacter pylori* isolates. FEMS Immunol. Med. Microbiol. *28*, 113–119.

Sakagami, T., Vella, J., Dixon, M.F., O'Rourke, J., Radcliff, F., Sutton, P., Shimoyama, T., Beagley, K., and Lee, A. (1997). The endotoxin of *Helicobacter pylori* is a modulator of host-dependent gastritis. Infect. Immun. *65*, 3310–3316.

Sakamoto, J., Yin, B.T., and Lloyd, K.O. (1984). Analysis of the expression of H, Lewis, X, Y and precursor blood group determinants in saliva and red cells using a panel of mouse monoclonal antibodies. Mol. Immunol. *21*, 1093–1098.

Sakamoto, J., Watanabe, T., Tokumare, T., Takagi, H., Nakazato, H., and Lloyd, K.O. (1989). Expression of Lewis^{a}, Lewis^{b}, Lewis^{x}, Lewis^{y}, silayl-Lewis^{a} and silayl-Lewis^{x} blood group antigens in human gastric carcinoma and in normal gastric tissue. Cancer Res. *49*, 745–752.

Schmausser, B., Andrulis, M., Endrich, S., Lee, S.K., Josenhans, C., Müller-Hermelink, H.K., and Eck, M. (2004). Expression and subcellular distribution of toll-like receptors TLR4, TLR5 and TLR9 on the gastric epithelium in *Helicobacter pylori* infection. Clin. Exp. Immunol. *136*, 521–526.

Schromm, A.B., Brandenburg, K., Loppnow, H., Moran, A.P., Koch, M.H.J., Rietschel, E.T., and Seydel, U. (2000). Biological activities of lipopolysaccharides are determined by the shape of their lipid A portion. Eur. J. Biochem. *267*, 2008–2013.

Semeraro, N., Montemurro, P., Piccoli, C., Muolo, V., Colucci, M., Giuliani, G., Fumarola, D., Pece, S., and Moran, A.P. (1996). Effect of *Helicobacter pylori* lipopolysaccharide (LPS) and LPS derivatives on the production of tissue factor and plasminogen activator inhibitor type 2 by human blood mononuclear cells. J. Infect. Dis. *174*, 1255–1260.

Senchenkova, S.N., Shashkov, A.S., Knirel, Y.A., McGovern, J.J., and Moran, A.P. (1996). The O-specific polysaccharide chain of *Campylobacter fetus* serotype B lipopolysaccharide is a D-rhamnan terminated with 3-*O*-methyl-D-rhamnose (D-acofriose). Eur. J. Biochem. *239*, 434–438.

Senchenkova, S.N., Shashkov, A.S., Knirel, Y.A., McGovern, J.J., and Moran, A.P. (1997). The O-specific polysaccharide chain of *Campylobacter fetus* serotype A lipopolysaccharide is a partially *O*-acetylated 1,3-linked α-D-mannan. Eur. J. Biochem. 245, 637–641.

Senchenkova, S.N., Zatonsky, G.V., Hynes, S.O., Widmalm, G., Andersen, L.P., Knirel, Y.A., Jansson, P.-E., and Moran, A.P. (2001). Structure of a D-*glycero*-D-*manno*-heptan from the lipopolysaccharide of *Helicobacter pylori*. Carbohydr. Res. *331*, 219–224.

Seydel, U., Oikawa, M., Fukase, K., Kusumoto, S., and Brandenburg, K. (2000). Intrinsic conformation of lipid A is responsiblke for aginostic and antagonistic activity. Eur. J. Biochem. *267*, 3032–3039.

Simoons-Smit, I.M., Appelmelk, B.J., Verboom, T., Negrini, R., Penner, J.L., Aspinall, G.O., Moran, A.P., Fei Fei S., Bi-shan, S., Rudnica, W., Savio, A., and De Graaff, J. (1996). Typing of *Helicobacter pylori* with monoclonal antibodies against Lewis antigens in lipopolysaccharide. J. Clin. Microbiol. *34*, 2196–2200.

Slomiany, B.L., Liau, Y.H., Lopez, R.A., Piotrowski, J., Czajowski, A., and Slomiany, A. (1992). Effect of *Helicobacter pylori* lipopolysaccharide on the synthesis of sulfated gastric mucin. Biochem. Int. *27*, 687–697.

Slomiany, B.L., Piotrowski, J., Rajiah, G., and Slomiany, A. (1991). Inhibition of gastric mucosal laminin receptor by *Helicobacter pylori* lipopolysaccharide: effect of nitecapone. Gen. Pharmacol. 22, 1063–1069.

Slomiany, B.L., and Slomiany, A. (2005). Endothelin-1-dependent leptin induction in gastric mucosal inflammatory responses to *Helicobacter pylori* lipopolysaccharide. Biochem. Biophys. Res. Commun. *336*, 1106–1111.

Slomiany, B.L., and Slomiany, A. (2006). Cytosolic phospholipase A2 activation in *Helciobacter pylori* lipopolysaccharide-induced interference with gastric mucin synthesis. IUBMB Life 58, 217–223.

Smith, G.V., Moran, A.P., Bajaj-Elliott, M., and Farthing, M.J.G. (2003b). Induction of cyclooxygenase 2 by *Escherichia coli* but not *Helicobacter pylori* lipopolysaccharide in gastric epithelial cells *in vitro*. Helicobacter *8*, 513–520.

Smith, M.F., Mitchell, A., Li, G., Ding, S., Fitzmaurice, A.M., Ryan, K., Crowe, S., and Goldberg, J.B. (2003a). Toll-like receptor (TLR) 2 and TLR5, but not TLR4, are required for *Helicobacter pylori*-induced NF-κB activation and chemokine expression by epithelial cells. J. Biol. Chem. *278*, 32552–32560.

Smith, S., Moran, A.P., and Kelleher, D.P. (2006). Lipopolysaccharides from *Helicobacter pylori* NCTC 11637 upregulate IL-8 and VEGF via Toll-like receptor 2 in an epithelial cell system. Gastroenterology *130* (*Suppl.* 2), A526.

Stead, C., Tran, A., Ferguson, D., McGrath, S., Cotter, R., and Trent, S. (2005). A novel 3-deoxy-D-manno-octulosonic acid (Kdo) hydrolase that removes the outer Kdo sugar of *Helicobacter pylori* lipopolysaccharide. J. Bacteriol. *187*, 3374–3383.

Sun, H.Y., Lin, S.W., Ko, T.P., Pan, J.F., Liu, C.L., Lin, C.N., Wang, A.H., and Lin, C.H. (2007). Structure and mechanism of *H. pylori* fucosyltransferase. A basis for lipopolysaccharide variation and inhibitor design. J. Biol. Chem. *282*, 9973–9982.

Suda, Y., Kim, Y.M., Ogawa, T., Yasui, N., Hasegawa, Y., Kashihara, W., Shimoyama, T., Aoyama, K., Nagat, K., Tamura, T., and Kusomoto, S. (2001). Chemical structure and biological activity of a lipid A component from *Helicobacter pylori* strain 206. J. Endotoxin Res. *7*, 95–104.

Suda, Y., Ogawa, T., Kashihara, W., Oikawa, M., Shimoyama, T., Hayashi, T., Tamura, T., and Kusumoto, S. (1997). Chemical structure of lipid A from *Helicobacter pylori* strain 206-1 lipopolysaccharide. J. Biochem (Tokyo) *121*, 1129–1133.

Suerbaum, S., Josenhans, C., Sterzenbach, T., Drescher, B., Brandt, P., Bell, M., Dröge, M., Fartmann, B., Fischer, H.-P., Ge, Z., Hörster, A., Holland, R., Klein, K., König, J., Macko, L., Mendz, G.L., Nyakatura, G., Schauer, D.P., Shen, Z., Weber, J., Frosch, M., and Fox, J.G. (2003). The complete genome sequence of the carcinogenic bacterium *Helicobacter hepaticus*. Proc. Natl. Acad. Sci. USA *100*, 7901–7906.

Takenaka, R., Yokota, K., Ayada, K., Mizuno, M., Zhao, Y., Fujinami, Y., Lin, S.N., Toyokawa, T., Okada, H., Shiratori, Y., and Oguma, K. (2004). *Helicobacter pylori* heat shock protein 60 induces inflammatory responses through the Toll-like receptor-triggered pathway in cultured human gastric epithelial cells. Microbiology *150*, 3913–3922.

Terrés, A.M., Pajares, J.M., Hopkins, A.M., Murphy, A., Moran, A., Baird, A.W., and Kelleher, D. (1998). *Helicobacter pylori* disrupts epithelial barrier function in a process inhibited by protein kinase C activators. Infect. Immun. *66*, 2943–2950.

Tomb, J.F., White, O., Kerlavage, A.R., Clayton, R.A., Sutton, G.G., Fleischmann, R.D., Ketchum, K.A., Klenk, H.P., Gill, S., Dougherty, B.A., Nelson, K., Quackenbush, J., Zhou, L. , Kirkness, E.F., Peterson, S., Loftus, B., Richardson, D., Dodson, R., Khalak, H.G., Glodek, A., McKenney, K., Fitzegerald, L.M., Lee, N., Adams M.D., Hickey, E.K., Berg, D.E., Gocayne, J.D., Utterback, T.R., Peterson, J.D., Kelley, J.M., Cotton, M.D., Weidman, J.M., Fujii, C., Bowman, C. , Watthey, L., Wallin, E., Hayes, W.S., Borodovsky, M., Karp, P.D., Smith, H.O., Fraser, C.M., and Venter, J.C. (1997). The complete genome sequence of the gastric pathogen *Helicobacter pylori*. Nature *388*, 539–547.

Tran, A.X., Karbarz, M.J., Wang, X., Raetz, C.R., McGrath, S.C., Cotter, R.J., and Trent, M.S. (2004). Periplasmic cleavage and modification of the 1-phosphate group of *Helicobacter pylori* lipid A. J. Biol. Chem. *279*, 55780–55791.

Tran, A.X, Stead, C.M., and Trent, M.S. (2005). Remodeling of *Helicobacter pylori* lipopolysaccharide. J. Endotoxin Res. *11*, 161–166.

Tran, A.X., Whittimore, J.D., Wyrick, P.B., McGrath, S.C., Cotter, R.J., and Trent, M.S. (2006). The lipid A 1-phosphatase of *Helicobacter pylori* is required for resistance to the antimicrobial peptide polymyxin. J. Bacteriol. *188*, 4531–4541.

Trent, M.S., and Raetz C.R.H. (2002). Cloning of EptA, the lipid A phosphoethanolamine transferase associated with polymyxin resistance. J. Endotoxin Res. *8*, 159.

Trent, M.S., Ribeiro, A.A., Lin, S., Cotter, R.J., and Raetz, C.R. (2001). An inner membrane enzyme in *Salmonella* and *Escherichia coli* that transfers 4-amino-4-deoxy-L-arabinose to lipid A: induction on polymyxin-resistant mutants and role of a novel lipid-linked donor. J. Biol. Chem. *276*, 43122–43131.

Trent, M.S., Stead, C.M., Tran, A.X., and Hankins, J.V. (2006). Diversity of endotoxin and its impact on pathogenesis. J. Endotoxin Res. *12*, 205–223.

Tzeng, Y.L., Ambrose, K.D., Zughaier, S., Zhou, X., Miller, Y.K., Shafer, W.M., and Stephens, D.S. (2005). Cationic antimicrobial peptide resistance in *Neisseria meningitidis*. J. Bacteriol. *187*, 5387–5396.

Vaara, M., and Vaara, T. (1981). Outer membrane permeability barrier disruption by polymyxin in polymyxin-susceptible and -resistant *Salmonella typhimurium*. Antimicrob. Agents Chemother. *19*, 578–583.

Vaara, M., and Vaara, T. (1983). Polycations sensitize enteric bacteria to antibiotics. Antimicrob. Agents Chemother. *24*, 107–113.

Valkonen, K.H., Wadström, T., and Moran, A.P. (1994). Interaction of lipopolysaccharides of *Helicobacter pylori* with basement membrane protein laminin. Infect. Immun. *62*, 3640–3648.

Valkonen, K.H., Wadström, T., and Moran, A.P. (1997). Identification of the *N*-acetylneuraminyllactose-specific laminin-binding protein of *Helicobacter pylori*. Infect. Immun. *65*, 916–923.

Viala, J., Chaput, C., Boneca, I.G., Cardona, A., Girardin, S.E., Moran, A.P., Athman, R., Mémet, S., Huerre, M.R., Coyle, A.J., DiStefano, P.S., Sansonetti, P.J., Labigne, A., Bertin, J., Philpott, D.J., and Ferrero, R.L. (2004). Nod1 responds to peptidoglycan delivered by the *Helicobacter pylori cag* pathogenicity island. Nat. Immunol. *5*, 1166–1174.

Wada, A., Mori, N., Oishi, K., Hojo, H., Nakahara, Y., Hamanaka, Y., Nagashima, M., Sekine, I., Ogushi, K., Niidome, T., Nagatake, T., Moss, J., and Hirayama, T. (1999). Induction of human β-defensin-2 mRNA expression by *Helicobacter pylori* in human gastric cell line MKN45 cells on *cag* pathogenicity island. Biochem. Biophys. Res. Commun. *263*, 770–774.

Walsh , E.J., and Moran, A.P. (1996). Serological analysis of the core region of *Helicobacter pylori* lipopolysaccharide. Gut 39 (Suppl. 2), A71–A72.

Walsh, E.J., and Moran, A.P. (1997). Influence of medium composition on the growth and antigen expression of *Helicobacter pylori*. J. Appl. Microbiol. *83*, 67–75.

Wang, G., Boulton, P.G., Chan, N.W., Palcic, M.M., and Taylor, D.E. (1999a). Novel *H. pylori* α1, 2-fucosyltransferase, a key enzyme in the synthesis of Lewis antigens. Microbiology *145*, 3245–3253.

Wang, G., Ge, Z., Rasko, D.A., and Taylor, D.E. (2000). Lewis antigens in *Helicobacter pylori*: biosynthesis and phase variation. Mol. Microbiol. *36*, 1187–1196.

Wang, G., Rasko, D.A., Sherburne, R., and Taylor, D.E. (1999b). Molecular genetic basis for the variable expression of Lewis Y antign in *Helicobacter pylori*: analysis of the α(1, 2) fucosyltransferase gene. Mol. Microbiol. *31*, 1265–1274.

Wang, X., Karbarz, M.J., McGrath, S.C., Cotter, R.J., and Raetz, C.R. (2004). MsbA transporter-dependent lipid A 1-dephosphorylation on the periplasmic surface of the inner membrane: topography of *Francisella novicida* LpxE expressed in *Escherichia coli*. J. Biol. Chem. *279*, 49470–49478.

Wang, X., McGrath, S.C., Cotter, R.J., and Raetz, C.R. (2006). Expression cloning and periplasmic orientation of the *Francisella novicida* lipid A 4′-phosphatase LpxF. J. Biol. Chem. *281*, 9321–9330.

White, K.A., Lin, S., Cotter, R.J., and Raetz, C.R. (1999). A *Haemophilus influenzae* gene that encodes a membrane bound 3-deoxy-D-*manno*-octulosonic acid (Kdo) kinase. Possible involvement of Kdo phosphorylation in bacterial virulence. J. Biol. Chem. *274*, 31391–31400.

Whittle, B.J.R., Morschl, É., Pozsár, J., Moran, A.P., and László, F. (2001). *Helicobacter pylori* lipopolysaccharide provokes iNOS-mediated acute systemic microvascular inflammatory responses in rat cardiac, hepatic, renal and pulmonary tissues. J. Physiol. (Paris) *95*, 257–259.

Wirth, H.P., Yang, M., Karita, M., and Blaser, M.J. (1996). Expression of the human cell surface glycoconjugates Lewis X and Lewis Y by *Helicobacter pylori* isolates is related to *cagA* status. Infect. Immun. 64, 4598–4605.

Wirth, H.P., Yang, M., Peek, R.M., Höök-Nikanne, J., Fried, M., and Blaser, M.J. (1999). Phenotypic diversity in Lewis expression of *Helicobacter pylori* isolates from the same strain. J. Lab. Clin. Med. *133*, 488–500.

Wirth, H.P., Yang, M., Peek, R.M., Tham, K.T., and Blaser, M.J. (1997). *Helicobacter pylori* Lewis expression is related to the host Lewis phenotype. Gastroenterology *113*, 1091–1098.

Wirth, H.P., Yang, M., Sanabria-Valentin, E., Berg, D.E., Dubois, A., and Blaser, M.J. (2006). Host Lewis phenotype-dependent *Helicobacter pylori* Lewis antigen expression in rhesus monkeys. FASEB J. *20*, 1534–1536.

Yethon, J.A., Gunn, J.S., Ernst, R.K., Miller, S.I., Laroche, L., Malo, D., and Whitfield, C. (2000). *Salmonella enterica* serovar Typhimurium *waaP* mutants show increased susceptibility to polymyxin and loss of virulence in vivo. Infect. Immun. *68*, 4485–4491.

Yethon, J.A., Heinrichs, D.E., Monteiro, M.A., Perry, M.B., and Whitfield, C. (1998). Involvement of *waaY*, *waaQ*, and *waaP* in the modification of *Escherichia coli* lipopolysaccharide and their role in the formation of a stable outer membrane. J Biol Chem *273*, 26310–26316.

Young, G.O., Stemmet, N., Lastovica, A., van der Merwe, E.L., Louw, J.A., Modlin, I.M., and Marks, I.N. (1992). *Helicobacter pylori* lipopolysaccharide stimulates gastric mucosal pepsinogen secretion. Aliment. Pharmacol. Ther. *6*, 169–177.

Zhou, Z., Ribeiro, A.A., Lin, S., Cotter, R.J., Miller, S.I., and Raetz, C.R. (2001). Lipid A modifications in polymyxin-resistant *Salmonella typhimurium*: PMRA-dependent 4-amino-4-deoxy-L-arabinose, and phosphoethanolamine incorporation. J. Biol. Chem. *276*, 43111–43121.

Zhou, Z., Ribeiro, A.A., and Raetz, C.R. (2000). High-resolution NMR spectroscopy of lipid A molecules containing 4-amino-4-deoxy-L-arabinose and phosphoethanolamine substituents. Different attachment sites on lipid A molecules from NH4VO3-treated *Escherichia coli* versus *kdsA* mutants of *Salmonella typhimurium*. J. Biol. Chem. *275*, 13542–13551.

Zhou, Z., White, K.A., Polissi, A., Georgopoulos, C., and Raetz, C.R. (1998). Function of *Escherichia coli* MsbA, an essential ABC family transporter, in lipid A and phospholipid biosynthesis. J. Biol. Chem. *273*, 12466–12475.

Helicobacter pylori Outer Membrane Proteins

3

Yoshio Yamaoka and Richard A. Alm

Abstract

Analysis of the three completed *Helicobacter pylori* genomes has confirmed the presence of five major outer membrane proteins (OMPs) families. However, there appears to be a trend within the protein families as to which two of the three orthologues are more closely related. *H. pylori* 26695 and HPAG1 proteins are the most closely related more often as *H. pylori* J99 and HPAG1 or *H. pylori* J99 and 26695. Whether this is because both HPAG1 and 26695 were isolated from gastritis patients and J99 from a duodenal ulcer patient is an intriguing, but untested possibility.

It has been demonstrated that several OMPs in the largest family act as adhesions, and these include BabA, SabA, OipA, AlpAB and HopZ. This unusual set of specialized OMPs may be a reflection of the adaptation of *H. pylori* to the unique gastric environment where it is found. Many of these genes undergo phase variation such that not all strains will produce functional proteins, and the stability of expression with passage varied with OipA > BabA > BabB > SabA. In this chapter, as well as describing the general features of the *H. pylori* OMP families, we describe the well-characterized OMPs (BabA, SabA and OipA) in detail. Each OMP appears to have specific functions, and further studies will be necessary for investigating the detailed analyses of these OMPs as well as many unstudied OMPs.

Introduction

Approximately 4% of the *Helicobacter pylori* genome encodes a diverse repertoire of outer membrane proteins (OMPs). The paralogous families of OMPs from the three fully sequenced *H. pylori* isolates were compared. These were strain J99, isolated from a patient with duodenal ulcer disease (Alm *et al.*, 1999); strain 26695, isolated from a patient with superficial gastritis (Tomb *et al.*, 1997); and strain HPAG1, isolated from a patient with chronic atrophic gastritis (Oh *et al.*, 2006). All three isolates contain similar members of the five major OMP families as described previously (Alm *et al.*, 2000). Significantly, given the high level of diversity seen between *H. pylori* isolates, the orthologous proteins (equivalent proteins between the three genomes, same predicted function) shared considerable identity between the three strains. However, there appears to be a trend within the protein families as to which two of the three orthologues are more closely related. *H. pylori* 26695 and HPAG1 proteins are the most closely related approximately three times more often as *H. pylori* J99 and HPAG1 and seven times more often as *H. pylori* J99 and 26695. Whether this is because both HPAG1 and 26695 came from gastritis patients and J99 from a duodenal ulcer patient is an intriguing, but untested hypothesis.

The adherence of *H. pylori* to the gastric mucosa is widely assumed to play an important role in the initial colonization and long-term persistence in the human gastric mucosa. Several putative adhesins of *H. pylori* have been identified, most of which were members of the largest OMP family and include the blood group antigen binding adhesin (BabA: HopS), sialic acid binding adhesin (SabA: HopP), adherence-associated lipoprotein (AlpA and AlpB: HopB and HopC, respectively), outer inflammatory protein (OipA: HopH) and HopZ. Lewis b antigen (Le^b) and related fucosylated ABO blood group antigens are recognized by BabA (Ilver *et al.*, 1998), whereas sialyl-Lewis x and sialyl-Lewis a antigens (sLe^x and sLe^a) are recognized by SabA (Mahdavi *et al.*, 2002). However, the corresponding receptors for AlpAB, OipA and HopZ remain unknown. Some OMPs also play important roles in the immune response, and the mechanisms by which OipA induces signal transduction pathways in the host cells are well studied.

In this chapter, we first describe the general features of five major OMP families, and then describe the three OMPs (BabA, SabA and OipA) in detail.

Hop (and Hor) group of outer membrane proteins

Members of the large Hop (*Helicobacter* outer membrane protein) family were among the first characterized OMPs in *H. pylori*. Five members (HopA–E) were purified and shown to have a highly conserved N-terminal motif (A↓EX(D,N)G where the ↓ represents the cleavage point), and were demonstrated to function as porins (Doig *et al.*, 1995; Exner *et al.*, 1995), transmembrane diffusion channels that allow small hydrophilic molecules, nutrients and even small antibiotics to passively diffuse across the outer membrane of Gram-negative bacteria. Analysis of the first completed *H. pylori* genome, strain 26695, demonstrated that these proteins were part of a much larger paralogous family of related proteins, all of which possessed the characteristic N-terminal motif. Importantly, it has been demonstrated that several other OMPs in this family act as adhesins, such as BabA, SabA, AlpAB, OipA and HopZ. This unusual set of specialized OMPs may be a reflection of the adaptation of *H. pylori* to the unique gastric environment where it is found.

Analysis of the three completed *H. pylori* genomes has confirmed the presence of this large paralogous family of proteins. Similar to *H. pylori* 26695, *H. pylori* HPAG1 contains 21 members of the Hop protein family, while *H. pylori* J99 lacks the HopU (or BabC) protein (Table 3.1). Most of the Hop proteins are predicted to contain anti-parallel amphipathic β-sheets that can be modelled into β-barrels. Typically, these β-strands are connected on the periplasmic side by short amino acid loops and by longer loops on the external side of the porin. The external loops can then fold back into the core of the β-barrel to effect the pore characteristics (size, selectivity) or can function in protein-protein interactions. *H. pylori* HPAG1, like J99 and 26695, also contains orthologues of the twelve Hor (hop related) proteins (HorA to L), that together with the Hop proteins constitute the major OMP family.

The sequence diversity of the orthologous groups of OMPs was compared between the three strains. These strains were isolated approximately a decade apart on three different continents from patients presenting different clinical symptoms, and thus are unlikely to be directly related. Of the 20 Hop proteins where there was an orthologue in all three strains, 12 have an identity >90%, with the remainder being between 80% and 90%. The pairwise comparison of the HopU (BabC) proteins from 26695 and HPAG1 showed 87% identity.

Many Gram-negative bacterial OMPs end with a C-terminal phenylalanine residue, predicted to be important for the proper insertion into the lipid bilayer (Struyve *et al.*, 1991). All the Hop and Hor proteins have the characteristic C-terminus with alternating hydrophobic and hydrophilic residues, with aromatic residues occupying the majority of the alternating positions from −1 to −11 (counting back from the C-terminus). The three C-terminal residues are identical for all the Hop and Hor orthologues from all three strains, with the exception of HorI, in which the −2 position is different (Table 3.1). The earlier observation that half of the Hop family proteins terminate with tyrosine rather than phenylalanine (Y-Hops) and have highly similar C-terminal domains also holds true for the HPAG1 orthologues.

The Y-Hop proteins are less divergent at their mature N-terminus leaving the central hypervariable domain containing the majority of the member-specific sequences. Indeed, for several Y-Hop proteins, the similarity of the C-terminal domain between paralogues within a strain is higher than the similarity of the orthologous proteins between isolates. This is true for SabB (HopO) and SabA (HopP) as well as HopM

Table 3.1 Comparison of OMPs from *H. pylori* J99 (JHP), 26695 (HP) and HPAG1

Gene number			Length (aa)						
J99	26695	HPAG1	JHP	HP	HPAG1	% Size variation (aa)	C-terminal residues	% ID*h*	Gene name*a*
Family I: major outer membrane protein family									
Hop proteins									
7	0009	0009	669[b]	672[b]	667[b]	0.5 (3[c])	FAY	93.3	*hopZ*
21	0025	0023	690	711	688	3.3 (23)	FAY	86.1	*hopD*
212	0227	0229/0230	696	691	694[e]	0.7 (5)	FAY	80.6	*hopM*
214	0229	0232	483	483	486	0.6 (3)	LAY	90.1	*hopA*
237	0252	0255	479	487	485	1.6 (8)	VGF	95.4	*hopF*
238	0253/0254	0256	471	471[e]	473	0.2 (1[f])	IGF	96.0	*hopG*
429	0477	0454	371	367	368	1.1 (4)	YSF	86.7	*hopJ*
581	0638	0621	307	305	307	0.7 (2[c])	NKH	92.8	*oipA (hopH)*
645	0706	0691	270	273	273	0[f]	YTF	97.8	*hopE*
659	0722	0706	638[b]	644[b]	632[b]	1.9 (12[c])	FAY	85.1	*sabB (hopO)*
662	0725	0709	651[b]	653[b]	651	0.3 (2[c])	FAY	89.6	*sabA (hopP)*
833	1243	0876	744	733	744	1.5 (11)	FAY	91.3	*babA(hopS)*
848	0912	0895	520	515	517	1 (5)	YSF	96.1	*alpA (hopC)*
849	0913	0896	527	529	529	0.4 (2)	YAF	94.7	*alpB (hopB)*
857	0923	0904	366	369	368	0.8 (3)	YSF	86.6	*hopK*
1083	1156	1095	697	696	696	0.1 (1)	IGF	94.7	*hopI*
1084	1157	1096	1237	1230	1240	0.8 (10)	MGF	91.2	*hopL*
1164	0896	0320	703	708	707	0.4 (3[c])	FAY	90.2	*babB (hopT)*
1103	1177	1115	643	641	641	0.3 (2)	FAY	85.3	*hopQ*
1261	1342	1289	696	691	692	0.7 (5)	FAY	80.9	*hopN*
NA[d]	0317	1186	NA[d]	745	738	NA[d]	FAY	87.0	*babC (hopU)*

Table 3.1 *continued*

Gene number			Length (aa)						
J99	26695	HPAG1	JHP	HP	HPAG1	% Size variation (aa)	C-terminal residues	% ID[h]	Gene name[a]
Hor proteins									
0073[k]	0079	0080	k	595	567	4.9 (28)	INF	89.6	*horA*
0117	0127	0126	286	286	286	0	VSF	98.6	*horB*
0307	0324	0327	245	254	254	0[f]	YHF	92.7	*horC*
0359	1066	0381	200	200	200	0	WHF	99.5	*horD*
0424	0472	0449	186	186	186	0	FTF	98.9	*horE*
0614	0671	0654	270	270	270	0	YNF	97.4	*horF*
0732	0796	0781	278	278	264	0[f]	YDF	90.2	*horG*
1034	1107	1046	220	230	230	0[f]	YNF	93.2	*horH*
1040	1113	1052	277	277	277	0	Y(I/S[g])F	90.6	*horI*
1362	1469	1446	248	248	248	0	RDF	91.9	*horJ*
1394	1501	1411	388	388	388	0	YTF	94.9	*horK*
1432	1395	1475	242	242	242	0	FTF	90.9	*horL*
Family 2: Hof family of outer membrane proteins									
0195	0209	0210	438	450	450	0[f]	YRF	92.7	*hofA*
0342	1083	0364	479	479	479	0	AKF	94.2	*hofB*
0438	0486	0462	528	528	528	0	YSF	94.1	*hofC*
0439	0487	0463	465	480	470	0[f]	RIY	98.5	*hofD*
0719	0782	0768	455	455	455	0	FFF	89.0	*hofE*
0725	0788	0774	499	499	499	0	WKL	97.0	*hofF*
0850	0914	0897	514	514	514	0	LKF	98.1	*hofG*
1094	1167	1106	471	471	471	0	ASF	95.8	*hofH*

Family 3: Hom family of outer membrane proteins									
0649	0710	NA[d]	657	660	NA[d]	NA[d]	WVF	85.7	*homA*
0870	NA[d]	0695	668	NA[d]	667	NA[d]	WVF	93.4	*homB*
1008	0373	1019	751	700	700	6.8 (51)	W(I/V[g])F	80.0	*homC*
1346	1453	1379	744	746	746	0.3 (2)	WIF	93.7	*homD*
Family 4: iron-regulated outer membrane proteins									
FecA-like proteins									
0626	0686	0669	767	767	767	0	YEF	92.4	*fecA-1*
0743	0807	0792	792	787	792	0*i*	YNF	93.1	*fecA-2*
1426	1400	1469	841	842	842	0*f*	YTF	98.1	*fecA-3*
FrpB-like proteins									
0810	0876	0859	791	791	791	0	YKW	96.7	*frpB-1*
0851	0915/0916	0898	815	812*e*	813	0.4 (3)	YKF	95.9	*frpB-2*
1405	1512	1400	879	877	863	1.8 (16)	YQF	93.2	*frpB-3*
Family 5: efflux pump outer membrane proteins									
0552	0605	0586	477	477	477	0	YVH	97.9	*hefA*
0905	0971	0952	431	413	431	0.4 (2[f])	(E/V[g])LH	95.4	*hefD*
1247	1327	1272	412	412	412	0	GLE	92.2	*hefG*
Other outer membrane proteins									
0456	0506	0480	406	403	403	0.7 (3)	EGF	94.8	
0600	0655	0640	906	916	916	1.1 (10)	TRF	96.9	
0634	0694[j]	0679	336	257[j]	333	0.9 (3)[j]	FAF	97.6[j]	
0663	0726	0710	305	305	305	0	FLF	90.8	

Table 3.1 *continued*

Gene number			Length (aa)						
J99	26695	HPAG1	JHP	HP	HPAG1	% Size variation (aa)	C-terminal residues	% ID[h]	Gene name[a]
1022	0358	1034	511	511	511	0	GLF	89.8	
1360	1467	1448	231	231	231	0	YKF	95.2	
0308	0325	0328	237	237	237	0	MPY	97.1	*flgH*
0777	0839	0324	587	587	587	0	YRW	97.3	
1054	1125	1064	179	179	179	0	L(V/M[g])K	95.0	*palA*
1349	1456	1459	175	175	175	0	VKK	99.4	*lpp20*

[a]Gene names have been assigned as follows: those with Hop-like motifs have been named *hop* genes with the original *hopA-E* gene names being assigned to those previous identified. Proteins related to the *hop* family but lacking the N-terminal motif have been called *hor* genes (hop related). The family of 50 kDa outer membrane protein genes has been called *hof* genes (*Helicobacter* OMP family). The smaller family of outer membrane protein genes has been called *hom* genes (*Helicobacter* outer membrane).

[b]These proteins are out-of-frame due to a CT dinucleotide repeat in their signal sequence. The protein sizes were determined by adjusting the coding sequence by the addition/removal of a single dinucleotide repeat.

[c]The size variation in the length of these proteins does not include the differences caused by the different numbers of CT dinucleotide repeats in the coding sequence.

[d]Not applicable.

[e]The HP0253 and HP0254 genes, the HP0915 and HP0916 genes, and the HPAG0229 and HPAG0230 genes were joined by the addition or removal of a single nucleotide for ease of comparison

[f]Size difference due to difference in prediction of initiation codons between *H. pylori* strains.

[g]The orthologues differ in their sequence in this position.

[h]The identity was calculated over the aligned portion of the proteins only and represented as a percentage to the shortest orthologue.

[i]There is significant difference in the C-terminal 20 amino acids. The C-terminus is found in a different reading frame in *H. pylori* 26695, and most likely represents a frameshift in HP0807.

[j]The remainder of the orthologue in *H. pylori* 26695 is found in a different reading frame after a potential frameshift. If a true frameshift, the orthologue would be 333 residues in length, with a three-way identity of 96.1%. The identity between JHP0634 and HPAG0679 is 97.6%

[k]The JHP0073 protein may represent a pseudogene caused by an internal deletion and subsequent fusion between the orthologous HP0078/HP0079 proteins in *H. pylori* 26695 and the HPAG0079/HPAG0080 proteins in *H. pylori* HPAG1 (Alm *et al.*, 2000).

and HopN. It is also true for the BabA (HopS) and BabB (HopT) proteins in J99, whereas the C-terminal domain of HopU (BabC) is more similar to BabA and BabB in 26695 and HPAG1, respectively (Fig. 3.1). However, this separation is only based on 1–2 residues that are different. The high level of similarity between paralogues and orthologues strongly suggests that this protein family has arisen due to ancestral gene duplication.

All three sequenced *H. pylori* genomes possess the duplicated members of the Hop family (HopM/N and HopJ/K), as shown by the initial genome comparison and subsequent PCR analyses on a range of geographically diverse *H. pylori* isolates (Alm *et al.*, 2000). Interestingly, while HopM and HopN are 100% identical to each other in J99 and 26695, the two paralogues only show 86.7% identity within *H. pylori* HPAG1, with HPAG1 HopM sharing highest similarity with the J99 proteins, and HPAG1 HopN protein sharing the highest similarity with the 26695 proteins. This separation is based on the member-specific central hypervariable domain, as alignments of the C-terminal domains show differentiation based on the specific strain. These family members have the lowest three-way similarity score of any of the 32 Hop/Hor proteins (Table 3.1). The other highly similar pair of Hop proteins, HopJ and HopK, are 100% identical in HPAG1, and ~99% identical in both J99 and 26695, and yet the three way similarities drop to approximately 86% (Table 3.1). The presence of the same genes being duplicated almost identically in multiple strains but with different sequences between strains is intriguing. This could suggest that divergence occurred in an ancestor and then each isolate has duplicated the genes that encode for the same proteins. However, it is more likely a duplication event occurred in the ancestral strain and that each strain now maintains almost perfect duplication by repeatedly taking up DNA from lysed surrounding cells and integrating this DNA into the two duplicated sites. In this manner, the gene sequences between *H. pylori* strains would be able to diverge significantly while still

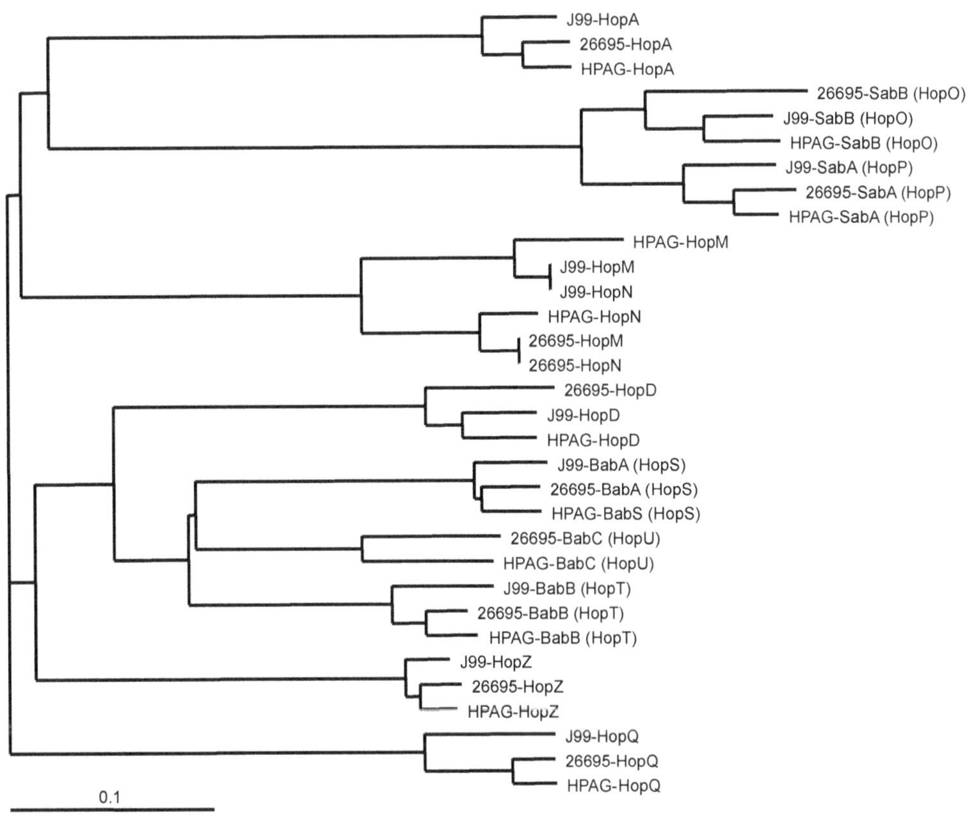

Figure 3.1 Phylogenic tree of the Y-Hop outer membrane protein family. Protein sequences were analysed using the ClustalX and PHYLIP programs.

maintaining a very high intra-strain identity. This model would also explain the higher C-terminal identity of some of the Hop proteins (e.g. BabA and BabB) within a given strain than either protein with its corresponding orthologue from another *H. pylori* strain.

The sequence identity of the Hor family of proteins between the three *H.* pylori strains is > 90% with the exception of HorA. In addition, the length of the HorB–HorL polypeptides is identical across all three isolates, a feature that is in contrast to the Hop proteins, where all but one protein differs in size, by up to 3% (Table 3.1). The HorA protein in *H. pylori* J99 is 255 amino acids in length, and appears to represent a gene-fusion between two adjacent genes in *H. pylori* 26695 as the N-terminus of JHP73 aligns with the N-terminus of HP0078, while the C-terminus of JHP73 aligns with the C-terminus of HP0079 (Table 3.1). When this region was analysed by PCR from a variety of strains there was considerable variability in the size of the HorA orthologue (Alm *et al.*, 2000). The HorA orthologue from *H. pylori* HPAG1 is 5% smaller than the 26695 protein, and shares almost 90% identity across the length of the protein.

Other families of outer membrane proteins

The Hof (*Helicobacter* OMP family) family of related proteins was identified from the *H. pylori* proteome based on the documentation of a 50 kDa non-heat-modifiable protein located in the outer membrane of strain CCUG17874 (Exner *et al.*, 1995). There are eight members in this paralogous family with molecular masses ranging from 51.2 to 59.7 kDa, and the overall similarity of the proteins from the three strains is quite high, with seven members having identities > 90% between the strains (Table 3.1). In addition, the predicted mature forms of all the orthologous pairs of proteins are identical in size between *H. pylori* J99, 26695, and HPAG1 (Table 3.1).

There is a smaller paralogous family of proteins that also contain the C-terminal alternating hydrophobic motif and characteristic signal sequences typical of outer membrane proteins that were called the Hom (*Helicobacter* outer membrane) family (Alm *et al.*, 2000) (Table 3.1). *H. pylori* 26695 and HPAG1 contain three members of this family, whereas *H. pylori* J99 contains four members. All of the Hom proteins have conserved N- and C- termini, while the central domain of the molecule displays greater variability. All three strains contain a HomC and HomD orthologue, with the HomC proteins displaying far greater variability, only 80% identical, than the HomD orthologues (Table 3.1). An orthologue of *H. pylori* J99 HomB protein was not found in strain 26695, and represented a single gene insertion in the J99 genome. However, it appears that a Hom protein orthologue once existed at that genomic location in strain 26695 as highly homologous DNA that would encode 73 amino acid residues of the C-terminal region is found there. Importantly, there is also a remnant of the C-terminal domain of a Hom orthologue in *H. pylori* HPAG1 in that same genomic location, and was identified as gene HPAG1_0916. Interestingly, the HPAG1 Hom protein found at the physical location of the HomA orthologue in both J99 and 26695, HPAG1_0695, bears higher identity (93.4%) to J99 HomB than to either of the HomA orthologues (87%) (Table 3.1). This is another example of *H. pylori* swapping the physical locations of paralogous genes, similar to that seen for *babA*, *babB*, and *babC*. Importantly however, the J99 HomA and HomB proteins share ~92% identity with all the differences being confined to the central domain, suggesting relatively recent gene duplication.

The two protein families that display homology to the iron-regulated outer membrane proteins from other bacteria (FecA-like, similarity to the ferric citrate receptor of *E. coli*, and FrpB-like, similarity to a major iron-regulated outer membrane protein in *Neisseria* spp.) are also represented in all three *H. pylori* strains. Both of these families contain three paralogous members, and all six protein share greater than 92% identity across the strains.

All three of the sequenced *H. pylori* strains contain three clusters (*hefA–C*, *hefD–F*, *hefG–I*) which encode homologues of resistance–nodulation–division (RND) efflux pump systems (Bina *et al.*, 2000). Each system contains an outer membrane component with some homology to the *E. coli* TolC protein, and these proteins (HefA, D and G) share > 92% amino acid identity between the three strains (Table 3.1). These proposed efflux systems have been shown to be highly conserved in sequence and organization between multiple *H. pylori* strains (Bina *et al.*, 2000), and the HefABC system has been shown to be an active multidrug efflux system (Kutschke and de Jonge, 2005).

Additional outer membrane protein singletons

Ten additional ORFs whose products are predicted to be located in the outer membrane were identified based on the C-terminal motif characteristic of OMPs and are not part of paralogous families are listed in Table 3.1. The FlgH homologue of *H. pylori* contains a 21 amino acid signal sequence but is smaller compared with homologues in other bacterial species. There are also several lipoproteins that may be associated with the outer membrane. Among these was a homologue of the peptidoglycan associated lipoprotein (PAL) family that includes the *Campylobacter jejuni* PAL protein (Burnens *et al.*, 1995) and the lpp20 lipoprotein that has been localized to the outer membrane, albeit not exclusively (Kostrzynska *et al.*, 1991) (Table 3.1).

BabA

Discovery of BabA

Lewis b (Le^b) is the dominant antigen in the gastric mucosa of secretor-positive individuals, and is characterized by the secretion of an α1,2-fucosyltransferase (Sakamoto *et al.*, 1989). In early studies, *H. pylori* were reported to be able to bind to fucosylated glycoconjugates containing Le^b structures on the surface of gastric epithelial cells in human biopsy specimens (Boren *et al.*, 1993; Falk *et al.*, 1993). Furthermore, monoclonal antibodies to Le^b or soluble Le^b antigens inhibited the binding of *H. pylori* to fixed gastric tissues (Boren *et al.*, 1993). Studies using transgenic mice expressing the human Le^b epitope in gastric epithelial cells showed that Le^b functions as a receptor for an *H. pylori*-specific adhesin and mediates its attachment to gastric pit and surface mucous cells (Falk *et al.*, 1995). Further studies using the same transgenic mice demonstrated that *H. pylori* was found in the mucous layer and adhered to the surface of mucus-secreting gastric epithelial cells, resulting in severe chronic gastritis and gastric atrophy; whereas *H. pylori* stayed in the mucous layer in non-transgenic control mice (Guruge *et al.*, 1998).

In 1998, Ilver *et al.* analysed the prevalence of blood group antigen-binding activity by measuring binding of *H. pylori* to ^{125}I-labelled fucosylated blood group antigens (Ilver *et al.*, 1998). Among 100 *H. pylori* isolates studied, 66% bound the Le^b antigen. In contrast, most of the strains (95%) did not bind the related Le^a, H-2, Le^x, or Le^Y antigens. The adhesin (78 kDa) recognizing the Le^b antigen was isolated by a combined ligand identification and purification technique, and designated as blood group antigen-binding adhesin (BabA) (Ilver *et al.*, 1998).

Clones were identified by subsequent molecular analyses that encoded two proteins with almost identical N-terminal domains and completely identical C-terminal domains (~300 amino acids), but with divergent central domains. The corresponding genes were designated *babA* and *babB*, and BabA, but not BabB had Le^b antigen-binding activity.

Location of the *babA* gene

The *babA* and *babB* coding sequences identified by Ilver *et al.* were highly similar to open reading frames in the sequenced genome of strain 26695 (HP1243 and HP0896, respectively) (Tomb *et al.*, 1997). The *bab* genes were initially cloned from strain CCUG17875, and this strain contains two *babA* genes and one *babB* gene. Gene inactivation experiments identified that only *babA2* had Le^b antigen-binding activity. The sequence of these two *babA* genes differed only by the presence of a 10 bp deletion in the extreme 5′ coding sequence of *babA1* that eliminates its translational initiation codon (Ilver *et al.*, 1998).

H. pylori 26695, J99 and HPAG1 all possess one *babA* allele (HP1243/JHP833/HPAG1_0876) and one *babB* allele (HP0896/JHP1164/HPAG1_0320) (Alm *et al.*, 1999; Oh *et al.*, 2006; Tomb *et al.*, 1997). Interestingly, the genomic locations of *babA* and *babB* genes are completely different among the three strains (Fig. 3.2). In *H. pylori* J99, *babA* (JHP833) is downstream of *hypD* (JHP835) with a J99-specific gene (JHP834) intervening, and *babB* (JHP1164) is downstream of *s18* (JHP1165). In strain 26695, the locations of *babA* (HP1243)

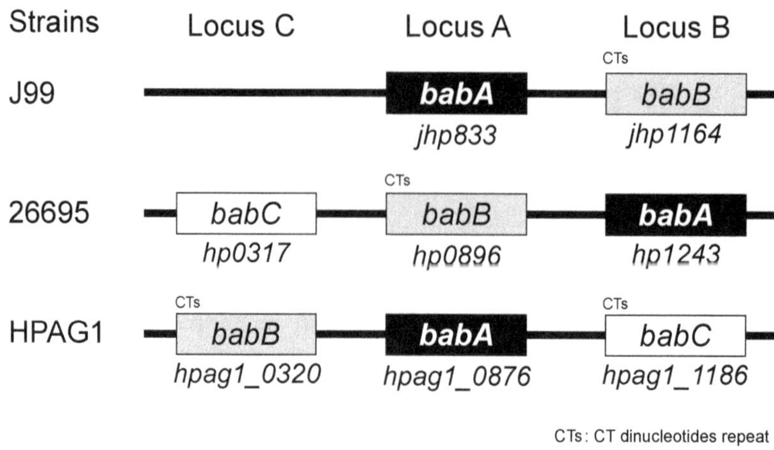

Figure 3.2 Genomic location of the *babA*, *babB* and *babC* genes in strains J99, 26695 and HPAG1.

and *babB* (HP0896) are reversed. For simplicity, the chromosomal locations downstream of *hypD* and *s18* will be referred to as locus A and locus B, respectively. *H. pylori* 26695 contains an orphan *hop* gene (HP0317: *hopU* or *babC*) whose function is unknown, and whose sequence is closely related to *babA* and *babB* and is called BabC (Alm *et al.*, 2000; Tomb *et al.*, 1997; Colbeck *et al.*, 2006). The chromosomal locations of *babC* in strain 26695 will be referred to as locus C. The *babC* gene does not exist in strain J99, and the *babC* gene in strain HPAG1 is located at locus B (Fig. 3.2). *H. pylori* CCUG17875 had two *babA* genes; *babA1* was located at locus B; however the locus of *babA2* was not determined (Ilver *et al.*, 1998).

Recently, two detailed studies investigating the location of *babA* and *babB* between different *H. pylori* isolates have been reported (Colbeck *et al.*, 2006; Hennig *et al.*, 2006). Hennig *et al.* (2006) analysed a panel of 35 *H. pylori* strains and found that *babA* was located at locus A for 19 strains (54%), at locus B for four strains (11%) and at locus C for three strains (9%). Four strains contained two copies of the *babA* gene similar to strain CCUG17875. In the four strains with two copies of the *babA* gene, the *babA* sequences found at two loci were identical in three strains and almost identical in one strain (i.e. three substitutions near the 5′ end of the genes in one strain), indicating that the multiple copies of *babA* presumably resulted from gene conversion (intragenomic non-reciprocal recombination) events. Importantly, two strains were shown to possess the *babA* gene; however the locus could not be identified, suggesting that there are additional unidentified chromosomal loci for *babA*. Overall, 24 (69%) contained *babA* sequences and 11 did not. In contrast, the *babB* gene was identified in 34 strains (97%).

Colbeck *et al.* (2006) also analysed a panel of 44 *H. pylori* strains and found that *babA* gene was located at locus A for 25 strains (57%) and at locus B for 18 strains (41%). Interestingly, although they used chromosomal DNA from low-passage-number, single-colony isolates, there was a mixed genotype in 30% (13/44) of the isolates, where the population of bacteria contained both *babA* and *babB* at the same locus, a phenomenon previously observed for strain J166 from the same group (Solnick *et al.*, 2004). The mixed genotype phenomenon was not rare, especially at locus B, and their further analyses showed that *babB* appeared to be predominant in these cases with the frequency of *babA* at locus B was between 10^{-3} and 10^{-4} (Colbeck *et al.*, 2006; Solnick *et al.*, 2004). As a result, 11 strains contain two copies of the *babA* gene including eight with mixed *babA* and *babB* at locus B. Among the 44 strains analysed, 32 (73%) contained *babA* sequences and in contrast, the *babB* gene was identified in 41 (95%) isolates.

In summary, it can be concluded that the *babA* gene prefers to be located at locus A and some strains do not contain the *babA* gene whereas other strains contain multiple copies. In addition, most strains possessed the *babB* gene, suggesting that the presence of *babB* confers a stronger selective advantage than does the presence of *babA*.

Regulation of BabA: translocation between *babA* and *babB*

Regulation of BabA expression is complicated, and several different mechanisms are predicted, including both at transcriptional and translational levels. For translational regulation, the formation of chimeric proteins appears to play an important role. Chimeric formations of *babA* and *babB* had initially been reported by Pride and Blaser (Pride and Blaser, 2002). They found that the 5′ regions of *babB* were replaced with the first 56 bp of *babA* (*babA/B* chimera) in two of 42 (5%) strains studied; one isolated in Japan and one in the USA. Furthermore, they showed that gene conversions can frequently (10^{-3}) occur in *H. pylori*, and the events are *recA* dependent and DNase resistant, indicating that their likely cause is intragenomic recombination. A similar phenomenon has been observed in an *in vivo* animal model of *H. pylori* infection (Solnick *et al.*, 2004). Rhesus monkeys were infected with a BabA-positive, Le^b-binding *H. pylori* J166. The inoculated strain J166 had both *babA* and *babB*; however most strains recovered after infection had replaced the *babA* gene with a chimeric *babA/B* allele, and subsequently lost their Le^b -binding ability.

In addition to the *babA/B* chimeras, *babB/A* chimeras have also been reported (Backstrom *et al.*, 2004). An isogenic *babA2* mutant from strain CCUG17875, defective in Le^b-binding, regained this activity by homologous recombination of a silent *babA1* gene into the *babB* locus, resulting in a chimeric *babB/A* gene, and the silent wild-type *babA1* gene was still maintained. The frequency of the *babA2* mutant regaining Le^b-binding was approximately 10^{-5}. The chimeric BabB/A protein binds Le^b, but its expression level was subject to phase variation through slipped-strand mispairing based on the number of cystidine–thymidine (CT) dinucleotide repeats in the 5′ region of the *babB* gene, with the frequencies of switching expression 'on' and 'off' at approximately 10^{-2}. Detailed analyses of the chimeric *babB/A* gene showed that first 47 bp were *babB* specific, the following 66 bp were shared between both *babA* and *babB*, and the remaining sequence was *babA*-specific. The second crossover event is predicted to have occurred within a region where the sequences of the *babA1* and the *babB* locus were identical.

As described above, Colbeck *et al.* recently identified many isolates with translocated *babA* and *babB* genes, especially at locus B, and the frequency of *babA* translocated at the *babB* locus was between 10^{-3} and 10^{-4} (Colbeck *et al.*, 2006), which is in agreement with the frequency in strain CCUG17875 (Backstrom *et al.*, 2004). Further investigation of the site of recombination using 10 representative strains found that the site of the recombination event was located approximately 50–200 bp downstream of the ATG in five strains (all recombination events occurred at locus B) and upstream of the ATG in the other five strains (Colbeck *et al.*, 2006). In the former cases, the resulting gene represents a *babB/A* chimera, whereas a complete recombination occurred in the latter case.

Phase variation had not been demonstrated as a method of regulating BabA expression until Solnick *et al.* recently reported that the *babA* gene in some variants of strain J166 possessed CT repeats (Solnick *et al.*, 2004). In their subsequent study, they found that CT repeats (minimum 5; maximum 11) were observed in *babA* [13/43 (30%)] and in *babB* [43/53 (81%)] and were more common at locus B [42/54 (78%)] than at locus A [14/42 (33%); $P < 0.001$) (Colbeck *et al.*, 2006). Hennig *et al.* also found hat 4 of 22 (18%) *babA* sequences, including strain J166, possessed CT repeats (eight or nine repeats) (Hennig *et al.*, 2006). Detailed analyses of the *babA* gene with CT repeats showed that the signal peptides are closely related to signal peptides of the paralogous BabB proteins, whereas sequences further downstream were typical BabA sequences (Hennig *et al.*, 2006). Taken together, these data suggest that a *babA* gene with CT repeats may be the result of the translocation of *babA* gene into the *babB* gene (chimeric *babB/A* gene). Overall, frequent translocation between *babA* and *babB* genes appears to be the predominant mechanism to regulate BabA expression. *H. pylori* can therefore use both antigenic variation and phase variation to regulate the expression of *babA*.

Regulation of BabA: point mutations, deletions and insertions

The *babA* gene was initially cloned from strain CCUG17875, which contains a silent *babA1* gene and an expressed *babA2* gene (Ilver *et al.*, 1998). The sequence of these two genes differed only by the presence of a 10-bp deletion in the signal peptide sequence of *babA1* that eliminates its translational initiation codon. However, we recently found that all 80 strains from a panel of Western and East Asian isolates contained an intact ATG start codon in the *babA* gene (Fujimoto *et al.*, 2007), which is in support of other groups that also reported the absence of the *babA1* type deletion (Colbeck *et al.*, 2006; Pride and Blaser, 2002; Pride *et al.*, 2001; Hennig *et al.*, 2006; Sheu *et al.*, 2003). Overall, the absence of a translational initiation codon, as described for *babA1* from CCUG17875 appears to be rare. Point mutation leading to stop codon, deletion and insertion in the other parts of the *babA* gene are also not common, and Hennig *et al.* found only one of 24 *babA*-positive strains (4%) contained a frameshift mutation that prevented expression of a full-length BabA protein (Hennig *et al.*, 2006).

Regulation of BabA: transcriptional regulation

Recently, transcriptional regulation of BabA expression has also been reported. Bäckström *et al.* found that only *babA2*, but not *babA1*, was transcribed using isogenic *babA1* or *babA2* mutants for strain CCUG17875 (Backstrom *et al.*, 2004). Their analyses showed that *babA* transcription seemed to be regulated by the number of Adenine [poly(A)] nucleotides within the −10 to −35 region of the *babA* promoter. The −10 and −35 region of the *babA2* sequences is highly homologous to the consensus for *E. coli* σ^{70} promoter sequences. This region is stable when the poly(A) stretch was 10 (*babA2*) but would become non-functional when the number was 14 (*babA1*). It was hypothesized that the poly(A) sequences between the −10 and the −35 motifs could be prone to slippage mutations which would subsequently affect the levels of transcription of downstream genes (Backstrom *et al.*, 2004). However, other studies could not confirm that the −10 to −35 spacing plays an important role in regulating *babA* expression (Fujimoto *et al.*, 2007; Hennig *et al.*, 2006). Further studies will be necessary to fully interrogate the roles of transcriptional regulation of BabA.

Analysis of the ribosome binding sites of the *hop* genes from the three sequenced *H. pylori* strains revealed the consensus of AAGGA-(5 to 9 nt)-ATG. While this is generally consistent with sequences observed in other bacteria, approximately half of the *omps* had the shorter spacing of 5 to 6 nt. This may either contribute to poorer expression for these genes or reflect a minor difference in the translation machinery of *H. pylori*. Several orthologous *hop* genes have nucleotide differences between the ribosome binding sites and initiation codon, however very few have insertion/deletion of bases that alters the spacing between the ribosome binding sites and the initiation codon, and such changes in spacing may affect the level of expression of these proteins. Significantly, differences in spacing are observed in both the *babA* and *babB* genes between the three *H. pylori* isolates. Several of the *hop* genes from these strains contain a poly(A) string between the ribosome binding sites and the initiation codon, and slipped-strand repair at these locations may play a role in the modulation of expression of these genes. Consistent with this notion is the difference in the number of Adenine nucleotides from GAAAAC to GAAAAAAC in the *babA* genes from *H. pylori* J99 and both HPAG1 and 26695, respectively.

There are several predicted mechanisms that may control BabA expression in some strains; however, there are many cases that remain unexplained. *H. pylori* strains that do not produce BabA can be divided into five types: (1) *babA* gene negative, which includes *babA/B* chimeras, (2) *babA* gene present; but is regulated by slipped strand repairing and the status is 'off', (3) *babA* gene present; but lacking of a translation initiation codon (single case of *babA1* in strain CCUG17875), (4) *babA* gene present; but has a frameshift mutation(s) causing non-productive translation, and (5) *babA* gene present without apparent mutations without a hypothesis for the lack of expression.

BabA protein levels

A panel of 80 strains (40 from Japan and 40 from Colombia) were recently examined for BabA protein levels and Le^b binding activity (Fujimoto *et al.*, 2007). BabA protein levels were measured by immunoblot analyses using anti-BabA antiserum (AK277), and Le^b binding activity was measuring binding of *H. pylori* to ^{125}I-labelled fucosylated blood group antigens. *H. pylori* strains were divided into two major groups: BabA-positive (76 strains) or BabA-negative (four strains). All four strains that failed to produce BabA did not possess the *babA* gene and were from Colombia. Semi-quantitative analyses of the BabA-positive strains allows BabA-positive strains to be further classified into two distinct groups: those with high levels of BabA expression (68 strains that express BabA to levels more than 20% of that detected in the strain J99) or those with low levels of BabA expression (eight strains that express BabA to levels less than 10% of that detected in the strain J99). All of the 68 strains that produced high levels of BabA exhibited Le^b binding activity. However, the strains that had no BabA or low levels of BabA did not exhibit Le^b binding activity (Fujimoto *et al.*, 2007). Based on this finding, *H. pylori* strains appear to be classifiable into three distinct groups based on their expression levels of BabA: (1) BabA-high producers (BabA-H), which produce BabA protein high enough levels to mediate Le^b binding, (2) BabA-low producers (BabA-L), which produce a small amount of BabA but not enough to mediate Le^b binding, and (3) BabA-negative strains, which do not produce any BabA protein.

BabA and clinical outcomes

Most studies evaluating BabA (*babA*) status have used PCR techniques based on detection of the 10 bp deletion to distinguish between the *babA2* and *babA1* genes (Gerhard *et al.*, 1999; Prinz *et al.*, 2001; Olfat *et al.*, 2005; Lehours *et al.*, 2004; Sheu *et al.*, 2003; Rad *et al.*, 2004; Rad *et al.*, 2002; Gatti *et al.*, 2006; Mattar *et al.*, 2005; Lai *et al.*, 2005; Han *et al.*, 2004; Oliveira *et al.*, 2003; Podzorski *et al.*, 2003; Zambon *et al.*, 2003; Lai *et al.*, 2002; Yu *et al.*, 2002; Mizushima *et al.*, 2001; Oleastro *et al.*, 2003). A number of studies have suggested a correlation between *babA2*-positive *H. pylori* and increased cellular mucosal inflammations as well as an increased risk of developing significant clinical outcomes (Gerhard *et al.*, 1999; Oleastro *et al.*, 2003; Olfat *et al.*, 2005; Oliveira *et al.*, 2003; Zambon *et al.*, 2003). However, as described above, strains carrying the prototypical silent *babA1* gene are very rare, and in addition, the BabA protein levels often does not match the presence of the *babA* (*babA2*) gene. Therefore, information gained from currently used PCR-based methods must be interpreted with caution (Fujimoto *et al.*, 2007).

There have been a few studies aimed at investigating the importance of BabA protein with respect to clinical outcomes using immunoblot analyses and Le^b binding activity (Fujimoto *et al.*, 2007; Yamaoka *et al.*, 2002c; Yamaoka *et al.*, 2006). A recent large-scale study using 520 geographically diverse patients presenting with different clinical symptoms was performed to evaluate BabA status by immunoblot analysis. A total of 250 isolates were studied from Western countries (150 strains from Colombia, 100 from the USA) in addition to 270 isolates from East Asia (150 from Korea and 120 from Japan) (Fujimoto *et al.*, 2007).

All strains from East Asia express the BabA protein. However, 24 (9.8%) of the Western strains were BabA negative and were associated with mild gastric injury and lower *H. pylori* density. Furthermore, BabA-negative strains were also infrequently associated with duodenal ulcer or gastric cancer. A small subset of strains were identified that were BabA-positive but produced low levels of the BabA protein and lacked Le^b-binding activity (BabA-L) (Fujimoto *et al.*, 2007). Although these strains were functionally BabA-negative, they were typically CagA-positive and were more likely to be associated with duodenal ulcer, gastric cancer, and increased mucosal inflammation and atrophy than BabA-positive strains that exhibit *in vitro* Le^b-binding activity (BabA-H strains) and BabA-negative strains. This suggests that either *in vitro* Le^b binding activity does not correlate with the severity of mucosal damage or that the clinical outcome or *in vitro* binding activity does not accurately reflect *in vivo* conditions. The underlying reason why strains with BabA-L status were more highly correlated with severe diseases than strains with BabA-H status is unknown, and it remains unclear whether expressing low levels of BabA has a direct role in the pathogenesis of duodenal ulcer or gastric cancer. It is possible that BabA expression is influenced by the intragastric environment and that the phenotype of the

BabA-L strains is a reflection rather than a cause of disease. It is possible that strong Le^b binding activity is associated with an enhanced immune response resulting in a severely inflamed mucosa. If so, the ability to change BabA status from a high producer to low producer (i.e. Le^b binding to Le^b non-binding) would be advantageous for the organism and low producers might reflect an adaptation of *H. pylori* that enhances survival in inflamed stomach tissue.

Recent discovered function of BabA

BabA was originally defined as an adhesin that binds to the Le^b antigen. The H-1 antigen is the carbohydrate structure that defines blood group O phenotype in the ABO blood group system. The Le^b, which is difucosylated, is formed by addition of a branched fucose (Fuc) residue to H-1. The antigens that define blood group A and B phenotypes and corresponding antigens in the Lewis blood group system are formed by terminal N-acetylgalactosamine (GalNAc) or galactose (Gal) substitutions of H-1 and Le^b [A-1 and A–Lewis b (ALe^b), and B-1 and B–Lewis b (BLe^b) antigens, respectively].

Recently, Aspholm-Hurtig *et al.* investigated the ability of BabA binding to Le^b, ALe^b and BLe^b (Aspholm-Hurtig *et al.*, 2004). Among 265 Le^b-binding *H. pylori* strains from various geographic regions, more than 95% *H. pylori* strains are 'generalists' (able to bind ALe^b and BLe^b in addition to Le^b), whereas a small subset of strains only bound to the naked Le^b structure and a few strains exclusively to ALe^b, and are called 'specialist' strains. Importantly, 'specialist' strains predominantly originated from South American individuals (where 60% were 'specialist'), which are known to express almost entirely the blood group O phenotype. They proposed that the central domain of BabA was responsible for determining the different binding patterns and transforming a 'generalist' BabA into a 'specialist' BabA; however the specific motifs could not be identified (Aspholm-Hurtig *et al.*, 2004).

The Le^b and ALe^b affinities were strongly correlated, although with about fivefold (median) higher affinity for Le^b. Aspholm-Hurtig *et al.* suggested that the higher affinity for Le^b may cause stronger gastric mucosal adherence of *H. pylori* in blood group O individuals and may thus contribute to their relatively higher risk of peptic ulcer disease in support of an earlier hypothesis (Clarke *et al.*, 1955). However, the binding affinities of BabA to Le^b in 'specialist' and 'generalist' strains were similar, suggesting no selective advantage of specialist binding activity and giving rise to the hypothesis that the emergence of specialists was due to repeated cycles of loss and regaining binding activity rather than to a one-step mutation process. Indeed, the rapid evolution of BabA in response to host mucosal glycosylation patterns would enable the pathogen to adapt to their individual hosts while avoiding host immune responses.

SabA

Discovery of SabA

Although BabA was believed to be the primary molecule involved in adherence to the gastric mucosa, it was significant that the *babA* deletion mutant that could not bind ABO/Le^b blood group antigens still bound to the gastric epithelium (Mahdavi *et al.*, 2002). Moreover, binding of the *babA* mutant could not be inhibited by pretreatment of bacteria with soluble Le^b antigen. Further analysis showed that the *babA* mutant preferentially bound to inflamed gastric mucosa. Taken together, these results implicated the presence of another adhesin, possibly specific to *H. pylori*-infected inflamed tissue, which binds to a receptor distinct from that of BabA.

Using thin layer chromatography, mass spectrometry and nuclear magnetic resonance, Mahdavi *et al.* identified sialyl-dimeric-Le^x ($sdiLe^x$) glycosphingolipid as a receptor for *H. pylori* and showed that *H. pylori* infection induced formation of sLe^x antigens in gastric epithelium in humans and in a Rhesus monkey (Mahdavi *et al.*, 2002). Using an affinity purification method, this sialyl-dimeric-Le^x-binding bacterial adhesin was identified as a 70 kDa OMP called SabA (Table 3.1). Importantly, a paralogous protein, called SabB, was also identified, and is located in all three sequenced isolates (Table 3.1). However, while both sLe^x and sLe^a antigen-binding activity was abolished in the *sabA* mutant, neither was affected in the *sabB* mutant (Mahdavi *et al.*, 2002).

Regulation of SabA

Based on the comparison of the three completed genomes of *H. pylori*, six *hop* genes (*oipA*, *sabA*, *sabB*, *babB*, *babC* and *hopZ*) are thought to undergo phase variation such that not all strains will produce functional proteins (Alm *et al.*, 1999; Oh *et al.*, 2006; Tomb *et al.*, 1997). The functional status is regulated by a slipped strand mispairing mechanism and is mediated by the number of CT dinucleotide repeats in the 5′ region of the genes.

The genome sequences of *H. pylori* J99 deposited in GenBank show that *sabA* is out of frame with nine CT repeats. However, Mahdavi *et al.* used J99 as the sLe^x-binding strain to identify *sabA*, but their isolate possessed 10 CT repeats, which would produce a functional SabA protein and could thus explain the sLe^x-binding (Mahdavi *et al.*, 2002). These results demonstrate the flexibility of the locus that subsequently confers 'on/off' binding properties. The metastability of the SabA adhesin might enable a subpopulation to detach from the gastric cells and the switch status of both the *sabA* and *sabB* genes has been demonstrated to change from 'on' to 'off' during *in vivo* passage in a murine model of *H. pylori* infection (Yamaoka *et al.*, 2002b). Interestingly, the bacterial density in mice whose expression switch status was 'off' was reduced compared with those of mice infected with *H. pylori* strains that had no change in the switch status.

The largest analysis of CT repeats in the *sabA* gene was recently performed in Taiwanese strains (Sheu *et al.*, 2006). Among the 145 *H. pylori* isolates analysed, 116 (80%) were *sabA*-genopositive by PCR, but only 45 (31%) of the isolates expressed SabA as detected by immunoblotting. Importantly, sequence of *sabA*-genopositive PCR products from 92 isolates identified that 60% ($n = 55$) had regular CT repeats whereas the other 40% ($n = 37$) had imperfect CT repeats (i.e. CT$\underline{\text{T}}$TCTCT$\underline{\text{A}}$TC$\underline{\text{C}}$) in the 5′ region of the *sabA* gene which in all isolates except one locked the expression status in the 'on' position. Interestingly, only 10 of these strains produced detectable SabA protein by immunoblotting. However, the expression status of *sabA* is not completely reliant on the number of CT dinucleotide repeats in the 5′ region of the *sabA* gene as there is also a report demonstrating that the *sabA* expression status can be varied by nucleotide changes after the dinucleotide repeats (Lehours *et al.*, 2004).

The presence of the *sabA* gene appears not to be universal among *H. pylori* isolates. Sheu *et al.* reported that 20% (29/145) of Taiwanese strains did not possess the *sabA* gene (Sheu *et al.*, 2006), whereas the prevalence of *sabA*-negative strains were 14% (6/43) in French strains (Lehours *et al.*, 2004) and 7% (7/96) in Dutch strains (de Jonge *et al.*, 2004), even though microarray analyses for 15 isolates showed that all strains possessed the *sabA* gene (Salama *et al.*, 2000). While the *sabA*-negative genotype may be false PCR negatives due to subtle mutations among strains in the primer regions, the use of several primer pairs in the Taiwanese and Dutch studies mitigates this risk.

Function of SabA

SabA was identified as an adhesin, with sLe^x and sLe^a serving as additional functional receptors for *H. pylori* adherence. Aspholm *et al.* recently reported that clinical *H. pylori* isolates showed 'polymorphism' in their abilities to bind various sialylated glycans including sdiLe^x, sLe^a and dialyl-lactosamine (sLn) (Aspholm *et al.*, 2006). They examined 99 Swedish clinical isolates, and found that 39 (39%) could bind to sialylated glycans, and the majority (87%) of these isolates could bind all three glycans. Preferential binding to sdiLe^x was the most common with 24 of 39 isolates demonstrating this phenotype although the binding patterns were not uniform among strains (e.g. some strains bind to sLe^a better than sdiLe^x and sLn). Further analyses using complementation of *sabA* mutants confirms that variation in sialyl binding affinity is an inherent property of the SabA protein, and likely is the result of subtle differences in sequence, although the specific motifs in SabA responsible for the affinity to various sialylated glycans are yet to be elucidated.

The minimal structure recognized by the SabA adhesin was determined to be α2–3-linked sialyllactose, irrespective of a terminal or subterminal fucosylation (Roche *et al.*, 2004;

Walz *et al.*, 2005). Fucosylation, sialylation, and length of the core chain are parameters that together form a high-affinity binding epitope for SabA. Recent studies using *sabA* mutants and sialidase treatment of glycoconjugate arrays showed that SabA is involved in the binding of the extracellular matrix protein laminin (Roche *et al.*, 2004).

The findings that SabA mediates binding to the sLe^x and sLe^a are important since *H. pylori*-induced gastric inflammation and gastric carcinoma are associated with the replacement of non-sialylated Lewis antigens by sLe^x and sLe^a (Mahdavi *et al.*, 2002; Ota *et al.*, 1998; Sakamoto *et al.*, 1989). Sialylated glycoconjugates are low in healthy gastric mucosa but *H. pylori* infection elicits gastric mucosal sialylation as part of the chronic inflammatory response. This sialylation was correlated with the capacity for SabA-dependent, but not BabA-dependent, binding in situ (Mahdavi *et al.*, 2002). Accordingly, high levels of sialylated glycoconjugates have been found in *H. pylori*-infected individuals, which decreased after eradication of infection and resolution of gastritis.

Recently, Unemo *et al.* found that the SabA adhesin, upon binding to sialic-acid containing neutrophil receptors, mediates the non-opsonic neutrophil oxidative burst induced by *H. pylori* (Unemo *et al.*, 2005). The specific binding to neutrophils through sialylated carbohydrates induces the oxidative burst resulting in oxidative damage of the gastric epithelium. In subsequent studies, SabA binding was identified not only as the key inducer of oxidative metabolism, but was also essential in phagocytosis induction (Petersson *et al.*, 2006).

H. pylori had been reported to agglutinate erythrocytes (Emody *et al.*, 1988), an activity proposed to be dependent on sialic acid since the activity was lost by prior sialidase treatment of erythrocytes (Evans *et al.*, 1988). Recently, Aspholm *et al.* demonstrated that adherence of *H. pylori* to erythrocytes and subsequent agglutination is mediated by SabA based on sialidase-sensitive haemagglutination, binding assays with sialylated glycoconjugates, and analysis of a series of isogenic *sabA* deletion mutants (Aspholm *et al.*, 2006).

SabA and pH

The stomach presents a variety of different microenvironments such that the ability to switch various adhesins and surface proteins 'on' and 'off' may offer survival advantages to the bacteria. For example, deep in the pits in the antrum the bacteria would be exposed to a different population of cells compared with those on the luminal surface as well as much lower fluctuations in pH. The finding that SabA expression was increased in strains obtained from patients with gastric cancer and the prevalence was reduced among those with duodenal ulcer suggested that alternating expression status may be related to intragastric pH (Yamaoka *et al.*, 2006).

To test this hypothesis, Yamaoka *et al.* compared cultures grown in high acidic conditions (pH 5.0) with those grown at pH 7.0 and SabA expression levels dramatically decreased following exposure of *H. pylori* to pH 5.0 for 2 hours (Yamaoka *et al.*, 2006). In contrast, the acid exposure did not result in any change in expression levels for BabA, BabB and OipA. They therefore further examined these three strains following an exposure to pH 5.5 for 20–24 hours, and found that SabA status changed from positive to negative in one strain and the expression of SabA decreased in the other two strains (Yamaoka *et al.*, 2006). These results are in agreement with a previous microarray study that demonstrated that the expression levels of *sabA* mRNA decreased in acidic conditions (Merrell *et al.*, 2003).

Overall, SabA-positive expression was inversely related to the ability of the stomach to secrete acid suggesting that its expression may be regulated by changes in acid secretion and/or in antigens expressed by the atrophic mucosa.

SabA and clinical outcomes

The relationship between the expression of SabA, BabA, BabB, and OipA and clinical outcomes was recently examined. A diverse cohort of strains from 200 patients including 120 from Colombia (40 with gastritis, 40 with duodenal ulcer and 40 with gastric cancer) and 80 from the USA (40 with gastritis and 40 with duodenal ulcer) was used (Yamaoka *et al.*, 2006). The prevalence of SabA-positive isolates was 66% (53/80) in gastritis, 88% (70/80) in duodenal ulcer, and 89% (36/40) in gastric cancer. Multiple logistic regression analysis among these OMPs showed that only SabA status was a predictor of gastric cancer vs. duodenal ulcer (OR 2.8; 95% CI = 1.2–6.7). Furthermore, SabA-positive status was also closely associated with decreased neutrophil infiltration in both the antrum and corpus and with severe intestinal metaplasia and gastric atrophy in the antrum (Yamaoka *et al.*, 2006). It was therefore proposed that SabA-positive status was associated with gastric cancer, intestinal metaplasia, and corpus atrophy and negatively associated with duodenal ulcer and neutrophil infiltration.

These results are in agreement with the pathogenesis scenario that Mahdavi *et al.* has proposed previously (Mahdavi *et al.*, 2002). Briefly, during persistent infection and chronic inflammation (gastritis), *H. pylori* triggers an alteration in the glycosylation patterns in the gastric mucosa including upregulation of the inflammation-associated sLe^x antigens, and that SabA binds the sialyl-(di)-$Le^{x/a}$ glycosphingolipids promoting membrane attachment and apposition. It is proposed that at sites of vigorous local inflammatory response (e.g. severe antral gastric inflammation in duodenal ulcer) *H. pylori* undergoes phase variation and turns 'off' SabA-mediated sLe^x-binding capacity, which in turn enables *H. pylori* cells to escape intimate contact with sialylated lymphocytes or other defensive immune cells. The ability of *H. pylori* strains to adhere to sialylated glycoconjugates expressed during chronic inflammation might therefore contribute to virulence and the extraordinary chronicity of *H. pylori* infection.

However, one study from Taiwan reported that the prevalence rate of SabA-positive isolates is similar between patients with different endoscopic diagnoses [33% (10/30) in duodenal ulcer, 22% (10/30) in gastric ulcer and 29% (25/85) in non-ulcer dyspepsia] by examining 145 *H. pylori*-infected patients (Sheu *et al.*, 2006). In addition, the prevalence rate of gastric atrophy and intestinal metaplasia were not statistically different between patients infected with SabA-positive and -negative isolates. Whether there are geographical differences in the importance of the relationship between SabA and clinical outcomes or histological changes warrants further attention.

OipA

Discovery of OipA

H. pylori infection of the gastric mucosa is characterized by an inflammatory response resulting in infiltration of neutrophils, lymphocytes, monocytes, and plasma cells. The initial migration and activation of inflammatory cells into the gastric mucosa is believed to depend on the production of various proinflammatory cytokines, especially IL-8, a potent neutrophil chemotactic and activating peptide produced by gastric epithelial cells (Yamaoka *et al.*, 1996; Yamaoka *et al.*, 1998a; Yamaoka *et al.*, 1999b; Yamaoka *et al.*, 2001). It is well known that many *H. pylori* strains contain a 40-kb genome segment encoding approximately 30 genes termed the *cag* pathogenicity island (PAI) (Censini *et al.*, 1996). *In vitro* experiments using gastric epithelial cancer cells co-cultured with *H. pylori* indicate that *cagE, cagH, cagI, cagM and cagT*, but not *cagA* and *cagN*, were involved in IL-8 induction (Censini *et al.*, 1996; Fischer *et al.*, 2001). There were additional reports that the IL-8 levels in the corporal gastric mucosa were increased in patients infected with *cag* PAI positive strains with serum antibody against a 33–35 kDa antigen than those without the antibody, suggesting that additional factors may be involved in IL-8 induction from gastric mucosal cells (Yamaoka *et al.*, 1998b). Further, *in vitro* experiments indicated that IL-8 is produced from the gastric epithelial cells after attachment of viable *H. pylori* (Aihara *et al.*, 1997). Thus, direct interaction of surface proteins with the host cells appears important and it was hypothesized that the OMPs with a molecular size of 33 to 35 kDa were excellent candidates for the proinflammatory virulence factor(s). Only two gene products, HopH (34 kDa) and HorG (33 kDa), were within that range of molecular weight among the OMP family, and inactivation of *hopH*, but not *horG*, reduced IL-8 induction by approximately 50% from AGS, MKN45 and KATOIII gastric epithelial cancer cells (Yamaoka *et al.*, 2000). Although inactivation of the *cagE* gene in the *cag* PAI caused a more dramatic reduction in IL-8 induction (approximately 80–90% reduction) than inactivation of the *hopH* gene, reduction was not absolute if the parental strain maintained a functional *hopH* gene. A double knockout mutant of the *cagE* and *hopH* genes reduced IL-8 production to background levels. The *hopH* gene was denoted as outer inflammatory protein (*oipA*) gene (Yamaoka *et al.*, 2000).

Regulation of OipA production

The expression of OipA is predicted to be regulated by a slipped strand mispairing system based on the number of CT dinucleotide repeats in the 5′ region of the gene (Alm *et al.*, 1999; Tomb *et al.*, 1997). This was experimentally confirmed since OipA production detected by immunoblot using polyclonal antisera against either a synthetic OipA peptide or a recombinant OipA protein correlated well with functional *oipA* status determined by PCR-based sequencing (sensitivity and specificity >94%) (Kudo *et al.*, 2004).

In a large population study, the 5′ region of the *oipA* gene was sequenced from 232 strains (111 from the U.S. and 121 from Colombia) and the number of CT dinucleotide repeats ranged from 4 to 12, with strains containing 6, 9 or 12 CT dinucleotide keeping the gene in-frame (Yamaoka *et al.*, 2002a). In 16 cases (7%), one or two CT dinucleotides in the repeats were changed to thymidine-thymidine (TT), which is similar to other reports including strains from Europe and India (Ando *et al.*, 2002b; de *et al.*, 2004c; Lehours *et al.*, 2004; Dossumbekova *et al.*, 2006). Interestingly, there are no CT repeats in the *oipA* gene in most East Asian strains and all East Asian strains reported so far possessed *oipA* with an 'on' status (Ando *et al.*, 2002b; Yamaoka *et al.*, 2000).

Although slipped strand mispairing can alter expression status during *in vitro* laboratory passages and over time *in vivo*, it was found that phase variation in *oipA* is rare when compared to other OMP genes (*sabA*, *babA* and *babB*) (Yamaoka *et al.*, 2006). Ando *et al.* also reported that the switch status of *oipA* was identical for the 15 strains obtained from a mother and her six daughters in The Netherlands, and for 9 of 11 pairs of strains obtained from Dutch individuals, on average 8 years apart, where one patient showed phase variation with the number of CT repeats shifting from 7 [off] to 6 [on] and another patient maintained the 'off' status although the number of CT repeats decreased from 8 to 7 (Ando *et al.*, 2002a). It has also been shown that the switch status of the *oipA* gene was stable and remained unchanged during the 12 weeks of infection in mice (Yamaoka *et al.*, 2002b).

Relationship between oipA status and other virulence factors

There have been many *H. pylori* putative virulence factors that might influence the risk of a patient developing different *H. pylori*-related gastroduodenal diseases. Furthermore, it is significant that the presence of these factors is often closely correlated. For example, clinical isolates that contain the *cag* PAI (CagA as a marker) typically also have an *oipA* 'on' status (Yamaoka *et al.*, 2006; Yamaoka *et al.*, 2002a; Ando *et al.*, 2002b; de Jonge *et al.*, 2004; Lehours *et al.*, 2004; Dossumbekova *et al.*, 2006) despite the *oipA* gene being physically located approximately 100 kbp from the *cag* PAI on the *H. pylori* chromosome (Alm *et al.*, 1999; Tomb *et al.*, 1997; Oh *et al.*, 2006). In one study, 95% (179 of 189) *cag* PAI-positive strains had *oipA* 'on' compared with 11% (6 of 54) *cag* PAI-negative strains (P <0.001) in the U.S.A. and Colombian populations (Yamaoka *et al.*, 2002a). The *oipA* functional status has also been strongly correlated with the *vacA* s1 genotype, and the presence of the *babA* gene in U.S.A., Colombian and East Asian strains (Yamaoka *et al.*, 2002a; Dossumbekova *et al.*, 2006).

Although the *oipA* functional status was closely correlated to the *babA* functional status, these have been reported to be independent of the expression status of other OMPs that are regulated by slipped strand mispairing (*babB*, *sabA*, *sabB* and *hopZ*) (Yamaoka *et al.*, 2006; de *et al.*, 2004a; Lehours *et al.*, 2004). There are several other *H. pylori* factors reported to be related to gastric injury and clinical outcomes including *iceA* (induced by contact with epithelium: *iceA1* or *iceA2*) (Peek, Jr. *et al.*, 1998) and *dupA* (duodenal ulcer promoting: positive or negative) (Lu *et al.*, 2005a); however, these are reported to be independent of *oipA* status ((Yamaoka *et al.*, 2002a) ; unpublished data).

Function of OipA

Although OipA was originally identified as a proinflammatory response-inducing protein given that *oipA* mutants reduced induction of IL-8 from gastric epithelial cells, there have been recent conflicting reports demonstrating that isogenic *oipA* mutants induce IL-8 similar to the parental strain from MKN45, AGS or KATO-III gastric epithelial cells (Odenbreit *et al.*, 2002; Ando *et al.*, 2002b; Dossumbekova *et al.*, 2006; Akanuma *et al.*, 2002). The conserved genetic organization of the *oipA* gene, as seen in the three genome sequences, limits the possibility that this being caused by polar effects on an unrelated gene.

It may be that the ability to induce IL-8 by *oipA* mutants is simply strain dependent. Such differences are not without precedence, as strain dependent differences in IL-8 induction from gastric epithelial cells has also been reported for CagA (Brandt *et al.*, 2005) and AlpAB (Lu *et al.*, 2007). Yamaoka *et al.* (Yamaoka, 2006) recently reported that a low percentage of *oipA* mutants from multiple *in vitro* passaged strains (i.e. 20 passages) induced IL-8 similar to their parental strains; however *oipA* mutants from minimal *in vitro* passaged strains (i.e. fewer than five passages) induced low IL-8 levels. This observation suggests that the subtle variations of IL-8 induction by *oipA* mutants may represent *in vitro* artefacts.

The mechanisms to induce IL-8 by OipA have been studied (Yamaoka *et al.*, 2004). OipA was involved in inducing interferon regulatory factor (IRF)-1 to bind and activate the interferon-stimulated responsive element (ISRE)-like element (−132 to −145) located in the IL-8 promoter. OipA was also involved in inducing the phosphorylation of signal transducers and activators of transcription (STAT)-1, an upstream signalling molecule of IRF-1, suggesting that OipA regulates STAT1→IRF1→ISRE-like element pathways. Recent studies also showed that OipA is involved in the activation of several other transcription factors including nuclear factor (NF)-κB, activator protein (AP)-1, ISRE, and cAMP response element (CRE) in a target gene-dependent and/or *H. pylori* strain-dependent manner (Kudo *et al.*, 2005; Lu *et al.*, 2005b; Wu *et al.*, 2006). An overview of the current understanding of the OipA signalling interactions are seen in Fig. 3.3.

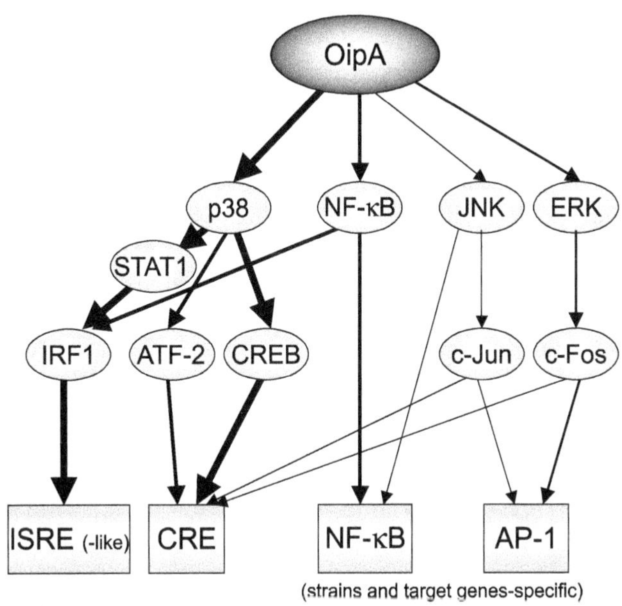

Figure 3.3 Hypothetical model for the signal transduction pathway in the host's *H. pylori* infected gastric epithelial cells caused by OipA. Thick lines represent the major pathways irrespective of the target genes and/or *H. pylori* strains. AP-1, activator protein-1; CRE: cAMP response element; CREB, CRE-binding protein; ERK, extracellular signal-regulated kinase; IRF, interferon regulatory factor; ISRE, interferon-stimulated responsive element; JNK, Jun-N-terminal kinase; NF, nuclear factor; STAT, signal transducers and activators of transcription

There is increasing evidence that OipA can also serve as an adhesin. An initial report that OipA was involved in attachment to the gastric epithelial cell line AGS (Yamaoka *et al.*, 2002b) was confirmed and extended using two additional gastric epithelial cell lines, AGS and KATOIII (Dossumbekova *et al.*, 2006). However, the affinity of OipA for gastric epithelial cells is not higher than that demonstrated for the HopZ and AlpAB adhesins. (Lu *et al.*, 2007; Yamaoka *et al.*, 2004). In the Mongolian gerbil model, Akanuma *et al.* reported that isogenic *oipA* mutants of the *H. pylori* TN2 strain failed to colonize gerbils following the 3-week observation period (Akanuma *et al.*, 2002), confirming the importance of OipA in gastric colonization.

OipA and clinical outcomes

Yamaoka *et al.* has examined the expression status or presence of multiple virulence factors (*cag*PAI, *vacA*, *iceA*, *oipA* and *babA*) in different clinical outcomes (Yamaoka *et al.*, 2002a). *H. pylori* isolates were obtained from 247 patients in the USA and Colombia that had been differentiated based on gastroduodenal disease; 86 with gastritis, 86 with duodenal ulcer and 75 with gastric carcinoma. Independent univariate analysis showed that *oipA* 'on', *cag* PAI-positive, *vacA* s1 genotypes or *babA*-positive types were all related to the risk of duodenal ulcer. However multiple logistic regression analysis showed that only *oipA* 'on' status was an independent determinant predictor of duodenal ulcer from gastritis [adjusted odds ratio (OR) = 5.0 and 95% confidence interval (CI) = 2.1–11.9). In addition, among the factors studied, multiple linear regression analysis showed that only *oipA* 'on' status was significantly associated with high *H. pylori* density and severe neutrophil infiltration and high *in vivo* gastric mucosal IL-8 production ($P < 0.001$). In contrast, *oipA* functional status was independent of the atrophy and intestinal metaplasia (Yamaoka *et al.*, 2002a).

This finding was confirmed and extended in a distinct study using a non-overlapping cohort of 200 patients (80 with gastritis, 80 with duodenal ulcer and 40 with gastric carcinoma) using immunoblot analysis for OipA, BabA, BabB and SabA (Yamaoka *et al.*, 2006). Multiple logistic regression analysis showed that only OipA-positive status was an independent determinant predictor of gastric cancer vs. gastritis (adjusted OR 4.8; 95% CI = 1.4–16.8) and duodenal ulcer vs. gastritis (OR 4.0; 95% CI = 1.6–10.2). OipA-positive status was significantly associated with high *H. pylori* density and the grade of neutrophil infiltration. Importantly, when CagA protein status was added as an explanatory variable for the analyses, the calculation could not be converged since OipA and CagA status were essentially identical.

From the two studies examining both gene and protein levels (Yamaoka *et al.*, 2002a; Yamaoka *et al.*, 2006), it may be possible, albeit unlikely, that the presence of *cag* PAI and CagA are important only as surrogate markers for OipA. However, since the status of both OipA and CagA (as well as VacA and BabA) are so strongly linked, it is more likely that all of these factors act synergistically to maximize gastric inflammation as well as cytokine expression. Importantly, this hypothesis does not appear to hold true for East Asian countries, as all strains reported contain a functional *oipA* gene, *cag* PAI and *vacA* s1 irrespective of the clinical outcomes (Yamaoka *et al.*, 1998c; Yamaoka *et al.*, 1999a; Yamaoka *et al.*, 2002c). It appears that other factors, such as host susceptibility or environmental factors, also play a role in the described geographical differences of *H. pylori* -related clinical outcomes.

Conclusion

Analysis of three *H. pylori* genome has confirmed the presence of the five major families of outer membrane proteins in *H. pylori*. However, there appears to be a trend at the protein level within the protein families as to which two of the three orthologues are more closely related with the orthologues from the gastritis isolates being the closest. A greater number of samples from a range of *H. pylori-induced* clinical diseases would be needed to determine if there was any correlation and whether the subtle protein differences could be responsible for any specialized adaptation or differences in gastric injury.

Several of these OMPs have been shown to act as adhesins and the specific functions of BabA, SabA and OipA are gradually becoming clearer; however there are still questions that remain unanswered or require further clarification, such as the relationship between BabA protein expression and Le^b binding ability, the regulation of SabA production by slipped strand mispairing, and the relationship between OipA and IL-8 induction from gastric epithelial cells. BabA and OipA, in cooperation with other virulence factors such as *cag* PAI induce gastric injury and cause gastroduodenal diseases. Although AlpAB and HopZ have been reported to be adhesins, the receptor(s) for these in addition to OipA in the host cells are still unknown. There are numerous other OMPs that appear conserved between isolates that have not been studied in detail, further studies will be needed to identify their precise roles in the disease process.

References

Aihara, M., Tsuchimoto, D., Takizawa, H., Azuma, A., Wakebe, H., Ohmoto, Y., Imagawa, K., Kikuchi, M., Mukaida, N., and Matsushima, K. (1997). Mechanisms involved in *Helicobacter pylori*-induced interleukin-8 production by a gastric cancer cell line, MKN45. Infect. Immun. *65*, 3218–3224.

Akanuma, M., Maeda, S., Ogura, K., Mitsuno, Y., Hirata, Y., Ikenoue, T., Otsuka, M., Watanabe, T., Yamaji, Y., Yoshida, H., Kawabe, T., Shiratori, Y., and Omata, M. (2002). The evaluation of putative virulence factors of *Helicobacter pylori* for gastroduodenal disease by use of a short-term Mongolian gerbil infection model. J. Infect. Dis *185*, 341–347.

Alm, R.A., Bina, J., Andrews, B.M., Doig, P., Hancock, R.E., and Trust, T.J. (2000). Comparative genomics of *Helicobacter pylori*: analysis of the outer membrane protein families. Infect. Immun. *68*, 4155–4168.

Alm, R.A., Ling, L.S., Moir, D.T., King, B.L., Brown, E.D., Doig, P.C., Smith, D.R., Noonan, B., Guild, B.C., de Jonge, B.L., Carmel, G., Tummino, P.J., Caruso, A., Uria-Nickelsen, M., Mills, D.M., Ives, C., Gibson, R., Merberg, D., Mills, S.D., Jiang, Q., Taylor, D.E., Vovis, G.F., and Trust, T.J. (1999). Genomic-sequence comparison of two unrelated isolates of the human gastric pathogen *Helicobacter pylori*. Nature *397*, 176–180.

Ando, T., Peek, R.M., Jr., Lee, Y.C., Krishna, U., Kusugami, K., and Blaser, M.J. (2002a). Host cell responses to genotypically similar *Helicobacter pylori* isolates from United States and Japan. Clin. Diagn. Lab Immunol. *9*, 167–175.

Ando, T., Peek, R.M., Pride, D., Levine, S.M., Takata, T., Lee, Y.C., Kusugami, K., van der, E.A., Kuipers, E.J., Kusters, J.G., and Blaser, M.J. (2002b). Polymorphisms of *Helicobacter pylori* HP0638 reflect geographic origin and correlate with *cagA* status. J. Clin. Microbiol. *40*, 239–246.

Aspholm, M., Olfat, F.O., Norden, J., Sonden, B., Lundberg, C., Sjostrom, R., Altraja, S., Odenbreit, S., Haas, R., Wadstrom, T., Engstrand, L., Semino-Mora, C., Liu, H., Dubois, A., Teneberg, S., Arnqvist, A., and Boren, T. (2006). SabA is the *H. pylori* hemagglutinin and is polymorphic in binding to sialylated glycans. PLoS. Pathog. *2*, e110.

Aspholm-Hurtig, M., Dailide, G., Lahmann, M., Kalia, A., Ilver, D., Roche, N., Vikstrom, S., Sjostrom, R., Linden, S., Backstrom, A., Lundberg, C., Arnqvist, A., Mahdavi, J., Nilsson, U.J., Velapatino, B., Gilman, R.H., Gerhard, M., Alarcon, T., Lopez-Brea, M., Nakazawa, T., Fox, J.G., Correa, P., Dominguez-Bello, M.G., Perez-Perez, G.I., Blaser, M.J., Normark, S., Carlstedt, I., Oscarson, S., Teneberg, S., Berg, D.E., and Boren, T. (2004). Functional adaptation of BabA, the *H. pylori* ABO blood group antigen binding adhesin. Science *305*, 519–522.

Backstrom, A., Lundberg, C., Kersulyte, D., Berg, D.E., Boren, T., and Arnqvist, A. (2004). Metastability of *Helicobacter pylori bab* adhesin genes and dynamics in Lewis b antigen binding. Proc. Natl. Acad. Sci. USA *101*, 16923–16928.

Bina, J.E., Alm, R.A., Uria-Nickelsen, M., Thomas, S.R., Trust, T.J., and Hancock, R.E. (2000). *Helicobacter pylori* uptake and efflux: basis for intrinsic susceptibility to antibiotics in vitro. Antimicrob. Agents Chemother. *44*, 248–254.

Boren, T., Falk, P., Roth, K.A., Larson, G., and Normark, S. (1993). Attachment of *Helicobacter pylori* to human gastric epithelium mediated by blood group antigens. Science *262*, 1892–1895.

Brandt, S., Kwok, T., Hartig, R., Konig, W., and Backert, S. (2005). NF-kappaB activation and potentiation of proinflammatory responses by the *Helicobacter pylori* CagA protein. Proc. Natl. Acad. Sci. USA *102*, 9300–9305.

Burnens, A., Stucki, U., Nicolet, J., and Frey, J. (1995). Identification and characterization of an immunogenic outer membrane protein of *Campylobacter jejuni*. J. Clin. Microbiol. *33*, 2826–2832.

Censini, S., Lange, C., Xiang, Z., Crabtree, J.E., Ghiara, P., Borodovsky, M., Rappuoli, R., and Covacci, A. (1996). *cag*, a pathogenicity island of *Helicobacter pylori*, encodes type I-specific and disease-associated virulence factors. Proc. Natl. Acad. Sci. USA *93*, 14648–14653.

Clarke, C.A., Cowan, W.K., Edwards, J.W., Howel-Evans, A.W., McConnell, R.B., Woodrow, J.C., and Sheppard, P.M. (1955). The relationship of the ABO blood groups to duodenal and gastric ulceration. Br. Med. J. *2*, 643–646.

Colbeck, J.C., Hansen, L.M., Fong, J.M., and Solnick, J.V. (2006). Genotypic profile of the outer membrane proteins BabA and BabB in clinical isolates of *Helicobacter pylori*. Infect. Immun. *74*, 4375–4378.

de Jonge, R., Pot, R.G., Loffeld, R.J., van Vliet, A.H., Kuipers, E.J., and Kusters, J.G. (2004). The functional status of the *Helicobacter pylori sabB* adhesin gene as a putative marker for disease outcome. Helicobacter. *9*, 158–164.

Doig, P., Exner, M.M., Hancock, R.E., and Trust, T.J. (1995). Isolation and characterization of a conserved porin protein from *Helicobacter pylori*. J. Bacteriol. *177*, 5447–5452.

Dossumbekova, A., Prinz, C., Mages, J., Lang, R., Kusters, J.G., van Vliet, A.H., Reindl, W., Backert, S., Saur, D., Schmid, R.M., and Rad, R. (2006). *Helicobacter pylori* HopH (OipA) and bacterial pathogenicity: genetic and functional genomic analysis of *hopH* gene polymorphisms. J. Infect. Dis *194*, 1346–1355.

Emody, L., Carlsson, A., Ljungh, A., and Wadstrom, T. (1988). Mannose-resistant haemagglutination by *Campylobacter pylori*. Scand. J. Infect. Dis. *20*, 353–354.

Evans, D.G., Evans, D.J., Jr., Moulds, J.J., and Graham, D.Y. (1988). N-acetylneuraminyllactose-binding fibrillar hemagglutinin of *Campylobacter pylori*: a putative colonization factor antigen. Infect. Immun. *56*, 2896–2906.

Exner, M.M., Doig, P., Trust, T.J., and Hancock, R.E. (1995). Isolation and characterization of a family of porin proteins from *Helicobacter pylori*. Infect. Immun. *63*, 1567–1572.

Falk, P., Roth, K.A., Boren, T., Westblom, T.U., Gordon, J.I., and Normark, S. (1993). An *in vitro* adherence assay reveals that *Helicobacter pylori* exhibits cell lineage-specific tropism in the human gastric epithelium. Proc. Natl. Acad. Sci. USA *90*, 2035–2039.

Falk, P.G., Bry, L., Holgersson, J., and Gordon, J.I. (1995). Expression of a human alpha-1, 3/4-fucosyltransferase in the pit cell lineage of FVB/N mouse stomach results in production of Leb-containing glycoconjugates: a potential transgenic mouse model for studying *Helicobacter pylori* infection. Proc. Natl. Acad. Sci. USA *92*, 1515–1519.

Fischer, W., Puls, J., Buhrdorf, R., Gebert, B., Odenbreit, S., and Haas, R. (2001). Systematic mutagenesis of the *Helicobacter pylori cag* pathogenicity island: essential genes for CagA translocation in host cells and induction of interleukin-8. Mol. Microbiol. *42*, 1337–1348.

Fujimoto, S., Olaniyi, O.O., Arnqvist, A., Wu, J.Y., Odenbreit, S., Haas, R., Graham, D.Y., and Yamaoka, Y. (2007). *Helicobacter pylori* BabA expression, gastric mucosal injury, and clinical outcome. Clin. Gastroenterol. Hepatol. *5*, 49–58.

Gatti, L.L., Modena, J.L., Payao, S.L., Smith, M.A., Fukuhara, Y., Modena, J.L., de Oliveira, R.B., and Brocchi, M. (2006). Prevalence of *Helicobacter pylori cagA*, *iceA* and *babA2* alleles in Brazilian patients with upper gastrointestinal diseases. Acta Trop. *100*, 232–240.

Gerhard, M., Lehn, N., Neumayer, N., Boren, T., Rad, R., Schepp, W., Miehlke, S., Classen, M., and Prinz, C. (1999). Clinical relevance of the *Helicobacter pylori* gene for blood-group antigen-binding adhesin. Proc. Natl. Acad. Sci. USA *96*, 12778–12783.

Guruge, J.L., Falk, P.G., Lorenz, R.G., Dans, M., Wirth, H.P., Blaser, M.J., Berg, D.E., and Gordon, J.I. (1998). Epithelial attachment alters the outcome of *Helicobacter pylori* infection. Proc. Natl. Acad. Sci. USA *95*, 3925–3930.

Han, Y.H., Liu, W.Z., Zhu, H.Y., and Xiao, S.D. (2004). Clinical relevance of *iceA* and *babA2* genotypes of *Helicobacter pylori* in a Shanghai population. Chin J. Dig. Dis *5*, 181–185.

Hennig, E.E., Allen, J.M., and Cover, T.L. (2006). Multiple chromosomal loci for the *babA* gene in *Helicobacter pylori*. Infect. Immun. *74*, 3046–3051.

Ilver, D., Arnqvist, A., Ogren, J., Frick, I.M., Kersulyte, D., Incecik, E.T., Berg, D.E., Covacci, A., Engstrand, L., and Boren, T. (1998). *Helicobacter pylori* adhesin binding fucosylated histo-blood group antigens revealed by retagging. Science *279*, 373–377.

Kostrzynska, M., Betts, J.D., Austin, J.W., and Trust, T.J. (1991). Identification, characterization, and spatial localization of two flagellin species in *Helicobacter pylori* flagella. J. Bacteriol. *173*, 937–946.

Kudo, T., Lu, H., Wu, J.Y., Graham, D.Y., Casola, A., and Yamaoka, Y. (2005). Regulation of RANTES promoter activation in gastric epithelial cells infected with *Helicobacter pylori*. Infect. Immun. *73*, 7602–7612.

Kudo, T., Nurgalieva, Z.Z., Conner, M.E., Crawford, S., Odenbreit, S., Haas, R., Graham, D.Y., and Yamaoka, Y. (2004). Correlation between *Helicobacter pylori* OipA protein expression and *oipA* gene switch status. J. Clin. Microbiol. *42*, 2279–2281.

Kutschke, A. and de Jonge, B.L. (2005). Compound efflux in *Helicobacter pylori*. Antimicrob. Agents Chemother. *49*, 3009–3010.

Lai, C.H., Kuo, C.H., Chen, Y.C., Chao, F.Y., Poon, S.K., Chang, C.S., and Wang, W.C. (2002). High prevalence of *cagA*- and *babA2*-positive *Helicobacter pylori* clinical isolates in Taiwan. J. Clin. Microbiol. *40*, 3860–3862.

Lai, C.H., Poon, S.K., Chen, Y.C., Chang, C.S., and Wang, W.C. (2005). Lower prevalence of *Helicobacter pylori* infection with *vacA* s1a, *cagA*-positive, and *babA2*-positive genotype in erosive reflux esophagitis disease. Helicobacter *10*, 577–585.

Lehours, P., Menard, A., Dupouy, S., Bergey, B., Richy, F., Zerbib, F., Ruskone-Fourmestraux, A., Delchier, J.C., and Megraud, F. (2004). Evaluation of the association of nine *Helicobacter pylori* virulence factors with strains involved in low-grade gastric mucosa-associated lymphoid tissue lymphoma. Infect. Immun. *72*, 880–888.

Lu, H., Hsu, P.I., Graham, D.Y., and Yamaoka, Y. (2005a). Duodenal ulcer promoting gene of *Helicobacter pylori*. Gastroenterology *128*, 833–848.

Lu, H., Wu, J.Y., Beswick, E.J., Ohno, T., Odenbreit, S., Haas, R., Reyes, V.E., Kita, M., Graham, D.Y., and Yamaoka, Y. (2007). Functional and intracellular signaling differences associated with the *Helicobacter pylori* AlpAB adhesin from Western and East Asian strains. J. Biol. Chem. *282*, 6242–6254.

Lu, H., Wu, J.Y., Kudo, T., Ohno, T., Graham, D.Y., and Yamaoka, Y. (2005b). Regulation of interleukin-6 promoter activation in gastric epithelial cells infected with *Helicobacter pylori*. Mol. Biol. Cell *16*, 4954–4966.

Mahdavi, J., Sonden, B., Hurtig, M., Olfat, F.O., Forsberg, L., Roche, N., Angstrom, J., Larsson, T., Teneberg, S., Karlsson, K.A., Altraja, S., Wadstrom, T., Kersulyte, D., Berg, D.E., Dubois, A., Petersson, C., Magnusson, K.E., Norberg, T., Lindh, F., Lundskog, B.B., Arnqvist, A., Hammarstrom, L., and Boren, T. (2002). *Helicobacter pylori* SabA adhesin in persistent infection and chronic inflammation. Science *297*, 573–578.

Mattar, R., dos Santos, A.F., Eisig, J.N., Rodrigues, T.N., Silva, F.M., Lupinacci, R.M., Iriya, K., and Carrilho, F.J. (2005). No correlation of *babA2* with *vacA* and *cagA* genotypes of *Helicobacter pylori* and grading of gastritis from peptic ulcer disease patients in Brazil. Helicobacter. *10*, 601–608.

Merrell, D.S., Goodrich, M.L., Otto, G., Tompkins, L.S., and Falkow, S. (2003). pH-regulated gene expression of the gastric pathogen *Helicobacter pylori*. Infect. Immun. *71*, 3529–3539.

Mizushima, T., Sugiyama, T., Komatsu, Y., Ishizuka, J., Kato, M., and Asaka, M. (2001). Clinical relevance of the *babA2* genotype of *Helicobacter pylori* in Japanese clinical isolates. J. Clin. Microbiol. *39*, 2463–2465.

Odenbreit, S., Kavermann, H., Puls, J., and Haas, R. (2002). CagA tyrosine phosphorylation and interleukin-8 induction by *Helicobacter pylori* are independent from *alpAB*, *hopZ* and *bab* group outer membrane proteins. Int. J. Med. Microbiol. *292*, 257–266.

Oh, J.D., Kling-Backhed, H., Giannakis, M., Xu, J., Fulton, R.S., Fulton, L.A., Cordum, H.S., Wang, C., Elliott, G., Edwards, J., Mardis, E.R., Engstrand, L.G., and Gordon, J.I. (2006). The complete genome sequence of a chronic atrophic gastritis *Helicobacter pylori* strain: evolution during disease progression. Proc. Natl. Acad. Sci. USA *103*, 9999–10004.

Oleastro, M., Gerhard, M., Lopes, A.I., Ramalho, P., Cabral, J., Sousa, G.A., and Monteiro, L. (2003). *Helicobacter pylori* virulence genotypes in Portuguese children and adults with gastroduodenal pathology. Eur. J. Clin. Microbiol. Infect. Dis *22*, 85–91.

Olfat, F.O., Zheng, Q., Oleastro, M., Voland, P., Boren, T., Karttunen, R., Engstrand, L., Rad, R., Prinz, C., and Gerhard, M. (2005). Correlation of the *Helicobacter pylori* adherence factor BabA with duodenal ulcer disease in four European countries. FEMS Immunol. Med. Microbiol. *44*, 151–156.

Oliveira, A.G., Santos, A., Guerra, J.B., Rocha, G.A., Rocha, A.M., Oliveira, C.A., Cabral, M.M., Nogueira, A.M., and Queiroz, D.M. (2003). *babA2*- and *cagA*-positive *Helicobacter pylori* strains are associated with duodenal ulcer and gastric carcinoma in Brazil. J. Clin. Microbiol. *41*, 3964–3966.

Ota, H., Nakayama, J., Momose, M., Hayama, M., Akamatsu, T., Katsuyama, T., Graham, D.Y., and Genta, R.M. (1998). *Helicobacter pylori* infection produces reversible glycosylation changes to gastric mucins. Virchows Arch. *433*, 419–426.

Peek, R.M., Jr., Thompson, S.A., Donahue, J.P., Tham, K.T., Atherton, J.C., Blaser, M.J., and Miller, G.G. (1998). Adherence to gastric epithelial cells induces expression of a *Helicobacter pylori* gene, *iceA*, that is associated with clinical outcome. Proc. Assoc. Am. Physicians *110*, 531–544.

Petersson, C., Forsberg, M., Aspholm, M., Olfat, F.O., Forslund, T., Boren, T., and Magnusson, K.E. (2006). *Helicobacter pylori* SabA adhesin evokes a strong inflammatory response in human neutrophils which is downregulated by the neutrophil-activating protein. Med. Microbiol. Immunol. (Berl) *195*, 195–206.

Podzorski, R.P., Podzorski, D.S., Wuerth, A., and Tolia, V. (2003). Analysis of the *vacA*, *cagA*, *cagE*, *iceA*, and *babA2* genes in *Helicobacter pylori* from sixty-one pediatric patients from the Midwestern United States. Diagn. Microbiol. Infect. Dis *46*, 83–88.

Pride, D.T. and Blaser, M.J. (2002). Concerted evolution between duplicated genetic elements in *Helicobacter pylori*. J. Mol. Biol. *316*, 629–642.

Pride, D.T., Meinersmann, R.J., and Blaser, M.J. (2001). Allelic Variation within *Helicobacter pylori babA* and *babB*. Infect. Immun. *69*, 1160–1171.

Prinz, C., Schoniger, M., Rad, R., Becker, I., Keiditsch, E., Wagenpfeil, S., Classen, M., Rosch, T., Schepp, W., and Gerhard, M. (2001). Key importance of the *Helicobacter pylori* adherence factor blood group antigen binding adhesin during chronic gastric inflammation. Cancer Res. *61*, 1903–1909.

Rad, R., Dossumbekova, A., Neu, B., Lang, R., Bauer, S., Saur, D., Gerhard, M., and Prinz, C. (2004). Cytokine gene polymorphisms influence mucosal cytokine expression, gastric inflammation, and host specific colonisation during *Helicobacter pylori* infection. Gut *53*, 1082–1089.

Rad, R., Gerhard, M., Lang, R., Schoniger, M., Rosch, T., Schepp, W., Becker, I., Wagner, H., and Prinz, C. (2002). The *Helicobacter pylori* blood group antigen-binding adhesin facilitates bacterial colonization and augments a nonspecific immune response. J. Immunol. *168*, 3033–3041.

Roche, N., Angstrom, J., Hurtig, M., Larsson, T., Boren, T., and Teneberg, S. (2004). *Helicobacter pylori* and complex gangliosides. Infect. Immun. *72*, 1519–1529.

Sakamoto, S., Watanabe, T., Tokumaru, T., Takagi, H., Nakazato, H., and Lloyd, K.O. (1989). Expression of Lewisa, Lewisb, Lewisx, Lewisy, siayl-Lewisa, and sialyl-Lewisx blood group antigens in human gastric carcinoma and in normal gastric tissue. Cancer Res. *49*, 745–752.

Salama, N., Guillemin, K., McDaniel, T.K., Sherlock, G., Tompkins, L., and Falkow, S. (2000). A whole-genome microarray reveals genetic diversity among *Helicobacter pylori* strains. Proc. Natl. Acad. Sci. USA *97*, 14668–14673.

Sheu, B.S., Odenbreit, S., Hung, K.H., Liu, C.P., Sheu, S.M., Yang, H.B., and Wu, J.J. (2006). Interaction between host gastric Sialyl-Lewis X and *H. pylori* SabA enhances *H. pylori* density in patients lacking gastric Lewis B antigen. Am. J. Gastroenterol. *101*, 36–44.

Sheu, B.S., Sheu, S.M., Yang, H.B., Huang, A.H., and Wu, J.J. (2003). Host gastric Lewis expression determines the bacterial density of *Helicobacter pylori* in *babA2* genopositive infection. Gut 52, 927–932.

Solnick, J.V., Hansen, L.M., Salama, N.R., Boonjakuakul, J.K., and Syvanen, M. (2004). Modification of *Helicobacter pylori* outer membrane protein expression during experimental infection of rhesus macaques. Proc. Natl. Acad. Sci. USA *101*, 2106–2111.

Struyve, M., Moons, M., and Tommassen, J. (1991). Carboxy-terminal phenylalanine is essential for the correct assembly of a bacterial outer membrane protein. J. Mol. Biol. *218*, 141–148.

Tomb, J.F., White, O., Kerlavage, A.R., Clayton, R.A., Sutton, G.G., Fleischmann, R.D., Ketchum, K.A., Klenk, H.P., Gill, S., Dougherty, B.A., Nelson, K., Quackenbush, J., Zhou, L., Kirkness, E.F., Peterson, S., Loftus, B., Richardson, D., Dodson, R., Khalak,

H.G., Glodek, A., McKenney, K., Fitzegerald, L.M., Lee, N., Adams, M.D., Hickey, E.K., Berg, D.E., Gocayne, J.D., Utterback, T.R., Peterson, J.D., Kelley, J.M., Cotton, M.D., Weidman, J.M., Fujii, C., Bowman, C., Watthey, L., Wallin, E., Hayes, W.S., Borodovsky, M., Karp, P.D., Smith, H.O., Fraser, C.M., and Venter, J.C. (1997). The complete genome sequence of the gastric pathogen *Helicobacter pylori*. Nature *388*, 539–547.

Unemo, M., spholm-Hurtig, M., Ilver, D., Bergstrom, J., Boren, T., Danielsson, D., and Teneberg, S. (2005). The sialic acid binding SabA adhesin of *Helicobacter pylori* is essential for nonopsonic activation of human neutrophils. J. Biol. Chem. *280*, 15390–15397.

Walz, A., Odenbreit, S., Mahdavi, J., Boren, T., and Ruhl, S. (2005). Identification and characterization of binding properties of *Helicobacter pylori* by glycoconjugate arrays. Glycobiology *15*, 700–708.

Wu, J.Y., Lu, H., Sun, Y., Graham, D.Y., Cheung, H.S., and Yamaoka, Y. (2006). Balance between polyoma enhancing activator 3 and activator protein 1 regulates *Helicobacter pylori*-stimulated matrix metalloproteinase 1 expression. Cancer Res. *66*, 5111–5120.

Yamaoka, Y. (2006). *Helicobacter pylori* outer membrane proteins and gastric inflammation: Author's reply. Gut *55*, 1361.

Yamaoka, Y., Kikuchi, S., El-Zimaity, H.M., Gutierrez, O., Osato, M.S., and Graham, D.Y. (2002a). Importance of *Helicobacter pylori oipA* in clinical presentation, gastric inflammation, and mucosal interleukin 8 production. Gastroenterology *123*, 414–424.

Yamaoka, Y., Kita, M., Kodama, T., Imamura, S., Ohno, T., Sawai, N., Ishimaru, A., Imanishi, J., and Graham, D.Y. (2002b). *Helicobacter pylori* infection in mice: Role of outer membrane proteins in colonization and inflammation. Gastroenterology *123*, 1992–2004.

Yamaoka, Y., Kita, M., Kodama, T., Sawai, N., and Imanishi, J. (1996). *Helicobacter pylori cagA* gene and expression of cytokine messenger RNA in gastric mucosa. Gastroenterology *110*, 1744–1752.

Yamaoka, Y., Kita, M., Kodama, T., Sawai, N., Tanahashi, T., Kashima, K., and Imanishi, J. (1998a). Chemokines in the gastric mucosa in *Helicobacter pylori* infection. Gut *42*, 609–617.

Yamaoka, Y., Kodama, T., Graham, D.Y., and Kashima, K. (1998b). Search for putative virulence factors of *Helicobacter pylori*: the low-molecular-weight (33–35 K) antigen. Dig. Dis Sci. *43*, 1482–1487.

Yamaoka, Y., Kodama, T., Gutierrez, O., Kim, J.G., Kashima, K., and Graham, D.Y. (1999a). Relationship between *Helicobacter pylori iceA*, *cagA*, and *vacA* status and clinical outcome: studies in four different countries. J. Clin. Microbiol. *37*, 2274–2279.

Yamaoka, Y., Kodama, T., Kita, M., Imanishi, J., Kashima, K., and Graham, D.Y. (1998c). Relationship of *vacA* genotypes of *Helicobacter pylori* to *cagA* status, cytotoxin production, and clinical outcome. Helicobacter. *3*, 241–253.

Yamaoka, Y., Kodama, T., Kita, M., Imanishi, J., Kashima, K., and Graham, D.Y. (1999b). Relation between clinical presentation, *Helicobacter pylori* density, interleukin 1beta and 8 production, and cagA status. Gut *45*, 804–811.

Yamaoka, Y., Kodama, T., Kita, M., Imanishi, J., Kashima, K., and Graham, D.Y. (2001). Relation between cytokines and *Helicobacter pylori* in gastric cancer. Helicobacter. *6*, 116–124.

Yamaoka, Y., Kudo, T., Lu, H., Casola, A., Brasier, A.R., and Graham, D.Y. (2004). Role of interferon-stimulated responsive element-like element in interleukin-8 promoter in *Helicobacter pylori* infection. Gastroenterology *126*, 1030–1043.

Yamaoka, Y., Kwon, D.H., and Graham, D.Y. (2000). A M(r) 34, 000 proinflammatory outer membrane protein (*oipA*) of *Helicobacter pylori*. Proc. Natl. Acad. Sci. USA *97*, 7533–7538.

Yamaoka, Y., Ojo, O., Fujimoto, S., Odenbreit, S., Haas, R., Gutierrez, O., El-Zimaity, H.M., Reddy, R., Arnqvist, A., and Graham, D.Y. (2006). *Helicobacter pylori* outer membrane proteins and gastroduodenal disease. Gut *55*, 775–781.

Yamaoka, Y., Souchek, J., Odenbreit, S., Haas, R., Arnqvist, A., Boren, T., Kodama, T., Osato, M.S., Gutierrez, O., Kim, J.G., and Graham, D.Y. (2002c). Discrimination between cases of duodenal ulcer and gastritis on the basis of putative virulence factors of *Helicobacter pylori*. J. Clin. Microbiol. *40*, 2244–2246.

Yu, J., Leung, W.K., Go, M.Y., Chan, M.C., To, K.F., Ng, E.K., Chan, F.K., Ling, T.K., Chung, S.C., and Sung, J.J. (2002). Relationship between *Helicobacter pylori babA2* status with gastric epithelial cell turnover and premalignant gastric lesions. Gut *51*, 480–484.

Zambon, C.F., Navaglia, F., Basso, D., Rugge, M., and Plebani, M. (2003). *Helicobacter pylori babA2, cagA*, and s1 *vacA* genes work synergistically in causing intestinal metaplasia. J. Clin. Pathol. *56*, 287–291.

Helicobacter Flagella, Motility and Chemotaxis

4

Melanie Rust, Tobias Schweinitzer and Christine Josenhans

Abstract

All gastric and enterohepatic *Helicobacter* species are highly motile. Comparative genomic analysis in different *Helicobacter* species and related bacteria in recent years has facilitated the analysis of genus-, species- and niche-specific properties. The characteristic sheathed flagellar filaments of helicobacters are composed of two copolymerized flagellins, FlaA and FlaB. Experiments in different animal models with *Helicobacter pylori*, *Helicobacter mustelae* and *Helicobacter felis* showed that flagellar motility is essential for *Helicobacter* species to colonize the gastric mucus. *H. pylori* has homologues of almost all flagellar structural proteins known from *Enterobacteriaceae*, but the regulatory network of *H. pylori* motility genes, which has been investigated in some detail, lacks a *flhCD* master operon and other regulatory factors, and relies on a three-tiered hierarchical regulatory system with RpoN as the central sigma factor. *H. pylori* shows taxis (directed motility) towards urea, amino acids, and bicarbonate, and moves away from low pH. Recently, it has been shown in several *in vivo* systems that chemotaxis in addition to motility is required for colonisation, and that *H. pylori* and *H. felis* reside in a very narrow layer of the gastric mucus, close to the gastric epithelium, where they are guided by the mucus pH gradient. So far, the mucus pH gradient seems to be the most important sensing stimulus *in vivo* for *H. pylori* orientation. However, it is still not completely clear which combinations of horizontal and vertical chemical gradients are used by *H. pylori in vivo* to maintain an optimal position in the gastric mucus layer. The chemotaxis systems of *H. pylori* and *H. hepaticus* are genetically similar to the *Salmonella* system as based on genomic analysis, but extensive functional analyses on the proteins involved have just been started. A dominant property of the *H. pylori* sensing system from recent *in vivo* and *in vitro* studies seems to be pH taxis.

Introduction

A prominent characteristic of all *Helicobacter* species identified so far is flagella-driven motility. Motility is essential for colonization by the three gastric *Helicobacter* species, *Helicobacter pylori*, *Helicobacter mustelae*, and *Helicobacter felis*, that have been tested in experimental and natural infection (Eaton *et al.*, 1996; Ottemann and Lowenthal, 2002; Terry *et al.*, 2005; Kavermann *et al.*, 2003; Andrutis *et al.*, 1997; Josenhans *et al.*, 1999). Flagellar motility is probably required for the initial stages of infection, when *H. pylori* must move from the acidic gastric lumen into the only slightly acidic layer of the gastric mucus close to the epithelium that is colonized during chronic infection (Schreiber *et al.*, 2004). Motility is also considered essential during persistent infection for maintaining a constant reservoir of bacteria deep in the gastric mucus which is subject to constant and rapid turnover (Suerbaum, 1995; Blaser, 1994).

Flagella-driven motility is common among both pathogenic and non-pathogenic bacteria. Bacteria may be propelled by one or multiple flagella alternatively. Flagella consist of three main structural parts: the flagellar motor plus other components of the basal body, the hook and the filament. The basal body spans the cell envelope, anchors the flagellum in the membranes, and is linked to the flagellar rotary motor at the base. The basal body is connected to the filament *via* the curved hook substructure. The filament is a long appendage that consists mainly of polymeric flagellin subunits. When the motor turns in the correct direction (e.g. counterclockwise for a left-handed helix), one or several filaments jointly form a spiral and propel the bacterium. In all known motile bacteria, motility is linked to a chemotactic signal transduction system that enables bacteria to move in a biased random walk towards attractants or away from repellents. Three known mechanisms for implementing this directed movement are alternation between periods of 'runs' and 'tumbles', backward movement after a change of flagellar rotary direction, and start-stop movements. The flagellar apparatus and the chemotaxis system have been studied thoroughly in several different organisms. The best studied bacterium in this regard is *Salmonella* Typhimurium, but extensive information is also available for the motility systems of *Pseudomonas aeruginosa*, *Vibrio* spp., *Bacillus subtilis* and *Caulobacter crescentus* (Aizawa, 1996; Prouty *et al.*, 2001; Macnab, 1996; Shapiro, 1995; Dasgupta *et al.*, 2003; Kim and McCarter, 2000).

The flagellar apparatus of *H. pylori* differs substantially from that of other motile bacteria. The investigation of *H. pylori* motility has already provided important clues about structural and regulatory adaptation of this common bacterial organelle to the particular requirements of the gastric niche. The chemotactic machinery that *H. pylori* uses to orient itself in the gastric mucus appears to be similar to that of other chemotactic eubacteria. It consists of transmembrane sensory systems (protein complexes), which can sense the presence of attractants and repellents in the outside milieu and transduce signals to the bacterial flagellar basal body *via* several steps of a molecular chain reaction involving protein phosphorylation (for reviews, see Stock and Surette, 1994; Bren and Eisenbach, 2000; Falke and Hazelbauer, 2001; Wadhams and Armitage, 2004; Parkinson, 1993).

Motility and flagella – molecular biology and structure

Most members of the growing group of gastric and enterohepatic *Helicobacter* species (Solnick and Schauer, 2001) possess multiple sheathed flagella (unsheathed in *H. rodentium* and *H. pullorum*, which form a separate phylogenetic branch called 'unsheathed'; see Dewhirst *et al.*, 2005) that are arranged in different patterns in various species, e.g. unipolar polytrichous (*H. pylori*; Fig. 4.1), bipolar polytrichous (*H. felis*), bipolar monotrichous (*H. hepaticus*) and peritrichous (*H. mustelae*). Together with slight

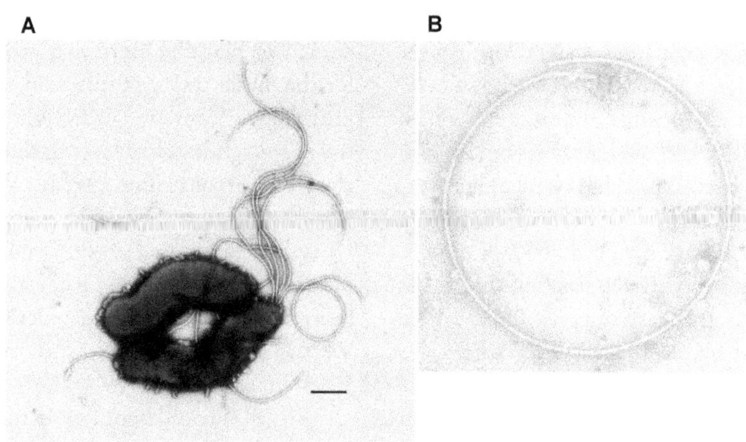

Figure 4.1 Transmission electron micrograph of *H. pylori* LSU1040–1 (A). Negative staining was performed with 1% phosphotungstate (pH 7.0). Note the curved bacteria with unipolar sheathed flagella. Panel B shows an isolated flagellum including a part of the basal body ring structure.

differences in cell body morphology (e.g. the possession of periplasmic fibrils in some species is observed), these different patterns are thought to offer optimum motility in the stomach and intestinal mucus of different hosts. Each *Helicobacter* flagellar filament is enveloped individually by a membranous sheath. This sheath is continuous with the outer membrane and, in *H. pylori*, has a similar composition of lipopolysaccharides, phospholipids and proteins (Geis *et al.*, 1993). The flagellar sheath-associated lipoprotein, HpaA, was localized in flagellar sheath membranes using immunogold electron microscopy (Jones *et al.*, 1997). A different study in an unrelated *H. pylori* strain reported the protein to be present also in the whole bacterial outer membrane (Lundstrom *et al.*, 2001). Uniquely sheath-specific protein components have not yet been identified. The sheath is thought to protect the acid-labile flagellar filaments against low pH in the stomach, because sheath vesicles are more resistant to dissociation under acidic conditions than are naked filaments (Geis *et al.*, 1993; Geis, Forsthoff and Suerbaum, unpublished).

All genes known to code for components of the flagellar substructures, i.e. the extended basal body (including the C-ring), the hook and the filament, are present in the so far available *H. pylori* genome sequences (Table 4.1) (Tomb *et al.*, 1997; Alm *et al.*, 1999; Oh *et al.*, 2006). Several genes encoding putative flagellar components that are not present in Enterobacteria were identified by homology to motility genes of other bacterial species (Table 4.1). The list of known flagella and motility genes lacks several genes that are known in *Salmonella*.

The flagellar filaments of all tested *Helicobacter* species are composed of two distinct flagellin subunits, FlaA and FlaB (Suerbaum *et al.*, 1993; Josenhans *et al.*, 1995; Josenhans *et al.*, 1999). These two flagellins share only 50% amino acid identity but probably evolved from a common ancestral flagellin molecule. Their molecular masses (53–54 kDa) are similar to those of enterobacterial flagellins. It has been suggested that *Helicobacter* FlaB is an exclusively hook proximal flagellin subunit that is solely expressed early during flagellar assembly (Kostrzynska *et al.*, 1991). Regulatory studies place FlaB among the middle flagellar genes (Niehus *et al.*, 2004) implying that this protein is expressed earlier during the cell cycle than FlaA (Niehus *et al.*, 2002), therefore it appears quite unlikely that both flagellin subunits are present in a mixed fashion throughout the length of the filaments. The ratio of FlaA to FlaB changes in a growth phase-dependent manner, and probably also in response to environmental conditions (Niehus *et al.*, 2002; Niehus *et al.*, 2004; Ye *et al.*, 2007). The two flagellins of *H. felis* and *H. pylori* (probably also those of other *Helicobacter* species) undergo post-translational glycosylation, but the function of this modification is not completely clear (Josenhans *et al.*, 1999; Josenhans *et al.*, 2002b; Schirm *et al.*, 2003; Schoenhofen *et al.*, 2006). Flagellin glycosylation has also been described for the closely related *Campylobacter* species (Thibault *et al.*, 2001; McNally *et al.*, 2006).

Mutants lacking the hook protein FlgE1 (O'Toole *et al.*, 1994), the flagellar cap protein FliD (Kim *et al.*, 1999), the export-associated proteins FlhA, FlhB1, FliQ (Schmitz *et al.*, 1997; Foynes *et al.*, 1999) or the energy-providing ATPase of the flagellar type III system FliI (Jenks *et al.*, 1997; Lane *et al.*, 2006; Porwollik *et al.*, 1999), were deficient in flagellar filament formation. Thus, all these proteins are also needed in *H. pylori* for important steps during the synthesis of flagella. (For a review of the functions of those proteins in *Enterobacteriaceae* see Macnab, 2003.)

The energy for flagellar rotation in *H. pylori*, as in *Salmonella*, is derived from the proton motive force, which is thought to be composed of (i) a membrane potential at the bacterial inner membrane and (ii) a pH gradient between the environment and the bacterial cytoplasm (Nakamura *et al.*, 1998).

The genomic sequences of *H. pylori* strains 26695, J99, HPAG1 (Alm *et al.*, 1999; Tomb *et al.*, 1997; Oh *et al.*, 2006), of *Helicobacter acynonychis* (Eppinger *et al.*, 2006), *Helicobacter hepaticus* (Suerbaum *et al.*, 2003), *H. mustelae* (Paul O'Toole and coworkers in collaboration with the Wellcome Trust Sanger Institute, unpublished), numerous closely related *Campylobacter* species, and *Wolinella succinogenes* (Baar *et al.*, 2003) have resulted in the tentative identification of many motility-associated genes,

Table 4.1 Motility- and flagella-associated genes in the complete genome sequences of *H. pylori* 26695 and J99

Gene number in TIGR genome	Gene number in J99 and HPAG genomes (JHP/AG)	Gene designation	Protein designation and function/homologue in *S.* Typhimurium yes/no (gene designation in *S.* Typhimurium, if different from HP nomenclature)	Promoter if known
I: genes involved in flagellar structure and function				
HP0114	jhp106/AG114	–	Flagellin glycosylation/no	?
HP0115	jhp107/AG115	*flaB*	Flagellin B/no	Sigma54
HP0170	jhp156/AG166	*cheZ**	Remote CheZ orthologue, possibly involved in phosphate flow in the chemotaxis system/yes	?
HP0173	jhp159/AG169	*fliR*	Regulatory protein of flagellar biogenesis flagellar basal body protein/yes	
HP0178	jhp166/AG175	–	Flagellin glycosylation/no	?
HP0232	jhp217/AG235	–	Secreted protein involved in motility/no	?
HP0246	jhp231/AG249	*flgI*	P ring forming protein of basal body/yes	Sigma70 (?)
HP0295	jhp280/AG297	*flgL*	HAP-3 homologue/yes	Sigma54 (?)
HP0325	jhp308/AG328	*flgH*	L ring forming protein of Basal body/yes	Sigma70 (?)
HP0326	jhp309/AG329	*neuA*	Flagellin glycosylation/no	?
HP0326b	jhp309b/AG330	*flmD*	Flagellin glycosylation/no	?
HP0327	jhp310/AG331	*flaG1* *(flmH)*	Flagella-associated protein – glycosylation/no	Sigma 70 (?)
HP0351	jhp325/AG346	*fliF*	M ring-forming protein of basal body/yes	Sigma70 (?)
HP0352	jhp326/AG347	*fliG*	Flagellar motor switch protein, putative C ring component/yes	Sigma70 (?)
HP0353	jhp327/AG348	*fliH*	Protein involved in flagellar export apparatus/yes	Sigma70 (?)
HP0366	jhp1015/AG1028	–	Involved in flagellin glycosylation/no	Sigma54 and Sigma28 (?)
HP0410	jhp971/AG982	*hpa2*	Paralogue of HP hemagglutinin protease – sheath associated/no protein/no	?
HP0472	jhp424/AG449	*omp11*	Coregulated with late flagellar genes/flagellar sheath protein?/no	Sigma28
HP0492	jhp444/AG468	*hpaA3*	Orthologue of HpaA/no	?
HP0584	jhp531/AG563	*fliN*	Flagellar motor switch protein, putative C ring component/yes	Sigma70 (?)

HP0601	jhp548/AG582	*flaA*	Major flagellin A/yes (*fliC*)	Sigma28
HP0684	jhp625/AG668	*fliP'*	N-terminal domain of basal body protein FliP involved in flagellar export apparatus/yes	?
HP0685	jhp625AG668	*fliP'*	Second half of basal body protein FliP[a]/yes	?
HP0751	jhp688/AG736	*flaG2*	Gene of unknown function involved in biogenesis of polar flagellum/no	Sigma70, sigma28 (?)
HP0752	jhp689/AG737	*fliD*	Hook-associated protein-2 (HAP-2), filament cap protein FliD/yes	Sigma70, sigma28 (?)
HP0753	jhp690/AG738	*fliS*	Flagellin chaperone protein FliS/yes	Sigma70, sigma28 (?)
HP0754	jhp691/AG739	*fliT*	Flagellar chaperone	Sigma70, sigma28 (?)
HP0770	jhp707/AG755	*flhB1*	Flagellar basal body protein involved in export/yes	Sigma70 (?)
HP0797	jhp733/AG782	*hpaA*	Flagellar sheath 'adhesin' protein -neuraminyllactose binding protein/no	?
HP0815	jhp751/AG800	*motA*	Flagellar motor protein A/yes	Sigma70 (?)
HP0816	jhp752/AG801	*motB*	Flagellar motor protein B/yes	Sigma70 (?)
HP0840	jhp778/AG825	*flaA*1	Flagella-associated protein involved in glycosylation/no	?
HP0870	jhp804/AG853	*flgE1*	Flagellar hook protein/yes	Sigma54
HP0906	jhp842/AG888	*fliK*	Flagellar substrate switch protein	Sigma70
HP0907	jhp843/AG889	*flgD*	Flagellar hook scaffolding protein/yes	Sigma54
HP0908	jhp844/AG890	*flgE2* (*flgF*)	Putative flagellar hook basal body protein homologue/no	Sigma54
HP0958	jhp892/AG942	–	Stabilizes flagellar s54 factor/no	?
HP1030	jhp394/AG417	*fliY*	FliY protein/no	Sigma70, sigma28 (?)
HP1031	jhp393/AG416	*fliM*	Flagellar motor switch protein putative C ring component/yes	Sigma70, sigma28 (?)
HP1034	jhp390/AG413	*flhG/ylxH*	Cooperates with FlhF in polar flagellar biosynthesis/no	Sigma70, sigma28 (?)
HP1035	jhp389/AG412	*flhF*	Flagellar biosynthesis protein of unknown function/no	Sigma70, sigma28 (?)

Table 4.1 *continued*

Gene number in TIGR genome	Gene number in J99 and HPAG genomes (JHP/AG)	Gene designation	Protein designation and function/homologue in *S.* Typhimurium yes/no (gene designation in *S.* Typhimurium, if different from HP nomenclature)	Promoter if known
HP1041	jhp383/AG406	*flhA*	Flagellar basal body protein involved in export/yes	No consensus, evtl. sigma70 without the –35 region
HP1051	jhp372/AG394	–	Hypothetical protein	Sigma28
HP1052	jhp373/AG395		Deacetylase EnvA (cell wall biosynthesis)	Sigma28
HP1076	jhp349/AG371	–	Hypothetical protein coregulated with middle flagellar genes	Sigma54
HP1092	jhp333/AG355	*flgG*1	Homologous to distal rod protein of flagellar basal body, putative functional FlgF substitute/no-yes or rod-hook junction protein	Sigma54
HP1119	jhp1047/AG1057	*FlgK*	Hook-associated protein 1 (HAP 1)/yes	Sigma54
HP1120	jhp1048/AG1058	–	Coregulated with *flgK*	Sigma54
HP1154	jhp1081/AG1093	–	Hypothetical protein	Sigma54
HP1155	jhp1082/AG1094	*murG*	Peptidoglycan biosynthesis	Sigma54
HP1192	jhp1117/AG1133	–	Secreted protein involved in motility/no	?
HP1233	jhp1154/AG1175	*flgJ*?*	Remote orthologue to flagellar muraminidase/yes	Sigma54
HP1419	jhp1314/AG1345	*FliQ*	Flagellar basal body protein involved in export/yes	?
HP1420	jhp1315/AG1346	*FliI*	Flagella-associated ATPase, energizer of flagellar export/yes	?
HP1462	jhp1355/AG1453	–	Secreted protein involved in motility/no	?
HP1557	jhp1465/AG1506	*fliE*	Flagellar basal body protein/yes	Sigma70 (?)
HP1558	jhp1466AG1507	*flgC*	Proximal rod protein of basal body/yes	Sigma54
HP1559	jhp1467/AG1508	*flgB*	Proximal rod protein of basal body/yes	Sigma54
HP1575	jhp1483/AG1524	*flhB2*	Homologue of FlhB protein basal body/no	?
HP1585	jhp1492/AG1533	*flgG2*	Distal rod protein of flagellar basal body/yes	Sigma70-like/ sigma54

II: genes involved in chemotaxis				
HP0019	jhp17/AG19	*cheV1*	CheW/CheY hybrid chemotaxis protein/no	?
HP0082	jhp75/AG83	*tlpC*	Receptor protein of chemotaxis signalling system/specificity unknown	?
HP0099	jhp91/AG99	*tlpA*	Chemotaxis receptor/specificity unknown	?
HP0103	jhp95/AG103	*tlpB*	Chemotaxis receptor/specificity unknown	?
HP0170	jhp156/AG166	*cheZ**	Remote CheZ orthologue, possibly involved in phosphate flow in the chemotaxis system/yes	?
HP0391	jhp990/AG1001	*chew*	Chemotaxis adaptor protein/yes	?
HP0392	jhp989/AG1000	*cheA*	Receptor-coupled histidine kinase/yes, but fused to CheY2 domain in HP	?
HP0393	jhp988/AG999	*cheV3*	Hybrid CheW–CheY chemotaxis protein/no	?
HP0599	jhp546/AG579	*tlp*	Soluble chemotaxis receptor/specificity unknown	?
HP0616	jhp559/AG597	*cheV2*	Hybrid CheW–CheY chemotaxis protein	?
HP1067	jhp358/AG380	*cheY1*	Chemotaxis effector protein – link to basal body motor/yes	Sigma70
III: genes involved in flagellar regulation				
HP0244	jhp229/AG247	*flgS*	Histidine kinase involved in activation of FlgR and RpoN/no	Sigma 70
HP0703	jhp643/AG688	*flgR*	NtrC homologue Transcriptional activator of sigma54-regulated motility genes (response regulator)/no	Sigma70
HP0714	jhp652/AG699	*rpoN*	Sigma54 subunit of DNA-dependent RNA polymerase	?
HP0958	jhp892/AG942	–	Activator of RpoN-dependent genes/no	?
HP1032	jhp392/AG415	*fliA*	Sigma28 subunit of DNA-dependent RNA Polymerase/'flagellar' sigma factor/yes	Sigma70, sigma28 (?)
HP1122	jhp1051/AG1060	*FlgM*	Flagellar anti-sigma28 factor	Sigma70, sigma54, sigma28

[a]HP0684 and HP0685 are one continuous ORF in strains J99 and HPAG1. See text for details.

and have facilitated comparative approaches to the motility and chemotaxis systems of those bacteria. Although some of the genome annotations have already been verified experimentally, functional assignments only based on sequence similarity should be treated with caution. An updated list of the putative motility-associated genes in the complete genome sequences of *H. pylori* (all of which are conserved and can be identified by sequence similarity in other genome sequences of *Helicobacter* spp.) is provided in Table 4.1, together with a note of whether the genes are present among the set of motility genes in *Salmonella*.

Including the tentative assignments, approximately 60 genes of the *H. pylori* genomes are related to motility and the structural assembly of flagella, and approximately 16 genes are involved in chemotaxis and the regulation of motility. Table 4.1 includes several genes that are not present or poorly characterized in the *Salmonella* or *B. subtilis* motility systems: (i) homologues of two *Caulobacter crescentus* flagella-associated genes, *flaG* (designated *flaG1* in Table 4.1) and *flaA1* (both involved in flagellin glycosylation), (ii) the gene encoding the putative flagellar sheath protein HpaA is present in three copies (Niehus *et al.*, 2004) and is unique to *H. pylori* and (iii) a gene that is weakly homologous to *Vibrio parahaemolyticus flaG* (*flaG2* in Table 4.1) and which is needed for the biogenesis of the polar flagellum in *V. parahaemolyticus*. Genes which play an important role for flagellar assembly in *Salmonella*, but which appear to be missing from the *H. pylori* genomes, include *flgN*, *fliJ*, and *flgA*, encoding chaperones for the hook protein FlgE, for hook/rod type proteins and P-ring assembly, respectively (Fraser *et al.*, 1999; Minamino and Macnab, 2000; Nambu and Kutsukake, 2000). Also missing from the genomes are genes encoding FlgJ, a flagella-specific muraminidase with cell wall lytic activity (Nambu *et al.*, 1999), and the master flagellar regulators FlhC and FlhD. Genes for *fliK*, involved in hook length control, *flgM* and *fliT*, which formerly were thought to be missing in helicobacters, have recently been identified in *H. pylori* and related bacteria either on the ground of weak homologies, protein interactions, or by coregulation with other flagellar genes (Rain *et al.*, 2001; Colland *et al.*, 2001; Josenhans *et al.*, 2002a; Lane *et al.*, 2006; Ryan *et al.*, 2005b).

The analogous regulatory functions of FlgM and FliK in *H. pylori* have also been confirmed experimentally in *H. pylori* and in a heterologous complementation system (for FlgM) in *Salmonella* (Colland *et al.*, 2001; Josenhans *et al.*, 2002a; Lane *et al.*, 2006). A mutation in FliK, which is supposed to regulate the switch from early flagellar export substrates ('rod-hook'-type) to the flagellin subunits ('filament'-type) resulted in the formation of polyhook structures (Lane *et al.*, 2006). The N-terminus of the *Helicobacter* anti-sigma factor FlgM is considerably shorter than in other bacteria, although its FliA binding and regulatory functions seem to be conserved (Josenhans *et al.*, 2002a). However, in contrast to FlgM in other bacteria, *H. pylori* FlgM seems not to be exported by the bacteria to exert its demonstrated function in regulation (see below; Rust and Josenhans, unpublished results). The gene encoding the hook protein FlgE, the export-linked gene *flhB*, which is involved in switching of export type specificities (Minamino and Macnab, 2000), the distal rod-associated gene *flgG* and the *hpaA* genes are each present in two copies (three related genes in the case of *hpaA*).

Recently, a study on the function of the two FlhB homologues (FlhB1 and FlhB2), which belong to the export substrate switching group of flagellar basal body proteins in *Salmonella*, suggested that the short FlhB2, which is homologous to the C terminus of FlhB1, can work as a functional substitute for this C terminal domain of FlhB1 (Wand *et al.*, 2006). Since a *flhB2* mutant did not show any particular phenotype, while *flhB1* mutants lacked flagella, the function of the second gene remains unclear. Novel flagella-associated genes have been found and experimentally characterized in *H. pylori*: The gene HP0958 was recently identified as a regulatory factor involved in the activation of RpoN-dependent genes and in RpoN binding (Pereira *et al.*, 2006; Ryan *et al.*, 2005a). A function of this protein in stabilization of RpoN and in increasing transcript abundance of several important flagellar genes was reported in those two studies. Further novel genes (HP0114, *neuA*, *flmD*, *flmH*, *flaA1*, HP0178[*neuB*], HP0366), some of which have homologies to glycan biosynthesis proteins of other bacteria, were found to be involved in the glycosylation of flagellar proteins in *H. pylori* (Schirm *et al.*, 2003; Josenhans *et al.*, 2002b; Schoenhofen *et al.*, 2006). In addition to molecular work on single motility-related proteins, the protein interaction map based on yeast two hybrid interactions, presented by Rain and coworkers (Rain *et al.*, 2001), has been a valuable resource in the recent past for identifying novel proteins associated to the flagellar functional network.

Regulation of motility genes

The assembly of flagella requires the coordinated synthesis and sequential assembly of more than 40 proteins. In *Salmonella typhimurium*, this is achieved by a hierarchical control system ensuring that the components of the filament are only synthesized if an intact basal body and hook can be assembled (Aizawa, 2000). Major components of the *Salmonella* motility regulon are (i) the master regulatory operon *flhCD* encoding transcriptional regulators that link motility to crucial cell functions like cell division (Kutsukake *et al.*, 1990; Aizawa and Kubori, 1998), (ii) the alternative sigma factor sigma28 (sigmaD), which controls the expression of most late flagellar genes (e.g. flagellin genes) and iii) the sigma28 antagonist FlgM (Hughes *et al.*, 1993), a stop valve to ensure that late flagellar genes are only transcribed after intact basal body-hook structures have been assembled. Recently, it has been established that other flagellated bacteria also have hierarchical systems of flagellar regulation, which are functioning quite differently from those of *Enterobacteriaceae* such as *Salmonella*. Those strongly rely on several sigma54-like sigma factors to establish the ordered expression of flagellar genes (Kim and McCarter, 2000; Prouty *et al.*, 2001; Niehus *et al.*, 2004; Dasgupta *et al.*, 2003; Jagannathan *et al.*, 2001).

Although the regulation of *Helicobacter* motility genes is not completely elucidated, some distinctive features have emerged. Many of the motility-associated genes of *H. pylori* are not organized in large operons, but are scattered around the whole chromosome, which makes the search for expression networks and coregulated genes more complicated (Fig. 4.2). Whole genome microarray analysis has therefore proven to be the method of choice to define flagellar regulons. A *flhCD* master operon is absent in *H. pylori*. Sigma28 (FliA, s28) as well as an anti sigma28 factor FlgM with a conserved function as a repressor of FliA are present (Josenhans *et al.*, 2002a). Expression of most flagellar genes in *H. pylori* have been shown to rely on the housekeeping sigma factor RpoD (early genes) and on RpoN (s54; middle genes) (Niehus *et al.*, 2004) (Fig. 4.3). Although RpoN is central to *H. pylori* flagellar regulation (Fig. 4.3), it is still a less dominant regulatory factor than in *Pseudomonas* spp., *Vibrio* spp., or *Caulobacter crescentus*. Using microarray analyses, the late flagellar sigma factor FliA has been shown to control few genes in *H. pylori*, which are termed late genes (Josenhans *et al.*, 2002a). A new group of flagellar genes has been defined in *H. pylori*, which are not completely dependent on either one of the flagellar sigma factors, but are controlled jointly either by two or three sigma factors (intermediate genes; see Fig. 4.3 and Niehus *et al.*, 2004). Those genes also occupy an intermediate structural or regulatory position in the *Helicobacter* flagellar hierarchy, as they belong to the middle to late genes. *H. pylori* middle to late flagellar protein synthesis is not completely dependent on the successful assembly of hook-basal bodies because an *H. pylori* mutant lacking the hook protein FlgE still expressed flagellin proteins (O'Toole *et al.*, 1994; Niehus *et al.*, 2004). Still, FlgM appears to control effectively the synthesis of late flagellar genes (Josenhans *et al.*, 2002a). Unlike in *Salmonella*, most of the *H. pylori* late flagellar genes are not exclusively under the control of s28. Most of the middle to late flagellar genes characterized (Table 4.1) are under the control of the s54 RNA polymerase cofactor, which works in concert with the two-component system FlgRS (Beier and Frank, 2000; Niehus *et al.*, 2004; Brahmachary *et al.*, 2004). Many of the early flagellar genes are controlled by sigma80-like (s80) housekeeping promoters, which seems to be the most common flagella-associated promoter in *H. pylori* (see below, Table 4.1 and Fig. 4.3). The sigma70 homologue of the *H. pylori* DNA-dependent RNA polymerase has been purified and found to be an 80 kDa protein (hence s80) (Beier *et al.*, 1998), which seems to have a slightly different DNA substrate specificity than does the

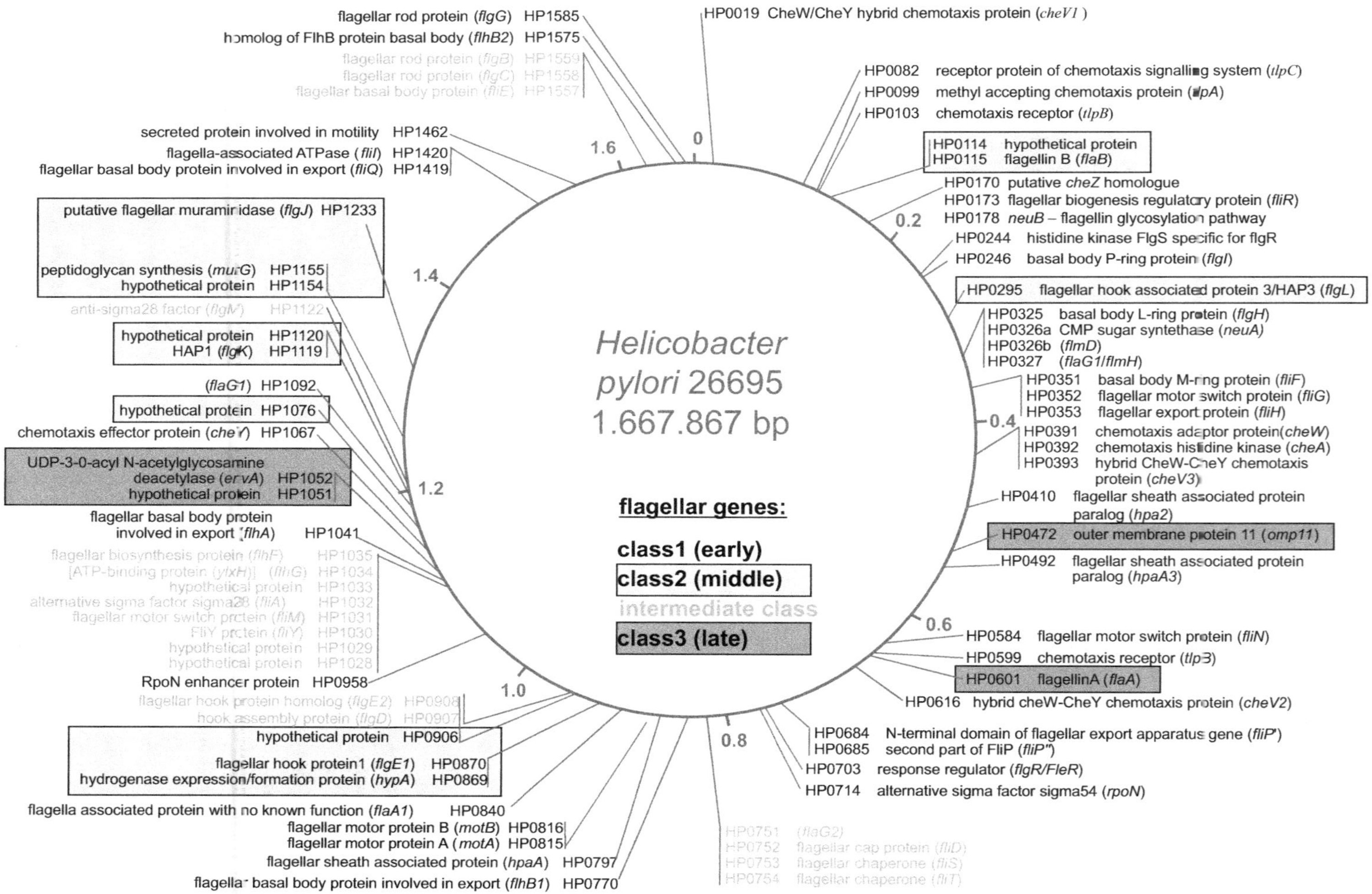

Figure 4.2 View of the whole genome of *H. pylori* (strain 26695, shown as a circular chromosome) demonstrating the scattered localization of all currently known genes involved in flagellar biosynthesis, motility and chemotaxis. (Modified according to supplementary Fig. 4.1 of Niehus *et al.*, 2004.)

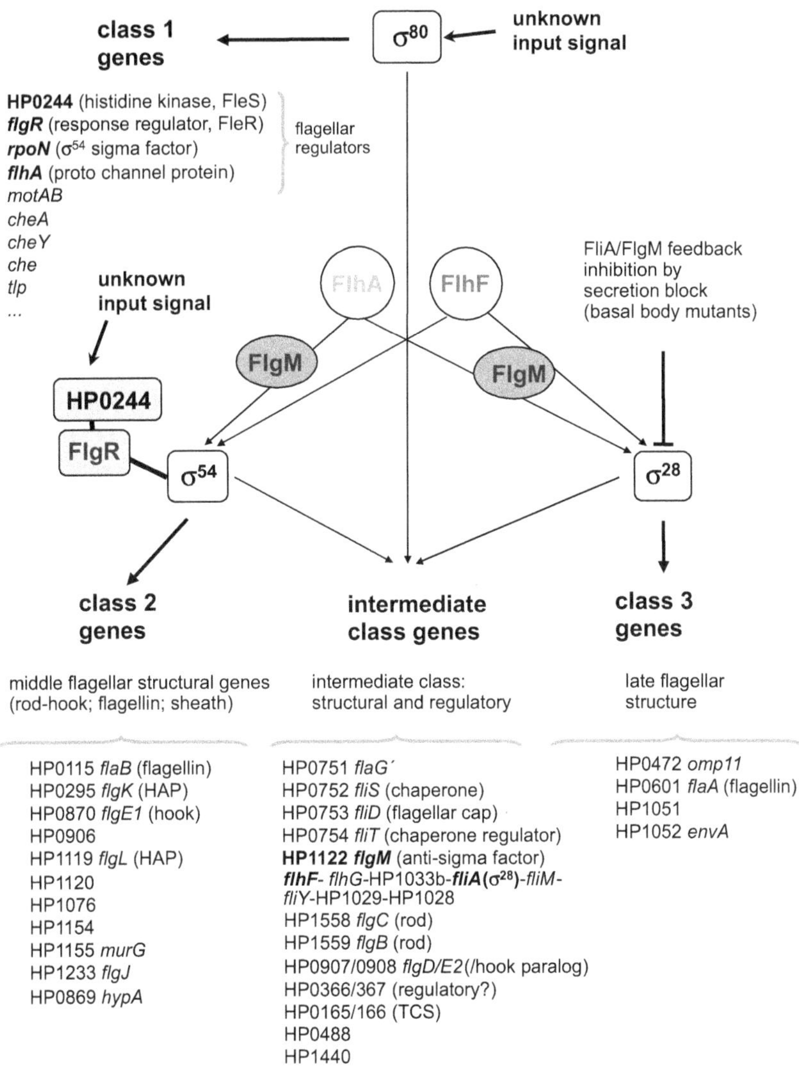

Figure 4.3 A model of the motility regulatory network in *H. pylori* (modified from Niehus *et al.*, 2004). The model integrates the three different sigma factors involved in regulation of motility associated genes and the genes they control into a tentative scheme together with factors which are known to play a role in expression of flagellar components (Schmitz *et al.*, 1997; Beier *et al.*, 1997; Spohn *et al.*, 1999; Vanet *et al.*, 2000; Niehus *et al.*, 2004; Ye *et al.*, 2007). *flhA* was previously named *flbA* (Schmitz *et al.*, 1997). The system is different from the hierarchical regulation present in *Salmonella*. Gene names and respective functions can be looked up in Table 4.1.

Escherichia coli sigma70 subunit. A consensus s80 promoter sequence was defined by computational analysis of the genomic sequence data (Vanet *et al.*, 2000). A great number of motility-associated genes in *H. pylori* are directly coregulated with genes involved in metabolic pathways by housekeeping promoters, unlike in other bacteria (Spohn and Scarlato, 1999; Niehus *et al.*, 2004) (Table 4.1 and Fig. 4.3). The transcriptional activator FlgR (response regulator), a homologue of the nitrogen-dependent regulator NtrC in *E. coli*, interlinks the expression of a number of s54-dependent motility genes (Spohn and Scarlato, 1999; Niehus *et al.*, 2004; Brahmachary *et al.*, 2004). FlaA, which is governed by a typical flagellar s28 promoter, is slightly repressed by RpoN and FlgR (Spohn and Scarlato, 1999). Various motility-related genes and operons have been shown to be expressed *in vivo* by screening for specific mRNA in biopsies of infected animal hosts and human patients (Porwollik *et al.*, 1999; Boonjakuakul *et al.*, 2004; Huff *et al.*, 2004; Boonjakuakul *et al.*, 2005). Several studies have also elucidated differential regulation of motility and chemotaxis genes *in vivo* as compared to *in vitro* in humans and animal models (Scott *et al.*, 2007; Boonjakuakul *et al.*, 2005), but in those studies, diverse gene groups have been found and it has been difficult to interpret these data within a well-defined common regulatory framework.

Despite the enormous progress in our understanding of global regulation in *H. pylori*, we still do not completely understand how the motility genes, which are distributed over the whole chromosome, can be coordinately regulated and synchronized with the bacterial cell cycle and other cell functions. The flagellar basal body membrane protein FlhA and the flagellar signal recognition particle homologue FlhF both control a large set of the *H. pylori* transcriptome (approximately 10% of the whole genome), encompassing the regulons dependent on the known flagellar transcription factors FliA and RpoN, but also genes not directly related to motility, and have therefore been termed functional flagellar master regulators in *H. pylori* (Niehus *et al.*, 2004). However, the mechanism by which FlhA and FlhF regulate their target genes is not clear, since there is no indication that they are working directly as DNA-binding transcription factors or enhancers. As one important dynamic factor, DNA structure, in particular the specific state of DNA superhelicity throughout the growth phases in *H. pylori*, was found to be strongly involved in the coordinated regulation of motility genes (Ye *et al.*, 2007). Among numerous other genes, the expression of both flagellin genes is influenced by local or global changes of supercoiling (Suerbaum *et al.*, 1998; Ye *et al.*, 2007; Spohn and Scarlato, 1999). The *flaB* gene maps next to the gene encoding topoisomerase I (*topA*), with promoters of both genes overlapping and oriented in the opposite direction. TopA is an enzyme which resolves negative supercoiling of DNA. The gene encoding the sigma54 activating protein FlgR is in the same operon with the *gyrA* gene (encoding gyrase, which introduces negative DNA supercoils). These genetic linkages provide clear support for a role of superhelicity in regulation of flagellar biosynthesis, and also give a hint about possible mechanisms of inverse regulation of middle (*flaB*) and late (*flaA*) flagellar genes. The gene linkage suggests a coregulation of FlgR activation (middle genes) with gyrase activation (increasing negative DNA supercoiling). Furthermore, activation of FlaB, which is the latest structural component encoded by a middle flagellar gene, would coincide with increased activation of TopA resulting in DNA relaxation, which would antagonize gyrase, and lead to increased *flaA* activation, as soon as TopA is sufficiently activated. Regarding *flaA*, the unusually short spacing of its s28 promoter (13 nucleotides as opposed to 15 nt for a standard s28 consensus promoter spacing in Enterobacteriaceae) is strongly involved in differential and growth phase regulation by DNA superhelicity, as has been shown by chromosomal *flaA* promoter mutagenesis in *H. pylori* (Ye *et al.*, 2007) The current model of how this regulation works is that binding of the sigma factor or other DNA binding proteins to regulatory DNA sequences, and/or open complex formation of the DNA may be modulated by winding and relaxation (unwinding) of the promoter segment and vice versa. Flagellar and chemotaxis genes *flaA*, *flaB*, *HP0472*, *cheV*, have been found to be modulated by DNA structure, as are numerous other housekeeping and virulence genes (Ye *et al.*, 2007). The exact processes by which the expression of each of the more than 100 superhelicity-sensitive genes in *H. pylori* may be modulated by supercoiling are unknown and certainly diverse, and may depend on various DNA binding proteins and transcription factors.

Expression of several *H. pylori* flagella-related genes is differentially regulated during the growth phases (Niehus *et al.*, 2002). This may be due, among other factors, to growth phase-dependent expression of transcription factors and growth-phase specific changes in DNA superhelicity (Ye *et al.*, 2007). One other superimposed factor recently reported to play a role in growth phase-related regulation of flagellar genes in *H. pylori* is quorum sensing, by a secreted autoinducer-2-like (AI-2) molecule produced through the enzymatic action of *H. pylori* LuxS (Rader *et al.*, 2007). Two reported effects of AI-2 were direct regulation of motility and of *flhA* transcript levels, but the latter appeared to have only little influence on flagellar morphology. Consistent with growth phase-dependent regulation of motility functions, *H. pylori* is only motile during the exponential growth phase, not in lag or stationary phase (Worku *et al.*, 1999). Evidence for differential regulation of *flaA* and *flaB* gene expression by specific environmental conditions is still lacking, but the transcription of an operon containing chemotaxis genes is influenced by the presence of copper in the medium and by temperature shifts from 37 to 45°C (Beier *et al.*, 1997). The role of metal ion (copper, iron – by the regulator Fur -, manganese) homeostasis and pH for gene regulation in *H. pylori* has been extensively investigated in the last few years by several groups. The bottom line of those studies is that the regulons dependent on environmental acidity, including the flagellar regulons, overlap partially with genes regulated by environmental metal ion availability which are governed by the two-component systems ArsRS and CrdRS, and by Fur (Gancz *et al.*, 2006; Pflock *et al.*, 2006; Waidner *et al.*, 2005; Danielli *et al.*, 2006; Loh and Cover, 2006). Since both acid-regulated and metal-regulated gene groups comprise motility and chemotaxis-related genes, we have to consider an extensive environmental influence overarching the regulons dependent on the above-described transcription factors and DNA superhelicity. DNA superhelicity may be the connecting factor, since it may depend on acidity and metal availability both, which are probably transmitted to the inside of the cell by two-component signal transduction systems such as ArsRS (Pflock *et al.*, 2006).

All clinical isolates of *H. pylori* are motile, but non-motile variants may accumulate rapidly during repeated *in vitro* passage. Non-motility in some strains results from the slipped-strand mispairing mutagenesis of a poly-cytidine repeat within the coding region of *fliP*, a gene that encodes a structural component of the export machinery in the basal body (Josenhans *et al.*, 2000). In the genome sequence of *H. pylori* strain 26695, the *fliP* repeat is C9 which leads to a premature stop codon. 26695 bacteria are non-motile and deficient in flagella production. Motile revertants contain a C8 repeat that restores the reading frame. Switching from non-motile to motile repeat length occurs with a frequency of approximately 10^{-4} to 10^{-5} per generation (Josenhans *et al.*, 2000). On–off switching regulation of the motility system warrants further investigation, because it might be important for *H. pylori* persistence and microecology in the host, but it has not yet been observed *in vivo*. Surprisingly, some variants of *H. pylori* and other *Helicobacter* species did not show phase variability in *fliP*, but in those cases, flagellar switching may occur by means of other flagellar genes such as *flhA* and *flgR*, as has been found in *Campylobacter jejuni* (Carrillo *et al.*, 2004; Hendrixson, 2006).

Posttranslational modification of flagellar proteins

Helicobacter flagellins are post-translationally modified by glycan residues which was first suspected since recombinantly expressed flagellins run faster in SDS-PAGE than the native proteins do (Josenhans *et al.*, 1999; Josenhans *et al.*, 2002b; Lee *et al.*, 2003). This was later confirmed by different biochemical methods, glycan staining (Josenhans *et al.*, 2002b) and mass spectrometry, analysis of glycosylation gene mutants and further biochemical characterization of isolated, protease-digested native flagellins (Schirm *et al.*, 2003). S. Logan and colleagues could show for the first time that *H. pylori* flagellins are modified by an unusual glycan, pseudaminic acid (Pse5Ac7Ac), and that only a few residues in the central portion (not in the terminal domains) of both flagellin proteins are variably modified (approximately 7–10 residues) by O-glycosylation, although numerous potential glycosylation sites are present throughout the whole molecule (Schirm *et al.*, 2003). This would suggest a glycosylation occurs after folding, which implies glycosylation coinciding with flagellar type III secretion, which has not been experimentally proven. A group of six enzymes encoded on the *H. pylori* chromosome which had been first annotated as capsule biosynthesis genes or flagellin-related genes have been confirmed to be involved in the glycosylation process (Josenhans *et al.*, 2002b) and (Schirm *et al.*, 2003; Schoenhofen *et al.*, 2006; indicated in Table 4.1). Protein glycosylation was earlier thought to be very rare in bacteria, but has since been shown to be quite frequent, in particular in *Campylobacter* spp. (Szymanski *et al.*, 2003). Also, flagellin glycosylation has meanwhile been demonstrated for many different bacteria (Eubacteria and Archaea), and is not specific for helicobacters (Logan, 2006; Verma *et al.*, 2006). In *H. pylori*, flagellin glycosylation is essential for flagellar assembly (Josenhans *et al.*, 2002b), which has also been demonstrated for a number of other bacteria (*Listeria, Caulobacter, Campylobacter* spp.). The exact function of the glycosylation for flagellar assembly in *H. pylori* is not known, and it is still possible that it serves more than one function, e.g. it may also be required for the stabilization of unfolded or folded flagellin subunits and their multimers in the cytoplasm or inside the flagellum.

Chemotaxis and behaviour

The chemotaxis system of *Helicobacter* spp. is still poorly characterized, although it has long been presumed that *H. pylori* must react to attractants and repellents in order to persist and to find its optimal niche within the stomach mucus and in relation to the gastric epithelium. In addition to a role in bacterial persistence, chemotactic abilities of the bacteria may also affect the host immune defence: various *H. pylori* chemotaxis mutants exerted distinct influences on the host immune system in a mouse model (Williams *et al.*, 2007).

The *Helicobacter* motility pattern has been studied in liquid culture and in motility agar plates by phase contrast microscopy and computerized motility tracking (so-called temporal assay), for example with a Hobson BacTracker (Eaton *et al.*, 1992; Foynes *et al.*, 2000; Hazell *et al.*, 1986; Karim *et al.*, 1998). *Helicobacter* chemotactic behaviour, similarly to the alternating phases of 'runs' and 'tumbling' observed in *E. coli* or *Salmonella*, consists of a combination of backward and forward movements, random darting movements with frequent changes in direction, and straight to curved runs, interspersed by stops. *In vitro*, *H. pylori* displays positive chemotaxis to moderate concentrations of urea, bicarbonate, and a structural analogue of urea, the urease inhibitor flurofamide (Mizote *et al.*, 1997). This positive chemotaxis behaviour may depend on a functional urease enzyme, especially in viscous media (Nakamura *et al.*, 1998). Intrabacterial but not extracellular urease activity was crucial for this function. A positive chemotactic movement towards hog gastric mucin, human plasma, and several amino acids has also been observed (Foynes *et al.*, 2000; Worku *et al.*, 2004). Human bile, aspartic acid and glutamic acid were reported to be repellents for *H. pylori* (Worku *et al.*, 2004).

A novel technology permits taking minute samples of mucus from defined positions within the mucus layer in order to study the distribution of bacteria within the mucus of anesthetized *H. felis*-infected mice or *H. pylori*-infected Mongolian gerbils. In the mouse model, the bacteria accumulated in a distinct horizontal layer of the stomach mucus, about 5–20 µm above the stomach epithelium (Schreiber *et al.*, 1999). The simplest explanation for such a specific location is the existence of one or several convergent vertical and horizontal chemical gradients that are detected by the bacteria to guide their accumulation. In a later study using the technique of nanosampling in the stomach of the Mongolian gerbil *in vivo*, the vertical mucus pH gradient above the gastric mucosa was shown to be required for proper orientation of *H. pylori* in a narrow layer (0– 15 µm) of the stomach mucus just above the mucosa (Schreiber *et al.*, 2004; see Fig. 4.4). Changes in other mucus vertical gradients, such as the bicarbonate or CO_2 gradients or a proposed urea gradient, did not affect the mucus distribution of *H. pylori* in the gerbil stomach. This suggests that horizontal orientation with respect to pH (pH taxis) is a dominant behaviour of *H. pylori in vivo*. Low pH in combination with pepsin, conditions which occur readily in the stomach lumen of healthy humans, quickly render *H. pylori* immotile (Schreiber *et al.*, 2005). This effect of low pH clearly highlights the necessity for *H. pylori* to avoid these conditions and to leave the luminal mucus layers as quickly as possibly to regain its niche within the deeper layers of the gastric mucus. Although this behaviour appears to make much sense for *H. pylori in vivo*, analysing and mechanistically explaining pH taxis as a bacterial behaviour is not a simple task. pH taxis as one particular mode of bacterial energy taxis (Edwards *et al.*, 2006) is closely linked to the energy status of the cell and to the proton gradient across the bacterial membrane which is generated by the respiratory chain and may be modulated by environmental

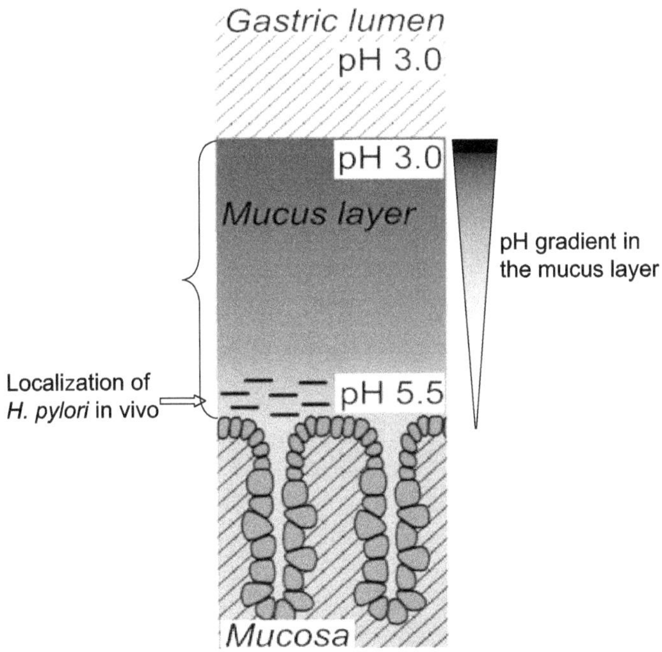

Figure 4.4 The spatial localization of *H. pylori* in the gastric mucus: role of the vertical pH gradient (modified from Schreiber *et al.*, 2004).

influences. pH sensing in *E. coli*, for example, is the result of sensing overlap between two distinct ways of bacterial energy sensing which are both metabolism dependent. In *E. coli*, pH or energetic changes would be sensed either as a change in proton motive force, or as a change in cellular redox potential (Edwards *et al.*, 2006). In *E. coli*, this is mediated by two different sensor proteins, Tsr and Aer, respectively, both of which are not present in *H. pylori*.

Genes in the complete *H. pylori* genomes have high homologies to chemotaxis genes from the model bacteria *Salmonella typhimurium* and *Caulobacter crescentus*. Homologues of all major components required for chemotaxis have been identified, including the receptor-coupled autokinase CheA, the linker protein CheW and the chemotaxis effector CheY (Tomb *et al.*, 1997; Alm *et al.*, 1999; Beier *et al.*, 1997; Foynes *et al.*, 2000; see also Table 4.2). However, chemotaxis in *H. pylori* does differ somewhat from other chemotaxis systems. A major difference from the *Salmonella* system is the lack of the CheB methylesterase and CheR methyltransferase. In *Salmonella*, these proteins regulate the chemotactic response and adaptation by methylation and demethylation of the membrane-bound receptors that are called methyl-accepting chemotaxis proteins (MCPs).

Among the helicobacters, *H. pylori* and *H. acinonychis* appear to be the only species lacking CheR and CheB. This lack indicates that the *H. pylori* chemotaxis cascade works without methylation of the receptor proteins, similar to *Sinorhizobium meliloti* (Sourjik and Schmitt, 1998) and *Rhodobacter sphaeroides* (Shah *et al.*, 1999). *H. pylori* possesses not only CheW and CheY (CheY1) homologues but also three *cheV* genes. In *B. subtilis*, *cheV* corresponds to a hybrid of CheW and CheY, implying that *H. pylori* has multiple CheW plus CheY equivalents. Furthermore, *H. pylori* expresses still another CheY (CheY2) orthologue which consists of a CheY domain fused to the C-terminus of CheA (Foynes *et al.*, 2000). Some of the five CheY domains might serve as high-affinity 'phosphate sinks' and act as terminators of the chemotaxis system (Foynes *et al.*, 2000; Jimenez-Pearson *et al.*, 2005). These are supposed to exert their function by dephosphorylating the specific CheY homologue(s) (which remains to be characterized) which may interact with the flagellar motor, thereby terminating the tumbling response and inducing the return to a straight run movement.

Table 4.2 Comparative genomics of the complete set of *H. pylori* chemotaxis-associated proteins, which are shown in the context of the complete sets of chemotaxis proteins of other epsilon-proteobacteria and of the enterobacterial species *E. coli* and *Salmonella* Typhimurium for comparison. The functional protein designations are included to the left of each column alongside the genome-specific gene numbers for easy identification in each bacterial species

Species	MCPs			Signal transduction proteins of chemotaxis pathway			Proteins involved In behavioural adaptation		
Helicobacter pylori	4	TlpA	HP0099	6	CheAY2	HP0392	1		
		TlpB	HP0103		CheY1	HP1067		CheZ*[a]	HP0170
		TlpC	HP0082		CheV1	HP0019			
		TlpD	HP0599		CheV2	HP0616			
					CheV3	HP0393			
					CheW	HP0391			
Helicobacter acinonychis	4	MCP	Hac_1409	6	CheA	Hac_0450	1		
			Hac_1462		CheY	Hac_1178		CheZ*[b]	Hac0351
			Hac_1478		CheV	Hac_0451			
			Hac_1524		CheW	Hac_0832			
						Hac_1689			
						Hac_0449			
Helicobacter hepaticus	12	TlpA	HH_0481	6	CheA	HH_0672	3		
		TlpB	HH_0725		CheY	HH_0825		CheB	HH_0456
		TlpC	NT02HP0764[c]		CheV1	HH_0819		CheR	HH_0455
		putative MCP	HH_0971		CheV2	HH_1802		CheZ*[b]	HH_0659
		TlpD	HH_1337		CheV3	HH_0671			
			HH_0513		CheW	HH_0673			
			HH_0736						
			HH_1088						
			HH_0004						
			HH_1155						
			HH_1310						
			HH_0891						
Campylobacter jejuni	12	Tlp1	Cj1506c	4	CheA	Cj0284	3		
		Tlp2	Cj0144		CheY	Cj1118		CheB	Cj0924
		Tlp3	Cj1564		CheV	Cj0285		CheR	Cj0925
		Tlp4	Cj0262c		CheW	Cj0283		CheZ*[b]	Cj0700
		Tlp5	Cj0246c						
		Tlp6	Cj0448c						
		Tlp7[d]	Cj0951c/Cj0952c						
		Tlp8	Cj1110c						
		Tlp9/CetA[e]	Cj1190c						
		Tlp10	Cj0019c						
		Aer1[g]	Cj1191c						
		Aer2/CetB[e,g]	Cj1189c						

Wolinella succinogenes	32	TlpA	WS0128	10	CheA2	WS1080	3		
		putative TlpB	WS0283		CheY	WS0619		CheYB[f]	WS1376
		TlpC	WS0843		CheY2	WS1077		CheR	WS1212
		TlpC2	WS0864		CheV	WS0257		CheZ*[b]	WS2123
		MCP	WS0929		CheV1	WS1916			
		putative MCP	WS0939		CheW	WS2084			
			WS1047		putative CheW	WS1808			
			WS1209			WS1078			
			WS1217			WS2082			
			WS1229			WS0095			
			WS1333						
			WS1845						
			WS1916						
			WS1951						
			WS1958						
			WS2163						
			WS0644						
			WS0761						
			WS1923						
			WS0432						
			WS0697						
			WS1520						
			WS0094						
			WS1956						
			WS2114						
			WS0281						
			WS0859						
			WS0949						
			WS1002						
			WS1406						
			WS1510						
			WS1527						

Table 4.2 *continued*

Species	MCPs			Signal transduction proteins of chemotaxis pathway			Proteins involved In behavioural adaptation	
Salmonella	13	Aer	STM3217	4	CheA	STM3670	3	
Typhimurium		CheM	STM1919		CheY	STM1916	CheB	STM1917
		LppB	STM1376		CheW	STM1920	CheR	STM1918
		Putative MCP	STM3138		CheV	STM2314	CheZ	STM1915
		Tcp	STM4210					
		Trg	STM4258					
		Tsr	STM3577					
		YeaJ	STM1626					
			STM1657					
			STM3216					
			STM3152					
			STM4533					
			STM1283					
Escherichia coli	5	Aer		3	CheA		3	
		Tsr			CheY		CheB	
		Trg			CheW		CheR	
		Tar					CheZ	
		Tap						

Abbreviations: Tlp, taxis sensor-like protein; MCP, methyl-accepting chemotaxis protein; Che, chemotaxis protein; Aer, aerotaxis-associated PAS domain-containing sensor protein. Except for the energy taxis proteins CetAB in *Campylobacter jejuni*, the sensing specificities of the diverse sensors in epsilon-proteobacteria are unknown.

[a]Found to be CheW mutant suppressor, remote homologue to enterobacterial CheZ.
[b]Identified by amino acid homology to *H. pylori* CheZ*/HP0170.
[c]Not identified in primary annotation.
[d]Tlp7 is depicted as one protein, assuming fusion of frameshifted ORFs Cj0951c and Cj0952c for a functional protein.
[e]CetA/CetB is a bipartite energy taxis system containing a PAS domain.
[f]WS1376 consists of a CheY and a CheB domain.
[g]*C. jejuni* Aer1 and Aer2 (CetB) do not contain a MCP signalling domain, but are annotated as PAS domain proteins.

Such a mechanism has been described for *S. meliloti* (Sourjik and Schmitt, 1998) and *R. sphaeroides* (Shah *et al.*, 2000) and could compensate for the apparent lack of an *H. pylori* homologue of the response terminator gene *cheZ*, whose gene product induces autodephosphorylation of CheY.

Recently, a novel chemotaxis-associated gene (HP0170/*cheZ**) which is a remote homologue of the missing *cheZ* was identified by a proteomics analysis of a *cheW* suppressor mutant (Terry *et al.*, 2006). Only four genes with homology to chemotaxis receptors and of unknown specificity have been identified in *H. pylori* (TlpA, TlpB, TlpC, TlpD). One of these (HP599/TlpD) seems to be a soluble cytoplasmic orthologue lacking the transmembrane domains. This paucity of chemotaxis receptors, which is quite unique in the Eubacterial kingdom, suggests that *H. pylori* uses few chemical gradients for orientation. The sensing specificities of the four annotated *H. pylori* chemosensors are not well studied, since few defined media for *H. pylori* are available and, since those are not minimal media, they are not well suited for taxis analyses. In confirmation of the above-mentioned result that *H. pylori* orients *in vivo* along a mucus pH gradient (Schreiber *et al.*, 2004), negative pH taxis *in vitro* (motility away from low pH) has also been discovered, and was found to be dependent on the specific sensor protein TlpB (Croxen *et al.*, 2006). The mechanistic basis of pH sensing by this sensor has not been clarified yet. Another possibility to perform energy or pH sensing are PAS domain proteins (such as Aer in *E. coli*). However, no member of this specific group of proteins has been identified in *H. pylori*. The *H. pylori* sensor TlpA has been subjected to heterologous expression in *E. coli*, and provided tactic movement towards serine, urea and bicarbonate (Cerda *et al.*, 2003).

Role of motility and flagella in colonization, pathogenesis and immune evasion

Motility is not a general requirement for successful gastrointestinal pathogens. For example, all Shigella are non-motile, as are some of the most widely spread serovars of enterohaemorrhagic *E. coli* (O157:H-, O26:H- etc.). Despite the lack of motility, the infectious doses of these bacteria are extremely low. Frequently, pathogenic bacteria express flagella only outside of the host and shut off flagellar biosynthesis once inside the host by specific regulatory mechanisms, sometimes dependent on the host's body temperature (Ottemann and Miller, 1997; Josenhans and Suerbaum, 2002). In contrast, results of three different animal models have shown that motility is essential for *H. pylori* infection. Early studies by Eaton and coworkers had demonstrated that a non-motile laboratory strain (Tx30a) was unable to colonize gnotobiotic piglets. Animal passage selects for motile and highly flagellated variants of *H. pylori* (Eaton *et al.*, 1992), and of *H. hepaticus*, accompanied by a concomitant increase in flagellin mRNA (Mankoski *et al.*, 1999; personal communication by J.G. Fox and coworkers; and own unpublished data).

Non-motile/aflagellate *H. pylori flaA/flaB* double mutants were unable to colonize gnotobiotic piglets for more than 2 days (Eaton *et al.*, 1996). *flaA/flaB* double mutants of *H. mustelae* were completely unable to colonize specific pathogen-free ferrets at any time point (Andrutis *et al.*, 1997).

The effects of the inactivation of single flagellin genes differed between *H. pylori* and *H. mustelae*. Both FlaA and FlaB flagellins are necessary for infectivity of *H. pylori* in a gnotobiotic piglet model. Single isogenic *flaA*- and *flaB*- mutants of the *H. pylori* strain N6 were recovered from gnotobiotic piglets 2 days and 4 days after inoculation, but not after 10 days, in contrast to the wild type. In contrast, single flagellin mutants of *H. mustelae* could establish persistent colonization in SPF ferrets. The density of colonization was initially much lower than that of the wild-type strain, but it increased with time. For *H. felis*, inactivation of the major flagellin gene *flaA* alone was sufficient to prevent colonization of mice (Josenhans *et al.*, 1999). A screen for *in vivo* essential genes in an infection model of *H. pylori* in the Mongolian gerbil yielded a wealth of motility and chemotaxis genes, indicating that in this model, which shares a lot of physiological and anatomical similarities with the human stomach, directed motility is a prime factor in colonization (Kavermann *et al.*, 2003). Infection experiments with signature-tagged mutant banks of *H. pylori* in two different strain backgrounds in mice confirmed numerous motility and taxis genes as *in vivo* essential genes (Baldwin *et al.*, 2007). In several mouse and gerbil infection experiments with *H. pylori* strains lacking in components of the chemotaxis system, poor colonization or lack of colonization in single strain infection or in competitive infection experiments with the wild type was observed (McGee *et al.*, 2005; Terry *et al.*, 2005).

The reasons for the differences between the diverse infection models for *H. pylori* are unknown. Since *H. felis* appears not to adhere to murine gastric epithelial cells, it makes sense that any reduction of motility should have pronounced effects on its ability to colonize. The differences between the *H. pylori*/piglet or mouse and the *H. mustelae*/ferret and *H. felis*/mouse models may simply reflect the fact that ferrets are the natural host of *H. mustelae*, as mice are with *H. felis*, whereas the piglet and mouse are experimental hosts for *H. pylori*. Hence, one of the major differences between the models is that all physiological adhesin receptors are present for the *H. mustelae*/ferret model but not necessarily for *H. pylori* in piglets, mice or gerbils. This interpretation is in agreement with mathematical models of *H. pylori* colonisation. Although the vast majority of the bacterial population is within the mucus, these models predict that adherent bacteria do make an important contribution to maintaining colonization (Blaser and Kirschner, 1999). It has been shown that motility is needed during different phases of the *H. pylori* infection, but not, whether there is also a phase of infection where motility is not required, as suggested by the *fliP* switch mechanism to shut off flagellar biosynthesis (Josenhans *et al.*, 2000).

Flagellar filaments do not seem to play a role in the adhesion of *Helicobacter*, because flagellin mutants of *H. pylori* and *H. mustelae* were unaffected in their ability to adhere to primary gastric epithelial cells, although a mutant in the pleiotropic flagellar master regulator, FlhA, showed diminished adherence (Clyne *et al.*, 2000). Secretion of virulence factors other than the flagellins via the flagellar type III secretion system in *H. pylori* seems to be possible, given recent results in *Campylobacter* (Poly *et al.*, 2007), but has not been demonstrated yet.

A new role of *Helicobacter* flagellar filaments and flagellins in the infection and probably pathogenesis emerged only recently: Since motility and flagella are required both for colonization and persistence of *H. pylori*, flagellar proteins need to be expressed continuously and they are in general not shut down during the infection as this may happen for numerous other pathogenic bacteria which undergo downregulation of motility functions *in vivo* (Josenhans and Suerbaum, 2002). Hence, one would expect that the continuous presence *in vivo* of *H. pylori* flagellins as potential ligands of the innate immune receptor TLR5 (Hayashi *et al.*, 2001; Smith *et al.*, 2003) would increase the inflammatory response to the *Helicobacter* infection greatly. However, this seems not to be the case, since all *H. pylori* and related bacteria seem to have developed flagellins which have a low activity on the innate immune system *via* TLR5 (Lee *et al.*, 2003; Gewirtz *et al.*, 2004; Andersen-Nissen *et al.*, 2005). This low activity seems to be intrinsic and provided by the protein sequence of the *Helicobacter* flagellins in their N and C termini, and not by variable post-translational modification (e.g. glycosylation). We hypothesize that this intrinsic low activity of flagellins may be one major mechanism of immune evasion by *H. pylori* and related bacteria, permitting persistent infection in the gastrointestinal tract. In addition, low-activity flagellins may not only be relevant as an immune evasion mechanism for persistence, but may even permanently suppress the immune response, thus contributing to chronic inflammation by these bacterial species and even to cancerogenesis.

Perspectives

Numerous questions regarding motility and chemotaxis in *Helicobacter* spp. remain open. Which effects on the immune system do the evasive properties of *Helicobacter* flagellins exert during the infection *in vivo*? What is the role(s) of the posttranslational modification of flagellins? Which substances does *H. pylori* detect for vertical and horizontal orientation in the mucus in addition to pH, and which chemotaxis sensor proteins are involved in sensing these gradients? Identification of these gradients and signal transduction pathways may provide an interesting new avenue for the specific treatment of *H. pylori* infection, which is inhibition of the motility and taxis system. Already now, with currently recommended treatment regimens, one of the components in the therapeutic regimen against *H. pylori*, omeprazole and related gastric proton pump inhibitors, may exert their assistant effect towards *H. pylori* antibiotic therapy by disturbing the pH gradient in the mucus and permitting a more efficient shedding of the bacteria from the stomach (Schreiber *et al.*, 2004). Another promising therapeutic approach against *H. pylori* in this context may be the specific inhibition of its flagellar type III secretion by chemical inhibitors, which may prevent bacterial persistence. In the near future, whole genome microarray hybridization technology, which has already been put to use successfully for elucidating major flagellar regulons, will allow the completion of the regulatory pathways in *H. pylori* including all regulons related to motility and chemotaxis. Microarray analyses will also further improve our understanding of how the *H. pylori* flagellar network is coordinated with other cell functions and integrates environmental signals without a master regulon. Finally, we still do not understand the selective advantage in certain environments of a complex filament composed of two or more flagellins, and the various *Helicobacter* species inhabiting different hosts and host niches provide good model systems to comparatively investigate this question.

Acknowledgements

We thank Sebastian Suerbaum for continuous collaboration and numerous valuable discussions. We are very grateful for collaboration and discussions concerning bacterial flagella, motility and taxis to Chi Aizawa, Mark Johnson, Kylie Watts, Barry Taylor, Soeren Schreiber, Kelly Hughes, Phil Aldridge, and Tomoko Mizote. Work in the laboratory on *H. pylori* motility and taxis was supported by grants from the Deutsche Forschungsgemeinschaft (Jo 344/2–2), the Federal Ministry for Education and Research (BMBF) and the European Union (INCA FP6, NoE EPG FP6) to C.J., and by a Bennigsen-Foerder award from the Ministry of Science and Education of North-Rhine Westphalia to C.J..

References

Aizawa, S.-I. (2000). Flagella. In Encyclopedia of Microbiology, J.Lederberg, ed. (New York: Academic Press).

Aizawa, S.I. (1996). Flagellar assembly in salmonella typhimurium [review]. Mol. Microbiol. *19*, 1–5.

Aizawa, S.I. and Kubori, T. (1998). Bacterial flagellation and cell division. Genes Cells 3, 625–634.

Alm, R.A., Ling, L.-S.L., Moir, D.T., King, B.L., Brown, E.D., Doig, P.C., Smith, D.R., Noonan B., Guild, B.C., deJonge, B.L., Carmel, G., Tummino, P.J., Caruso, A., Uria-Nickelsen M., Mills, D.M., Ives, C., Gibson, R., Merberg, D., Mills, S.D., Jiang, Q., Taylor, D.E., Vovis, G.F., and Trust, T.J. (1999). Genomic-sequence comparison of two unrelated isolates of the human gastric pathogen *Helicobacter pylori*. Nature *397*, 176–180.

Andersen-Nissen, E., Smith, K.D., Strobe, K.L., Barrett, S.L., Cookson, B.T., Logan, S.M., and Aderem, A. (2005). Evasion of Toll-like receptor 5 by flagellated bacteria. Proc. Natl. Acad. Sci. USA *102*, 9247–9252.

Andrutis, K.A., Fox, J.G., Schauer, D.B., Marini, R.P., Li, X., Yan, L., Josenhans, C., and Suerbaum, S. (1997). Infection of the ferret stomach by isogenic flagellar mutant strains of *Helicobacter mustelae*. Infect. Immun. *65*, 1962–1966.

Baar, C., Eppinger, M., Raddatz, G., Simon, J., Lanz, C., Klimmek, O., Nandakumar, R., Gross, R., Rosinus, A., Keller, H., Jagtap, P., Linke, B., Meyer, F., Lederer, H., and Schuster, S.C. (2003). Complete genome sequence and analysis of *Wolinella succinogenes*. Proc. Natl. Acad. Sci. USA *100*, 11690–11695.

Baldwin, D.N., Shepherd, B., Kraemer, P., Hall, M.K., Sycuro, L.K., Pinto-Santini, D.M., and Salama, N.R. (2007). Identification of *Helicobacter pylori* genes that contribute to stomach colonization. Infect. Immun. *75*, 1005–1016.

Beier, D. and Frank, R. (2000). Molecular characterization of two-component systems of *Helicobacter pylori*. J. Bacteriol. *182*, 2068–2076.

Beier, D., Spohn, G., Rappuoli, R., and Scarlato, V. (1997). Identification and characterization of an operon of *Helicobacter pylori* that is involved in motility and stress adaptation. J. Bacteriol. *179*, 4676–4683.

Beier, D., Spohn, G., Rappuoli, R., and Scarlato, V. (1998). Functional analysis of the *Helicobacter pylori* principal sigma subunit of RNA polymerase reveals that the spacer region is important for efficient transcription. Mol. Microbiol. *30*, 121–134.

Blaser, M.J. (1994). *Helicobacter pylori*: microbiology of a 'slow' bacterial infection. Trends Microbiol. *1*, 255–260.

Blaser, M.J. and Kirschner, D. (1999). Dynamics of *Helicobacter pylori* colonization in relation to the host response. Proc. Natl. Acad. Sci. USA *96*, 8359–8364.

Boonjakuakul, J.K., Canfield, D.R., and Solnick, J.V. (2005). Comparison of *Helicobacter pylori* virulence gene expression *in vitro* and in the Rhesus macaque. Infect. Immun. 73, 4895–4904.

Boonjakuakul, J.K., Syvanen, M., Suryaprasad, A., Bowlus, C.L., and Solnick, J.V. (2004). Transcription profile of *Helicobacter pylori* in the human stomach reflects its physiology in vivo. J. Infect. Dis. *190*, 946–956.

Brahmachary, P., Dashti, M.G., Olson, J.W., and Hoover, T.R. (2004). *Helicobacter pylori* FlgR is an enhancer-independent activator of sigma54-RNA polymerase holoenzyme. J. Bacteriol. *186*, 4535–4542.

Bren, A. and Eisenbach, M. (2000). How signals are heard during bacterial chemotaxis: protein-protein interactions in sensory signal propagation. J. Bacteriol. *182*, 6865–6873.

Carrillo, C.D., Taboada, E., Nash, J.H., Lanthier, P., Kelly, J., Lau, P.C., Verhulp, R., Mykytczuk, O., Sy, J., Findlay, W.A., Amoako, K., Gomis, S., Willson, P., Austin, J.W., Potter, A., Babiuk, L., Allan, B., and Szymanski, C.M. (2004). Genome-wide expression analyses of *Campylobacter* jejuni NCTC11168 reveals coordinate regulation of motility and virulence by FlhA. J. Biol. Chem. *279*, 20327–20338.

Cerda, O., Rivas, A., and Toledo, H. (2003). *Helicobacter pylori* strain ATCC700392 encodes a methyl-accepting chemotaxis receptor protein (MCP) for arginine and sodium bicarbonate. FEMS Microbiol. Lett. *224*, 175–181.

Clyne, M., O'Croinin, T., Suerbaum, S., Josenhans, C., and Drumm, B. (2000). Adherence of isogenic flagella negative mutants of *Helicobacter pylori* and *Helicobacter mustelae* to primary human and ferret gastric epithelial cells cells. Infect. Immun. *68*, 4335–4339.

Colland, F., Rain, J.C., Gounon, P., Labigne, A., Legrain, P., and de Reuse, H. (2001). Identification of the *Helicobacter pylori* anti-sigma28 factor. Mol. Microbiol. *41*, 477–487.

Croxen, M.A., Sisson, G., Melano, R., and Hoffman, P.S. (2006). The *Helicobacter pylori* chemotaxis receptor TlpB (HP0103) is required for pH taxis and for colonization of the gastric mucosa. J. Bacteriol. *188*, 2656–2665.

Danielli, A., Roncarati, D., Delany, I., Chiarini, V., Rappuoli, R., and Scarlato, V. (2006). In vivo dissection of the *Helicobacter pylori* Fur regulatory circuit by genome-wide location analysis. J. Bacteriol. *188*, 4654–4662.

Dasgupta, N., Wolfgang, M.C., Goodman, A.L., Arora, S.K., Jyot, J., Lory, S., and Ramphal, R. (2003). A four-tiered transcriptional regulatory circuit controls flagellar biogenesis in *Pseudomonas aeruginosa*. Mol. Microbiol. *50*, 809–824.

Dewhirst, F.E., Shen, Z., Scimeca, M.S., Stokes, L.N., Boumenna, T., Chen, T., Paster, B.J., and Fox, J.G. (2005). Discordant 16S and 23S rRNA gene phylogenies for the genus *Helicobacter*: implications for phylogenetic inference and systematics. J. Bacteriol. *187*, 6106–6118.

Eaton, K.A., Morgan, D.R., and Krakowka, S. (1992). Motility as a factor in the colonisation of gnotobiotic piglets by *Helicobacter pylori*. J. Med. Microbiol. *37*, 123–127.

Eaton, K.A., Suerbaum, S., Josenhans, C., and Krakowka, S. (1996). Colonization of gnotobiotic piglets by *Helicobacter pylori* deficient in two flagellin genes. Infect. Immun. *64*, 2445–2448.

Edwards, J.C., Johnson, M.S., and Taylor, B.L. (2006). Differentiation between electron transport sensing and proton motive force sensing by the Aer and Tsr receptors for aerotaxis. Mol. Microbiol. *62*, 823–837.

Eppinger, M., Baar, C., Linz, B., Raddatz, G., Lanz, C., Keller, H., Morelli, G., Gressmann, H., Achtman, M., and Schuster, S.C. (2006). Who ate whom? Adaptive *Helicobacter* genomic changes that accompanied a host jump from early humans to large felines. PLoS Genet. *2*, e120.

Falke, J.J. and Hazelbauer, G.L. (2001). Transmembrane signaling in bacterial chemoreceptors. Trends Biochem. Sci. *26*, 257–265.

Foynes, S., Dorrell, N., Ward, S.J., Stabler, R.A., McColm, A.A., Rycroft, A.N., and Wren, B.W. (2000). *Helicobacter pylori* possesses two CheY response regulators and a histidine kinase sensor, CheA, which are essential for chemotaxis and colonization of the gastric mucosa. Infect. Immun. *68*, 2016–2023.

Foynes, S., Dorrell, N., Ward, S.J., Zhang, Z.W., McColm, A.A., Farthing, M.J., and Wren, B.W. (1999). Functional analysis of the roles of FliQ and FlhB in flagellar expression in *Helicobacter pylori*. FEMS Microbiol. Lett. *174*, 33–39.

Fraser, G.M., Bennett, J.C., and Hughes, C. (1999). Substrate-specific binding of hook-associated proteins by FlgN and FliT, putative chaperones for flagellum assembly. Mol. Microbiol. *32*, 569–580.

Gancz, H., Censini, S., and Merrell, D.S. (2006). Iron and pH homeostasis intersect at the level of Fur regulation in the gastric pathogen *Helicobacter pylori*. Infect. Immun. *74*, 602–614.

Geis, G., Suerbaum, S., Forsthoff, B., Leying, H., and Opferkuch, W. (1993). Ultrastructure and biochemical studies of the flagellar sheath of *Helicobacter pylori*. J. Med. Microbiol. *38*, 371–377.

Gewirtz, A.T., Yu, Y., Krishna, U.S., Israel, D.A., Lyons, S.L., and Peek, R.M., Jr. (2004). *Helicobacter pylori* flagellin evades toll-like receptor 5-mediated innate immunity. J. Infect. Dis. *189*, 1914–1920.

Hayashi, F., Smith, K.D., Ozinsky, A., Hawn, T.R., Yi, E.C., Goodlett, D.R., Eng, J.K., Akira, S., Underhill, D.M., and Aderem, A. (2001). The innate immune response to bacterial flagellin is mediated by Toll-like receptor 5. Nature *410*, 1099–1103.

Hazell, S.L., Lee, A., Brady, L., and Hennessy, W. (1986). *Campylobacter* pyloridis and gastritis: association with intercellular spaces and adaptation to an environment of mucus as important factors in colonization of the gastric epithelium. J. Infect. Dis. *153*, 658–663.

Hendrixson, D.R. (2006). A phase-variable mechanism controlling the *Campylobacter jejuni* FlgR response regulator influences commensalism. Mol. Microbiol. *61*, 1646–1659.

Huff, J.L., Hansen, L.M., and Solnick, J.V. (2004). Gastric transcription profile of *Helicobacter pylori* infection in the rhesus macaque. Infect. Immun. *72*, 5216–5226.

Hughes, K.T., Gillen, K.L., Semon, M.J., and Karlinsey, J.E. (1993). Sensing structural intermediates in bacterial flagellar assembly by export of a negative regulator. Science *262*, 1277–1280.

Jagannathan, A., Constantinidou, C., and Penn, C.W. (2001). Roles of *rpoN*, *fliA*, and *flgR* in expression of flagella in *Campylobacter jejuni*. J. Bacteriol. *183*, 2937–2942.

Jenks, P.J., Foynes, S., Ward, S.J., Constantinidou, C., Penn, C.W., and Wren, B.W. (1997). A flagellar-specific ATPase (FliI) is necessary for flagellar export in *Helicobacter pylori*. FEMS. Microbiol. Lett. *152*, 205–211.

Jimenez-Pearson, M.A., Delany, I., Scarlato, V., and Beier, D. (2005). Phosphate flow in the chemotactic response system of *Helicobacter pylori*. Microbiology *151*, 3299–3311.

Jones, A.C., Logan, R.P.H., Foynes, S., Cockayne, A., Wren, B.W., and Penn, C.W. (1997). A flagellar sheath protein of *Helicobacter pylori* is identical to HpaA, a putative *N*-Acetylneuraminyllactose-Binding Hemagglutinin, but is not an adhesin for AGS Cells. J. Bacteriol. *179*, 5643–5647.

Josenhans, C., Eaton, K.A., Thevenot, T., and Suerbaum, S. (2000). Switching of flagellar motility in *Helicobacter pylori* by reversible length variation of a short homopolymeric sequence repeat in *fliP*, a gene encoding a basal body protein. Infect. Immun. *68*, 4598–4603.

Josenhans, C., Ferrero, R.L., Labigne, A., and Suerbaum, S. (1999). Cloning and allelic exchange mutagenesis of two flagellin genes from *Helicobacter felis*. Mol. Microbiol. *33*, 350–362.

Josenhans, C., Labigne, A., and Suerbaum, S. (1995). Comparative ultrastructural and functional studies of *Helicobacter pylori* and *Helicobacter mustelae* flagellin mutants: Both flagellin subunits, FlaA and FlaB, are necessary for full motility in *Helicobacter* species. J. Bacteriol. *177*, 3010–3020.

Josenhans, C., Niehus, E., Amersbach, S., Hörster, A., Betz, C., Drescher, B., Hughes, K.T., and Suerbaum, S. (2002a). Functional characterization of the antagonistic flagellar late regulators FliA and FlgM of *Helicobacter pylori* and their effects on the *H. pylori* transcriptome. Mol. Microbiol. *43*, 307–322.

Josenhans, C. and Suerbaum, S. (2002). The role of motility as a virulence factor in bacteria. Int. J. Med. Microbiol. *291*, 605–614.

Josenhans, C., Vossebein, L., Friedrich, S., and Suerbaum, S. (2002b). The *neuA/flmD* gene cluster of *Helicobacter pylori* is involved in flagellar biosynthesis and flagellin glycosylation. FEMS Microbiol. Lett. *210*, 165–172.

Karim, Q.N., Logan, R.P.P., Puels, J., Karnholz, A., and Worku, M.L. (1998). Measurement of motility of *Helicobacter pylori*, *Campylobacter* jejuni, and *Escherichia coli* by real time computer tracking using the Hobson BacTracker. J. Clin. Pathol. *51*, 623–628.

Kavermann, H., Burns, B.P., Angermuller, K., Odenbreit, S., Fischer, W., Melchers, K., and Haas, R. (2003). Identification and characterization of *Helicobacter pylori* genes essential for gastric colonization. J. Exp. Med. *197*, 813–822.

Kim, J.S., Chang, J.H., Chung, S.I., and Yum, J.S. (1999). Molecular cloning and characterization of the *Helicobacter pylori fliD* gene, an essential factor in flagellar structure and motility. J Bacteriol. *181*, 6969–6976.

Kim, Y.K. and McCarter, L.L. (2000). Analysis of the polar flagellar gene system of *Vibrio parahaemolyticus*. J. Bacteriol. *182*, 3693–3704.

Kostrzynska, M., Betts, J.D., Austin, J.W., and Trust, T.J. (1991). Identification, characterization, and spatial localization of two flagellin species in *Helicobacter pylori* flagella. J. Bacteriol. *173*, 937–946.

Kutsukake, K., Ohya, Y., and Iino, T. (1990). Transcriptional analysis of the flagellar regulon of *Salmonella typhimurium*. J. Bacteriol. *172*, 741–747.

Lane, M.C., O'Toole, P.W., and Moore, S.A. (2006). Molecular basis of the interaction between the flagellar export proteins FliI and FliH from *Helicobacter pylori*. J. Biol. Chem. *281*, 508–517.

Lee, S.K., Stack, A., Katzowitsch, E., Aizawa, S.I., Suerbaum, S., and Josenhans, C. (2003). *Helicobacter pylori* flagellins have very low intrinsic activity to stimulate human gastric epithelial cells via TLR5. Microbes. Infect. *5*, 1345–1356.

Logan, S.M. (2006). Flagellar glycosylation – a new component of the motility repertoire? Microbiology *152*, 1249–1262.

Loh, J.T. and Cover, T.L. (2006). Requirement of histidine kinases HP0165 and HP1364 for acid resistance in *Helicobacter pylori*. Infect. Immun. *74*, 3052–3059.

Lundstrom, A.M., Blom, K., Sundaeus, V., and Bolin, I. (2001). HpaA shows variable surface localization but the gene expression is similar in different *Helicobacter pylori* strains. Microb. Pathog. *31*, 243–253.

Macnab, R.M. (1996). Flagella and Motility. In *Escherischia coli* and *Salmonella*: cellular and molecular biology, F.C.Neidhardt, ed. (Washington D. C.: ASM Press), pp. 123–145.

Macnab, R.M. (2003). How bacteria assemble flagella. Annu. Rev. Microbiol. *57*, 77–100.

Mankoski, R., Hoepf, T., Krakowka, S., and Eaton, K.A. (1999). *flaA* mRNA transcription level correlates with *Helicobacter pylori* colonisation efficiency in gnotobiotic piglets. J Med Microbiol 48, 395–399.

McGee, D.J., Langford, M.L., Watson, E.L., Carter, J.E., Chen, Y.T., and Ottemann, K.M. (2005). Colonization and inflammation deficiencies in Mongolian gerbils infected by *Helicobacter pylori* chemotaxis mutants. Infect. Immun. *73*, 1820–1827.

McNally, D.J., Hui, J.P., Aubry, A.J., Mui, K.K., Guerry, P., Brisson, J.R., Logan, S.M., and Soo, E.C. (2006). Functional characterization of the flagellar glycosylation locus in *Campylobacter jejuni* 81–176 using a focused metabolomics approach. J. Biol. Chem. *281*, 18489–18498.

Minamino, T. and Macnab, R.M. (2000). Interactions among components of the *Salmonella* flagellar export apparatus and its substrates. Mol Microbiol 35, 1052–1064.

Mizote, T., Yoshiyama, H., and Nakazawa, T. (1997). Urease-independent chemotactic responses of *Helicobacter pylori* to urea, urease inhibitors, and sodium bicarbonate. Infect Immun. 65, 1519–1521.

Nakamura, H., Yoshiyama, H., Takeuchi, H., Mizote, T., Okita, K., and Nakazawa, T. (1998). Urease plays an important role in the chemotactic motility of *Helicobacter pylori* in a viscous environment. Infect Immun. 66, 4832–4837.

Nambu, T. and Kutsukake, K. (2000). The *Salmonella* FlgA protein, a putative periplasmic chaperone essential for flagellar P ring formation. Microbiology *146*, 1171–1178.

Nambu, T., Minamino, T., Macnab, R.M., and Kutsukake, K. (1999). Peptidoglycan-hydrolyzing activity of the FlgJ protein, essential for flagellar rod formation in *Salmonella typhimurium*. J Bacteriol *181*, 1555–1561.

Niehus, E., Gressmann, H., Ye, F., Schlapbach, R., Dehio, M., Dehio, C., Stack, A., Meyer, T.F., Suerbaum, S., and Josenhans, C. (2004). Genome-wide analysis of transcriptional hierarchy and feedback regulation in the flagellar system of *Helicobacter pylori*. Mol. Microbiol. *52*, 947–961.

Niehus, E., Ye, F., Suerbaum, S., and Josenhans, C. (2002). Growth phase-dependent and differential transcriptional control of flagellar genes in *Helicobacter pylori*. Microbiology *148*, 3827–3837.

O'Toole, P.W., Kostrzynska, M., and Trust, T.J. (1994). Non-motile mutants of *Helicobacter pylori* and *Helicobacter mustelae* defective in flagellar hook production. Mol. Microbiol. *14*, 691–703.

Oh, J.D., Kling-Backhed, H., Giannakis, M., Xu, J., Fulton, R.S., Fulton, L.A., Cordum, H.S., Wang, C., Elliott, G., Edwards, J., Mardis, E.R., Engstrand, L.G., and Gordon, J.I. (2006). The complete genome sequence of a chronic atrophic gastritis *Helicobacter pylori* strain: evolution during disease progression. Proc. Natl. Acad. Sci. USA *103*, 9999–10004.

Ottemann, K.M. and Lowenthal, A.C. (2002). *Helicobacter pylori* uses motility for initial colonization and to attain robust infection. Infect. Immun. *70*, 1984–1990.

Ottemann, K.M. and Miller, J.F. (1997). Roles for motility in bacterial-host interactions. Mol. Microbiol. *24*, 1109–1117.

Parkinson, J.S. (1993). Signal transduction schemes of bacteria. Cell *73*, 857–871.

Pereira, L.E., Brahmachary, P., and Hoover, T.R. (2006). Characterization of *Helicobacter pylori* sigma54 promoter-binding activity. FEMS Microbiol. Lett. *259*, 20–26.

Pflock, M., Finsterer, N., Joseph, B., Mollenkopf, H., Meyer, T.F., and Beier, D. (2006). Characterization of the ArsRS regulon of *Helicobacter pylori*, involved in acid adaptation. J. Bacteriol. *188*, 3449–3462.

Poly, F., Ewing, C., Goon, S., Hickey, T.E., Rockabrand, D., Majam, G., Lee, L., Phan, J., Savarino, N.J., and Guerry, P. (2007). Heterogeneity of a *Campylobacter jejuni* protein that is secreted through the flagellar filament. Infect. Immun. *75*, 3859–3867.

Porwollik, S., Noonan, B., and O'Toole, P.W. (1999). Molecular characterization of a flagellar export locus of *Helicobacter pylori*. Infect. Immun. *67*, 2060–2070.

Prouty, M.G., Correa, N.E., and Klose, K.E. (2001). The novel sigma54- and sigma28-dependent flagellar gene transcription hierarchy of *Vibrio cholerae*. Mol. Microbiol. *39*, 1595–1609.

Rain, J.C., Selig, L., de Reuse, H., Battaglia, V., Reverdy, C., Simon, S., Lenzen, G., Petel, F., Wojcik, J., Schachter, V., Chemama, Y., Labigne, A., and Legrain, P. (2001). The protein-protein interaction map of *Helicobacter pylori*. Nature *409*, 211–215.

Ryan, K.A., Karim, N., Worku, M., Moore, S.A., Penn, C.W., and O'Toole, P.W. (2005a). HP0958 is an essential motility gene in *Helicobacter pylori*. FEMS Microbiol. Lett. *248*, 47–55.

Ryan, K.A., Karim, N., Worku, M., Penn, C.W., and O'Toole, P.W. (2005b). *Helicobacter pylori* flagellar hook-filament transition is controlled by a FliK functional homolog encoded by the gene HP0906. J. Bacteriol. *187*, 5742–5750.

Schirm, M., Soo, E.C., Aubry, A.J., Austin, J., Thibault, P., and Logan, S.M. (2003). Structural, genetic and functional characterization of the flagellin glycosylation process in *Helicobacter pylori*. Mol. Microbiol. *48*, 1579–1592.

Schmitz, A., Josenhans, C., and Suerbaum, S. (1997). Cloning and characterization of the *Helicobacter pylori flbA* gene, which codes for a membrane protein involved in coordinated expression of flagellar genes. J. Bacteriol. *179*, 987–997.

Schoenhofen, I.C., McNally, D.J., Brisson, J.R., and Logan, S.M. (2006). Elucidation of the CMP-pseudaminic acid pathway in *Helicobacter pylori*: synthesis from UDP-N-acetylglucosamine by a single enzymatic reaction. Glycobiology *16*, 8C-14C.

Schreiber, S., Konradt, M., Groll, C., Scheid, P., Hanauer, G., Werling, H.O., Josenhans, C., and Suerbaum, S. (2004). The spatial orientation of *Helicobacter pylori* in the gastric mucus. Proc. Natl. Acad. Sci. USA *101*, 5024-5029.

Schreiber, S., Stüben, M., Josenhans, C., Scheid, P., and Suerbaum, S. (1999). *In vivo* distribution of *Helicobacter felis* in the gastric mucus of the mouse: Experimental method and results. Infect. Immun. *67*, 5151–5156.

Scott, D.R., Marcus, E.A., Wen, Y., Oh, J., and Sachs, G. (2007). Gene expression in vivo shows that *Helicobacter pylori* colonizes an acidic niche on the gastric surface. Proc. Natl. Acad. Sci. USA *104*, 7235–7240.

Shah, D.S., Porter, S.L., Harris, D.C., Wadhams, G.H., Hamblin, P.A., and Armitage, J.P. (2000). Identification of a fourth *cheY* gene in *Rhodobacter sphaeroides* and interspecies interaction within the bacterial chemotaxis signal transduction pathway. Mol. Microbiol. *35*, 101–112.

Shah, S., Qaqish, R., Patel, V., and Amiji, M. (1999). Evaluation of the factors influencing stomach-specific delivery of antibacterial agents for *Helicobacter pylori* infection. J. Pharm. Pharmacol *51*, 667–672.

Shapiro, L. (1995). The bacterial flagellum: from genetic network to complex architecture. Cell *80*, 525–527.

Smith, K.D., Andersen-Nissen, E., Hayashi, F., Strobe, K., Bergman, M.A., Barrett, S.L., Cookson, B.T., and Aderem, A. (2003). Toll-like receptor 5 recognizes a conserved site on flagellin required for protofilament formation and bacterial motility. Nat. Immunol. *4*, 1247–1253.

Solnick, J.V. and Schauer, D.B. (2001). Emergence of diverse *Helicobacter* species in the pathogenesis of gastric and enterohepatic diseases. Clin. Microbiol. Rev. *14*, 59–97.

Sourjik, V. and Schmitt, R. (1998). Phosphotransfer between CheA, CheY1, and CheY2 in the chemotaxis signal transduction chain of *Rhizobium meliloti*. Biochemistry *37*, 2327–2335.

Spohn, G. and Scarlato, V. (1999). Motility of *Helicobacter pylori* is coordinately regulated by the transcriptional activator FlgR, an NtrC homolog. J. Bacteriol. *181*, 593–599.

Stock, J. and Surette, M. (1994). Bacterial chemotaxis. The motor connection. Curr. Biol. *4*, 143–144.

Suerbaum, S. (1995). The complex flagella of gastric *Helicobacter* species. Trends Microbiol. 3, 168–170.

Suerbaum, S., Brauer-Steppkes, T., Labigne, A., Cameron, B., and Drlica, K. (1998). Topoisomerase I of *Helicobacter pylori*: juxtaposition with a flagellin gene (*flaB*) and functional requirement of a fourth zinc finger motif. Gene *210*, 151–161.

Suerbaum, S., Josenhans, C., and Labigne, A. (1993). Cloning and genetic characterization of the *Helicobacter pylori* and *Helicobacter mustelae flaB* flagellin genes and construction of *H. pylori flaA*- and *flaB*-negative mutants by electroporation-mediated allelic exchange. J. Bacteriol. *175*, 3278–3288.

Suerbaum, S., Josenhans, C., Sterzenbach, T., Drescher, B., Brandt, P., Bell, M., Droege, M., Fartmann, B., Fischer, H.-P., Ge, Z., Hörster, A., Holland, R., Klein, K., König, J., Macko, L., Mendz, G.L., Nyakatura, G., Schauer, D.B., Shen, Z., Weber, J., Frosch, M., and Fox, J.G. (2003). The complete genome sequence of the carcinogenic bacterium *Helicobacter hepaticus*. Proc. Natl. Acad. Sci. USA *100*, 7901–7906.

Szymanski, C.M., Logan, S.M., Linton, D., and Wren, B.W. (2003). *Campylobacter*-a tale of two protein glycosylation systems. Trends Microbiol. *11*, 233–238.

Terry, K., Go, A.C., and Ottemann, K.M. (2006). Proteomic mapping of a suppressor of non-chemotactic *cheW* mutants reveals that *Helicobacter pylori* contains a new chemotaxis protein. Mol. Microbiol. *61*, 871–882.

Terry, K., Williams, S.M., Connolly, L., and Ottemann, K.M. (2005). Chemotaxis plays multiple roles during *Helicobacter pylori* animal infection. Infect. Immun. *73*, 803–811.

Thibault, P., Logan, S.M., Kelly, J.F., Brisson, J.R., Ewing, C.P., Trust, T.J., and Guerry, P. (2001). Identification of the carbohydrate moieties and glycosylation motifs in *Campylobacter* jejuni flagellin. J. Biol. Chem. *276*, 34862–34870.

Tomb, J.-F., White, O., Kerlavage, A.R., Clayton, R.A., Sutton, G.G., Fleischmann, R.D., Ketchum, K.A., Klenk, H.P., Gill, S., Dougherty, B.A., Nelson, K., Quackenbush, J., Zhou, L., Kirkness, E.F., Peterson, S., Loftus, B., Richardson, D., Dodson, R., Khalak, H.G., Glodek, A., McKenney, K., Fitzegerald, L.M., Lee, N., Adams, M.D., Hickey, E.K., Berg, D.E., Gocayne, J.D., Utterback, T.R., Peterson, J.D., Kelley, J.M., Cotton, M.D., Weidman, J.M., Fujii, C., Bowman, C., Watthey, L., Wallin, E., Hayes, W.S., Borodovsky, M., Karp, P.D., Smith, H.O., Fraser, C.M., and Venter, J.C. (1997). The complete genome sequence of the gastric pathogen *Helicobacter pylori*. Nature *388*, 539–547.

Vanet, A., Marsan, L., Labigne, A., and Sagot, M.F. (2000). Inferring regulatory elements from a whole genome. An analysis of *Helicobacter pylori* sigma(80) family of promoter signals. J. Mol. Biol. *297*, 335–353.

Verma, A., Schirm, M., Arora, S.K., Thibault, P., Logan, S.M., and Ramphal, R. (2006). Glycosylation of b-Type flagellin of *Pseudomonas aeruginosa*: structural and genetic basis. J. Bacteriol. *188*, 4395–4403.

Wadhams, G.H. and Armitage, J.P. (2004). Making sense of it all: bacterial chemotaxis. Nat. Rev. Mol. Cell Biol. *5*, 1024–1037.

Waidner, B., Melchers, K., Stahler, F.N., Kist, M., and Bereswill, S. (2005). The *Helicobacter pylori* CrdRS two-component regulation system (HP1364/HP1365) is required for copper-mediated induction of the copper resistance determinant CrdA. J. Bacteriol. *187*, 4683–4688.

Wand, M.E., Sockett, R.E., Evans, K.J., Doherty, N., Sharp, P.M., Hardie, K.R., and Winzer, K. (2006). *Helicobacter pylori* FlhB function: the FlhB C-terminal homologue HP1575 acts as a 'spare part' to permit flagellar export when the HP0770 FlhBCC domain is deleted. J. Bacteriol. *188*, 7531–7541.

Williams, S.M., Chen, Y.T., Andermann, T., Carter, J.E., McGee, D.J., and Ottemann, K.M. (2007). *Helicobacter pylori* chemotaxis modulates inflammation and gastric-epithelium interactions in infected mice. Infect. Immun.

Worku, M.L., Karim, Q.N., Spencer, J., and Sidebotham, R.L. (2004). Chemotactic response of *Helicobacter pylori* to human plasma and bile. J. Med. Microbiol. *53*, 807–811.

Worku, M.L., Sidebotham, R.L., Walker, M.M., Keshavarz, T., and Karim, Q.N. (1999). The relationship between *Helicobacter pylori* motility, morphology and phase of growth: implications for gastric colonization and pathology. Microbiology *145*, 2803–2811.

Ye, F., Brauer, T., Niehus, E., Drlica, K., Josenhans, C., and Suerbaum, S. (2007). Flagellar and global gene regulation in *Helicobacter pylori* modulated by changes in DNA supercoiling. Int. J. Med. Microbiol. *297*, 65–81.

Helicobacter pylori Vacuolating Toxin

5

Steven R. Blanke and Timothy L. Cover

Abstract

Helicobacter pylori VacA is a protein toxin that is secreted into the extracellular space by a type V autotransporter mechanism. Specific allelic variants of *vacA* exhibit different levels of toxin activity *in vitro* and are associated with different risks of gastroduodenal disease in *H. pylori*-infected humans. The structural features and the cell-modulating activities of VacA are distinct from those of any other known bacterial toxin. VacA can exert a wide array of effects on gastric epithelial cells, including alteration of endocytic compartments, alteration of mitochondrial membrane permeability, and activation of signal transduction pathways. Moreover, VacA inhibits activation and proliferation of T-cells and can have effects on several other types of immune cells. VacA contributes to the ability of *H. pylori* to colonize the stomach and also contributes to the pathogenesis of *H. pylori*-associated peptic ulcer disease and gastric cancer. VacA displays both immunosuppressive as well as proinflammatory activities *in vitro*, emphasizing the potentially complex role of this toxin *in vivo*.

Introduction

In 1988, Leunk and colleagues reported that a proteinaceous factor present in broth culture filtrates from *Helicobacter pylori* induced the formation of cytoplasmic vacuoles in several cultured cell lines (Leunk *et al.*, 1988). Since the initial discovery of this protein, which was designated vacuolating toxin or VacA, there has been a high level of interest in this toxin for several reasons. First, the structural features and the cell-modulating activities of this toxin are distinct from those of any other known bacterial toxin (de Bernard *et al.*, 2004; Cover and Blanke, 2005; Rieder *et al.*, 2005). Second, there is evidence that VacA contributes to the ability of *H. pylori* to colonize the stomach and contributes to the pathogenesis of *H. pylori*-associated gastroduodenal diseases. Third, the gene encoding VacA (*vacA*) is characterized by a high degree of genetic variation; specific allelic variants of *vacA* exhibit different levels of toxin activity *in vitro* and are associated with different risks of gastroduodenal disease in *H. pylori*-infected humans. Finally, VacA is a candidate antigen for inclusion in a vaccine against *H. pylori* (Marchetti *et al.*, 1995; Marchetti *et al.*, 1998).

Expression, secretion and maturation of *VacA*

Regulation of vacA expression

A single chromosomal copy of the *vacA* gene is present in essentially all *H. pylori* strains (Cover *et al.*, 1994; Phadnis *et al.*, 1994; Schmitt and Haas, 1994; Telford *et al.*, 1994). Among strains that have been characterized, the gene encoding VacA varies in length from approximately 3860 to 3940 nucleotides (Cover *et al.*, 1994; Phadnis *et al.*, 1994; Schmitt and Haas, 1994; Telford *et al.*, 1994; Atherton *et al.*, 1995; Ito *et al.*, 1998; Pagliaccia *et al.*, 1998). The *vacA* transcript is monocistronic, with a transcriptional start point located approximately 120 nucleotides upstream from the ATG start site (Schmitt and Haas, 1994; Forsyth *et al.*, 1998; Forsyth and Cover,

1999). The initiation of *vacA* transcription apparently involves a promoter that contains both an extended −10 sequence and a −35 binding site, despite the lack of a −35 consensus sequence for *H. pylori* promoters (Forsyth and Cover, 1999). There is considerable variation among *H. pylori* strains in levels of *vacA* transcription (Forsyth and Cover, 1999; Ayala *et al.*, 2004).

VacA transcription has been studied both *in vitro* and *in vivo*. When *H. pylori* is cultured in liquid medium, the level of *vacA* transcription increases about twofold as the cultures enter stationary phase, compared with the level of *vacA* transcription during log phase (Forsyth and Cover, 2000; Boonjakuakul *et al.*, 2004). Transcriptional analysis of gastric biopsy specimens collected from *H. pylori*-infected human patients revealed that *vacA* was among the most highly expressed of the genes screened, which included 18 putative virulence factors (Boonjakuakul *et al.*, 2004). In studies of gastric biopsies from *H. pylori*-infected rhesus macaques, higher levels of *vacA* transcripts were detected at 1 week post-infection than at 4–6 weeks post infection, suggesting that there may be an important role of VacA during the early stages of colonization (Boonjakuakul *et al.*, 2005).

In comparison to pathogens such as *Vibrio cholerae* or *Pseudomonas aeruginosa*, which boast numerous and overlapping regulatory transcriptional networks, *H. pylori* strains possess relatively few transcriptional regulatory systems. Nonetheless, several lines of evidence suggest that *vacA* transcription may be regulated. *vacA* transcription is altered when *H. pylori* is co-cultured with AGS (Gieseler *et al.*, 2005) or HM02 (van Amsterdam *et al.*, 2003) gastric epithelial cells. The expression of *vacA* is also altered when *H. pylori* is co-cultured with granulocytes, but the observed changes are strain-dependent (Gieseler *et al.*, 2005). Two studies reported that exposure of *H. pylori* to acidic pH results in alteration of *vacA* transcription (Merrell *et al.*, 2003a; Barnard *et al.*, 2004), but other studies of acid-induced gene transcription in *H. pylori* have not detected any changes in *vacA* expression (Allan *et al.*, 2001; Ang *et al.*, 2001). Several independent studies indicated that *vacA* transcription is upregulated by iron depletion *in vitro* (Szczebara *et al.*, 1999; Merrell *et al.*, 2003b), suggesting the possibility that during conditions of iron starvation *in vivo*, increased VacA production might allow *H. pylori* to more readily access iron released from toxin-damaged cells.

Very little is known about which regulatory elements might influence VacA expression. Iron-induced transcriptional regulation of *vacA* was reported to be Fur-independent (Szczebara *et al.*, 1999). One study reported that mutagenesis of the global regulatory protein CsrA (carbon storage regulator A) resulted in the deregulation of the acid induction of *vacA* (Barnard *et al.*, 2004). The effects of CsrA in *H. pylori* seem to occur at the post-transcriptional level by influencing the processing and translation of target transcripts, with minimal effects on transcript stability.

Secretion and proteolytic processing

The *vacA* gene of *H. pylori* strain 60190 encodes a 1287-amino-acid protoxin (Cover *et al.*, 1994), which, analogous to several other bacterial protein toxins, must undergo maturation to yield the extracellular active form of the toxin (Fig. 5.1A). In the case of VacA, the protoxin undergoes at least two distinct proteolytic processing steps during the secretion process, to yield a mature toxin with a mass of 88.2 kDa, consisting of approximately 821 amino acids (Cover and Blaser, 1992; Nguyen *et al.*, 2001).

The first processing step is proteolytic cleavage of a 33-amino-acid amino-terminal signal sequence, which presumably occurs as the toxin is transported across the cytoplasmic membrane and into the periplasmic space. After entering the periplasm, VacA is believed to be secreted across the outer membrane by a mechanism idiosyncratic to members of the autotransporter family of Gram-negative bacterial secreted proteins (Dautin and Bernstein, 2006). During the secretion process, VacA undergoes further proteolytic processing by one or more unidentified proteases, resulting in an 88 kDa amino-terminal 'passenger domain' (which is the mature form of the toxin), a 12 kDa secreted peptide, and a ~33 kDa carboxy-terminal domain that remains associated with the bacterial outer membrane (Schmitt and Haas, 1994; Telford *et al.*, 1994; Nguyen *et al.*, 2001; Bumann *et al.*, 2002). Analogous to other members of the autotransporter protein family, the VacA carboxyl-terminal domain contains a

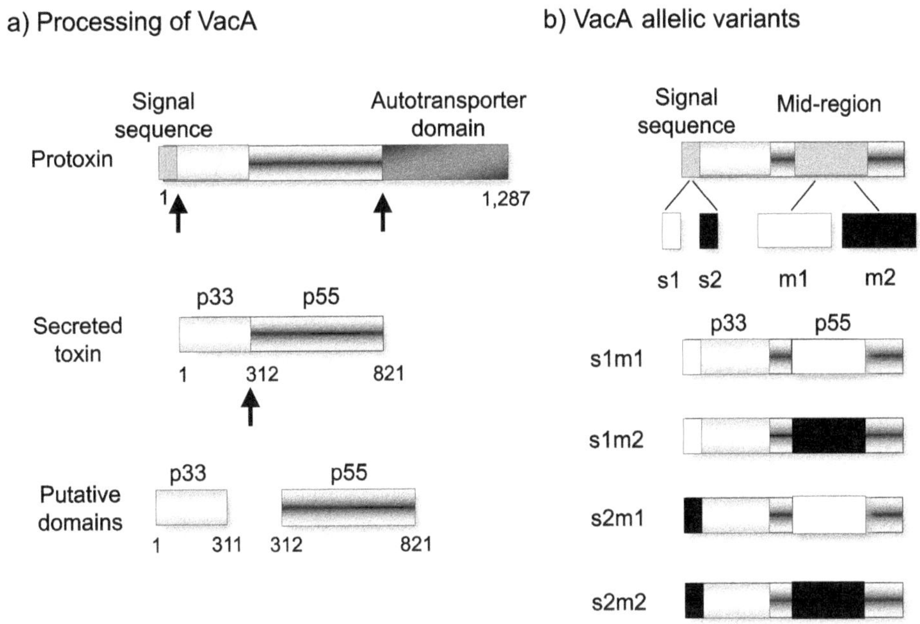

Figure 5.1 *vacA* gene structure and allelic diversity. (A) The 140 kDa VacA protoxin undergoes cleavage of an amino-terminal signal sequence and cleavage of a carboxyl-terminal domain to yield an 88 kDa mature secreted toxin that is secreted into the extracellular space. The structure of the *vacA* gene and the secretion and proteolytic processing of the VacA protein are characteristic of a family of proteins secreted by Gram-negative bacteria, known as 'autotransporters'. A 33 kDa C-terminal beta-barrel domain of VacA is predicted to insert into the outer membrane and form a channel, through which the mature VacA toxin is secreted. (B) There is a high level of diversity among *vacA* alleles from different *H. pylori* strains. Allelic diversity is particularly striking near the 5′ end of *vacA* (termed the s-region) and in the mid-region of the gene (termed the m-region). Two main families of s-region sequences (s1 and s2) and two main families of m-region sequences (m1 and m2) have been described. Type s1 and s2 VacA proteins differ in the site at which amino-terminal signal sequences are cleaved, and consequently, the mature secreted type s2 VacA toxin contains a 12-amino-acid hydrophilic extension at its amino-terminus which is absent from type s1 VacA toxins. VacA toxins that contain this 12-amino-acid hydrophilic extension fail to include cell vacuolation *in vitro*. The amino acid sequences of type m1 and m2 VacA proteins are about 65% identical within a region comprising 250 amino acids. Mosaic forms of *vacA* are thought to arise via homologous recombination among *vacA* alleles from different strains (adapted from Cover and Blanke, 2005).

canonical phenylalanine-containing motif and is predicted to have a beta-barrel structure (Cover *et al.*, 1994; Schmitt and Haas, 1994). Moreover, similar to several other passenger domains, the secreted VacA passenger domain contains a single pair of cysteine residues, which has been demonstrated to be required for high levels of VacA secretion, but not for VacA activity (Letley *et al.*, 2006). The most compelling evidence that VacA is an autotransporter protein is the demonstration that the predicted beta-barrel carboxyl-terminal domain of VacA can transport the cholera toxin B subunit to the external face of the *H. pylori* outer membrane (Fischer *et al.*, 2001).

Maturation of VacA culminates with the release of 88 kDa VacA toxin molecules from the bacteria as soluble proteins into the extracellular space (Cover and Blaser, 1992). In addition to secretion as a soluble protein, VacA can remain localized on the surface of *H. pylori*, spatially organized into distinct toxin-rich domains (Ilver *et al.*, 2004). Upon bacterial contact with host cells, toxin clusters can be transferred directly from the bacterial surface to host cells at the bacteria-cell interface (Ilver *et al.*, 2004). The toxin can also be released from the bacteria in the form of membrane blebs (Fiocca *et al.*, 1999; Keenan *et al.*, 2000). The mechanisms that govern release of VacA as a soluble protein and retention of VacA on the bacterial surface have not been elucidated. Presumably the capacity of either released or bacterial-associated VacA to interact with cells provides a means for *H. pylori* to modulate host cells adjacent to the bacteria as well as cells located at more distant sites. In addition, the presence of VacA on the bacterial surface may promote *H. pylori* adherence to gastric epithelial cells.

Properties of the 88 kDa secreted protein

The secreted VacA protein is an 88 kDa monomer under denaturing conditions (Cover and Blaser, 1992), but under non-denaturing conditions, VacA monomers can assemble into large water-soluble oligomeric structures. By using a variety of microscopic techniques, it has been possible to visualize multiple types of oligomeric VacA structures, including single layered forms (containing 6–9 subunits), bilayered forms (containing 12 or 14 subunits), and two-dimensional crystals (Lupetti *et al.*, 1996; Cover *et al.*, 1997; Lanzavecchia *et al.*, 1998; Czajkowsky *et al.*, 1999; Adrian *et al.*, 2002; El-Bez *et al.*, 2005). Upon exposure to either acidic or alkaline pH, VacA oligomeric complexes dissociate into monomeric components, but can reassemble into oligomeric structures following neutralization of the pH (Cover *et al.*, 1997; Molinari *et al.*, 1998a; Yahiro *et al.*, 1999).

The secreted 88-kDa toxin can undergo limited proteolytic nicking to yield two fragments, designated p33 and p55 (Telford *et al.*, 1994; Cover *et al.*, 1997; Ye *et al.*, 1999; Nguyen *et al.*, 2001; Willhite *et al.*, 2002; Torres *et al.*, 2004; Torres *et al.*, 2005) (Fig. 5.1A). There is no evidence that proteolytic cleavage into p33 and p55 fragments is required for VacA activity, but these two fragments are considered to represent two domains or subunits of VacA. Several studies indicate that the p55 domain has an important role in binding of VacA to host cells (see 'Interaction of VacA with host cells (binding, internalization, trafficking, localization)', below) (Garner and Cover, 1996; Pagliaccia *et al.*, 1998; Reyrat *et al.*, 1999; Wang and Wang, 2000; Wang *et al.*, 2001). About 422 amino acids at the amino-terminal end of VacA (which includes the p33 domain and about 100 amino acids of the p55 domain) are sufficient to induce cell vacuolation if VacA is expressed intracellularly in transiently transfected cells (de Bernard *et al.*, 1997; de Bernard *et al.*, 1998; Ye *et al.*, 1999). The p33 domain is predicted to contain a strongly hydrophobic region near the amino-terminus, and this region is required for membrane channel formation and many of the toxin's cell-modulating effects (see 'Membrane channel formation by VacA', 'Interaction of VacA with host cells (binding, internalization, trafficking, localization)' and 'Effects of VacA on host cells *in vitro*', below) (de Bernard *et al.*, 1998; Vinion-Dubiel *et al.*, 1999; Ye and Blanke, 2000; McClain *et al.*, 2003).

vacA genetic diversity

Diversity among *vacA* alleles

Although essentially all *H. pylori* strains have a copy of the *vacA* gene, early studies revealed marked differences in the levels of vacuolating toxin activity present in culture filtrates generated from different *H. pylori* strains (Leunk *et al.*, 1988; Figura *et al.*, 1989; Cover *et al.*, 1990; Cover and Blaser, 1992; Cover *et al.*, 1993a). Subsequent investigations have suggested that there are likely multiple explanations underlying the observed variation in vacuolating toxin activity. Some *H. pylori* isolates fail to express a functionally active VacA protein due internal duplications, deletions, 1-bp insertions, or nonsense mutations (Ito *et al.*, 1998). Among strains that express a VacA protein, the level of vacuolating toxin activity present in culture filtrates may be influenced by differences in *vacA* transcription or the efficiency of VacA secretion (Forsyth *et al.*, 1998). Variation among VacA amino acid sequences also influences the level of vacuolating toxic activity exhibited by VacA proteins (Atherton *et al.*, 1995). Analysis of *vacA* sequences from many different *H. pylori* strains has revealed several families of *vacA* alleles, which, as discussed below, are associated with differences in vacuolating toxic activity (Atherton *et al.*, 1995; Strobel *et al.*, 1998; van Doorn *et al.*, 1998a).

The section of *vacA* that exhibits maximum sequence diversity is an approximately 800-base-pair region located in the middle of the gene. This so-called 'm (for middle) region' encodes part of the 55 kDa domain (p55) of VacA (Fig. 5.1B). Two primary allelic families have been differentiated within the m region, and are designated m1 and m2 (Atherton *et al.*, 1995; Pagliaccia *et al.*, 1998; Atherton *et al.*, 1999; Ji *et al.*, 2002).

Substantial sequence diversity is also located at the 5′ end of *vacA*, referred to as the 's region,' which includes the amino-terminal VacA signal sequence and the amino terminus of the processed, mature toxin (Fig. 5.1B) (Atherton *et al.*, 1995; Strobel *et al.*, 1998; van Doorn *et al.*, 1998a). Two main allelic families are recognized, designated s1 and s2, and the s1 allelic type can be further divided into s1a, s1b, and s1c subtypes. Certain subtypes are found in limited geographic distributions (Van Doorn *et al.*, 1999). For example, s1c subtypes are found mainly in *H. pylori* strains isolated in East Asia (Van Doorn *et al.*, 1999).

vacA genes comprising all possible combinations of s and m allelic types (s1a/m1, s1b/m1, s1c/m1, s1a/m2, etc.) can theoretically arise, as a result of recombination among *vacA* alleles. Most such combinations have been detected commonly in *H. pylori* strains isolated from infected humans, although s2/m1 forms of *vacA* are quite rare (van Doorn *et al.*, 1998a; Letley *et al.*, 1999).

vacA polymorphisms and toxin activity

The *vacA* genotype is recognized as an important determinant of vacuolating activity. Specifically, strains containing a type s2 *vacA* allele consistently fail to induce cellular vacuolation (Atherton *et al.*, 1995). VacA proteins with a s2 signal sequence undergo amino-terminal signal peptide cleavage at a different site compared with *vacA* proteins with a s1 signal sequence (Atherton *et al.*, 1995). As a consequence, type s2 VacA proteins retain a 12-amino-acid hydrophilic peptide at the amino-terminus, which blocks toxin-medicated cell vasculation (Letley and Atherton, 2000; McClain *et al.*, 2001b; Letley *et al.*, 2003).

Diversity in the mid-region of *vacA* alleles has been linked to differences among VacA proteins in cell tropism (see 'Binding of VacA to the cell surface,' below) (Pagliaccia *et al.*, 1998; Ji *et al.*, 2000; Wang *et al.*, 2001). Type s1/m1 VacA proteins induce cell vacuolation of many epithelial cell lines, including HeLa cells. In contrast, several s1/m2 VacA proteins exhibit very low levels of vacuolating activity on HeLa cells, but cause vacuolation of RK-13 cells and primary gastric epithelial cells (Pagliaccia *et al.*, 1998; Ji *et al.*, 2000; Wang *et al.*, 2001). The difference in cell type specificity of s1/m1 and s1/m2 VacA proteins appears attributable, at least in part, to different cell-binding properties (Pagliaccia *et al.*, 1998; Ji *et al.*, 2000; Wang *et al.*, 2001). The VacA sequences that determine the cell type specificity of type m1 and m2 VacA proteins have been mapped to a 148-amino-acid region within the p55 domain (Ji *et al.*, 2000; Skibinski *et al.*, 2006), although the complete m-region is apparently required for VacA binding to cells and vacuolating toxin activity (Wang and Wang, 2000; Skibinski et *al.*, 2006).

Strains containing type s1/m1 and s2/m2 *vacA* alleles typically differ not only in VacA amino acid sequences but also in the level of *vacA* transcription (Forsyth *et al.*, 1998). The levels of *vacA* transcription are typically higher in strains containing s1/m1 alleles than in strains containing s2/m2 alleles. Sequence analysis and promoter exchange experiments indicated that the different levels of *vacA* transcription in these two groups of strains cannot be explained solely by a difference in promoter strength (Forsyth *et al.*, 1998). Similarly, an analysis of wild-type *H. pylori* strains detected up to 28-fold differences in levels of *vacA* expression among different strains, but the differences in expression were unrelated to detectable differences in putative −35 and −10 motifs or other untranslated sequences upstream of the ATG start site (Ayala *et al.*, 2004). The lack of correlation between *vacA* promoter sequences and levels of *vacA* transcription suggests the potential existence of a strain-specific regulatory factor that modulates *vacA* transcription.

Associations of *vacA* genotypes with disease

Efforts to evaluate potential roles of VacA in the pathogenesis of human gastroduodenal diseases have largely focused on searching for correlations between *vacA* allelic types and the development of disease in *H. pylori*-infected humans. These studies have used a broad range of sampling and analytical methodologies, and interpretation of the results has been limited by the potential for numerous confounding factors. Nevertheless, many of these studies have demonstrated a correlation between particular *vacA* m- and s-region allelic types and occurrence of gastroduodenal disease in *H. pylori*- infected humans.

Numerous studies conducted in Europe, North America, and South America have concluded that strains containing type s1 *vacA* alleles are more commonly associated with peptic ulceration than are strains containing type s2 *vacA* alleles (Atherton *et al.*, 1995; Atherton *et al.*, 1997; Han *et al.*, 1998; Rudi *et al.*, 1998; Strobel *et al.*, 1998; van Doorn *et al.*, 1998b; Gerhard *et al.*, 1999; Kidd *et al.*, 1999; Van Doorn *et al.*, 1999; De Gusmao *et al.*, 2000). Similarly, individuals infected with *H. pylori* harbouring s1 alleles are at higher risk for developing gastric carcinoma than are those with strains harbouring *vacA* s2 alleles (Miehlke *et al.*, 2000; Ashour *et al.*, 2002; Figueiredo *et al.*, 2002).

Several studies have shown that strains harbouring m1 *vacA* alleles are more highly associated with gastric carcinoma than are those harbouring m2 *vacA* alleles (Miehlke *et al.*, 2000; Ashour *et al.*, 2002; Figueiredo *et al.*, 2002). In comparison with *H. pylori* strains containing m2 *vacA* alleles, strains with an m1 allele are also more highly associated with histological abnormalities preceding the onset of gastric cancer, including epithelial damage, atrophic gastritis, and intestinal metaplasia (Atherton *et al.*, 1997; Nogueira *et al.*, 2001). The m-type of *vacA* alleles is not considered to be a strong independent predictor for development of peptic ulcer disease (van Doorn *et al.*, 1998b).

Because nearly all *H. pylori* clinical isolates from Asian countries harbour *vacA* s1 allele types (Van Doorn *et al.*, 1999), an association between type s1 *vacA* and disease is not readily apparent in many East Asian countries (Pan *et al.*, 1998). Nevertheless, it is striking that in Japan, a country with a very high incidence of gastric adenocarcinoma, nearly all *H. pylori* strains contain type s1/m1 *vacA* alleles (which encode forms of VacA that exhibit high levels of toxin activity *in vitro*).

The pathophysiological mechanism(s) underlying the observed associations between *vacA* genotypes and gastroduodenal disease are not yet entirely clear. However, the demonstration that s2 forms of VacA lack vacuolating toxin activity in cell culture assays (Atherton *et al.*, 1995; Letley and Atherton, 2000; McClain *et al.*, 2001b; Letley *et al.*, 2003) provides a possible functional explanation for the rarity of gastric disease and lack of gastric epithelial damage in patients infected with *H. pylori* strains containing s2 forms of VacA.

Several putative *H. pylori* virulence factors, including the *cag* pathogenicity island, an adhesin known as BabA, in-frame *oipA* (*hopH*) alleles, and type I *hopQ* alleles, are present in *H. pylori* strains containing type s1 *vacA* alleles more frequently than in *H. pylori* strains containing type s2 *vacA* alleles (Atherton *et al.*, 1995; van Doorn *et al.*, 1998b; Gerhard *et al.*, 1999; Cao and Cover, 2002; Yamaoka *et al.*, 2002; Dossumbekova *et al.*, 2006). It is difficult to rank the relative importance of these individual bacterial factors in the pathogenesis of human gastroduodenal disease. The development of gastroduodenal disease in *H. pylori*-infected humans is probably not attributable to the actions of a single bacterial virulence factor, but, instead, multiple bacterial factors may contribute to the development of disease. In addition to a role of bacterial virulence factors, various host and environmental factors are also likely to have important roles in the pathogenesis of *H. pylori*-associated diseases (Figueiredo *et al.*, 2002).

Membrane channel formation by VacA

Like many other bacterial toxins, VacA is able to insert into membranes and form channels. Hence, VacA has been classified as a pore-forming bacterial toxin. The formation of membrane channels by VacA has been studied in detail via the use of planar lipid bilayers (Czajkowsky *et al.*, 1999; Iwamoto *et al.*, 1999; Tombola *et al.*, 1999a; Tombola *et al.*, 1999b; Tombola *et al.*, 2001b; Czajkowsky *et al.*, 2005). The channels formed by VacA exhibit a moderate selectivity for anions over cations (Czajkowsky *et al.*, 1999; Iwamoto *et al.*, 1999; Tombola *et al.*, 1999a; Tombola *et al.*, 1999b; Tombola *et al.*, 2001b; Czajkowsky *et al.*, 2005). VacA channel activity can be blocked by several chemical inhibitors, including 5-nitro-2-(3-phenylpropylamino) benzoic acid (NPPB) (Tombola *et al.*, 1999b; Tombola *et al.*, 2000). VacA channels exhibit several electrophysiological features that are characteristics of anion-selective channels found in human cells (Campello *et al.*, 2002; Czajkowsky *et al.*, 2005).

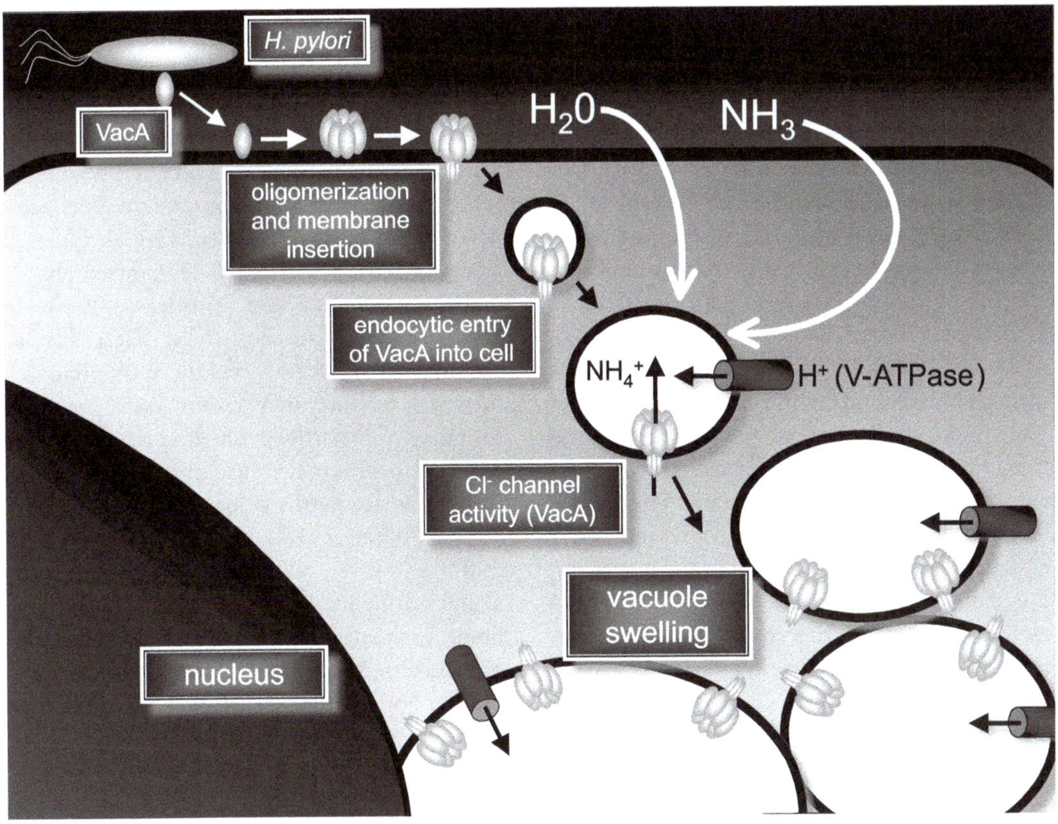

Figure 5.2 Cellular vacuolation induced by VacA. Schematic illustration of a model for the mechanism of VacA-induced vacuole formation (Czajkowsky et al., 1999; Szabo *et al.*, 1999; Tombola *et al.*, 1999a; Morbiato *et al.*, 2001; de Bernard *et al.*, 2004a). Following binding of VacA to the cell surface, VacA is internalized and forms anion-selective VacA channels in the membranes of late endocytic compartments. Conductance of chloride through these channels results in an increased intraluminal chloride concentration. To compensate for this increased anion concentration, vacuolar ATPase activity increases, resulting in increased proton pumping and a reduction in intraluminal pH. Membrane-permeable weak bases such as ammonia diffuse into late endocytic compartments, and become protonated and trapped within these compartments. Osmotic swelling of these compartments ultimately results in cell vacuolation (adapted from Cover and Blanke, 2005).

Images of membrane-associated VacA indicate that the toxin forms hexagonal ring-shaped structures (Czajkowsky *et al.*, 1999; Adrian *et al.*, 2002; Geisse *et al.*, 2004), similar in appearance to water-soluble VacA oligomers, except that the membrane-associated VacA oligomers are probably single-layered (Czajkowsky *et al.*, 1999). Presumably these membrane-associated oligomeric structures correspond to VacA membrane channels. Binding of VacA to lipid-containing membranes is enhanced by acid-activation of the water-soluble oligomeric toxin (Czajkowsky *et al.*, 1999) and, similarly, membrane channel formation by VacA requires acid-activation of the toxin (Czajkowsky *et al.*, 1999; Iwamoto *et al.*, 1999; Geisse *et al.*, 2004). These observations suggest that VacA monomers bind more avidly to lipid membranes than do water-soluble oligomeric forms of VacA. Channel formation probably results from interaction of VacA monomers with the membrane, followed by subsequent oligomerization and membrane insertion.

The VacA channel structure has not yet been analysed in any detail, but several residues near the amino terminus of the protein are known to be important for channel function (McClain *et al.*, 2003). The mature 88 kDa toxin is predicted to contain only one strongly hydrophobic region, located within the p33 domain near the amino-terminus of VacA (Vinion-Dubiel *et al.*, 1999). This region contains three tandem GXXXG motifs (defined by glycines at positions 14, 18,

22 and 26) (McClain *et al.*, 2001a; McClain *et al.*, 2003; Kim *et al.*, 2004), which are characteristic of transmembrane dimerization sequences. Mutagenesis of several residues within the amino-terminal hydrophobic region of VacA (including G14 and G18) abolishes the capacity of VacA to form membrane channels in planar lipid bilayers (McClain *et al.*, 2003) and also abolishes vacuolating cytotoxin activity (de Bernard *et al.*, 1998; Ye and Blanke, 2000; McClain *et al.*, 2003). These data provide evidence that the unique hydrophobic region at the amino-terminus of VacA plays a role in membrane channel formation, and that membrane channel formation is required for VacA-induced cell vacuolation (see 'Binding of VacA to the cell surface' and 'VacA as a modulator of epithelial cell function', below). Based on analysis of VacA associated with a lipid membrane, one study detected multiple protease-protected regions, comprising residues 40–66, 111–169, 205–266, 548–574 and 723–767. These may represent additional regions of VacA that interact with the membrane (Wang *et al.*, 2000).

Many of the cellular effects of VacA can be attributed to the capacity of this toxin to insert into membranes and form anion-selective channels. As discussed further below, the effects of VacA on human cells are dependent in part on the cellular site where these channels form. Several VacA effects, including leakage of ions and other small molecules from gastric epithelial cells, result from formation of VacA channels in the plasma membrane (Szabo *et al.*, 1999; Tombola *et al.*, 2001a). VacA-induced cell vacuolation can be attributed to VacA channel formation in the membranes of late endocytic compartments (Montecucco and Rappuoli, 2001) (Fig. 5.2). VacA-induced alterations in mitochondrial membrane permeability may be due to formation of VacA channels in mitochondrial membranes (Willhite *et al.*, 2003; Willhite and Blanke, 2004). Thus, the formation of VacA channels in multiple different cellular sites may account for several different VacA activities. Although many cellular effects of VacA are attributable to membrane channel formation, several VacA effects are probably the consequences of channel-independent actions (see 'Effects of VacA on host cells *in vitro*', below).

Interaction of VacA with host cells (binding, internalization, trafficking, localization)

The cell lines used most commonly to study VacA activity have been HeLa cells and cell lines derived from human gastric tumours, such as AGS and AZ-521 cells. Various types of immune cells, including T-cells, macrophages, and mast cells, are also sensitive to VacA. To effectively modulate host cells, VacA, like all bacterial toxins, must access the appropriate cellular targets. Entry of VacA into cells is required for many of the cellular activities that have been reported for this toxin, including vacuolation and mitochondrial perturbation (de Bernard *et al.*, 1997; Willhite *et al.*, 2003; Willhite and Blanke, 2004). However, VacA also can cause cellular alterations by interacting with the plasma membrane, without a requirement for internalization of the toxin (Nakayama *et al.*, 2004). Thus, VacA may act both at the cell surface and in at least one intracellular location.

Binding of VacA to the cell surface

Binding of VacA to the surface of cells is the first step in the intoxication pathway. The binding of VacA to cells is increased if water-soluble oligomeric forms of the toxin are acid- or alkaline-activated prior to contact with cells (de Bernard *et al.*, 1995; Yahiro *et al.*, 1999; McClain *et al.*, 2000; Schraw *et al.*, 2002; Willhite and Blanke, 2004; De Guzman *et al.*, 2005). Similarly, acid- or alkaline-activated VacA exhibits markedly increased cell-modulating activity compared to non-activated VacA (de Bernard *et al.*, 1995; Molinari *et al.*, 1998a; Yahiro *et al.*, 1999; McClain *et al.*, 2000; Willhite *et al.*, 2003; Willhite and Blanke, 2004; De Guzman *et al.*, 2005). Activation-dependent increases in VacA cellular binding and modulatory activities are observed for both m1 and m2 forms of the toxin (De Guzman *et al.*, 2005). Although the biochemical and structural changes accompanying VacA activation are not entirely understood, it is noteworthy that water-soluble VacA oligomers disassemble into monomeric components under conditions of acid or alkaline activation (Cover *et al.*, 1997; Molinari *et al.*, 1998a; Yahiro *et al.*, 1999). Therefore, the interaction of VacA monomers with the cell surface presumably is important for VacA intoxication of cells, although this has not been directly demonstrated.

Many intracellular-acting toxins interact with specific receptors on the plasma membrane of sensitive cells. Studies with radiolabeled VacA revealed that toxin binding to HeLa cells is not saturable, and is not entirely blocked by the presence of a molar excess of non-radiolabeled VacA (McClain *et al.*, 2000; Ricci *et al.*, 2000). On the other hand, indirect immunofluorescence and flow cytometry-based studies indicated that VacA binding to HeLa cells is saturable (Massari *et al.*, 1998), and binding of the native *H. pylori* VacA protein to RK13 cells was competitively inhibited by a GST-VacA fusion protein (Wang *et al.*, 2001). One interpretation of these results is that VacA may bind to specific receptors, and in addition, there may be low affinity and/or non-specific interactions between VacA and other plasma membrane components.

Several putative receptors for VacA have been described. One study suggested that VacA may bind to heparan sulfate, which is a sulfated polysaccharide found on the surface of most cells (Utt *et al.*, 2001), while another study implicated the epidermal growth factor receptor as a cell-surface receptor for VacA (Seto *et al.*, 1998). However, the most extensive studies of VacA binding have focused on interactions of VacA with two receptor protein tyrosine phosphatases (RPTP), α and β. RPTP-α and RPTP-β were first identified as proteins on the surface of AZ-521 gastric cells or G401 kidney cells that co-immunoprecipitated with VacA (Yahiro *et al.*, 1997; Yahiro *et al.*, 1999; Yahiro *et al.*, 2003).

In support of a role of RPTP-β in VacA cellular intoxication, the knockdown of RPTP-β expression in HL-60 leukaemic cells using antisense methodology resulted in a decrease in VacA-induced vacuolation (Padilla *et al.*, 2000). Moreover, BHK-21 cells, which are resistant to VacA cytotoxicity, became sensitive when they were transfected with expression plasmids encoding RPTP-β (Padilla *et al.*, 2000). In addition to a potential role of RPTP-β in VacA-induced vacuolation, binding of VacA to RPTP-β is reported to trigger activation of a signalling pathway involving tyrosine phosphorylation of the G protein-coupled receptor kinase interactor (Git1) (Fujikawa *et al.*, 2003). Further characterization revealed that pre-treatment of RPTP-β with neuraminidase and O-glycosidase diminished VacA binding (Yahiro *et al.*, 2004). Mutagenesis of the RPTP-β extracellular domain defined a region (including the sequence QTTQP at positions 747–751) that is critical for VacA binding (Yahiro *et al.*, 2004). Binding of pleiotrophin, a natural RPTP-β ligand, was not antagonistic for VacA binding, suggesting that VacA and pleiotrophin bind to RPTP-β at different sites (Yahiro *et al.*, 2004).

Although RPTP-β seems to be important for VacA-induced vacuolation of AZ-521 cells (Yahiro *et al.*, 1997; Yahiro *et al.*, 1999; Padilla *et al.*, 2000) and several other VacA-induced cellular alterations (Fujikawa *et al.*, 2003), RPTP-β is apparently not essential for VacA-mediated vacuolation of HeLa cells or G401 human kidney cells, which do not express RPTPβ (De Guzman *et al.*, 2005; Skibinski *et al.*, 2006). In addition, VacA can cause vacuolation of gastric epithelial cells from RPTP-β −/− mice, which do not express RPTP-β (De Guzman *et al.*, 2005). These data suggest that the essentiality of RPTP-β as a receptor for VacA may be cell-line specific, and support the view that VacA probably binds multiple cell-surface receptors.

VacA causes vacuolation of G401 cells, a human kidney tumour cell line, but these cells do not express RPTPβ. Treatment of G401 cells with RPTP-α antisense oligonucleotides resulted in decreased RPTP-α expression and inhibition of VacA-induced vacuolation, which provided evidence that RPTP-α has an important role in VacA intoxication of this cell line (Yahiro *et al.*, 2003). Notably, although both HeLa cells and G401 cells express RPTP-α, these two cell types differ markedly in their sensitivity to m2 VacA (De Guzman *et al.*, 2005). Differences in post-translational modifications of RPTP-α may possibly account for the different susceptibilities of HeLa cells and G401 cells to m2 VacA forms of VacA (De Guzman *et al.*, 2005).

VacA interacts with detergent-resistant membrane domains called lipid rafts (Patel *et al.*, 2002; Schraw *et al.*, 2002; Kuo and Wang, 2003; Gauthier *et al.*, 2004; Geisse *et al.*, 2004). Acid-activation of water-soluble VacA oligomers is required for association of the toxin with lipid rafts (Schraw *et al.*, 2002; Geisse *et al.*, 2004). The exact components of lipid rafts to which VacA binds are not entirely clear, but it is possible that VacA may interact with both lipid components (Schraw *et al.*, 2002; Geisse *et al.*, 2004) and protein components (including GPI-anchored proteins) (Ricci *et al.*, 2000). In AZ-521 cells, VacA initially binds to RPTP-β in non-lipid raft microdomains on the cell surface, and then localizes with RPTP-β in lipid rafts (Nakayama *et al.*, 2006). Disruption of lipid rafts by cholesterol depletion abrogates many VacA-induced cellular effects, which provides evidence that lipid rafts have an important role in VacA-mediated cellular alterations (Patel *et al.*, 2002; Schraw *et al.*, 2002; Kuo and Wang, 2003; Nakayama *et al.*, 2006). As discussed further in the section on VacA internalization and intracellular trafficking, the interaction of VacA with lipid rafts seems to be important for entry of the toxin into host cells.

VacA can bind to lipids in artificial membranes (Moll *et al.*, 1995; Czajkowsky *et al.*, 1999; Pagliaccia *et al.*, 2000; Wang *et al.*, 2000; Geisse *et al.*, 2004), and may bind to either raft-associated lipids or raft-independent lipids on the cell surface (Patel *et al.*, 2002; Schraw *et al.*, 2002; Kuo and Wang, 2003; Nakayama *et al.*, 2006). VacA association with human gastric glycosphingolipids was investigated by analysing VacA binding to these potential receptors on thin-layer chromatograms (Roche *et al.*, 2007). The VacA-binding glycosphingolipids were isolated and characterized by mass spectrometry and proton NMR as galactosylceramide (Galbeta1Cer) and galabiosylceramide (Galalpha4Galbeta1Cer). Analysis of the binding of VacA to reference glycosphingolipids from other sources demonstrated an interaction of VacA with glucosylceramide (Glcbeta1Cer), lactosylceramide (Galbeta4Glcbeta1Cer) and globotriaosylceramide (Galalpha4Galbeta4Glcbeta1Cer). It seems possible that VacA may interact with glycosphingolipids within lipid rafts, but this has not yet been investigated.

Depletion of sphingomyelin from the plasma membrane results in a reduction of VacA binding to cells as well as reduced VacA internalization (Gupta *et al.*, submitted 2007). VacA appears to associate with both raft-associated and non-raft-associated sphingomyelin. Manipulation of the levels of sphingomyelin in the plasma membrane (by pre-treatment of cells with sphingomyelinase, or supplementation with exogenous sphingomyelin) results in altered levels of VacA binding and alterations in VacA-induced cell vacuolation, suggesting that sphingomyelin might function as a receptor for VacA.

Finally, VacA has been reported to bind to the extracellular matrix component fibronectin (Hennig *et al.*, 2005). The adhesion of HeLa cells to fibronectin is decreased in the presence of VacA (Hennig *et al.*, 2005). Potentially VacA-fibronectin interactions might have effects on integrin receptor-induced cell signalling and cytoskeleton-dependent cell functions, but these topics have not yet been evaluated in detail.

Specific VacA residues that mediate toxin binding to receptors on the surface of host cells have not been identified. However, deletion of a ~100 amino acid region at the C-terminus of VacA results in decreased binding of the toxin to cells (Wang and Wang, 2000). Antiserum against p55 inhibits binding of VacA to cells (Garner and Cover, 1996), suggesting the importance of determinants within p55 for VacA binding to host cells. The p88 VacA holotoxin binds to glycosphingolipids, but a mutant toxin comprised mainly of p55 (lacking the p33 domain) does not exhibit detectable binding to glycosphingolipids (Roche *et al.*, 2007). Individual recombinant p33 and p55 domains can each bind to cells (Reyrat *et al.*, 1999; Torres *et al.*, 2005), and individual p33 and p55 fragments each demonstrate affinity for lipids (Moll *et al.*, 1995). A mixture of p33 and p55 proteins exhibits markedly enhanced binding to the plasma membrane of mammalian cells in comparison to the individual p33 or p55 domains (Torres *et al.*, 2005).

As noted in the section on *vacA* genetic diversity, type s1/m1 and s1/m2 forms of VacA differ in cell type specificity, and this is attributable to differences in the binding properties of these VacA proteins (Pagliaccia *et al.*, 1998; Ji *et al.*, 2000; Wang *et al.*, 2001). Accordingly, several studies have localized VacA cell-binding determinants to the m-region of VacA, located within the p55 domain (Reyrat *et al.*, 1999; Wang and Wang, 2000; Wang *et al.*, 2001). Binding studies have shown that m1 forms of VacA bind to HeLa cells more strongly than do m2 forms of VacA, whereas m1 and m2 forms of VacA bind similarly to RK-13 cells (Reyrat *et al.*, 1999; Wang *et al.*, 2001). Immunoprecipitation experiments have shown that m1 and m2 forms of VacA each bind to both RPTP-α and RPTP-β in AZ-521 cells (De Guzman *et al.*, 2005). A recent study, in which *vacA* m-region sequences were systematically swapped to create m1/m2 chimeric VacA toxins, demonstrated that both the m1 and m2 forms of VacA are able to bind RPTP-β, implying that the cell-specific phenotype of the mid-region is not dependent on VacA interactions with RPTP-β (Skibinski *et al.*, 2006).

The binding of VacA to the surface of host cells may be followed by three possible overall outcomes. One possibility is that binding of VacA to the cell surface can result in activation of signal transduction pathways (see 'Effects of VacA on host cells *in vitro*', below). Alternatively, VacA may insert into the plasma membrane to form channels (see 'Membrane channel formation by VacA' and below). Finally, VacA may internalize into an intracellular compartment, as discussed further below.

Membrane insertion and formation of VacA channels in the plasma membrane

As described in 'Membrane channel formation by VacA', VacA can insert into membranes and can form membrane channels. Electrophysiological and pharmacological approaches suggested that VacA forms low-conductance pores in the cell plasma membrane of HeLa cells, resulting in increased anion permeability and membrane depolarization (Szabo *et al.*, 1999). Other probable consequences of VacA pore formation in the plasma membrane include transepithelial flux of urea (Tombola *et al.*, 2001a) and bicarbonate efflux (Debellis *et al.*, 2001). Formation of VacA channels in the plasma membrane is inhibited by the removal of GPI-anchored proteins from the surface of HeLa cells (Gauthier *et al.*, 2004).

VacA internalization and intracellular trafficking

The first indication that VacA might function from within an intracellular site came from fluorescence microscopy studies showing that VacA is internalized by cells and localizes in a punctate manner to a perinuclear location (Garner and Cover, 1996). Internalization of VacA into HeLa or AGS cells occurs by a temperature-dependent, ATP-dependent process that requires actin filaments (Garner and Cover, 1996; McClain *et al.*, 2000; Ricci *et al.*, 2000; Gauthier *et al.*, 2004; Gauthier *et al.*, 2005; Gauthier *et al.*, 2006). This clathrin-independent process has been termed pinocytosis (Ricci *et al.*, 2000; Gauthier *et al.*, 2005; Gauthier *et al.*, 2006). Presumably the internalization of VacA is a consequence of toxin binding to one or more specific receptors on the cell surface, but the identity of VacA receptors required for internalization is not known. Treatment of cells with methyl-beta cyclodextrin (a cholesterol-depleting agent) blocks internalization of VacA, which suggests that interaction of VacA with lipid rafts may be required for internalization (Patel *et al.*, 2002; Schraw *et al.*, 2002; Kuo and Wang, 2003; Nakayama *et al.*, 2006).

VacA enters HeLa or AGS cells via peripheral early endocytic compartments that are characterized by the lack of caveolin 1, Rab5, early endosome antigen-1, and transferrin, but able to take up fluid-phase markers and GPI-anchored proteins (Kuo and Wang, 2003; Gauthier *et al.*, 2005). Pinocytosis of VacA is Cdc42-dependent, but is independent of cellular tyrosine kinases, dynamin 2, ADP-ribosylating factor 6, and RhoA GTPase activities. Prior to uptake, VacA interacts with detergent-resistant membrane domains localized above filamentous actin (F-actin) structures that are controlled by Rac1 (Gauthier *et al.*, 2005). Following uptake, VacA traffick from early endosomes to late endosomes by a mechanism requiring polymerized actin (Gauthier *et al.*, 2007). Throughout the uptake process, VacA is continuously associated with detergent-resistant membrane domains (Gauthier *et al.*, 2005). The CD2-associated protein (CD2AP), which is a docking factor likely to be involved in intracellular trafficking, bridges the F-actin structures with VacA-containing early endosomes, and is required to transfer VacA from GPI-AP-enriched early endosomal compartments to late endosomes (Gauthier *et al.*, 2007).

Water-soluble oligomeric forms of VacA are not efficiently internalized into cells unless the oligomers are first activated by exposure to low pH or high pH (McClain *et al.*, 2000). Recently, a mutant form of VacA (VacAΔ49–57) was demonstrated to be defective in oligomerization, but was able to be internalized normally by HeLa cells, suggesting that, at least in these cells, VacA oligomerization is not required for internalization (Genisset *et al.*, 2006). Individual p33 or p55 domains can bind to the cell surface, but are not efficiently internalized (Reyrat *et al.*, 1999; Torres *et al.*, 2005). In contrast, if p33 and p55 domains are independently expressed and a mixture of the two domains is added to cells, both domains are internalized (Torres *et al.*, 2005).

VacA can assemble into water-soluble oligomeric structures and can form membrane channels in the plasma membrane of cells. However, very little is known about the oligomeric state of VacA following its internalization into cells. Several lines of evidence suggest that VacA has membrane channel activity within cells (McClain *et al.*, 2003), which suggests that the intracellular toxin associates with membranes in an oligomeric state. In support of this view, FRET studies indicate that when VacA is expressed intracellularly in transiently transfected cells, it can assemble into oligomeric structures (Willhite *et al.*, 2002). In addition, a mutant form of VacA that lacks the capacity to form membrane channels can block the activity of wild-type VacA in a dominant-negative fashion (Vinion-Dubiel *et al.*, 1999). This dominant negative phenotype is associated with the formation of mixed oligomeric structures, comprised of both wild-type and mutant VacA (McClain *et al.*, 2001b). The dominant negative mutant exhibits inhibitory activity regardless whether it is added to the surface of cells or expressed intracellularly (Vinion-Dubiel *et al.*, 1999). The capacity of a dominant negative mutant toxin to block wild-type VacA activity provides evidence that an intracellular oligomeric form of VacA is required for VacA cytotoxicity.

Following internalization, VacA can localize to a variety of endocytic compartments, including the large vacuoles that form in response to VacA (Ricci *et al.*, 1997; Fiocca *et al.*, 1999; Gauthier *et al.*, 2005; Yamasaki *et al.*, 2006; Gauthier *et al.*, 2007). In addition, when fluorescently labelled VacA is added to cells, a fraction of the fluorescently labelled VacA localizes in association with mitochondria (Willhite and Blanke, 2004). Similarly, VacA expressed intracellularly can localize to mitochondria (Galmiche *et al.*, 2000). The trafficking pathway that allows VacA to move from the cell surface to mitochondria has not yet been elucidated.

In summary, VacA interactions with host cells are complex, and the toxin is apparently able to engage multiple plasma membrane components. The relationship between these different putative receptors, as well as their role in VacA cellular entry, remains poorly defined. Recent work has defined a novel pinocytic-like pathway by which VacA is internalized into several epithelial cell lines. It will be interesting to determine whether VacA utilizes the same mechanism for entry into other cell lines, including those originating from immune cells, which, as described below, VacA modulates in several different ways.

Effects of VacA on host cells *in vitro*

VacA is able to intoxicate a wide range of cell types, including gastric epithelial cells and several types of immune cells. In response to VacA, several cell types are known to undergo an array of multiple different alterations (Table 5.1 and Fig. 5.3).

VacA as a modulator of epithelial cell function

Effects of VacA on endosomal compartments

The first documented effect of VacA was cellular vacuolation (Leunk *et al.*, 1988). Many different types of cultured cells, including primary human gastric epithelial cells, transformed cell lines derived from multiple different types of human tissue, and cell lines derived from multiple mammalian species, undergo vacuolation in response to VacA (Leunk *et al.*, 1988; Smoot *et al.*, 1996; Pagliaccia *et al.*, 1998). The intraluminal pH of VacA-induced vacuoles is acidic, and the vacuoles are endocytically active (Cover *et al.*, 1991; Catrenich and Chestnut, 1992; Cover *et al.*, 1992; Papini *et al.*, 1994). The membranes of these vacuoles are enriched in Rab7 and other markers for late endocytic compartments (Papini *et al.*, 1994; Papini *et al.*, 1996; Molinari *et al.*, 1997; Papini *et al.*, 1997; Li *et al.*, 2004), which suggests the vacuoles are derived from late endocytic compartments. Vacuolated cells exclude trypan blue, which indicates that VacA-induced

Table 5.1. Effects of VacA on mammalian cells *in vitro*

Effects on gastric epithelial cells	
Alterations of endocytic compartments	
	Cell vacuolation
Alterations of mitochondria	
	Cytochrome c release
	Reduction in mitochondrial transmembrane potential ($\Delta\Psi_m$)
	Apoptosis
Formation of channels in plasma membrane	
Activation of signal transduction pathways (p38 and Erk1/2 MAP kinases, Git1)	
Increased permeability of epithelial monolayers	
Cytoskeletal alterations	
Enhanced intracellular survival of *H. pylori*	
Effects on immune cells	
T-cells	
	Inhibition of NFAT activation
	Inhibition of activation-induced proliferation
	Activation of signal transduction pathways (p38 MAP kinase, MKK3/6, Vav)
	Alterations of mitochondria
B-cells	
	Inhibition of antigen presentation
	Inhibition of activation-induced proliferation
Macrophages	
	Formation of megasomes
Mast cells	
	Stimulation of cytokine production
	Alterations in intracellular calcium
	Degranulation
Eosinophils	
	Stimulation of cytokine production
Neutrophils	
	Increased COX-2 expression

cell vacuolation is not a cytolethal phenomenon (Leunk *et al.*, 1988).

Like many other cellular alterations caused by VacA, VacA-induced cell vacuolation is markedly enhanced if water-soluble VacA oligomers are activated by low pH or high pH prior to contact with cells (de Bernard *et al.*, 1995; Molinari *et al.*, 1998a). The formation of intracellular vacuoles in response to purified VacA is enhanced by the presence of weak bases such as ammonium chloride (Cover and Blaser, 1992; Cover *et al.*, 1992; Li *et al.*, 2004). In the absence of weak bases, relatively high concentrations of VacA are required to cause cell vacuolation. VacA can cause detectable alterations in late endocytic compartments even in the absence of visible vacuolation, resulting in inhibition of the intracellular degradation of epidermal growth factor (Satin *et al.*, 1997), and inhibition of procathepsin D maturation (associated with mis-targeting of procathepsin D outside the cell) (Satin *et al.*, 1997).

A current model for VacA-induced vacuolation proposes that the toxin is internalized into

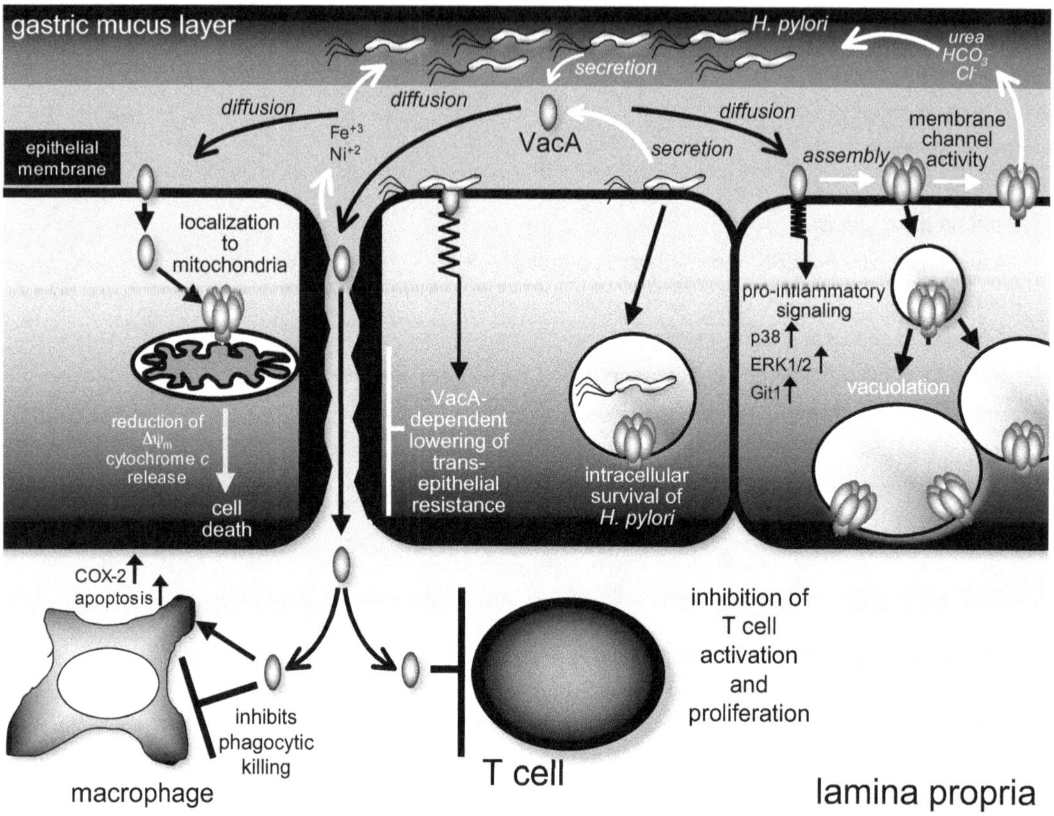

Figure 5.3. Multiple actions of VacA contribute to *H. pylori* colonization of the stomach. Effects of VacA on gastric epithelial cells include alterations in mitochondrial membrane permeability ($\Delta\Psi m$ indicates mitochondrial transmembrane potential) and apoptosis, stimulation of proinflammatory signaling (as indicated by upward-pointing arrows; squiggly arrows indicate the involvement of multistep pathways), increased permeability of the plasma membrane, and alterations in endocytic compartments. Multiple *H. pylori* factors, including VacA, have been reported to decrease trans-epithelial resistance, and may thereby facilitate passage of VacA through the epithelial layer. At the same time, disruption of the membrane may allow *H. pylori* to access $Fe^{+}3$ and $Ni^{+}2$. Within the lamina propria, VacA interferes with the activation and proliferation of T lymphocytes. Moreover, VacA has been reported to increase both apoptosis and pro-inflammatory signaling in macrophages (as indicated by upward-pointing arrows). VacA has been reported to enhance intracellular survival of *H. pylori* within epithelial cells and macrophages. Many, but not all of these effects of VacA are attributable to the formation of VacA membrane channels (adapted from Cover and Blanke, 2005).

cells and forms anion-selective channels in the membranes of late endocytic compartments, and that, in the presence of permeant weak bases, vacuoles arise due to swelling of late endosomal compartments (Czajkowsky *et al.*, 1999; Szabo *et al.*, 1999; Tombola *et al.*, 1999a; Morbiato *et al.*, 2001; de Bernard *et al.*, 2004) (Fig. 5.2). Several lines of evidence support this model. First, VacA has been localized to the membranes of VacA-induced vacuoles (Ricci *et al.*, 1997; Fiocca *et al.*, 1999). Intracellular expression of VacA in transiently transfected cells results in cell vacuolation (de Bernard *et al.*, 1997; Ye *et al.*, 1999), which provides additional evidence that VacA can act in an intracellular site. VacA-induced vacuolation requires the influx of extracellular chloride (Morbiato *et al.*, 2001), which presumably enters the cell through VacA channels in the plasma membrane. An inhibitor of VacA channel activity, 5-nitro-2-(3-phenylpropylamino) benzoic acid (NPPB), blocks VacA-induced cell vacuolation (Szabo *et al.*, 1999; Tombola *et al.*, 1999b). VacA mutant toxins that are deficient in membrane channel formation lack the capacity to cause cell vacuolation, regardless whether they are added to the surface of cells or expressed intracellularly in transiently transfected cells (Vinion-Dubiel *et al.*, 1999; Ye and Blanke, 2000; McClain *et al.*, 2003). Finally, wild-type VacA causes swelling of isolated endosomes and enhances the v-ATPase proton pump activity of endosomes in a chloride-dependent manner, whereas mutant toxins lacking channel activity do not induce swelling of isolated endosomes (Genisset *et al.*, 2007).

The complexity of VacA-induced vacuole formation is underscored by the apparent involvement of multiple cellular factors, including vacuolar ATPase, Rab7, Rac1, syntaxin 7, dynamin, and GPI-anchored proteins (Cover *et al.*, 1993b; Papini *et al.*, 1993; Papini *et al.*, 1996; Papini *et al.*, 1997; Hotchin *et al.*, 2000; Suzuki *et al.*, 2001; Suzuki *et al.*, 2003; Li *et al.*, 2004). Several of these factors may be required for VacA internalization or the process of vesicle swelling. Overexpression of PIKfyve kinase has been reported to inhibit VacA-induced vacuolation, which suggests that VacA-induced cellular alterations may somehow be related to changes in cellular phosphatidylinositol metabolism (Ikonomov *et al.*, 2002).

Several aspects of VacA-induced vacuolation remain poorly understood. One unresolved question is whether VacA inserts directly into endosomal membranes to form channels, or, whether the channels form in the plasma membrane and are then taken up by endocytosis. Another uncertainty involves the source (or sources) of membrane required for the formation of intracellular vacuole compartments. The progressive enlargement of pre-existing vesicular compartments seems unlikely, because the compartments would be expected to undergo lysis during the swelling process required for vacuole formation. It has been suggested that vacuoles could arise from late endosomes without a requirement for fusion of different compartments, via a process involving fusion of late endosomal internal membranes with the late endosomal limiting membrane (de Bernard *et al.*, 2002). In support of the latter model, a recent study reported that VacA could induce swelling of isolated endosomes (Genisset *et al.*, 2007). Alternatively, VacA-induced vacuoles may result from the fusion of multiple smaller endocytic compartments. One study reported that the SNARE protein syntaxin 7, which is involved in intracellular membrane fusion events, is localized to the membranes of VacA-induced vacuoles, and that intracellular expression of a dominant-negative mutant form of syntaxin 7 blocks VacA-induced vacuolation (Suzuki *et al.*, 2003). However, another study did not detect a role of SNARE proteins in VacA-induced vacuolation (de Bernard *et al.*, 2002).

Effects of VacA on mitochondria

In addition to its effects on late endocytic compartments, VacA causes alterations in mitochondria. Effects of VacA on mitochondria include reduction of the mitochondrial transmembrane potential and release of cytochrome *c*, both of which are consistent with alterations in mitochondrial membrane permeability (Kimura *et al.*, 1999; Galmiche *et al.*, 2000; Willhite *et al.*, 2003; Willhite and Blanke, 2004; Blanke, 2005). VacA-induced alterations of mitochondria have been reported to result in a reduction in cellular ATP levels and impaired cell cycle progression (Kimura *et al.*, 1999). VacA-induced reductions in intracellular ATP levels may have multiple consequences, including impairment of glutathione metabolism and impaired resistance of cells against oxidative stress (Kimura *et al.*, 2001). The development of mitochondrial alterations in response to exogenously added VacA requires longer time periods and higher doses than are required for VacA-induced cell vacuolation, which suggests that these two processes occur via separate pathways (Willhite *et al.*, 2003). Effects of VacA on mitochondria have been detected following addition of VacA to the surface of cells, and also following intracellular expression of VacA in transiently transfected cells (Kimura *et al.*, 1999; Galmiche *et al.*, 2000; Willhite *et al.*, 2003; Willhite and Blanke, 2004). VacA expression in transiently transfected cells results in the cleavage of poly(ADP-ribose) polymerase (PARP), an event consistent with upstream executioner caspase 3 activation (Galmiche *et al.*, 2000). Overexpression of Bcl-2, an anti-apoptotic protein that is believed to interact with and block the action of pro-apoptotic proteins that act at mitochondria, inhibits VacA-induced PARP cleavage (Galmiche *et al.*, 2000). Release of cytochrome *c* from mitochondria and activation of caspase 3 are pro-apoptotic phenomena, and accordingly, prolonged exposure of AGS gastric cells to exogenously added VacA can result in apoptosis (Peek *et al.*, 1999; Kuck *et al.*, 2001; Boquet *et al.*, 2003; Cho *et al.*, 2003; Cover *et al.*, 2003). Similarly, microinjection of DNA encoding VacA into cells can induce apoptosis (Galmiche *et al.*, 2000).

Several studies have reported localization of VacA to mitochondria (Galmiche *et al.*, 2000; Willhite and Blanke, 2004), which suggests that VacA might act directly on mitochondria. Both the p33 fragment of VacA expressed directly within mammalian cells, as well as full-length fluorescently labelled VacA applied to the surface of cells, have been reported to localize to mitochondria (Galmiche *et al.*, 2000; Willhite and Blanke, 2004). In addition, VacA is translocated into isolated yeast mitochondria *in vitro* (Galmiche *et al.*, 2000). Mutant forms of VacA defective in the capacity to form membrane channels fail to cause cytochrome *c* release, and NPPB, an agent that blocks formation of VacA membrane channels, inhibits VacA-induced cytochrome *c* release (Willhite *et al.*, 2003; Willhite and Blanke, 2004). Based on these data, it seems possible that VacA-induced effects on mitochondria may result from the formation of VacA channels in mitochondrial membranes. However, alternate mechanisms can also be considered. For example, it is possible that the observed changes in mitochondrial membrane permeability may result from activation of endogenous channels found in mitochondria. A third possibility is that VacA-induced mitochondrial alterations and cytochrome c release may occur via activation of the proapoptotic proteins Bax and Bak (Yamasaki *et al.*, 2006).

Effects of VacA on cellular signal transduction pathways

The effects of VacA on late endosomal compartments and mitochondria are typically observed within several hours after addition of VacA to cells. However, VacA causes several other cellular effects that can be detected at much earlier time points. Within 10 min after addition of VacA to a human gastric adenocarcinoma cell line (AZ-521), two classes of mitogen-activated protein (MAP) kinases (p38 and ERK1/2) and the activating transcription factor 2 (ATF-2) signalling pathway are activated (Nakayama *et al.*, 2004; Hisatsune *et al.*, 2007). An inhibitor of p38 kinase activity (SB203580) does not block VacA-induced vacuolation or VacA-induced cytochrome *c* release, which indicates that VacA-induced activation of the p38/ATF-2 signal pathway is independent of VacA effects on late endocytic compartments and mitochondria (Nakayama *et al.*, 2004). One potential consequence of VacA-induced p38 activation is upregulation of cyclooxygenase-2 (COX-2) expression, leading to increased PGE2 production (Hisatsune *et al.*, 2007). In addition to effects of VacA on MAP kinases, VacA is reported to activate a signalling pathway involving tyrosine phosphorylation of the G protein-coupled receptor kinase interactor (Git1) (Fujikawa *et al.*, 2003) and a signalling pathway that leads to upregulated expression of vascular endothelial growth factor (Caputo *et al.*, 2003).

In contrast to effects of VacA on endocytic compartments and mitochondria, the relatively rapid cellular responses to VacA described above are likely to be the consequences of VacA interactions with specific cell surface components, without a requirement for internalization of the toxin. The cell surface receptors for VacA that are involved in activating MAP kinase signalling pathways have not yet been characterized. Activation of the Git1 pathway is dependent on VacA binding to RPTP-β (Fujikawa *et al.*, 2003). VacA-induced upregulation of vascular endothelial growth factor reportedly occurs through a pathway involving activation of epidermal growth factor receptor, MAP kinases, and COX-2 (Caputo *et al.*, 2003).

Effects of VacA on epithelial monolayer permeability

VacA lowers the transepithelial electrical resistance (TER) of monolayers formed by several different types of polarized epithelial cells, an effect that is attributable to increased paracellular epithelial permeability of the monolayers to low molecular mass molecules and ions (Papini *et al.*, 1998; Pelicic *et al.*, 1999). Selective permeabilization of epithelial monolayers by VacA has been proposed to support the growth of *H. pylori* in the gastric mucus layer by promoting the release of factors such as Fe^{3+}, Ni^{2+}, sugars, and amino acids (Papini *et al.*, 1998). The mechanisms by which VacA alters paracellular permeability are not yet well understood. This activity does not require acid-activation of purified oligomeric VacA, and is not inhibited by bafilomycin A1.

In addition to altering paracellular permeability, VacA also increases the transepithelial flux of certain molecules, including urea and bicarbonate (Guarino *et al.*, 1998; Szabo *et al.*, 1999; Debellis *et al.*, 2001; Tombola *et al.*, 2001a). Transepithelial flux of these molecules is attributed to the formation of VacA channels in the plasma membrane (Szabo *et al.*, 1999). Release of bicarbonate may have a favourable effect on survival of *H. pylori* within the acidic gastric mucus layer, and release of urea may optimize the enzymatic activity of *H. pylori* urease.

Effects of VacA on the cytoskeleton

Several studies have reported that VacA causes alterations in the cytoskeleton of intoxicated cells (Pai *et al.*, 1999; Ashorn *et al.*, 2000; Pai *et al.*, 2000; Tabel *et al.*, 2003; Yuan *et al.*, 2004; Hennig *et al.*, 2005; Wang *et al.*, 2005). Alterations in both actin filaments and the microtubule system have been reported. The mechanisms underlying these cytoskeletal alterations are not clear. However, one study noted that VacA interacted with an intermediate filament interacting protein in a yeast two hybrid system (de Bernard *et al.*, 2000), and it is possible that VacA may alter the expression of multiple cytoskeletal-related genes (Pai *et al.*, 2000; Yuan *et al.*, 2004; Wang *et al.*, 2005).

Role of VacA in promoting intracellular survival of H. pylori *in gastric epithelial cells.*

Within the human stomach, *H. pylori* is found mainly in the mucus layer overlying gastric epithelial cells, but there is evidence that the bacteria may occasionally localize within gastric epithelial cells. VacA may promote survival of *H. pylori* within gastric epithelial cells (Bjorkholm *et al.*, 2000; Petersen *et al.*, 2001). A recent study suggested that VacA promotes intracellular survival of *H. pylori* via a mechanism involving VacA-dependent retention of Rab7, fusion of late endocytic compartments, and sequestration *H. pylori*-containing cell compartments away from the lysosomal degradative pathway (Terebiznik *et al.*, 2006). However, another study did not detect a role of VacA in intracellular survival of *H. pylori* (Amieva *et al.*, 2002).

VacA as a modulator of immune cell function

Effects of VacA on lymphocytes

VacA causes multiple alterations in T lymphocytes (Boncristiano *et al.*, 2003; Gebert *et al.*, 2003; Sundrud *et al.*, 2004; Oswald-Richter *et al.*, 2006; Algood *et al.*, 2007). When added to Jurkat T-cells, VacA inhibits production of interleukin 2 (IL-2) (a factor required for T-cell viability and proliferation) and downregulates surface expression of the IL-2 receptor (Boncristiano *et al.*, 2003; Gebert *et al.*, 2003; Sundrud *et al.*, 2004). These effects are attributable to the capacity of VacA to inhibit activation of nuclear factor of T-cells (NFAT) (Boncristiano *et al.*, 2003; Gebert *et al.*, 2003), a transcription factor that acts as a global regulator of immune response genes and that is required for optimal T-cell activation. The mechanism by which VacA inhibits NFAT activation may involve blocking calcium influx into cells from the extracellular milieu, thereby inhibiting the activity of the Ca^{2+}-calmodulin-dependent phosphatase calcineurin (an enzyme that dephosphorylates NFAT) (Boncristiano *et al.*, 2003; Gebert *et al.*, 2003). Intoxication of Jurkat T-cells by VacA alters the expression of more than 100 genes, including many that are affected by FK506 (tacrolimus) and cyclosporine A, two immunosuppressive drugs that inhibit calcineurin activity (Gebert *et al.*, 2003). Other effects of VacA on T lymphocytes include activation of signalling pathways that include mitogen-activated protein kinases (p38 and MKK3/6) and the Rac-specific nucleotide exchange factor, Vav (Boncristiano *et al.*, 2003; Oswald-Richter *et al.*, 2006). All of the effects of VacA described above are reported to occur without a substantial increase in apoptosis or cell death.

VacA treatment of primary human CD4+ T-cells results in inhibition of activation-induced proliferation, mitochondrial depolarization, ATP depletion, and cell cycle arrest (Sundrud *et al.*, 2004; Oswald-Richter *et al.*, 2006). Several differences have been noted when comparing the effects of VacA on primary human T-cells versus cultured Jurkat cells. For example, the inhibitory effects of VacA on IL-2 secretion are much more prominent in Jurkat cells than in primary cells (Sundrud *et al.*, 2004; Oswald-Richter *et al.*, 2006). In addition to having effects on primary human CD4+ T-cells, VacA also inhibits stimulation-induced proliferation of human CD8+ T-cells and B-cells (Torres *et al.*, in press).

The effects of VacA on T-cells may occur via more than one mechanism. Some effects on T-cells are dependent on the formation of VacA channels in cell membranes (Boncristiano *et al.*, 2003; Sundrud *et al.*, 2004; Oswald-Richter *et al.*, 2006). Other effects are the result of activation of intracellular signalling in T-cells, via a channel-independent mechanism (Boncristiano *et al.*, 2003).

It is hypothesized that VacA enters the lamina propria via disruptions in the gastric epithelial layer, and thereby intoxicates T-cells and various other types of immune cells. In addition, it is possible that VacA may target intraepithelial T lymphocytes. Most of the effects of VacA on T lymphocytes described above are expected to result in localized immunosuppression. VacA also stimulates expression of cyclooxygenase-2, a proinflammatory enzyme, in T lymphocytes, which is expected to have a proinflammatory effect (Boncristiano *et al.*, 2003). Thus, the effects of VacA on immune cells are complex, and are characterized by both immunostimulatory and immunosuppressive actions.

Effects of VacA on antigen presentation

VacA has been reported to interfere with the process of antigen presentation by B lymphocytes. In one model system, VacA interfered with proteolytic processing of tetanus toxoid and inhibited the invariant chain (Ii)-dependent pathway of antigen presentation mediated by newly synthesized major histocompatibility complex (MHC) class II molecules (Molinari *et al.*, 1998b). This effect of VacA on antigen-presenting cells is likely due to VacA-induced effects on endocytic compartments, resulting in alterations in endocytic trafficking.

Effects of VacA on macrophages

Two reports provided evidence that VacA contributes to the formation of large vesicular compartments (termed megasomes) in *H. pylori*-infected macrophages by stimulating homotypic phagosome fusion (Allen *et al.*, 2000; Zheng and Jones, 2003). VacA stimulates the recruitment and retention of the tryptophan aspartate-containing coat protein (TACO or coronin 1) to phagosomes, thereby disrupting proper vesicular maturation (Zheng and Jones, 2003), which in turn, may impair phagocytic killing of *H. pylori*. Notably, one report did not detect any effects of VacA on phagosome formation or intracellular survival of *H. pylori* in human monocytes (Rittig *et al.*, 2003). In addition to having effects on vesicular compartments in macrophages, VacA also is reported to stimulate activation of p38 MAP kinase and cause increased expression of COX-2 in macrophages (Boncristiano *et al.*, 2003). Finally, it has been reported that VacA can stimulate macrophages to undergo apoptosis (Menaker *et al.*, 2004).

Effects of VacA on mast cells, eosinophils and neutrophils

Incubation of VacA with mast cells stimulates production of proinflammatory cytokines and induces mast cell chemotaxis (Supajatura *et al.*, 2002; de Bernard *et al.*, 2005). Binding of VacA to a mast cell line (RBL-2H3 cells) induces an oscillation in levels of cytosolic calcium and exocytosis of secretory granules (de Bernard *et al.*, 2005). These effects of VacA on mast cells do not require acid-activation of purified oligomeric VacA, which contrasts the requirement of acid-activation for cell-vacuolating activity of VacA. The rapid changes in calcium concentration that follow VacA interaction with mast cells may result from VacA interactions with cell surface receptors and activation of specific signal transduction pathways, without a requirement for membrane channel formation.

VacA is reported to have an effect on eosinophils, manifested as upregulated expression of chemokines (Kim *et al.*, 2007). This effect is reported to occur via a pathway involving calcium influx, mitochondrial generation of reactive oxygen intermediates, and NF-κB activation (Kim *et al.*, 2007). VacA also is reported to have effects on neutrophils, including activation of p38 MAP kinase and increased expression of COX-2 (Boncristiano *et al.*, 2003; Brest *et al.*, 2006).

Role of VacA *in vivo*

Role of VacA in *H. pylori* colonization of the stomach

The actions of bacterial protein toxins can result in extensive remodelling of host cells or tissues, resulting in a more suitable colonization niche for the bacteria. Toxin production may be absolutely required for bacterial colonization of the host, or alternatively, toxin-producing bacteria may have a selective advantage for colonization compared with strains that do not produce a particular toxin. All human isolates of *H. pylori* possess *vacA*, which suggests that VacA production is important for colonization or persistence. A degenerate *vacA* gene has been detected in *H. acinonychis* (a gastric pathogen of cheetahs and other large cats) (Dailidiene *et al.*, 2004). In contrast, *vacA* orthologues have not been detected in most other species of *Helicobacter* that colonize non-human mammalian stomachs (Schmitt and Haas, 1994). The presence of *vacA* in *H. pylori* and its absence from other *Helicobacter* species suggests that VacA may interact with host molecules found uniquely in humans.

A potential role for VacA in *H. pylori* colonization of the stomach of several animal models has been experimentally evaluated. Mice, gerbils and gnotobiotic piglets can be colonized by *H. pylori vacA*-null mutant strains (Eaton *et al.*, 1997; Wirth *et al.*, 1998; Ogura *et al.*, 2000; Salama *et al.*, 2001; Guo and Mekalanos, 2002), which indicates that, for these particular animal models, VacA production is not an absolute requirement for gastric colonization. In a Mongolian gerbil model, a wild-type *H. pylori* strain colonized only slightly better than a *vacA* null mutant (Wirth *et al.*, 1998). However, when mice were co-infected with a mixture of strains, the wild-type VacA-producing *H. pylori* strain had a selective colonization advantage compared with an isogenic *vacA* null mutant strain (Salama *et al.*, 2001). In addition, mice that were immunized with VacA were protected against subsequent challenges with *H. pylori* (Marchetti *et al.*, 1995; Ghiara *et al.*, 1997; Marchetti *et al.*, 1998). These studies provide evidence that VacA may contribute to the ability of *H. pylori* to colonize the stomach.

A hallmark of *H. pylori* infection in humans is the persistence of the bacteria in the stomach, despite the development of gastric inflammation and local and systemic humoral immune responses (Dunn *et al.*, 1997; Suerbaum and Michetti, 2002; Blaser and Atherton, 2004; Monack *et al.*, 2004; Algood and Cover, 2006). As discussed above, *in vitro* studies indicate that VacA has several immunosuppressive properties. Therefore, it seems plausible that VacA might have a role in enabling *H. pylori* to persistently colonize human hosts. The possibility that VacA facilitates *H. pylori* persistence has not yet been rigorously tested in animal models. Notably, murine T-cells are resistant to the effects of VacA (Algood *et al.*, 2007), and this will hinder efforts to study the role of VacA in *H. pylori* persistence in mouse models.

Role of VacA in gastroduodenal disease

As discussed in the section on *vacA* genetic diversity, strains of *H. pylori* that contain certain allelic forms of *vacA* are associated with an increased risk of symptomatic gastroduodenal disease compared with strains containing other allelic forms of *vacA*. In particular, *H. pylori* strains that contain *vacA* alleles of the s1 type are associated with an increased risk for development of peptic ulcer disease and gastric cancer compared with strains containing *vacA* alleles of the s2 type (Atherton *et al.*, 1995; Rudi *et al.*, 1998; Strobel *et al.*, 1998; van Doorn *et al.*, 1998b; Gerhard *et al.*, 1999; Kidd *et al.*, 1999; Van Doorn *et al.*, 1999; Figueiredo *et al.*, 2002). These data can be correlated with the failure of type s2 forms of VacA to cause detectable cytotoxicity using *in vitro* assays (Atherton *et al.*, 1995; Letley and Atherton, 2000; McClain *et al.*, 2001b; Letley *et al.*, 2003). Strains containing type m1 *vacA* alleles are associated with an increased risk of gastric epithelial injury and gastric cancer compared with strains containing *vacA* alleles of the m2 type (Atherton *et al.*, 1997; Figueiredo *et al.*, 2002). Although drawing conclusions from these epidemiological studies is limited by the possibility of numerous confounding variables, these data suggest that certain forms of VacA contribute to the pathogenesis of *H. pylori*-associated peptic ulcer disease and gastric cancer.

Several different experimental strategies have been used to investigate whether VacA contributes to gastric disease. Direct administration of VacA protein into the stomachs of mice results in gastric mucosal injury and gastric inflammation (Telford *et al.*, 1994; Supajatura *et al.*, 2002; Fujikawa *et al.*, 2003). Expression of RPTPβ (a VacA receptor) is required for the development of VacA-induced gastric ulcers in a mouse model (Fujikawa *et al.*, 2003). The ability of VacA to cause gastric damage in the mouse model through an RPTPβ-dependent pathway correlates with the finding that VacA interactions with RPTPβ result in detachment of primary mouse gastric epithelial cells from a reconstituted basement membrane (Fujikawa *et al.*, 2003). As a consequence of VacA-mediated damage to the gastric epithelium, a gastric mucosal inflammatory response may occur. In addition, several lines of evidence indicate that VacA may directly induce proinflammatory signalling within intoxicated cells. As discussed above, VacA induces p38 activation in several cell types, including gastric epithelial cells, T-cells, macrophages, and neutrophils (Boncristiano *et al.*, 2003). One potential consequence of p38 activation is increased expression of COX-2, which can contribute to inflammation via the generation of proinflammatory prostanoids (Boncristiano *et al.*, 2003; Hisatsune *et al.*, 2007). VacA can also contribute to inflammation by stimulating various cell types, including mast cells and eosinophils, to produce chemokines and proinflammatory cytokines (Supajatura *et al.*, 2002; Kim *et al.*, 2007).

Another approach for investigating a potential role of VacA in the pathogenesis of gastric disease has been to infect animals with isogenic wild type and *vacA*-null mutant strains, and compare the histological alterations produced by these strains. The results of one study suggested that VacA may have a small but detectable effect on development of gastric ulceration in gerbils (Ogura *et al.*, 2000), but VacA does not significantly contribute to changes in gastric histology in gnotobiotic piglets, mice, or gerbils (Eaton *et al.*, 1997; Ogura *et al.*, 2000; Salama *et al.*, 2001). Mouse T-cells are resistant to VacA activity (Algood *et al.*, 2007), which suggest that caution must be used when choosing appropriate animal models to study VacA activity. If VacA interacts specifically with cellular components found exclusively in human cells, most currently available animal models may not be suitable for studies of VacA activity *in vivo*.

Conclusions and future directions

Nearly 20 years of VacA research have revealed that this toxin has many novel attributes. Some traits of VacA are closely related to those of classic pore-forming bacterial toxins, whereas other characteristics are more closely related to those of bacterial toxins that act at intracellular sites. The structural features of VacA are unique among bacterial protein toxins, and intoxication of various types of human cells with VacA results in a wide range of different effects *in vitro*. Many cellular consequences of VacA intoxication are dependent on the ability of VacA to form membrane channels, whereas other cellular effects occur via channel-independent mechanisms.

A variety of evidence indicates that VacA may contribute to *H. pylori* colonization of the stomach, *H. pylori* persistence in the stomach, and gastroduodenal disease. Understanding the mechanisms by which VacA modulates the gastric environment represents a challenging topic for future investigation. Further study of VacA is likely to lead to important new findings that are relevant for diverse fields of biology, and may lead to a more clear understanding of how *H. pylori* interacts with the human stomach.

Acknowledgements

We thank members of the Blanke and Cover labs for helpful discussions. This work was supported by NIH R01 AI45928, NIH R01 AI39657, and the Department of Veterans Affairs.

References

Adrian, M., Cover, T.L., Dubochet, J., and Heuser, J.E. (2002). Multiple oligomeric states of the *Helicobacter pylori* vacuolating toxin demonstrated by cryo-electron microscopy. J. Mol. Biol. *318*, 121–133.

Algood, H.M., and Cover, T.L. (2006). *Helicobacter pylori* persistence: an overview of interactions between *H. pylori* and host immune defenses. Clin. Microbiol. Rev. *19*, 597–613.

Algood, H.M., Torres, V.J., Unutmaz, D., and Cover, T.L. (2007). Resistance of primary murine CD4+ T-cells to *Helicobacter pylori* vacuolating cytotoxin. Infect. Immun. *75*, 334–341.

Allan, E., Clayton, C.L., McLaren, A., Wallace, D.M., and Wren, B.W. (2001). Characterization of the low-pH responses of *Helicobacter pylori* using genomic DNA arrays. Microbiology *147*, 2285–2292.

Allen, L.A., Schlesinger, L.S., and Kang, B. (2000). Virulent strains of *Helicobacter pylori* demonstrate delayed phagocytosis and stimulate homotypic phagosome fusion in macrophages. J. Exp. Med. *191*, 115–128.

Amieva, M.R., Salama, N.R., Tompkins, L.S., and Falkow, S. (2002). *Helicobacter pylori* enter and survive within multivesicular vacuoles of epithelial cells. Cell Microbiol. *4*, 677–690.

Ang, S., Lee, C.Z., Peck, K., Sindici, M., Matrubutham, U., Gleeson, M.A., and Wang, J.T. (2001). Acid-induced gene expression in *Helicobacter pylori*: study in genomic scale by microarray. Infect. Immun. *69*, 1679–1686.

Ashorn, M., Cantet, F., Mayo, K., and Megraud, F. (2000). Cytoskeletal rearrangements induced by *Helicobacter pylori* strains in epithelial cell culture: possible role of the cytotoxin. Dig Dis. Sci. *45*, 1774–1780.

Ashour, A.A., Magalhaes, P.P., Mendes, E.N., Collares, G.B., de Gusmao, V.R., Queiroz, D.M., Nogueira, A.M., Rocha, G.A., and de Oliveira, C.A. (2002). Distribution of *vacA* genotypes in *Helicobacter pylori* strains isolated from Brazilian adult patients with gastritis, duodenal ulcer or gastric carcinoma. FEMS Immunol. Med. Microbiol. *33*, 173–178.

Atherton, J.C., Cao, P., Peek, R.M., Jr., Tummuru, M.K., Blaser, M.J., and Cover, T.L. (1995). Mosaicism in vacuolating cytotoxin alleles of *Helicobacter pylori*. Association of specific *vacA* types with cytotoxin production and peptic ulceration. J. Biol. Chem. *270*, 17771–17777.

Atherton, J.C., Peek, R.M., Jr., Tham, K.T., Cover, T.L., and Blaser, M.J. (1997). Clinical and pathological importance of heterogeneity in *vacA*, the vacuolating cytotoxin gene of *Helicobacter pylori*. Gastroenterology *112*, 92–99.

Atherton, J.C., Sharp, P.M., Cover, T.L., Gonzalez-Valencia, G., Peek, R.M., Jr., Thompson, S.A., Hawkey, C.J., and Blaser, M.J. (1999). Vacuolating cytotoxin (*vacA*) alleles of *Helicobacter pylori* comprise two geographically widespread types, m1 and m2, and have evolved through limited recombination. Curr. Microbiol. *39*, 211–218.

Ayala, G., Chihu, L., Perales, G., Fierros-Zarate, G., Hansen, L.M., Solnick, J.V., and Sanchez, J. (2004). Quantitation of *H. pylori* cytotoxin mRNA by real-time RT-PCR shows a wide expression range that does not correlate with promoter sequences. Microb. Pathog. *37*, 163–167.

Barnard, F.M., Loughlin, M.F., Fainberg, H.P., Messenger, M.P., Ussery, D.W., Williams, P., and Jenks, P.J. (2004). Global regulation of virulence and the stress response by CsrA in the highly adapted human gastric pathogen *Helicobacter pylori*. Mol. Microbiol. *51*, 15–32.

Bjorkholm, B., Zhukhovitsky, V., Lofman, C., Hulten, K., Enroth, H., Block, M., Rigo, R., Falk, P., and Engstrand, L. (2000). *Helicobacter pylori* entry into human gastric epithelial cells: A potential determinant of virulence, persistence, and treatment failures. *Helicobacter* *5*, 148–154.

Blanke, S.R. (2005). Micro-managing the executioner: pathogen targeting of mitochondria. Trends Microbiol. *13*, 64–71.

Blaser, M.J., and Atherton, J.C. (2004). *Helicobacter pylori* persistence: biology and disease. J. Clin. Invest *113*, 321–333.

Boncristiano, M., Paccani, S.R., Barone, S., Ulivieri, C., Patrussi, L., Ilver, D., Amedei, A., D'Elios, M.M., Telford, J.L., and Baldari, C.T. (2003). The *Helicobacter pylori* vacuolating toxin inhibits T-cell activation by two independent mechanisms. J. Exp. Med. *198*, 1887–1897.

Boonjakuakul, J.K., Canfield, D.R., and Solnick, J.V. (2005). Comparison of *Helicobacter pylori* virulence gene expression *in vitro* and in the Rhesus macaque. Infect. Immun. *73*, 4895–4904.

Boonjakuakul, J.K., Syvanen, M., Suryaprasad, A., Bowlus, C.L., and Solnick, J.V. (2004). Transcription profile of *Helicobacter pylori* in the human stomach reflects its physiology in vivo. J. Infect. Dis. *190*, 946–956.

Boquet, P., Ricci, V., Galmiche, A., and Gauthier, N.C. (2003). Gastric cell apoptosis and *H. pylori*: has the main function of VacA finally been identified? Trends Microbiol. *11*, 410–413.

Brest, P., Hofman, V., Lassalle, S., Cesaro, A., Ricci, V., Selva, E., Auberger, P., and Hofman, P. (2006). Human polymorphonuclear leukocytes are sensitive *in vitro* to *Helicobacter pylori* vaca toxin. *Helicobacter* *11*, 544–555.

Bumann, D., Aksu, S., Wendland, M., Janek, K., Zimny-Arndt, U., Sabarth, N., Meyer, T.F., and Jungblut, P.R. (2002). Proteome analysis of secreted proteins of the gastric pathogen *Helicobacter pylori*. Infect. Immun. *70*, 3396–3403.

Campello, S., Tombola, F., Cabrini, G., and Zoratti, M. (2002). The vacuolating toxin of *Helicobacter pylori* mimicks the CFTR-mediated chloride conductance. FEBS Lett. *532*, 237–240.

Cao, P., and Cover, T.L. (2002). Two different families of hopQ alleles in *Helicobacter pylori*. J. Clin. Microbiol. *40*, 4504–4511.

Caputo, R., Tuccillo, C., Manzo, B.A., Zarrilli, R., Tortora, G., Blanco Cdel, V., Ricci, V., Ciardiello, F., and Romano, M. (2003). *Helicobacter pylori* VacA toxin upregulates vascular endothelial growth factor expression in MKN 28 gastric cells through an epidermal growth factor receptor-, cyclooxygenase-2-dependent mechanism. Clin. Cancer Res. *9*, 2015–2021.

Catrenich, C.E., and Chestnut, M.H. (1992). Character and origin of vacuoles induced in mammalian cells by the cytotoxin of *Helicobacter pylori*. J. Med. Microbiol. *37*, 389–395.

Cho, S.J., Kang, N.S., Park, S.Y., Kim, B.O., Rhee, D.K., and Pyo, S. (2003). Induction of apoptosis and expression of apoptosis related genes in human epithelial carcinoma cells by *Helicobacter pylori* VacA toxin. Toxicon *42*, 601–611.

Cover, T.L., and Blanke, S.R. (2005). *Helicobacter pylori* VacA, a paradigm for toxin multifunctionality. Nat. Rev. Microbiol. 3, 320–332.

Cover, T.L., and Blaser, M.J. (1992). Purification and characterization of the vacuolating toxin from *Helicobacter pylori*. J. Biol. Chem. *267*, 10570–10575.

Cover, T.L., Cao, P., Lind, C.D., Tham, K.T., and Blaser, M.J. (1993a). Correlation between vacuolating cytotoxin production by *Helicobacter pylori* isolates *in vitro* and in vivo. Infect. Immun. *61*, 5008–5012.

Cover, T.L., Dooley, C.P., and Blaser, M.J. (1990). Characterization of and human serologic response to proteins in *Helicobacter pylori* broth culture supernatants with vacuolizing cytotoxin activity. Infect. Immun. 58, 603–610.

Cover, T.L., Hanson, P.I., and Heuser, J.E. (1997). Acid-induced dissociation of VacA, the *Helicobacter pylori* vacuolating cytotoxin, reveals its pattern of assembly. J. Cell Biol. *138*, 759–769.

Cover, T.L., Krishna, U.S., Israel, D.A., and Peek, R.M., Jr. (2003). Induction of gastric epithelial cell apoptosis by *Helicobacter pylori* vacuolating cytotoxin. Cancer Res. *63*, 951–957.

Cover, T.L., Puryear, W., Pérez-Pérez, G.I., and Blaser, M.J. (1991). Effect of urease on HeLa cell vacuolation induced by *Helicobacter pylori* cytotoxin. Infect. Immun. 59, 1264–1270.

Cover, T.L., Reddy, L.Y., and Blaser, M.J. (1993b). Effects of ATPase inhibitors on the response of HeLa cells to *Helicobacter pylori* vacuolating toxin. Infect. Immun. *61*, 1427–1431.

Cover, T.L., Tummuru, M.K.R., Cao, P., Thompson, S.A., and Blaser, M.J. (1994). Divergence of genetic sequences for the vacuolating cytotoxin among *Helicobacter pylori* strains. J. Biol. Chem. *269*, 10566–10573.

Cover, T.L., Vaughn, S.G., Cao, P., and Blaser, M.J. (1992). Potentiation of *Helicobacter pylori* vacuolating toxin activity by nicotine and other weak bases. J. Infect. Dis. *166*, 1073–1078.

Czajkowsky, D.M., Iwamoto, H., Cover, T.L., and Shao, Z. (1999). The vacuolating toxin from *Helicobacter pylori* forms hexameric pores in lipid bilayers at low pH. Proc. Natl. Acad. Sci. USA *96*, 2001–2006.

Czajkowsky, D.M., Iwamoto, H., Szabo, G., Cover, T.L., and Shao, Z. (2005). Mimicry of a host anion channel by a *Helicobacter pylori* pore-forming toxin. Biophys. J. *89*, 3093–3101.

Dailidiene, D., Dailide, G., Ogura, K., Zhang, M., Mukhopadhyay, A.K., Eaton, K.A., Cattoli, G., Kusters, J.G., and Berg, D.E. (2004). *Helicobacter acinonychis*: genetic and rodent infection studies of a *Helicobacter pylori*-like gastric pathogen of cheetahs and other big cats. J. Bacteriol. *186*, 356–365.

Dautin, N., and Bernstein, H. (2006). Protein secretion in Gram-negative bacteria via the autotransporter pathway. Annu. Rev. Microbiol.

de Bernard, M., Arico, B., Papini, E., Rizzuto, R., Grandi, G., Rappuoli, R., and Montecucco, C. (1997). *Helicobacter pylori* toxin VacA induces vacuole formation by acting in the cell cytosol. Mol. Microbiol. *26*, 665–674.

de Bernard, M., Burroni, D., Papini, E., Rappuoli, R., Telford, J., and Montecucco, C. (1998). Identification of the *Helicobacter pylori* VacA toxin domain active in the cell cytosol. Infect. Immun. *66*, 6014–6016.

de Bernard, M., Cappon, A., Del Giudice, G., Rappuoli, R., and Montecucco, C. (2004). The multiple cellular activities of the VacA cytotoxin of *Helicobacter pylori*. Int. J. Med. Microbiol. *293*, 589–597.

de Bernard, M., Cappon, A., Pancotto, L., Ruggiero, P., Rivera, J., Del Giudice, G., and Montecucco, C. (2005). The *Helicobacter pylori* VacA cytotoxin activates RBL-2H3 cells by inducing cytosolic calcium oscillations. Cell Microbiol. *7*, 191–198.

de Bernard, M., Moschioni, M., Habermann, A., Griffiths, G., and Montecucco, C. (2002). Cell vacuolization induced by *Helicobacter pylori* VacA cytotoxin does not depend on late endosomal SNAREs. Cell Microbiol. *4*, 11–18.

de Bernard, M., Moschioni, M., Napolitani, G., Rappuoli, R., and Montecucco, C. (2000). The VacA toxin of *Helicobacter pylori* identifies a new intermediate filament-interacting protein. Embo J. *19*, 48–56.

de Bernard, M., Papini, E., de Filippis, V., Gottardi, E., Telford, J., Manetti, R., Fontana, A., Rappuoli, R., and Montecucco, C. (1995). Low pH activates the vacuolating toxin of *Helicobacter pylori*, which becomes acid and pepsin resistant. J. Biol. Chem. *270*, 23937–23940.

De Gusmao, V.R., Nogueira Mendes, E., De Magalhaes Queiroz, D.M., Aguiar Rocha, G., Camargos Rocha, A.M., Ramadan Ashour, A.A., and Teles Carvalho, A.S. (2000). *vacA* genotypes in *Helicobacter pylori* strains isolated from children with and without duodenal ulcer in brazil. J. Clin. Microbiol. *38*, 2853–2857.

De Guzman, B.B., Hisatsune, J., Nakayama, M., Yahiro, K., Wada, A., Yamasaki, E., Nishi, Y., Yamazaki, S., Azuma, T., Ito, Y., Ohtani, M., van der Wijk, T., den Hertog, J., Moss, J., and Hirayama, T. (2005). Cytotoxicity and recognition of receptor-like protein tyrosine phosphatases, RPTPalpha and RPTPbeta, by *Helicobacter pylori* m2 VacA. Cell Microbiol. *7*, 1285–1293.

Debellis, L., Papini, E., Caroppo, R., Montecucco, C., and Curci, S. (2001). *Helicobacter pylori* cytotoxin VacA increases alkaline secretion in gastric epithelial cells. Am. J. Physiol Gastrointes Liver Physiol *281*, G1440–1448.

Dossumbekova, A., Prinz, C., Mages, J., Lang, R., Kusters, J.G., van Vliet, A.H.M., Reindl, W., Backert, S., Saur, D., Schmid, R.M., and Rad, R. (2006). *Helicobacter pylori* HopH (OipA) and bacterial pathogenicity: genetic and functional genomic analysis of hopH gene polymorphisms. J. Infect. Dis., in press.

Dunn, B.E., Cohen, H., and Blaser, M.J. (1997). *Helicobacter pylori*. Clin. Microbiol. Rev. *10*, 720–741.

Eaton, K.A., Cover, T.L., Tummuru, M.K., Blaser, M.J., and Krakowka, S. (1997). Role of vacuolating cytotoxin in gastritis due to *Helicobacter pylori* in gnotobiotic piglets. Infect. Immun. *65*, 3462–3464.

El-Bez, C., Adrian, M., Dubochet, J., and Cover, T.L. (2005). High resolution structural analysis of *Helicobacter pylori* VacA toxin oligomers by cryo-negative staining electron microscopy. J. Struc Biol. *151*, 215–228.

Figueiredo, C., Machado, J.C., Pharoah, P., Seruca, R., Sousa, S., Carvalho, R., Capelinha, A.F., Quint, W., Caldas, C., van Doorn, L.J., Carneiro, F., and Sobrinho-Simoes, M. (2002). *Helicobacter pylori* and interleukin 1 genotyping: an opportunity to identify high-risk individuals for gastric carcinoma. J. Natl. Cancer Inst *94*, 1680–1687.

Figura, N., Guglielmetti, P., Rossolini, A., Barberi, A., Cusi, G., Musmanno, R.A., Russi, M., and Quaranta, S. (1989). Cytotoxin production by *Campylobacter pylori* strains isolated from patients with peptic ulcers and from patients with chronic gastritis only. J. Clin. Microbiol. *27*, 225–226.

Fiocca, R., Necchi, V., Sommi, P., Ricci, V., Telford, J., Cover, T.L., and Solcia, E. (1999). Release of *Helicobacter pylori* vacuolating cytotoxin by both a specific secretion pathway and budding of outer membrane vesicles. Uptake of released toxin and vesicles by gastric epithelium. J. Pathol. *188*, 220–226.

Fischer, W., Buhrdorf, R., Gerland, E., and Haas, R. (2001). Outer membrane targeting of passenger proteins by the vacuolating cytotoxin autotransporter of *Helicobacter pylori*. Infect. Immun. *69*, 6769–6775.

Forsyth, M.H., Atherton, J.C., Blaser, M.J., and Cover, T.L. (1998). Heterogeneity in levels of vacuolating cytotoxin gene (*vacA*) transcription among *Helicobacter pylori* strains. Infect. Immun. *66*, 3088–3094.

Forsyth, M.H., and Cover, T.L. (1999). Mutational analysis of the *vacA* promoter provides insight into gene transcription in *Helicobacter pylori*. J. Bacteriol. *181*, 2261–2266.

Forsyth, M.H., and Cover, T.L. (2000). Intercellular communication in *Helicobacter pylori*: luxS is essential for the production of an extracellular signaling molecule. Infect. Immun. *68*, 3193–3199.

Fujikawa, A., Shirasaka, D., Yamamoto, S., Ota, H., Yahiro, K., Fukada, M., Shintani, T., Wada, A., Aoyama, N., Hirayama, T., Fukamachi, H., and Noda, M. (2003). Mice deficient in protein tyrosine phosphatase receptor type Z are resistant to gastric ulcer induction by VacA of *Helicobacter pylori*. Nat. Genet. *33*, 375–381.

Galmiche, A., Rassow, J., Doye, A., Cagnol, S., Chambard, J.C., Contamin, S., de Thillot, V., Just, I., Ricci, V., Solcia, E., Van Obberghen, E., and Boquet, P. (2000). The N-terminal 34 kDa fragment of *Helicobacter pylori* vacuolating cytotoxin targets mitochondria and induces cytochrome c release. EMBO J. *19*, 6361–6370.

Garner, J.A., and Cover, T.L. (1996). Binding and internalization of the *Helicobacter pylori* vacuolating cytotoxin by epithelial cells. Infect. Immun. *64*, 4197–4203.

Gauthier, N.C., Monzo, P., Gonzalez, T., Doye, A., Oldani, A., Gounon, P., Ricci, V., Cormont, M., and Boquet, P. (2007). Early endosomes associated with dynamic F-actin structures are required for late trafficking of *H. pylori* VacA toxin. J. Cell Biol. *177*, 343–354.

Gauthier, N.C., Monzo, P., Kaddai, V., Doye, A., Ricci, V., and Boquet, P. (2005). *Helicobacter pylori* VacA cytotoxin: a probe for a clathrin-independent and Cdc42-dependent pinocytic pathway routed to late endosomes. Mol. Biol. Cell *16*, 4852–4866.

Gauthier, N.C., Ricci, V., Gounon, P., Doye, A., Tauc, M., Poujeol, P., and Boquet, P. (2004). Glycosylphosphatidylinositol-anchored proteins and actin cytoskeleton modulate chloride transport by channels formed by the *Helicobacter pylori* vacuolating cytotoxin VacA in HeLa cells. J. Biol. Chem. *279*, 9481–9489.

Gauthier, N.C., Ricci, V., Landraud, L., and Boquet, P. (2006). *Helicobacter pylori* VacA toxin: a tool to study novel early endosomes. Trends Microbiol. *14*, 292–294.

Gebert, B., Fischer, W., Weiss, E., Hoffman, R., and Haas, R. (2003). *Helicobacter pylori* vacuolating cytotoxin inhibits T lymphocyte activation. Science *301*, 1099–1102.

Geisse, N.A., Cover, T.L., Henderson, R.M., and Edwardson, J.M. (2004). Targeting of *Helicobacter pylori* vacuolating toxin to lipid raft membrane domains analysed by atomic force microscopy. Biochem. J. *381*, 911–917.

Genisset, C., Galeotti, C.L., Lupetti, P., Mercati, D., Skibinski, D.A., Barone, S., Battistutta, R., de Bernard, M., and Telford, J.L. (2006). A *Helicobacter pylori* vacuolating toxin mutant that fails to oligomerize has a dominant negative phenotype. Infect. Immun. *74*, 1786–1794.

Genisset, C., Puhar, A., Calore, F., de Bernard, M., Dell'Antone, P., and Montecucco, C. (2007). The concerted action of the *Helicobacter pylori* cytotoxin VacA and of the v-ATPase proton pump induces swelling of isolated endosomes. Cell Microbiol. *9*, 1481–1490.

Gerhard, M., Lehn, N., Neumayer, N., Boren, T., Rad, R., Schepp, W., Miehlke, S., Classen, M., and Prinz, C. (1999). Clinical relevance of the *Helicobacter pylori* gene for blood-group antigen-binding adhesin. Proc. Natl. Acad. Sci. USA *96*, 12778–12783.

Ghiara, P., Rossi, M., Marchetti, M., Di Tommaso, A., Vindigni, C., Ciampolini, F., Covacci, A., Telford, J.L., De Magistris, M.T., Pizza, M., Rappuoli, R., and Del Giudice, G. (1997). Therapeutic intragastric vaccination against *Helicobacter pylori* in mice eradicates an otherwise chronic infection and confers protection against reinfection. Infect. Immun. *65*, 4996–5002.

Gieseler, S., Konig, B., Konig, W., and Backert, S. (2005). Strain-specific expression profiles of virulence genes in *Helicobacter pylori* during infection of gastric epithelial cells and granulocytes. Microbes Infect. *7*, 437–447.

Guarino, A., Bisceglia, M., Canani, R.B., Boccia, M.C., Mallardo, G., Bruzzese, E., Massari, P., Rappuoli, R., and Telford, J. (1998). Enterotoxic effect of the vacuolating toxin produced by *Helicobacter pylori* in Caco-2 cells. J. Infect. Dis. *178*, 1373–1378.

Guo, B.P., and Mekalanos, J.J. (2002). Rapid genetic analysis of *Helicobacter pylori* gastric mucosal colonization in suckling mice. Proc. Natl. Acad. Sci. USA *99*, 8354–8359.

Han, S.R., Schreiber, H.J., Bhakdi, S., Loos, M., and Maeurer, M.J. (1998). vacA genotypes and genetic diversity in clinical isolates of *Helicobacter pylori*. *Clin. Diagn.* Lab. Immunol. 5, 139–145.

Hennig, E.E., Godlewski, M.M., Butruk, E., and Ostrowski, J. (2005). *Helicobacter pylori* VacA cytotoxin interacts with fibronectin and alters HeLa cell adhesion and cytoskeletal organization in vitro. FEMS Immunol. Med. Microbiol. *44*, 143–150.

Hisatsune, J., Yamasaki, E., Nakayama, M., Shirasaka, D., Kurazono, H., Katagata, Y., Inoue, H., Han, J., Sap, J., Yahiro, K., Moss, J., and Hirayama, T. (2007). *Helicobacter pylori* VacA enhances PGE2 production through induction of COX-2 expression via a p38 MAP kinase/ATF-2 cascade in AZ-521 cells. Infect. Immun. *75*, 4472–4481.

Hotchin, N.A., Cover, T.L., and Akhtar, N. (2000). Cell vacuolation induced by the VacA cytotoxin of *Helicobacter pylori* is regulated by the rac1 GTPase. J. Biol. Chem. *275*, 14009–14012.

Ikonomov, O.C., Sbrissa, D., Yoshimori, T., Cover, T.L., and Shisheva, A. (2002). PIKfyve Kinase and SKD1 AAA ATPase define distinct endocytic compartments. Only PIKfyve expression inhibits the cell-vacoulating activity of *Helicobacter pylori* VacA toxin. J. Biol. Chem. *277*, 46785–46790.

Ilver, D., Barone, S., Mercati, D., Lupetti, P., and Telford, J.L. (2004). *Helicobacter pylori* toxin VacA is transferred to host cells via a novel contact-dependent mechanism. Cell Microbiol. *6*, 167–174.

Ito, Y., Azuma, T., Ito, S., Suto, H., Miyaji, H., Yamazaki, Y., Kohli, Y., and Kuriyama, M. (1998). Full-length sequence analysis of the vacA gene from cytotoxic and noncytotoxic *Helicobacter pylori*. *J.* Infect. Dis. *178*, 1391–1398.

Iwamoto, H., Czajkowsky, D.M., Cover, T.L., Szabo, G., and Shao, Z. (1999). VacA from *Helicobacter pylori*: a hexameric chloride channel. FEBS Lett. *450*, 101–104.

Ji, X., Fernandez, T., Burroni, D., Pagliaccia, C., Atherton, J.C., Reyrat, J.M., Rappuoli, R., and Telford, J.L. (2000). Cell specificity of *Helicobacter pylori* cytotoxin is determined by a short region in the polymorphic midregion. Infect. Immun. *68*, 3754–3757.

Ji, X., Frati, F., Barone, S., Pagliaccia, C., Burroni, D., Xu, G., Rappuoli, R., Reyrat, J.M., and Telford, J.L. (2002). Evolution of functional polymorphism in the gene coding for the *Helicobacter pylori* cytotoxin. FEMS Microbiol. Lett. *206*, 253–258.

Keenan, J., Day, T., Neal, S., Cook, B., Perez-Perez, G., Allardyce, R., and Bagshaw, P. (2000). A role for the bacterial outer membrane in the pathogenesis of *Helicobacter pylori* infection. FEMS Microbiol. Lett. *182*, 259–264.

Kidd, M., Lastovica, A.J., Atherton, J.C., and Louw, J.A. (1999). Heterogeneity in the *Helicobacter pylori vacA* and *cagA* genes: association with gastroduodenal disease in South Africa? Gut 45, 499–502.

Kim, J.M., Kim, J.S., Lee, J.Y., Kim, Y.J., Youn, H.J., Kim, I.Y., Chee, Y.J., Oh, Y.K., Kim, N., Jung, H.C., and Song, I.S. (2007). Vacuolating cytotoxin in *Helicobacter pylori* water-soluble proteins upregulates chemokine expression in human eosinophils via Ca2+ influx, mitochondrial reactive oxygen intermediates, and NF-kappaB activation. Infect. Immun. *75*, 3373–3381.

Kim, S., Chamberlain, A.K., and Bowie, J.U. (2004). Membrane channel structure of *Helicobacter pylori* vacuolating toxin: role of multiple GXXXG motifs in cylindrical channels. Proc. Natl. Acad. Sci. USA. *101*, 5988–5991.

Kimura, M., Goto, S., Ihara, Y., Wada, A., Yahiro, K., Niidome, T., Aoyagi, H., Hirayama, T., and Kondo, T. (2001). Impairment of glutathione metabolism in human gastric epithelial cells treated with vacuolating cytotoxin from *Helicobacter pylori*. Microb Pathog *31*, 29–36.

Kimura, M., Goto, S., Wada, A., Yahiro, K., Niidome, T., Hatakeyama, T., Aoyagi, H., Hirayama, T., and Kondo, T. (1999). Vacuolating cytotoxin purified from *Helicobacter pylori* causes mitochondrial damage in human gastric cells. Microb .Pathog. *26*, 45–52.

Kuck, D., Kolmerer, B., Iking-Konert, C., Krammer, P.H., Stremmel, W., and Rudi, J. (2001). Vacuolating cytotoxin of *Helicobacter pylori* induces apoptosis in the human gastric epithelial cell line AGS. Infect. Immun. *69*, 5080–5087.

Kuo, C.H., and Wang, W.C. (2003). Binding and internalization of *Helicobacter pylori* VacA via cellular lipid rafts in epithelial cells. Biochem. Biophys Res. Commun. *303*, 640–644.

Lanzavecchia, S., Bellon, P.L., Lupetti, P., Dallai, R., Rappuoli, R., and Telford, J.L. (1998). Three-dimensional reconstruction of metal replicas of the *Helicobacter pylori* vacuolating cytotoxin. J. Struct Biol. *121*, 9–18.

Letley, D.P., and Atherton, J.C. (2000). Natural diversity in the N. terminus of the mature vacuolating cytotoxin of *Helicobacter pylori* determines cytotoxin activity. J. Bacteriol. *182*, 3278–3280.

Letley, D.P., Lastovica, A., Louw, J.A., Hawkey, C.J., and Atherton, J.C. (1999). Allelic diversity of the *Helicobacter pylori* vacuolating cytotoxin gene in South Africa: rarity of the *vacA* s1a genotype and natural occurrence of an s2/m1 allele. J. Clin. Microbiol. *37*, 1203–1205.

Letley, D.P., Rhead, J.L., Bishop, K., and Atherton, J.C. (2006). Paired cysteine residues are required for high levels of the *Helicobacter pylori* autotransporter VacA. Microbiology *152*, 1319–1325.

Letley, D.P., Rhead, J.L., Twells, R.J., Dove, B., and Atherton, J.C. (2003). Determinants of non-toxicity in the gastric pathogen *Helicobacter pylori*. J. Biol. Chem. *278*, 26734–26741.

Leunk, R.D., P.T., J., David, B.C., Kraft, W.G., and Morgan, D.R. (1988). Cytotoxic activity in broth-culture filtrates of *Campylobacter pylori*. J. Med. Microbiol. *26*, 93–99.

Li, Y., Wandinger-Ness, A., Goldenring, J.R., and Cover, T.L. (2004). Clustering and redistribution of late endocytic compartments in response to *Helicobacter pylori* vacuolating toxin. Mol. Biol. Cell, *15*, 1946–1959.

Lupetti, P., Heuser, J.E., Manetti, R., Massari, P., Lanzavecchia, S., Bellon, P.L., Dallai, R., Rappuoli, R., and Telford, J.L. (1996). Oligomeric and subunit structure of the *Helicobacter pylori* vacuolating cytotoxin. J. Cell Biol. *133*, 801–807.

Marchetti, M., Aricò, B., Burroni, D., Figura, N., Rappuoli, R., and Ghiara, P. (1995). Development of a mouse model of *Helicobacter pylori* infection that mimics human disease. Science *267*, 1655–1658.

Marchetti, M., Rossi, M., Giannelli, V., Giuliani, M.M., Pizza, M., Censini, S., Covacci, A., Massari, P., Pagliaccia, C., Manetti, R., Telford, J.L., Douce, G., Dougan, G., Rappuoli, R., and Ghiara, P. (1998). Protection against *Helicobacter pylori* infection in mice by intragastric vaccination with *H. pylori* antigens is achieved using a non-toxic mutant of *E. coli* heat-labile enterotoxin (LT) as adjuvant. Vaccine *16*, 33–37.

Massari, P., Manetti, R., Burroni, D., Nuti, S., Norais, N., Rappuoli, R., and Telford, J.L. (1998). Binding of the *Helicobacter pylori* vacuolating cytotoxin to target cells. Infect. Immun. *66*, 3981–3984.

McClain, M.S., Cao, P., and Cover, T.L. (2001a). Amino-terminal hydrophobic region of *Helicobacter pylori* vacuolating cytotoxin (VacA) mediates transmembrane protein dimerization. Infect. Immun. *69*, 1181–1184.

McClain, M.S., Cao, P., Iwamoto, H., Vinion-Dubiel, A.D., Szabo, G., Shao, Z., and Cover, T.L. (2001b). A 12-Amino-acid segment, present in type s2 but not type s1 *Helicobacter pylori* VacA proteins, abolishes cytotoxin activity and alters membrane channel formation. J. Bacteriol. *183*, 6499–6508.

McClain, M.S., Iwamoto, H., Cao, P., Vinion-Dubiel, A.D., Li, Y., Szabo, G., Shao, Z., and Cover, T.L. (2003). Essential role of a GXXXG motif for membrane channel formation by *Helicobacter pylori* vacuolating toxin. J. Biol. Chem. *278*, 12101–12108.

McClain, M.S., Schraw, W., Ricci, V., Boquet, P., and Cover, T.L. (2000). Acid-activation of *Helicobacter pylori* vacuolating cytotoxin (VacA) results in toxin internalization by eukaryotic cells. Mol. Microbiol. *37*, 433–442.

Menaker, R.J., Ceponis, P.J., and Jones, N.L. (2004). *Helicobacter pylori* induces apoptosis of macrophages in association with alterations in the mitochondrial pathway. Infect. Immun. *72*, 2889–2898.

Merrell, D.S., Goodrich, M.L., Otto, G., Tompkins, L.S., and Falkow, S. (2003a). pH-regulated gene expression of the gastric pathogen *Helicobacter pylori*. Infect. Immun. *71*, 3529–3539.

Merrell, D.S., Thompson, L.J., Kim, C.C., Mitchell, H., Tompkins, L.S., Lee, A., and Falkow, S. (2003b). Growth phase-dependent response of *Helicobacter pylori* to iron starvation. Infect. Immun. *71*, 6510–6525.

Miehlke, S., Kirsch, C., Agha-Amiri, K., Gunther, T., Lehn, N., Malfertheiner, P., Stolte, M., Ehninger, G., and Bayerdorffer, E. (2000). The *Helicobacter pylori vacA* s1, m1 genotype and *cagA* is associated with gastric carcinoma in Germany. Int. J. Cancer *87*, 322–327.

Molinari, M., Galli, C., de Bernard, M., Norais, N., Ruysschaert, J.M., Rappuoli, R., and Montecucco, C. (1998a). The acid activation of *Helicobacter pylori* toxin VacA: structural and membrane binding studies. Biochem. Biophys Res. Commun. *248*, 334–340.

Molinari, M., Galli, C., Norais, N., Telford, J.L., Rappuoli, R., Luzio, J.P., and Montecucco, C. (1997). Vacuoles induced by *Helicobacter pylori* toxin contain both late endosomal and lysosomal markers. J. Biol. Chem. *272*, 25339–25344.

Molinari, M., Salio, M., Galli, C., Norais, N., Rappuoli, R., Lanzavecchia, A., and Montecucco, C. (1998b). Selective inhibition of Ii-dependent antigen presentation by *Helicobacter pylori* toxin VacA. J. Exp. Med. *187*, 135–140.

Moll, G., Papini, E., Colonna, R., Burroni, D., Telford, J., Rappuoli, R., and Montecucco, C. (1995). Lipid interaction of the 37-kDa and 58-kDa fragments of the *Helicobacter pylori* cytotoxin. Eur. J. Biochem. *234*, 947–952.

Monack, D.M., Mueller, A., and Falkow, S. (2004). Persistent bacterial infections: the interface of the pathogen and the host immune system. Nat. Rev. Microbiol. *2*, 747–765.

Montecucco, C., and Rappuoli, R. (2001). Living dangerously: how *Helicobacter pylori* survives in the human stomach. Nat. Rev. Mol. Cell Biol. *2*, 457–466.

Morbiato, L., Tombola, F., Campello, S., Del Giudice, G., Rappuoli, R., Zoratti, M., and Papini, E. (2001). Vacuolation induced by VacA toxin of *Helicobacter pylori* requires the intracellular accumulation of membrane permeant bases, Cl(-) and water. FEBS Lett. *508*, 479–483.

Nakayama, M., Hisatsune, J., Yamasaki, E., Nishi, Y., Wada, A., Kurazono, H., Sap, J., Yahiro, K., Moss, J., and Hirayama, T. (2006). Clustering of *Helicobacter pylori* VacA in lipid rafts, mediated by its receptor, receptor-like protein tyrosine phosphatase beta, is required for intoxication in AZ-521 Cells. Infect. Immun. *74*, 6571–6580.

Nakayama, M., Kimura, M., Wada, A., Yahiro, K., Ogushi, K.I., Niidome, T., Fujikawa, A., Shirasaka, D., Aoyama, N., Kurazono, H., Noda, M., Moss, J., and Hirayama, T. (2004). *Helicobacter pylori* VacA activates the p38/ATF-2-mediated signal pathway in AZ-521 cells. J. Biol. Chem. *279*, 7024–7028.

Nguyen, V.Q., Caprioli, R.M., and Cover, T.L. (2001). Carboxy-terminal proteolytic processing of *Helicobacter pylori* vacuolating toxin. Infection and Immunity *69*, 543–546.

Nogueira, C., Figueiredo, C., Carneiro, F., Gomes, A.T., Barreira, R., Figueira, P., Salgado, C., Belo, L., Peixoto, A., Bravo, J.C., Bravo, L.E., Realpe, J.L., Plaisier, A.P., Quint, W.G., Ruiz, B., Correa, P., and van Doorn, L.J. (2001). *Helicobacter pylori* genotypes may determine gastric histopathology. Am. J. Pathol. *158*, 647–654.

Ogura, K., Maeda, S., Nakao, M., Watanabe, T., Tada, M., Kyutoku, T., Yoshida, H., Shiratori, Y., and Omata, M. (2000). Virulence factors of *Helicobacter pylori* responsible for gastric diseases in mongolian gerbil. J. Exp. Med. *192*, 1601–1610.

Oswald-Richter, K., Torres, V.J., Sundrud, M.S., VanCompernolle, S.E., Cover, T.L., and Unutmaz, D. (2006). *Helicobacter pylori* VacA toxin inhibits human immunodeficiency virus infection of primary human T-cells. J. Virol. *80*, 11767–11775.

Padilla, P.I., Wada, A., Yahiro, K., Kimura, M., Niidome, T., Aoyagi, H., Kumatori, A., Anami, M., Hayashi, T., Fujisawa, J., Saito, H., Moss, J., and Hirayama, T. (2000). Morphologic differentiation of HL-60 cells is associated with appearance of RPTPbeta and induction of *Helicobacter pylori* VacA sensitivity. J. Biol. Chem. *275*, 15200–15206.

Pagliaccia, C., de Bernard, M., Lupetti, P., Ji, X., Burroni, D., Cover, T.L., Papini, E., Rappuoli, R., Telford, J.L., and Reyrat, J.M. (1998). The m2 form of the *Helicobacter pylori* cytotoxin has cell type-specific vacuolating activity. Proc. Natl. Acad. Sci. USA *95*, 10212–10217.

Pagliaccia, C., Wang, X.M., Tardy, F., Telford, J.L., Ruysschaert, J.M., and Cabiaux, V. (2000). Structure and interaction of VacA of *Helicobacter pylori* with a lipid membrane. Eur. J. Biochem. *267*, 104–109.

Pai, R., Cover, T.L., and Tarnawski, A.S. (1999). *Helicobacter pylori* vacuolating cytotoxin (VacA) disorganizes the cytoskeletal architecture of gastric epithelial cells. Biochem. Biophys Res. Commun. *262*, 245–250.

Pai, R., Sasaki, E., and Tarnawski, A.S. (2000). *Helicobacter pylori* vacuolating cytotoxin (VacA) alters cytoskeleton-associated proteins and interferes with re-epithelialization of wounded gastric epithelial monolayers. Cell Biol. Int. *24*, 291–301.

Pan, Z.J., Berg, D.E., van der Hulst, R.W., Su, W.W., Raudonikiene, A., Xiao, S.D., Dankert, J., Tytgat, G.N., and van der Ende, A. (1998). Prevalence of vacuolating cytotoxin production and distribution of distinct *vacA* alleles in *Helicobacter pylori* from China. J. Infect. Dis. *178*, 220–226.

Papini, E., Bugnoli, M., De Bernard, M., Figura, N., Rappuoli, R., and Montecucco, C. (1993). Bafilomycin A1 inhibits *Helicobacter pylori*-induced vacuolization of HeLa cells. Mol. Microbiol. *7*, 323–327.

Papini, E., de Bernard, M., Milia, E., Bugnoli, M., Zerial, M., Rappuoli, R., and Montecucco, C. (1994). Cellular vacuoles induced by *Helicobacter pylori* originate from late endosomal compartments. Proc. Natl. Acad. Sci. USA. *91*, 9720–9724.

Papini, E., Gottardi, E., Satin, B., de Bernard, M., Massari, P., Telford, J., Rappuoli, R., Sato, S.B., and Montecucco, C. (1996). The vacuolar ATPase proton pump is present on intracellular vacuoles induced by *Helicobacter pylori*. J. Med. Microbiol. *45*, 84–89.

Papini, E., Satin, B., Bucci, C., de Bernard, M., Telford, J.L., Manetti, R., Rappuoli, R., Zerial, M., and Montecucco, C. (1997). The small GTP binding protein rab7 is essential for cellular vacuolation induced by *Helicobacter pylori* cytotoxin. Embo J. *16*, 15–24.

Papini, E., Satin, B., Norais, N., de Bernard, M., Telford, J.L., Rappuoli, R., and Montecucco, C. (1998). Selective increase of the permeability of polarized epithelial cell monolayers by *Helicobacter pylori* vacuolating toxin. J. Clin. Invest *102*, 813–820.

Patel, H.K., Willhite, D.C., Patel, R.M., Ye, D., Williams, C.L., Torres, E.M., Marty, K.B., MacDonald, R.A., and Blanke, S.R. (2002). Plasma membrane cholesterol modulates cellular vacuolation induced by the *Helicobacter pylori* vacuolating cytotoxin. Infect. Immun. *70*, 4112–4123.

Peek, R.M., Jr., Blaser, M.J., Mays, D.J., Forsyth, M.H., Cover, T.L., Song, S.Y., Krishna, U., and Pietenpol, J.A. (1999). *Helicobacter pylori* strain-specific genotypes and modulation of the gastric epithelial cell cycle. Cancer Res. *59*, 6124–6131.

Pelicic, V., Reyrat, J.M., Sartori, L., Pagliaccia, C., Rappuoli, R., Telford, J.L., Montecucco, C., and Papini, E. (1999). *Helicobacter pylori* VacA cytotoxin associated with the bacteria increases epithelial permeability independently of its vacuolating activity. Microbiology *145*, 2043–2050.

Petersen, A.M., Sorensen, K., Blom, J., and Krogfelt, K.A. (2001). Reduced intracellular survival of *Helicobacter pylori vacA* mutants in comparison with their wild-types indicates the role of VacA in pathogenesis. FEMS Immunol. Med. Microbiol. *30*, 103–108.

Phadnis, S.H., Ilver, D., Janzon, L., Normark, S., and Westblom, T.U. (1994). Pathological significance and molecular characterization of the vacuolating toxin gene of *Helicobacter pylori*. Infect. Immun. *62*, 1557–1565.

Reyrat, J.M., Lanzavecchia, S., Lupetti, P., de Bernard, M., Pagliaccia, C., Pelicic, V., Charrel, M., Ulivieri, C., Norais, N., Ji, X., Cabiaux, V., Papini, E., Rappuoli, R., and Telford, J.L. (1999). 3D imaging of the 58 kDa cell binding subunit of the *Helicobacter pylori* cytotoxin. J. Mol. Biol. *290*, 459–470.

Ricci, V., Galmiche, A., Doye, A., Necchi, V., Solcia, E., and Boquet, P. (2000). High cell sensitivity to *Helicobacter pylori* VacA toxin depends on a GPI-anchored protein and is not blocked by inhibition of the clathrin- mediated pathway of endocytosis. Mol. Biol. Cell *11*, 3897–3909.

Ricci, V., Sommi, P., Fiocca, R., Romano, M., Solcia, E., and Ventura, U. (1997). *Helicobacter pylori* vacuolating toxin accumulates within the endosomal- vacuolar compartment of cultured gastric cells and potentiates the vacuolating activity of ammonia. J. Pathol. *183*, 453–459.

Rieder, G., Fischer, W., and Haas, R. (2005). Interaction of *Helicobacter pylori* with host cells: function of secreted and translocated molecules. Curr. Opin. Microbiol. *8*, 67–73.

Rittig, M.G., Shaw, B., Letley, D.P., Thomas, R.J., Argent, R.H., and Atherton, J.C. (2003). *Helicobacter pylori*-induced homotypic phagosome fusion in human monocytes is independent of the bacterial *vacA* and *cag* status. Cell Microbiol. *5*, 887–899.

Roche, N., Ilver, D., Angstrom, J., Barone, S., Telford, J.L., and Teneberg, S. (2007). Human gastric glycosphingolipids recognized by *Helicobacter pylori* vacuolating cytotoxin VacA. Microbes Infect. *9*, 605–614.

Rudi, J., Kolb, C., Maiwald, M., Kuck, D., Sieg, A., Galle, P.R., and Stremmel, W. (1998). Diversity of *Helicobacter pylori vacA* and *cagA* genes and relationship to VacA and CagA protein expression, cytotoxin production, and associated diseases. J. Clin. Microbiol. *36*, 944–948.

Salama, N.R., Otto, G., Tompkins, L., and Falkow, S. (2001). Vacuolating cytotoxin of *Helicobacter pylori* plays a role during colonization in a mouse model of infection. Infect. Immun. *69*, 730–736.

Satin, B., Norais, N., Telford, J., Rappuoli, R., Murgia, M., Montecucco, C., and Papini, E. (1997). Effect of *Helicobacter pylori* vacuolating toxin on maturation and extracellular release of procathepsin D and on epidermal growth factor degradation. J. Biol. Chem. *272*, 25022–25028.

Schmitt, W., and Haas, R. (1994). Genetic analysis of the *Helicobacter pylori* vacuolating cytotoxin: structural similarities with the IgA protease type of exported protein. Mol. Microbiol. *12*, 307–319.

Schraw, W., Li, Y., McClain, M.S., van der Goot, F.G., and Cover, T.L. (2002). Association of *Helicobacter pylori* vacuolating toxin (VacA) with lipid rafts. J. Biol. Chem. *277*, 34642–34650.

Seto, K., Hayashi-Kuwabara, Y., Yoneta, T., Suda, H., and Tamaki, H. (1998). Vacuolation induced by cytotoxin from *Helicobacter pylori* is mediated by the EGF receptor in HeLa cells. FEBS Lett. *431*, 347–350.

Skibinski, D.A., Genisset, C., Barone, S., and Telford, J.L. (2006). The cell-specific phenotype of the polymorphic *vacA* midregion is independent of the appearance of the cell surface receptor protein tyrosine phosphatase beta. Infect. Immun. *74*, 49–55.

Smoot, D.T., Resau, J.H., Earlington, M.H., Simpson, M., and Cover, T.L. (1996). Effects of *Helicobacter pylori* vacuolating cytotoxin on primary cultures of human gastric epithelial cells. Gut *39*, 795–799.

Strobel, S., Bereswill, S., Balig, P., Allgaier, P., Sonntag, H.G., and Kist, M. (1998). Identification and analysis of a new *vacA* genotype variant of *Helicobacter pylori* in different patient groups in Germany. J. Clin. Microbiol. *36*, 1285–1289.

Suerbaum, S., and Michetti, P. (2002). *Helicobacter pylori* infection. N. Engl. J. Med. *347*, 1175–1186.

Sundrud, M.S., Torres, V.J., Unutmaz, D., and Cover, T.L. (2004). Inhibition of primary human T-cell proliferation by *Helicobacter pylori* vacuolating toxin (VacA) is independent of VacA effects on IL-2 secretion. Proc. Natl. Acad. Sci. USA *101*, 7727–7732.

Supajatura, V., Ushio, H., Wada, A., Yahiro, K., Okumura, K., Ogawa, H., Hirayama, T., and Ra, C. (2002). Cutting edge: VacA, a vacuolating cytotoxin of *Helicobacter pylori*, directly activates mast cells for migration and production of proinflammatory cytokines. J. Immunol. *168*, 2603–2607.

Suzuki, J., Ohnishi, H., Shibata, H., Wada, A., Hirayama, T., Iiri, T., Ueda, N., Kanamaru, C., Tsuchida, T., Mashima, H., Yasuda, H., and Fujita, T. (2001). Dynamin is involved in human epithelial cell vacuolation caused by the *Helicobacter pylori*-produced cytotoxin VacA. J. Clin. Invest *107*, 1329.

Suzuki, J., Ohnishi, H., Wada, A., Hirayama, T., Ohno, H., Ueda, N., Yasuda, H., Iiri, T., Wada, Y., Futai, M., and Mashima, H. (2003). Involvement of syntaxin 7 in human gastric epithelial cell vacuolation induced by the *Helicobacter pylori*-produced cytotoxin VacA. J. Biol. Chem. *278*, 25585–25590.

Szabo, I., Brutsche, S., Tombola, F., Moschioni, M., Satin, B., Telford, J.L., Rappuoli, R., Montecucco, C., Papini, E., and Zoratti, M. (1999). Formation of anion-selective channels in the cell plasma membrane by the toxin VacA of *Helicobacter pylori* is required for its biological activity. EMBO J. *18*, 5517–5527.

Szczebara, F., Dhaenens, L., Armand, S., and Husson, M.O. (1999). Regulation of the transcription of genes encoding different virulence factors in *Helicobacter pylori* by free iron. FEMS Microbiol. Lett. *175*, 165–170.

Tabel, G., Hoa, N.T., Tarnawski, A., Chen, J., Domek, M., and Ma, T.Y. (2003). *Helicobacter pylori* infection inhibits healing of the wounded duodenal epithelium in vitro. J. Lab. Clin. Med. *142*, 421–430.

Telford, J.L., Ghiara, P., Dell'Orco, M., Comanducci, M., Burroni, D., Bugnoli, M., Tecce, M.F., Censini, S., Covacci, A., Xiang, Z., and *et al.* (1994). Gene structure of the *Helicobacter pylori* cytotoxin and evidence of its key role in gastric disease. J. Exp. Med. *179*, 1653–1658.

Terebiznik, M.R., Vazquez, C.L., Torbicki, K., Banks, D., Wang, T., Hong, W., Blanke, S.R., Colombo, M.I., and Jones, N.L. (2006). *Helicobacter pylori* VacA toxin promotes bacterial intracellular survival in gastric epithelial cells. Infect. Immun. *74*, 6599–6614.

Tombola, F., Carlesso, C., Szabo, I., de Bernard, M., Reyrat, J.M., Telford, J.L., Rappuoli, R., Montecucco, C., Papini, E., and Zoratti, M. (1999a). *Helicobacter pylori* vacuolating toxin forms anion-selective channels in planar lipid bilayers: possible implications for the mechanism of cellular vacuolation. Biophys. J. *76*, 1401–1409.

Tombola, F., Del Giudice, G., Papini, E., and Zoratti, M. (2000). Blockers of VacA provide insights into the structure of the pore. Biophys. J. *79*, 863–873.

Tombola, F., Morbiato, L., Del Giudice, G., Rappuoli, R., Zoratti, M., and Papini, E. (2001a). The *Helicobacter pylori* VacA toxin is a urea permease that promotes urea diffusion across epithelia. J. Clin. Invest. *108*, 929–937.

Tombola, F., Oregna, F., Brutsche, S., Szabo, I., Del Giudice, G., Rappuoli, R., Montecucco, C., Papini, E., and Zoratti, M. (1999b). Inhibition of the vacuolating and anion channel activities of the VacA toxin of *Helicobacter pylori*. FEBS Lett. *460*, 221–225.

Tombola, F., Pagliaccia, C., Campello, S., Telford, J.L., Montecucco, C., Papini, E., and Zoratti, M. (2001b). How the loop and middle regions influence the properties of *Helicobacter pylori* VacA channels. Biophys J. *81*, 3204–3215.

Torres, V.J., Ivie, S.E., McClain, M.S., and Cover, T.L. (2005). Functional properties of the p33 and p55 domains of the *Helicobacter pylori* vacuolating cytotoxin. J. Biol. Chem. *280*, 21107–21114.

Torres, V.J., McClain, M.S., and Cover, T.L. (2004). Interactions between p-33 and p-55 domains of the *Helicobacter pylori* vacuolating cytotoxin (VacA). J. Biol. Chem. *279*, 2324–2331.

Utt, M., Danielsson, B., and Wadstrom, T. (2001). *Helicobacter pylori* vacuolating cytotoxin binding to a putative cell surface receptor, heparan sulfate, studied by surface plasmon resonance. FEMS Immunol. Med. Microbiol. *30*, 109–113

van Amsterdam, K., van Vliet, A.H., Kusters, J.G., Feller, M., Dankert, J., and van der Ende, A. (2003). Induced *Helicobacter pylori* vacuolating cytotoxin VacA expression after initial colonisation of human gastric epithelial cells. FEMS Immunol. Med. Microbiol. *39*, 251–256.

Van Doorn, L.J., Figueiredo, C., Megraud, F., Pena, S., Midolo, P., Queiroz, D.M., Carneiro, F., Vanderborght, B., Pegado, M.D., Sanna, R., De Boer, W., Schneeberger, P.M., Correa, P., Ng, E.K., Atherton, J., Blaser, M.J., and Quint, W.G. (1999). Geographic distribution of *vacA* allelic types of *Helicobacter pylori*. Gastroenterology *116*, 823–830.

van Doorn, L.J., Figueiredo, C., Sanna, R., Pena, S., Midolo, P., Ng, E.K., Atherton, J.C., Blaser, M.J., and Quint, W.G. (1998a). Expanding allelic diversity of *Helicobacter pylori vacA*. J. Clin. Microbiol. *36*, 2597–2603.

van Doorn, L.J., Figueiredo, C., Sanna, R., Plaisier, A., Schneeberger, P., de Boer, W., and Quint, W. (1998b). Clinical relevance of the *cagA*, *vacA*, and *iceA* status of *Helicobacter pylori*. Gastroenterology *115*, 58–66.

Vinion-Dubiel, A.D., McClain, M.S., Czajkowsky, D.M., Iwamoto, H., Ye, D., Cao, P., Schraw, W., Szabo, G., Blanke, S.R., Shao, Z., and Cover, T.L. (1999). A dominant negative mutant of *Helicobacter pylori* vacuolating toxin (VacA) inhibits VacA-induced cell vacuolation. J. Biol. Chem. *274*, 37736–37742.

Wang, H.J., and Wang, W.C. (2000). Expression and binding analysis of GST-vacA fusions reveals that the C- terminal approximately 100-residue segment of exotoxin is crucial for binding in HeLa cells. Biochem. Biophys Res. Commun. *278*, 449–454.

Wang, H.T., Li, Z.H., Yuan, J.P., Zhao, W., Shi, X.D., Tong, S.Q., and Guo, X.K. (2005). Effect of *Helicobacter pylori* VacA on gene expression of gastric cancer cells. World J. Gastroenterol. *11*, 109–113.

Wang, W.-C., Wang, H.-J., and Kuo, C.-H. (2001). Two distinctive cell binding patterns by vacuolating toxin fused with glutathione S-transferase: one high-affinity m1-specific binding and the other lower-affinity binding for variant m forms. Biochemistry *40*, 11887–11896.

Wang, X., Wattiez, R., Paggliacia, C., Telford, J.L., Ruysschaert, J., and Cabiaux, V. (2000). Membrane topology of VacA cytotoxin from H. pylori. FEBS Lett. *481*, 96–100.

Willhite, D.C., and Blanke, S.R. (2004). *Helicobacter pylori* vacuolating cytotoxin enters cells, localizes to the mitochondria, and induces mitochondrial membrane permeability changes correlated to toxin channel activity. Cell Microbiol. *6*, 143–154.

Willhite, D.C., Cover, T.L., and Blanke, S.R. (2003). Cellular vacuolation and mitochondrial cytochrome c release are independent outcomes of *Helicobacter pylori* vacuolating cytotoxin activity that are each dependent on membrane channel formation. J. Biol. Chem. *278*, 48204–48209.

Willhite, D.C., Ye, D., and Blanke, S.R. (2002). Fluorescence resonance energy transfer microscopy of the *Helicobacter pylori* vacuolating cytotoxin within mammalian cells. Infect. Immun. *70*, 3824–3832.

Wirth, H.P., Beins, M.H., Yang, M., Tham, K.T., and Blaser, M.J. (1998). Experimental infection of Mongolian gerbils with wild-type and mutant *Helicobacter pylori* strains. Infect. Immun. *66*, 4856–4866.

Yahiro, K., Niidome, T., Hatakeyama, T., Aoyagi, H., Kurazono, H., Padilla, P.I., Wada, A., and Hirayama, T. (1997). *Helicobacter pylori* vacuolating cytotoxin binds to the 140-kDa protein in human gastric cancer cell lines, AZ-521 and AGS. Biochem. Biophys Res. Commun. *238*, 629–632.

Yahiro, K., Niidome, T., Kimura, M., Hatakeyama, T., Aoyagi, H., Kurazono, H., Imagawa, K., Wada, A., Moss, J., and Hirayama, T. (1999). Activation of *Helicobacter pylori* VacA toxin by alkaline or acid conditions increases its binding to a 250-kDa receptor protein-tyrosine phosphatase beta. J. Biol. Chem. *274*, 36693–36699.

Yahiro, K., Wada, A., Nakayama, M., Kimura, T., Ogushi, K., Niidome, T., Aoyagi, H., Yoshino, K., Yonezawa, K., Moss, J., and Hirayama, T. (2003). Protein-tyrosine phosphatase alpha, RPTP alpha, is a *Helicobacter pylori* VacA receptor. J. Biol. Chem. *278*, 19183–19189.

Yahiro, K., Wada, A., Yamasaki, E., Nakayama, M., Nishi, Y., Hisatsune, J., Morinaga, N., Sap, J., Noda, M., Moss, J., and Hirayama, T. (2004). Essential domain of receptor tyrosine phosphatase beta (RPTPbeta) for interaction with *Helicobacter pylori* vacuolating cytotoxin. J. Biol. Chem. *279*, 51013–51021.

Yamaoka, Y., Kikuchi, S., el-Zimaity, H.M., Gutierrez, O., Osato, M.S., and Graham, D.Y. (2002). Importance of *Helicobacter pylori oipA* in clinical presentation, gastric inflammation, and mucosal interleukin 8 production. Gastroenterology *123*, 414–424.

Yamasaki, E., Wada, A., Kumatori, A., Nakagawa, I., Funao, J., Nakayama, M., Hisatsune, J., Kimura, M., Moss, J., and Hirayama, T. (2006). *Helicobacter pylori* vacuolating cytotoxin induces activation of the proapoptotic protein Bax and Bak, leading to cytochrome c release and cell death, independent of vacuolation. J. Biol. Chem. *281*, 11250–11259.

Ye, D., and Blanke, S.R. (2000). Mutational analysis of the *Helicobacter pylori* vacuolating toxin amino terminus: identification of amino acids essential for cellular vacuolation. Infect. Immun. *68*, 4354–4357.

Ye, D., Willhite, D.C., and Blanke, S.R. (1999). Identification of the minimal intracellular vacuolating domain of the *Helicobacter pylori* vacuolating toxin. J. Biol. Chem. *274*, 9277–9282.

Yuan, J.P., Li, T., Chen, H.B., Li, Z.H., Yang, G.Z., Hu, B.Y., Shi, X.D., Tong, S.Q., Li, Y.X., and Guo, X.K. (2004). Analysis of gene expression profile in gastric cancer cells stimulated with *Helicobacter pylori* isogenic strains. J. Med. Microbiol. *53*, 965–974.

Zheng, P.Y., and Jones, N.L. (2003). *Helicobacter pylori* strains expressing the vacuolating cytotoxin interrupt phagosome maturation in macrophages by recruiting and retaining TACO (coronin 1) protein. Cell Microbiol. *5*, 25–40.

Type IV Secretion Systems in *Helicobacter pylori*

6

Wolfgang Fischer, Arno Karnholz, Luisa F. Jimenez-Soto and Rainer Haas

Abstract

Type IV secretion systems are widely distributed in prokaryotes, where they are used to deliver DNA or protein substrates to bacterial or eukaryotic target cells. They are structurally complex molecular machines, typically composed of a cell envelope-spanning translocation channel, three cytoplasmic ATPases, and potentially an extracellular pilus structure. The *Agrobacterium tumefaciens* VirB/D4 type IV secretion system serves as a prototype system for which detailed structural and functional data are available. In *H. pylori*, three type IV secretion systems have been identified, the ComB, Cag and Tfs3 systems. The ComB system is found in all *H. pylori* strains and is involved in DNA uptake during natural transformation competence. The Cag type IV secretion system is restricted to a subset of more pathogenic *H. pylori* strains and mediates transfer of the effector protein CagA into different cell types of the host. The distribution and function of the Tfs3 system is unknown. This review summarizes recent progress in our understanding of the structure, function and regulation of the different type IV secretion systems in *H. pylori*.

Introduction

Similar to many other pathogenic bacteria, *Helicobacter pylori* possesses a relatively small genome of approximately 1.7 Mio base pairs (Bukanov and Berg, 1994; Tomb *et al.*, 1997). This implicates a specialized metabolism that is well adapted to its habitat within the human stomach (Marais *et al.*, 1999; Schilling *et al.*, 2002). Nevertheless, early comparative studies revealed a very high diversity among *H. pylori* isolates on the genetic level (Akopyanz *et al.*, 1992a; Akopyanz *et al.*, 1992b; Go *et al.*, 1996). The available complete genome sequences of *H. pylori* strains 26695, J99 and HPAG1 (Tomb *et al.*, 1997; Alm *et al.*, 1999; Oh *et al.*, 2006) confirm these early findings. A comparison between strains 26695 and J99 disclosed a considerable variability in the arrangement of common genes, and the presence of about 7% strain-specific genes, many of which are clustered in so-called plasticity regions (Alm *et al.*, 1999).

The plasticity regions often contain genes with homologies to type IV secretion systems (termed *tfs3* genes for type IV secretion system 3), but their function is so far unknown (Kersulyte *et al.*, 2003). In addition, some genes, such as the virulence factor genes *vacA* (encoding the vacuolating cytotoxin A) and *cagA* (encoding the cytotoxin associated antigen A), have a mosaic structure, consisting of variable and conserved parts (Marshall *et al.*, 1998). The *cagA* gene is located on the *cag* pathogenicity island, a 37 kb genomic island that is present in more virulent *H. pylori* strains and encodes a type IV secretion system as well (Censini *et al.*, 1996; Tomb *et al.*, 1997; Akopyants *et al.*, 1998). Accordingly, *H. pylori* is considered as one of the genetically most diverse bacteria (Suerbaum and Josenhans, 2007).

This diversity results from an elevated mutation rate and frequent recombination (Suerbaum and Achtman, 1999), which requires an intensive horizontal gene transfer, possibly even across the species barrier. In general, horizontal gene

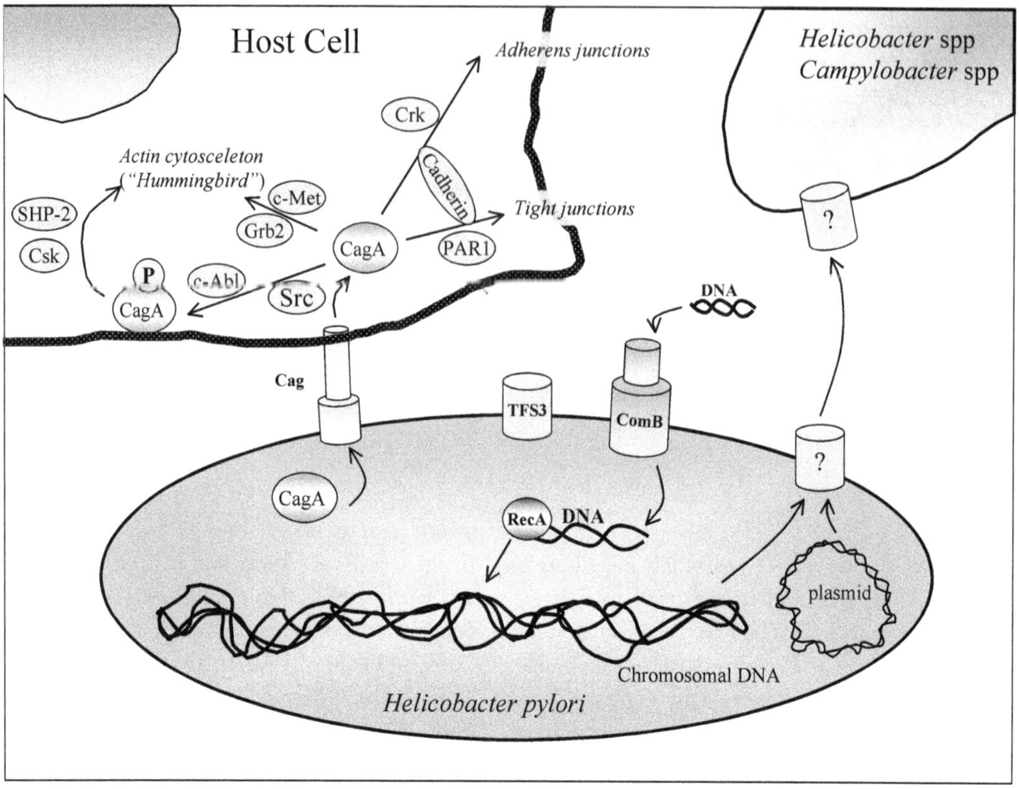

Figure 6.1 Type IV secretion systems in *H. pylori* and associated (putative) functions. Three type IV secretion systems have been described in *H. pylori*: the Cag system mediates translocation of the effector protein CagA into eukaryotic cells, the ComB system facilitates DNA uptake during transformation. Evidence for conjugation-like processes from *H. pylori* to *H. pylori* or to *C. jejuni* suggests that a third, conjugative type IV secretion system must exist; however, it is currently unclear whether the Tfs3 system or an as yet unknown type IV secretion system is involved in these processes. Putative interaction partners of the CagA protein in eukaryotic cells and some target structures are indicated; see text for further details.

transfer is an important factor with respect to the evolution and stability of microbial communities (Gogarten and Townsend, 2005). An exchange of single alleles, but also the transfer of complete metabolic pathways may take place. For pathogenic bacteria, this provides the possibility of a fast and flexible adaptation to the host and also facilitates the transfer of antibiotic resistances. Especially the latter becomes an increasing problem for treatment of *H. pylori* infections due to the rising number of resistant strains (Mégraud, 2004).

The known pathways of horizontal gene transfer include transduction via phages, conjugation and natural transformation (Mazodier and Davies, 1991; Davison, 1999). There is not very much known about phages in *H. pylori* (Schmid *et al.*, 1990; Heintschel von Heinegg *et al.*, 1993), and the structural basis of a conjugation-like mechanism is so far not clear (see '*H. pylori* plasmids and conjugation', below). Most certainty exists about the occurrence of natural competence for genetic transformation. The uptake of exogenous DNA in bacteria via natural transformation can be divided into several steps. These involve DNA binding on the cell surface, passing the murein sacculus or periplasm, degradation of one DNA strand and translocation of the complementary strand across the cytoplasmic membrane into the cytoplasm, and finally a stable integration into the chromosome via homologous recombination (Dubnau, 1999).

The ability to take up free DNA was early discovered in *H. pylori* (Nedenskov-Sorensen *et al.*, 1990) and is regarded as the main reason for the high diversity of *H. pylori*. In 1996, the first proteins of the transport system involved, designated the ComB system, were identified

(Odenbreit *et al.*, 1996). Meanwhile, the ComB system has been characterized as another type IV secretion system operating in the reverse orientation as compared with typical type IV protein or DNA secretion systems. Thus, some *H. pylori* strains contain at least three different type IV secretion systems (Fig. 6.1).

The ComB system – a modified type IV secretion system mediates transformation competence

General structure of the ComB system

Natural competence for genetic transformation is widespread among Gram-negative as well as Gram-positive bacteria (Solomon and Grossman, 1996; Dubnau, 1999; Chen and Dubnau, 2004). Generally, DNA transformation systems show homology with secretion systems involved in the assembly of type IV pili or type II secretion (Chen and Dubnau, 2004). The ComB system is unique among DNA competence systems, since it is related to type IV secretion systems, a feature that is unique to *H. pylori* and its close relative, *Campylobacter jejuni* (Bacon *et al.*, 2000; Hofreuter *et al.*, 2001). Although other known type IV secretion systems of *H. pylori*, the *cag* and the *tfs3* system, show homology to components of the ComB system, they seem to work independently and are unable to complement each other functionally (Hofreuter *et al.*, 2001).

A number of proteins that constitute the ComB transport apparatus have been named according to the nomenclature used for the VirB type IV secretion system of *Agrobacterium tumefaciens*, which is considered as a prototype system. The VirB system consists of 11 VirB proteins (VirB1–11) and the substrate recognition factor VirD4, which build up the transport apparatus necessary for the delivery of proteins or nucleoprotein complexes into target cells. For the *H. pylori* ComB system, all orthologues of the VirB type IV secretion system, except VirB1, VirB5, VirB11 and VirD4, have been identified, with homologies of up to 37% on the protein level (Hofreuter *et al.*, 2001). In contrast to the *virB* genes of *A. tumefaciens*, the *comB* genes are not arranged in a single operon, but are distributed over two different loci in the genome. One locus contains the genes *comB2*, *comB3*, and *comB4* organized in an operon, and the second locus the genes *comB6* to *comB10*.

With the exception of ComB7, which seems to have a modulating effect on the transformation efficiency, each ComB protein is essential for the DNA transformation process, as demonstrated by the generation of knockout mutants and genetic complementation studies (Hofreuter *et al.*, 2001; Karnholz *et al.*, 2006). It is supposed that the structural proteins ComB2, ComB3, ComB6 to ComB10, and possibly also ComB4, constitute a membrane-spanning protein complex, involved in bridging the cytoplasmic membrane with the outer membrane, in analogy to the *A. tumefaciens* VirB system (Christie *et al.*, 2005). Although the topology and localization of the ComB proteins seem to be similar to that of the corresponding VirB proteins in most cases (Hofreuter *et al.*, 2003; Karnholz *et al.*, 2006), some interesting differences exist in the predicted secondary structure, as detailed in the next section.

Components of the ComB system

ComB2

Despite rather low sequence identities between the pilin subunit VirB2 and the equivalent protein ComB2, the structural similarities are significant. ComB2 possesses some of the rather specific characteristics of these proteins, such as an extended signal sequence, two transmembrane domains and a high proportion of hydrophobic amino acids in relation to its small size. VirB2 is a small protein, which assembles in association with VirB5 and VirB7 as a surface-exposed pilus structure, the T pilus (Lai and Kado, 2000). No typical pili have been described for *H. pylori*, but ComB2 might be part of a 'stump structure' in the cell envelope that facilitates DNA uptake but does not form an extended extracellular pilus, as also suggested for VirB2 (Schröder and Lanka, 2005). Such a scenario is also known from type IV pili. For example, in *Neisseria gonorrhoeae*, the formation of long surface type IV pili is neither necessary nor sufficient for transformation competence, whereas the pilus forming subunit PilE is essential (Aas *et al.*, 2002). Due to the lack of a conserved AEIA motif (Kalkum *et al.*, 2002), a C-terminal processing of ComB2, resulting in the cyclization of the protein, as reported for certain TrbC-like pilins (Eisenbrandt *et al.*, 1999), is unlikely.

ComB3

Although VirB3 homologues are essential components of type IV secretion systems, their precise function remains unknown. Originally the protein was suggested to localize at the periplasmic face of the outer membrane, but recent analyses suggest that VirB3 proteins assemble as a functional complex with the VirB4 ATPase at the inner membrane (Christie *et al.*, 2005). For ComB3, the arrangement of a single hydrophobic transmembrane helix indicates the relation to VirB3 homologues (Karnholz *et al.*, 2006), but the localization of the protein and its function for DNA transformation competence is so far unknown.

ComB4

In the *A. tumefaciens* VirB type IV secretion system, three ATPases, VirB4, VirB11 and VirD4, are necessary for the correct assembly of the secretion apparatus or the delivery of the effector molecules, a nucleoprotein complex. In the ComB system, only a single putative ATPase essential for the transformation process has been identified, ComB4, a homologue to VirB4 (Hofreuter *et al.*, 2001). It is unclear whether ComB4 is necessary to assemble the structure of the channel or to energize DNA transport. Besides its function in DNA transformation, ComB4 seems to have other important functions in *H. pylori*, since a *comB4* knockout *H. pylori* strain, but not a *comB7–10* deletion mutant was unable to colonize the stomach mucosa in the Mongolian gerbil model (Kavermann *et al.*, 2003).

ComB6

Although the position of the *comB6* gene directly upstream of *comB7* to *comB10* is identical to the arrangement in other type IV secretion systems, the identification of ComB6 as a VirB6-homologous protein was not obvious. The gene product was originally annotated as an 'NADH-ubiquinone oxidoreductase subunit' in the *H. pylori* 26695 genome sequence (Tomb *et al.*, 1997). A deletion of the gene caused a complete loss of transformation competence of the corresponding mutant strain, and genetic complementation resulted in a restoration of the phenotype (Karnholz *et al.*, 2006). The secondary structure of ComB6 reveals a five-transmembrane-spanning topology with the processed N-terminus in the periplasm and the C-terminal part in the cytoplasm, similar to the topology known from VirB6 of *A. tumefaciens* (Jakubowski *et al.*, 2004; Judd *et al.*, 2005). The polytopic cytoplasmic membrane protein VirB6 is a core component of the transformation apparatus and is considered to form a pore at the cytoplasmic membrane. It catalyses the translocation of the DNA substrate coming from the ATPases VirB11 and VirD4 into the membrane-spanning channel (Jakubowski *et al.*, 2004). Since the direction of DNA transport in *H. pylori* is inverse, as compared to the VirB system, interaction of ComB6 with proteins in the cytoplasm might be different to other type IV secretion systems. In this respect it is interesting to note that the C-terminus of ComB6 is extended, as compared to VirB6 and this extension might well have some function in respect to protein-protein interaction.

ComB7

ComB7, a putative lipoprotein, is found membrane-attached, probably by its lipid anchor. Similar to VirB7 of *A. tumefaciens*, ComB7 possesses a conserved cysteine residue (C27), which might facilitate the interaction between ComB7 and ComB9 by formation of a disulfide bond to the N-terminal cysteine of ComB9 (C316) (Hofreuter *et al.*, 2001). This interaction seems to stabilize the transformation channel, although it is not essential for the DNA substrate transport, since a *comB7* deletion mutant was not completely defective for natural transformation (Hofreuter *et al.*, 2001). In the VirB system, suppression of disulfide bond formation between VirB7 and VirB9 does not prevent the induction of tumour formation in plants (Das *et al.*, 1997). The structural basis for the interaction of VirB7 and VirB9 homologues from plasmid pKM101 has been solved recently by NMR (Bayliss *et al.*, 2007). VirB9 consists of a β-sandwich, forming a conserved heterodimer interface, around which VirB7 winds.

ComB8

ComB8 is a bitopic membrane protein with a short N-terminal portion in the cytoplasm and the remainder of the protein expanding into the periplasmic space (Hofreuter *et al.*, 2003), and thus has a similar topology to VirB8 (Das and Xie, 1998). In analogy to the VirB system, ComB8 is supposed to interact with itself and with ComB9 and ComB10 to stabilize the secretion channel, possibly by conserved periplasmic domains (Hofreuter *et al.*, 2001). In contrast to VirB8, the C-terminal half of which consists mainly of β-barrels, ComB8 has significantly more α-helical domains.

ComB9

ComB9 is a periplasmic protein, which is tightly attached to the membrane. The strong membrane association might be mediated by disulfide interaction with ComB7 (Hofreuter *et al.*, 2001). However, intensive treatment of the membrane fractions with rather stringent extraction methods did not release the ComB9 protein, indicating that in addition to the putative disulfide bond, a strong anchoring of the protein to the outer membrane seems to exist (Hofreuter *et al.*, 2003). Recent work on the membrane interaction of VirB9 and the structure of the VirB7/VirB9 complex suggested that VirB9 inserts into the outer membrane and protrudes extracellularly, and that it has a role in regulating substrate selection and T-pilus biogenesis (Jakubowski *et al.*, 2005; Bayliss *et al.*, 2007). Whether ComB9 of *H. pylori* shows a similar outer membrane insertion and thus a similar function as VirB9, remains to be shown.

ComB10

Similar to the ComB8 protein, the N-terminus of ComB10 is anchored in the cytoplasmic membrane and the major portion of the protein, including a putative coiled-coil domain, is located in the periplasmic space (Hofreuter *et al.*, 2003). This α-helical coiled-coil domain, which is missing in VirB10 of *A. tumefaciens*, might be involved in homomultimerization of the ComB10 protein, or in interaction with other ComB proteins. In addition to the topology data obtained by creating fusions with reporter proteins, the crystal structure for the C-terminal portion of the ComB10 protein has been determined, which supported a dimerization of the subunits (Terradot *et al.*, 2005). The structure comprises an extensively modified β-barrel, open on one side and extended by a flexible α-helical antenna at the top. The sequence alignment of ComB10 and its VirB10 homologues indicates that the overall fold is likely conserved among VirB10 proteins. Thus, the structure of ComB10 can be seen as a prototype for all periplasmic domains of VirB10 proteins. Together with computer predictions and experimental topology determination, a low resolution model for the ComB system was established (Hofreuter *et al.*, 2001; Smeets and Kusters, 2002; Karnholz *et al.*, 2006) (Fig. 6.2a).

Other VirB-homologous proteins?

As outlined above, the ComB system lacks VirB1 and VirB5 homologues as well as VirB11- and VirD4-like ATPases. It cannot be excluded that orthologues of the latter proteins with low primary sequence homology might be encoded in the *H. pylori* genome, but for the basic function of the DNA transformation system, no further ComB proteins seem to be obligatory. This might be due to the inverse direction of substrate transport as compared to type IV secretion systems. For example, the ATPases VirB11 and VirD4 are responsible for energizing the assembly of the apparatus and the subsequent substrate transport (Schröder and Lanka, 2005). Using a transfer DNA immunoprecipitation (TrIP) assay, VirD4 and VirB11 were shown to be involved in the initial steps of DNA transfer, delivering the DNA substrate into the translocation channel (Cascales and Christie, 2004; Atmakuri *et al.*, 2004). Since DNA import via the ComB system functions in the inverse direction as compared to the VirB system, it is assumed that these coupling and delivering processes at the inner membrane are not necessary. Interestingly, *A. tumefaciens* strains harbouring an intact VirB system are more efficient recipients for VirB-mediated plasmid transfer, and for this recipient function of a type IV system, VirB5, VirB11 and VirD4 are not required either (Bohne *et al.*, 1998).

VirB1 is a lytic transglycosylase, capable of localized disruption of the peptidoglycan layer

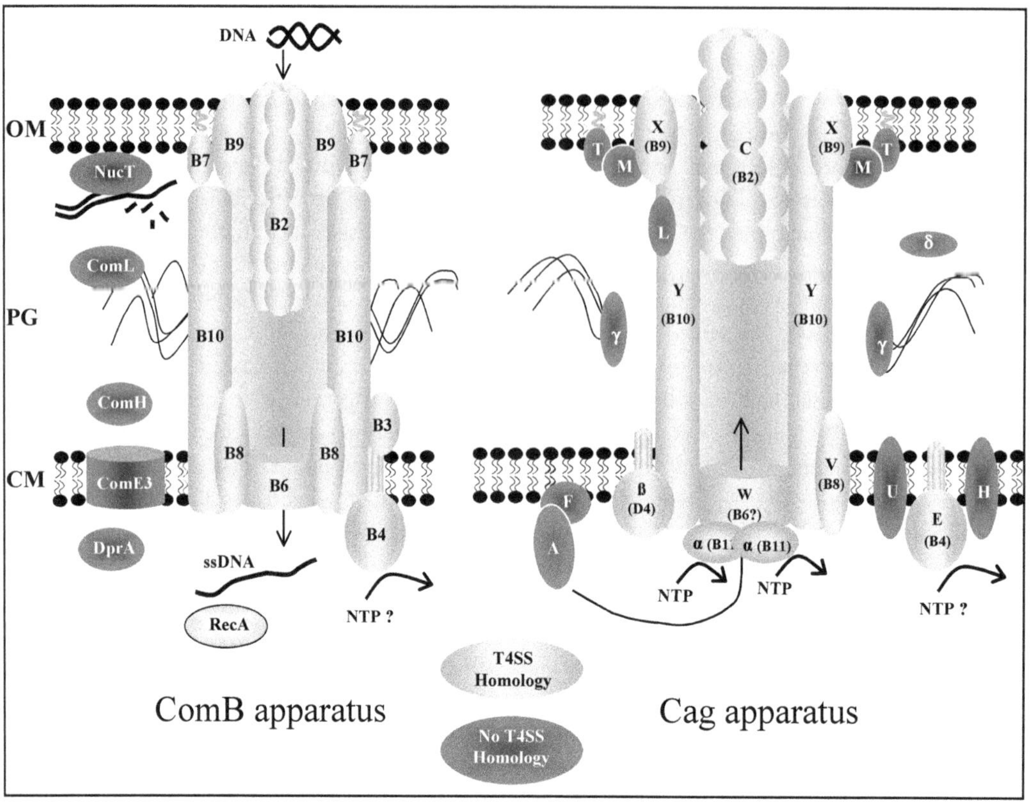

Figure 6.2 Structural models of the ComB and Cag type IV secretion complexes. The putative structures of the ComB (left) and the Cag (right) type IV secretion apparatus are drawn according to available localization and interaction data, or in analogy to the VirB system, as far as homologues are present (light grey boxes). Components that are involved in DNA uptake or CagA translocation, but have no counterparts in other type IV secretion systems, are drawn in dark grey. Depicted interactions between CagA and CagF, CagL and CagY, CagM and CagX, and between CagM and CagT are based on experimental data; see text for details. T4SS, type IV secretion system; OM, outer membrane, CM, cytoplasmic membrane, PG, peptidoglycan.

(Mushegian *et al.*, 1996). Although VirB1 is required for efficient type IV secretion system function in *A. tumefaciens*, it is not absolutely essential for secretion apparatus assembly and substrate translocation (Berger and Christie, 1994). It was therefore speculated that the relatively thin murein layer of Gram-negative bacteria may also permit a limited type IV secretion system assembly in the absence of VirB1 or that other cellular enzymes may substitute for this function (Koraimann, 2003). Such an alternative enzyme in *H. pylori* could be a gene product homologous to ComL of *Neisseria gonorrhoeae*, which in the latter bacteria is involved in the transformation-associated hydrolysis of murein (Fussenegger *et al.*, 1996). Unfortunately, a possible involvement of the putative ComL protein in DNA transformation of *H. pylori* could not be addressed genetically, as inactivation of the gene was lethal (Smeets *et al.*, 2000b; Yeh *et al.*, 2003).

VirB5 of *A. tumefaciens* is a secreted protein that localizes mainly to the T pilus, where it forms complexes with VirB2 (Yuan *et al.*, 2005). In addition, an adhesion-like function was suggested for its homologue, TraC (Yeo *et al.*, 2003). While a direct cell-to-cell contact is required for T-DNA transfer and bacterial conjugation, this is unlikely for DNA uptake, which might explain the absence of a VirB5 homologue in the *H. pylori* ComB system. Consistent with this, the Ptl type IV secretion system of *Bordetella pertussis*, which is involved in the contact-independent secretion of the pertussis toxin, also lacks a VirB5 homologue (Burns, 2003).

Other proteins involved in the natural transformation process

A number of other factors that are not typical type IV secretion system proteins have been found to be involved in the process of natural transformation in *H. pylori*. RecA-mediated homologous recombination is an essential step for incoming DNA to be integrated into the *H. pylori* genome (Schmitt *et al.*, 1995; Thompson and Blaser, 1995). The knockout of the *dprA* gene in *H. pylori* resulted in a strongly reduced transformation efficiency, as compared to the wild-type strain (Ando *et al.*, 1999; Smeets *et al.*, 2000b). DprA is found in many naturally competent bacteria, but also in non-competent species (Takata *et al.*, 2005). In transformation-competent *Streptococcus pneumoniae*, incoming single-stranded DNA is immediately degraded in the absence of DprA or RecA (Berge *et al.*, 2003). Whether DprA in *H. pylori* also mediates active protection of DNA prior to the RecA-driven search for homology is unknown.

A knockout mutant in the *comH* gene, for which no orthologue is known, is completely defective in natural transformation competence, although it is still able to integrate DNA introduced via electroporation (Smeets *et al.*, 2000a). This suggests that the protein, which possesses an N-terminal signal sequence, is probably active in the periplasm or the outer membrane, where it might be involved in binding or processing of incoming DNA. Furthermore, a protein with sequence homology to *B. subtilis* ComE3 (also known as ComEC) was identified (Yeh *et al.*, 2003). ComE3 is a multispanning transmembrane protein for DNA binding and uptake in *B. subtilis* and other genetically transformable bacteria, a function similar to that of VirB6 (Hahn *et al.*, 1993; Draskovic and Dubnau, 2005). *H. pylori* therefore possesses both types of proteins, a ComE3 homologue and the ComB6 protein, both absolutely essential for transformation competence, although they do not complement each other. In contrast to a *comB6* knockout mutant, a *H. pylori* ComE3 mutant was resistant to transformation via electroporation, a method that should be independent of DNA binding or a DNA uptake channel in the bacterial membrane (Yeh *et al.*, 2003). This indicates that both proteins seem to fulfil different functions in the transformation process.

The membrane-bound endonuclease NucT was also found to be essential for the transformation process (O'Rourke *et al.*, 2004). According to its activity to hydrolyze single- and double-stranded DNA it is reasonable to assume that the nuclease is responsible for the processing of the incoming DNA (O'Rourke *et al.*, 2004). Based on a random transposon mutagenesis approach, some additional genes, including *hp1089*, *hp1424* and *hp1473*, have been identified as being involved in the transformation process, since corresponding mutants are partially or completely defective in natural transformation, but no further functional data are available so far (Chang *et al.*, 2001).

Regulation of the ComB system

The activity of the ComB system *in vivo* is mainly inferred from the increasing genetic diversity of bacteria reisolated from the same patients over time (Kersulyte *et al.*, 1999; Falush *et al.*, 2001; Smeets *et al.*, 2003). Most conclusions about ComB system regulation are derived from *in vitro* assays in liquid cultures, by analysing the transformation rates during bacterial growth. One study found an increase of transformation rates at the early and late logarithmic phases (Israel *et al.*, 2000), whereas recent data suggest a complex, and strain-specific, transformation pattern during the entire growth phase (Baltrus and Guillemin, 2006). For other naturally competent bacteria, it is known that DNA uptake is subject to a regulated release of DNA from a subset of the population (Moscoso and Claverys, 2004; Guiral *et al.*, 2005). For *H. pylori*, neither a crosstalk within a population, nor any regulatory circuit within the cell concerning natural transformation capacities is known. Natural transformation is a fast and saturable process, and double-stranded DNA seems to be the preferred substrate (Levine *et al.*, 2007). It is unclear if there are any recognition sequences which control the DNA before entering the recipient cells, as known from other Gram-negative competent species, such as *N. gonorrhoeae* or *H. influenzae* (Goodman and Scocca, 1988). In *H. pylori*, restriction/modification (R/M) systems occur in a particularly high number, as compared to other bacteria, and their composition and activity is strain-specific (Xu *et al.*, 2000; Lin *et al.*, 2001). These R/M systems are involved in restricting transformation with plasmid DNA, but also with chromosomal DNA, depending on the length of the transforming DNA fragment (Ando *et al.*, 2000; Donahue *et al.*, 2000; Aras *et al.*, 2002). Thus, transfer of DNA between different *H. pylori* strains is usually less efficient than within the same strain (Israel *et al.*, 2000; Gorrell *et al.*, 2005), and DNA fragments larger than 1 kb are effectively transferred between isogenic strains only, whereas frequencies of point mutation transfer are independent of the R/M system equipment (Aras *et al.*, 2002). Generally, transformation of *H. pylori* cells with short DNA fragments seems to be much more efficient than in other competent bacteria (Levine *et al.*, 2007). Together with the average size of DNA exchanged by recombination, which has been determined to be about 400 bp (Falush *et al.*, 2001), this supports a role for R/M systems to prevent uncontrolled genome transformation without influencing the generation of a high microdiversity.

The biological impact of the ComB system for *H. pylori*

Horizontal DNA exchange by natural transformation is considered beneficial for bacteria living in changing environments, for example to manage host adaptation and immune modulation (Solomon and Grossman, 1996). The unusually high frequency of DNA exchange and microdiversity in *H. pylori* is attributed to natural transformation. The importance of the ComB system for *H. pylori* is also supported by the fact that it is found in all strains (Hofreuter *et al.*, 2001), in contrast to the other type IV secretion systems of *H. pylori*, which are limited to subgroups of strains only. This high degree of preservation raises the question if the ComB system might fulfil additional functions. As discussed for other transformation-competent bacteria, such additional benefits might be the use of DNA from the environment as a nutrient source (Finkel and Kolter, 2001), or as a reservoir for gene repair (Tønjum *et al.*, 2004). Since *H. pylori* lives in its niche under conditions of chronic gastric inflammation (chronic gastritis), that is under DNA-damaging conditions (O'Rourke *et al.*, 2003), the ComB system might offer a means to provide the cells with intact chromosomal DNA, and might thus enhance their fitness and allow them to persist within the host. Although recombination is usually involved in the generation of new variants by allowing the incorporation of exogenous DNA or the reshuffling of endogenous sequences, the MutS2 system of *H. pylori* (HPMutS2) inhibits homologous and homeologous recombination (recombination between closely related but non-identical DNA sequences). Disruption of HpMutS2 leads to a more efficient incorporation of exogenous DNA (Pinto *et al.*, 2005). Whether the ComB system is involved in providing the DNA is unclear. Except for a colonization defect of a *comB4* mutant in the Mongolian gerbil model (Kavermann *et al.*, 2003), no experimental data supporting a role of the ComB system for long-term survival *in vivo* are available.

The Cag system – type IV secretion as a virulence determinant

Composition of the cag pathogenicity island and variability of *cag* genes

Originally, the *cag* pathogenicity island was identified by sequencing regions up- and downstream from *cagA*, a gene that was already known to be correlated with the development of more severe disease and also with the presence of an active (genotype s1/m1) VacA cytotoxin (hence the designation *cytotoxin-associated gene A*) (Censini *et al.*, 1996; Tomb *et al.*, 1997; Akopyants *et al.*, 1998). It was also known that a pronounced IL-8 induction in gastric epithelial cells depends on the presence of the *cagA* gene (Crabtree *et al.*, 1995). Mutational analysis indicated that various genes on the *cag* pathogenicity island, but not *cagA* itself, are responsible for IL-8 induction (Tummuru *et al.*, 1995; Censini *et al.*, 1996; Segal *et al.*, 1997). The *cag* pathogenicity island contains roughly 30 genes that encode the CagA protein and a type IV secretion apparatus, which is able to translocate CagA from the bacteria to several host cells (Odenbreit *et al.*, 2000; Asahi *et al.*, 2000; Stein *et al.*, 2000; Backert *et al.*, 2000; Odenbreit *et al.*, 2001). To date, complete *cag* PAI sequences of 19 strains are available in the databases. The variations among these are limited, however, the most pronounced differences being the deletion of *cagε* or the existence of an alternative *cagε* gene (termed *hp521b*), an inversion and sequence divergence of *cagQ*, and sequence variations in the *cagA* and *cagY* genes (Azuma *et al.*, 2004a; Blomstergren *et al.*, 2004).

Interestingly, however, the *cag* island is split by IS605 insertion elements into two parts that are about 500 kb apart in the genome of strain NCTC11638 (Akopyants *et al.*, 1998), and other types of rearrangements may occur. The percentage of *cag* pathogenicity island-positive strains varies considerably between different ethnic groups, ranging from complete absence in certain South African strains to ubiquitous presence in Asian strains (Gressmann *et al.*, 2005).

For the *cagA* gene, variations exist in the 5′ (van Doorn *et al.*, 1999) and the 3′ parts (Yamaoka *et al.*, 1998; Yamaoka *et al.*, 1999), and these regions are divergent between Western and East Asian strains. The 3′ variable region encodes the CagA tyrosine phosphorylation motifs (EPIYA motifs) which can, according to differences in the flanking amino acid residues, be classified as EPIYA-A, B, C, and D motifs (Hatakeyama, 2003), with motif ABC-type CagA molecules occurring in Western and ABD-type CagA molecules in East Asian strains (Higashi *et al.*, 2002a; Naito *et al.*, 2006). Different *cagA* genotypes, notably variations in the 3′ repeat region, are associated with the severity of diseases (Yamaoka *et al.*, 1999; Azuma *et al.*, 2002; Azuma *et al.*, 2004b; Argent *et al.*, 2004), and a correlation of the *vacA* s1c/m1b genotype with ABD-type CagA and *vacA* s2 or s1a/m1a genotypes with ABC-type CagA has been reported (Yamazaki *et al.*, 2005).

Regulation of cag pathogenicity island gene expression

The transcriptional organization of the *cag* pathogenicity island and the localization of promoters is not completely understood, except in the region upstream of the *cagA* gene, where several transcriptional start sites have been identified (Spohn *et al.*, 1997). Different putative promoter regions have been identified with a transcriptional reporter (Joyce *et al.*, 2001). Several genes of the *cag* pathogenicity island have been reported to be upregulated *in vitro* under acidic conditions in one study (Wen *et al.*, 2003), whereas other studies did not find significant inductions (Merrell *et al.*, 2003; Bury-Moné *et al.*, 2004). Inconclusive data have also been obtained concerning acid regulation of the *cagA* gene itself (Karita *et al.*, 1996; Merrell *et al.*, 2003; Barnard *et al.*, 2004), but *cagA* transcription seems to be upregulated between logarithmic and stationary phase (Thompson *et al.*, 2003; Boonjakuakul *et al.*, 2005) and may also depend on salt concentration (Loh *et al.*, 2007). CagA expression may be influenced by the global regulator CsrA (Barnard *et al.*, 2004), but is not regulated by the acid-responsive ArsRS two-component system (Pflock *et al.*, 2006). Several *cag* pathogenicity island genes are also induced during infection in Rhesus macaques or Mongolian gerbils (Boonjakuakul *et al.*, 2005; Scott *et al.*, 2007).

Essential and supportive components of the Cag type IV secretion apparatus

Among the components of the Cag secretion system, only five have significant homologies to known type IV secretion apparatus components (CagE/HP544 to VirB4, CagX/HP528 to VirB9, CagY/HP527 to VirB10, Cagα/HP525 to VirB11 and Cagβ/HP524 to VirD4). The first of these homologues to be described was CagE, originally designated PicB, which was shown to be necessary for IL-8 induction (Tummuru *et al.*, 1995). A comprehensive survey of the requirement of *cag*-encoded factors for the ability of *H. pylori* to induce IL-8 secretion or to translocate the CagA protein resulted in identification of 14 gene products as necessary for IL-8 induction as well as CagA translocation. Three gene products contribute to, but are not essential for both phenotypes, and four further gene products are necessary for CagA translocation only (Fischer *et al.*, 2001). The latter gene products were considered as substrate recognition or translocation factors, and the former as components of the secretion apparatus.

In most type IV secretion systems, as also seen in the ComB system (see above), the gene order of *virB2*, *virB3* and *virB4* homologues is conserved. Since *cagC* and *cagD* are upstream genes of the *virB4* homologue *cagE*, CagC (HP546) might represent a VirB2 homologue of the Cag system. Indeed, it has certain similarities to the major pilus component VirB2 (Kalkum *et al.*, 2002; Andrzejewska *et al.*, 2006). The fact that CagC is the most abundant protein encoded on the *cag* pathogenicity island (Boonjakuakul *et al.*, 2005), is also consistent with its putative role as a pilus subunit. In contrast to CagC, CagD is not essential for the function of the Cag type IV secretion system, and it does not have similarities to VirB3 or its homologues. Computer predictions rather suggest that the N-terminal portion of CagE (HP544), which is unrelated to VirB4 and its homologues, might represent a VirB3-like protein domain. Similar protein fusions between VirB3 and VirB4 homologues have also been found in conjugation system of *Campylobacter* species (Batchelor *et al.*, 2004). Apart from its N-terminal 120 amino acids, CagE displays a good homology to VirB4, and contains conserved motifs identified for this protein family (Rabel *et al.*, 2003).

No homologies were found for the essential apparatus components CagH (HP541) and CagL (HP539). CagH is a putative inner membrane protein, and CagL is located in the periplasm. Interestingly, CagL stabilizes the VirB10-homologous component CagY, suggesting an interaction between these proteins (Rohde *et al.*, 2003). The non-essential, but supportive protein CagN (HP538) has been characterized and was found to be processed at its C-terminus; however, its function for type IV secretion remains unclear (Bourzac *et al.*, 2006). CagM (HP537) is an essential apparatus component with an N-terminal signal sequence that is associated with the outer membrane. Determination of protein–protein interactions by co-immunoprecipitation in *H. pylori* and localization studies by bacterial cell subfractionation revealed the presence of an outer membrane-associated subcomplex composed of CagM, CagT, and CagX (S. Kutter, R. Buhrdorf, R. Haas and W. Fischer, in press).

CagT (HP532) is a putative lipoprotein, which has been postulated to be an equivalent to VirB7 (Akopyants *et al.*, 1998). However, VirB7 and its homologues are small lipoproteins of 45–70 amino acids and form disulphide bonds with VirB9, whereas CagT has a size of 280 amino acids and does not contain further cysteine residues apart from the ones within its signal sequence. Nevertheless, some type IV secretion systems might contain a second class of lipoproteins with a similar or an additional function. For example, plasmid conjugation of RP4 requires the outer membrane lipoprotein TrbH with 160 amino acids (Grahn *et al.*, 2000), and F-like conjugation systems also contain essential outer membrane lipoproteins such as TraV with a size of 171 amino acids (Harris *et al.*, 2001; Lawley *et al.*, 2003). CagU (HP531) is an inner membrane protein which is essential for secretion apparatus function, but does not have any homologies.

The CagV (HP530) protein has been shown to adopt a topology and to contain sequence motifs typical for VirB8-type proteins, and was thus suggested to be a VirB8 homologue (Buhrdorf *et al.*, 2003). CagW (HP529) is an integral inner membrane protein with six predicted transmembrane helices, which is a common feature of the otherwise only weakly conserved group of VirB6 homologues (Christie *et al.*, 2005). VirB6 homologues are essentially involved in forming the translocation channel and in pilus assembly and/or elongation (Schröder and Lanka, 2005). Experimental evidence for an analogous function of CagW is lacking so far.

CagX (HP528) is homologous to the outer membrane protein VirB9. It is associated with the *H. pylori* outer membrane and forms a complex there with the essential periplasmic component CagM and the lipoprotein CagT (S. Kutter, R. Buhrdorf, R. Haas and W. Fischer, in press; see above). Since parts of the CagX and CagT proteins are exposed on the bacterial surface (Tanaka *et al.*, 2003; Rohde *et al.*, 2003), it seems reasonable to speculate that these proteins form a secretin-like pore in the outer membrane, as also suggested for the VirB7-VirB9 or TraN–TraO heterodimers (Jakubowski *et al.*, 2005; Bayliss *et al.*, 2007). This pore may allow the passage of the observed pilus-like surface appendages and/or translocated substrates across the outer membrane.

In marked contrast to ComB10, the VirB10 homologue CagY (HP527) is a very unusual protein in that it contains two distinct repeat sections in addition to its C-terminal VirB10 homology region, one in the N-terminal part of the protein and one in its middle region (Liu *et al.*, 1999; Aras *et al.*, 2003). Remarkably, the middle repeat region consists of a regular array of 74 sequence repeats composed of six different consensus segments and contains 58 cysteine residues in strain 26695. A part of the protein, which is probably the middle repeat region, seems to be surface-exposed (Rohde *et al.*, 2003;

Tanaka *et al.*, 2003). This region is subject to antigenic variation that has been associated with immune evasion (Aras *et al.*, 2003).

The VirB11 homologue Cagα forms hexamers, as shown by biochemical experiments, electron microscopy and crystallization of the purified protein (Krause *et al.*, 2000a; Yeo *et al.*, 2000). *In vitro*, the purified protein has an ATPase activity (Krause *et al.*, 2000b), and its conformation changes according to the binding of ADP or ATP, indicating that a dynamic cyclic process occurs during substrate transport or apparatus assembly (Savvides *et al.*, 2003).

The essential secretion apparatus component Cagγ (HP523) contains sequence motifs that are characteristic for lytic transglycosylases, and it was reported to possess a corresponding enzymatic activity (Höppner *et al.*, 2004; Zahrl *et al.*, 2005), suggesting that it has a VirB1-like function in local degradation of the peptidoglycan. This function would require a periplasmic localization, but Cagγ does not have an N-terminal signal sequence. Interestingly, however, biochemical data indicated that Cagγ is necessary for proper folding and/or processing of the putative bridging component CagY (Rohde *et al.*, 2003), also indicating a VirB1-like function. Finally, Cagδ (HP522) is an essential periplasmic component of the secretion apparatus without any apparent homologies or sequence motifs. Interactions between Cagδ and CagV, and also between Cagδ and CagM, were identified in a yeast two-hybrid screen and confirmed biochemically (Busler *et al.*, 2006).

Taken together, the Cag system seems to contain homologues to most prototypical type IV secretion components, with the exception of VirB5, but also several additional secretion apparatus components that are unique for the Cag system. It is presently unclear whether the components encoded on the *cag* pathogenicity island are sufficient for type IV secretion apparatus function, or whether other factors are required. Interestingly, a genome-wide yeast two-hybrid study identified several interactions between Cag components and proteins encoded outside of the *cag* pathogenicity island (Rain *et al.*, 2001). Some of these, including interactions between Cagα and HP1451, and between CagA and HP496, could be confirmed biochemically by coexpression in *E. coli* (Terradot *et al.*, 2004); however, their functional significance remains to be determined.

Type IV secretion signals and secretion mechanism

Recognition mechanisms for type IV secretion substrates and the secretion signals involved have gained much attention recently (Cambronne and Roy, 2006). With the exception of the pertussis toxin subunits of *B. pertussis*, which are exported by the Sec apparatus to the periplasmic space (Burns, 2003), effector proteins of type IV secretion systems have no N-terminal, but C-terminal signal sequences (Nagai *et al.*, 2005; Vergunst *et al.*, 2005; Schulein *et al.*, 2005). In the case of the *A. tumefaciens* VirB system, positively charged amino acids near the C-terminus play an important role in effector protein recognition (Vergunst *et al.*, 2005). Whereas translocation of these effector proteins was measured by fusion with adenylate cyclase or Cre recombinase reporter proteins, a novel reporter system that relies on N-terminal fusions with the phosphorylatable GSK tag was used for determination of CagA translocation (Torruellas-Garcia *et al.*, 2006; Hohlfeld *et al.*, 2006). For translocation of CagA, 20 C-terminal amino acids, containing positively charged amino acids, are essential; however, positive charges do not seem to be as important as for *A. tumefaciens* VirB substrates. Nevertheless, the similarity to other type IV substrates is supported by the fact that C-terminal regions can be functionally exchanged. In contrast to other type IV effector proteins, N-terminal regions are required for CagA translocation as well, suggesting a more complex substrate recognition mechanism in the Cag system (Hohlfeld *et al.*, 2006).

Recognition of type IV secretion substrates is thought to be accomplished by the coupling proteins, cytoplasmic membrane proteins with ATP-binding capacity that are not required for secretion apparatus assembly (Gomis-Rüth *et al.*, 2004; Christie *et al.*, 2005; Cambronne and Roy, 2006). Consistent with this notion, the coupling protein homologue Cagβ is dispensable for IL-8 induction, but required for CagA translocation (Fischer *et al.*, 2001). Like other coupling protein homologues, Cagβ was shown to form oligomers and to bind DNA without sequence specificity, but it lacks the postulated ATPase activity (Schröder *et al.*, 2002). The exact function of Cagβ in the CagA signal sequence recognition process remains unclear. Apart from Cagβ, three further *cag*-encoded proteins, CagF, CagI and CagZ, are specifically required for CagA translocation without being secretion apparatus components (Fischer *et al.*, 2001). The three-dimensional structure of CagZ has been determined to consist of a single L-shaped domain without structural homologues (Cendron *et al.*, 2004). CagZ has been reported to interact with many Cag proteins in a yeast two-hybrid assay (Busler *et al.*, 2006), but its function and also that of the translocation factor CagI remain unclear.

The CagA translocation factor CagF is associated with the inner bacterial membrane and was shown to interact directly with CagA (Couturier *et al.*, 2006). Molecular characteristics of CagF as well as biochemical and genetic evidence indicate that CagF has secretion chaperone-like features, although a stabilizing function for CagA could not be demonstrated (Pattis *et al.*, 2007). Since the interaction between CagF and CagA takes place at the cytoplasmic membrane and is independent of signal sequence recognition, CagF probably acts by recruiting CagA to the secretion apparatus. It is also interesting in this regard that CagF interacts with the secretion apparatus component CagY (Busler *et al.*, 2006). The CagF-binding region is adjacent to, but distinct from the C-terminal CagA secretion signal (Pattis *et al.*, 2007). It may overlap with a domain required for dimerization in eukaryotic cells, which is adjacent to the EPIYA-C or EPIYA-D motifs (Ren *et al.*, 2006).

Pathogenicity-associated activities of the Cag type IV secretion system

CagA translocation and induction of a proinflammatory response

Since the cellular consequences of Cag type IV secretion are not the main focus of this review, we will discuss *cag*-associated phenotypes only briefly. For a more detailed discussion of these phenotypes, the reader is referred to several recent reviews (Blaser and Atherton, 2004; Bourzac and Guillemin, 2005; Naumann, 2005; Hatakeyama, 2006). Some of the phenotypes related to Cag type IV secretion system activity have been observed after infection of different cancer cell lines with *H. pylori* strains, and some after transfection with *cagA*-expressing constructs. CagA-independent effects caused by modulation of different signalling pathways may depend on the cell type used, and CagA-dependent effects may also be indirect due to the massive impact of translocated CagA on gene expression profiles (El Etr *et al.*, 2004). Thus, the relevance of the molecular interactions and cellular consequences under *in vivo* conditions remains to be determined.

CagA is the only effector protein that was reported to be translocated by the Cag type IV secretion system. Bacterial peptidoglycan fragments transferred by the type IV secretion apparatus may subsequently stimulate the intracellular pattern recognition receptor Nod1, resulting in NF-κB activation and induction of chemokines such as IL-8 (Viala *et al.*, 2004). However, different signalling pathways are involved in the proinflammatory response elicited by *H. pylori* (Naumann, 2005), and it is unlikely that Nod1 activation represents the only *cag*-dependent stimulation mechanism. For example, an activation of Nod2 seems to be possible as well (Rosenstiel *et al.*, 2006). Translocated CagA proteins of some *H. pylori* strains are able to contribute to IL-8 expression (Brandt *et al.*, 2005), and transfection of epithelial cells with *cagA* expression constructs has been shown to result in IL-8 induction as well (Kim *et al.*, 2006). Moreover, it should be noted that other bacterial factors such as the outer inflammatory protein OipA may act synergistically with the Cag type IV secretion system to stimulate a proinflammatory response (Yamaoka *et al.*, 2004).

CagA interference with host cell functions

The hallmark of CagA translocation into eucaryotic cells is its tyrosine phosphorylation by kinases of the Src family (Selbach *et al.*, 2002; Stein *et al.*, 2002), and, at later time points, the c-Abl kinase (Poppe *et al.*, 2007; Tammer *et al.*, 2007). Apart from Src and Abl kinases, CagA has been reported to interact in eucaryotic cells with a plethora of target molecules, including the tyrosine phosphatase SHP-2 (Higashi *et al.*, 2002b), the tyrosine kinase Csk (Tsutsumi *et al.*, 2003), the receptor tyrosine kinase c-Met (Churin *et al.*, 2003), adaptor molecules such as Grb2 (Mimuro *et al.*, 2002) and Crk (Suzuki *et al.*, 2005), the junction protein E-cadherin (Murata-Kamiya *et al.*, 2007), and the serine/threonine kinase PAR1/MARK (Saadat *et al.*, 2007). Some of these interactions are dependent on, and others independent of, CagA tyrosine phosphorylation. Tyrosine phosphorylation of CagA, on the other hand, results in dephosphorylation of several cellular proteins including cortactin, ezrin, and vinculin (Selbach *et al.*, 2003; Selbach *et al.*, 2004; Moese *et al.*, 2007), which is due to a feedback inactivation of Src kinase (Tsutsumi *et al.*, 2003; Selbach *et al.*, 2003).

The most prominent phenotypes associated with CagA translocation *in vitro* include a characteristic cell elongation of epithelial cells referred to as the 'hummingbird' phenotype (Segal *et al.*, 1999), a disruption of epithelial tight junctions (Amieva *et al.*, 2003; Bagnoli *et al.*, 2005) and adherens junctions (Suzuki *et al.*, 2005), and consequently an induction of epithelial cell motility (scattering phenotype). Development of the hummingbird phenotype may represent a lack of focal contact disassembly during migration (Bourzac *et al.*, 2007). It requires CagA-dependent Erk1/2 activation and other signalling pathway modulations (Higashi *et al.*, 2004; Suzuki *et al.*, 2005) and was explained by SHP-2-dependent dephosphorylation and inhibition of focal adhesion kinase (Tsutsumi *et al.*, 2006), by deregulation of Crk adaptor protein activity leading to adherens junction protein redistribution (Suzuki *et al.*, 2005), or by a loss of cell-matrix adhesion caused by dephosphorylation of vinculin, followed by a redistribution of focal contacts (Moese *et al.*, 2007). Furthermore, inhibition of the PAR1/MARK kinase by interaction with CagA and a concomitant loss of epithelial cell polarity are required for induction of the hummingbird and motility phenotypes (Saadat *et al.*, 2007).

The Cag type IV secretion system and gastric cancer

The Cag type IV secretion system and particularly the CagA protein have been associated with the development of cancer (Peek and Blaser, 2002). In Mongolian gerbils, a corpus-predominant infection with *H. pylori*, which is considered as a precursor of atrophic gastritis, and also *H. pylori* adaptation to a high carcinogenic potential, are dependent on CagA (Rieder *et al.*, 2005; Franco *et al.*, 2005). The activity of the Cag type IV secretion system and CagA translocation result in massive changes in gene expression profiles of infected cells. Upregulated genes include catenin, matrix metalloproteinase, and other genes which may play a role in carcinogenesis (Cox *et al.*, 2001; El Etr *et al.*, 2004). Many studies have addressed the upregulation of matrix metalloproteinases. Matrix metalloproteinase-7 (MMP-7) is induced in AGS cells, depending on a functional Cag secretion system and on Erk activation (Crawford *et al.*, 2003). MMP-1 upregulation in AGS cells upon infection is also Erk- and JNK-dependent and relies partly on a functional type IV system and on translocated and phosphorylated CagA (Krueger *et al.*, 2006; Pillinger *et al.*, 2007). Increased MMP-2 and MMP-9 activity can be stimulated in AGS cells in a CagA-dependent manner and leads to an invasive phenotype (Oliveira *et al.*, 2006). In mice, however, MMP-2 and MMP-9 are also upregulated by Cag$^-$ strains (Kundu *et al.*, 2006), possibly reflecting the dispensability of the Cag system for mouse infections.

In epithelial cells that were sensitized to carcinogenic agents, transfection of CagA leads to an abnormal activation of MAP kinase pathways which results in cell transformation (Zhu *et al.*, 2005). CagA transfection into B lymphocytes leads to an inhibition of cell proliferation, but also to prevention of p53 accumulation or Erk1/2-mediated Bad phosphorylation and thus an inhibition of apoptosis (Umehara *et al.*, 2003; Zhu *et al.*, 2007). A decrease in p53 phosphorylation and inhibition of apoptosis, which is dependent on CagA, was also observed in epithelial cells, but indirectly via stimulation of macrophage migration inhibitory factor (Beswick *et al.*, 2006). Infection of gastric epithelial cells with an adapted *H. pylori* strain displaying a high carcinogenic potential was characterized by activation of gene transcription via β-catenin in a CagA-dependent manner (Franco *et al.*, 2005). Activation of β-catenin was found to involve a tyrosine phosphorylation-independent interaction between CagA and E-cadherin and to result in upregulation of genes involved in intestinal differentiation (Murata-Kamiya *et al.*, 2007). This effect correlated with the ability of transfected CagA to induce the cyclin-dependent kinase inhibitor p21$^{\mathrm{WAF1/Cip1}}$, thus causing a G1 cell-cycle arrest (Yokoyama *et al.*, 2005). However, after infection of epithelial cells with *H. pylori*, CagA-dependent cyclin D1 induction resulted in enhanced cell cycle progression, rather than cell cycle arrest (Chang *et al.*, 2006). In polarized epithelial cells, disruption of E-cadherin-dependent adherens junctions upon *H. pylori* infection was found to be independent of CagA (Weydig *et al.*, 2007). In conclusion, although several interaction partners of CagA and associated effects with regard to the development of gastric cancer have been characterized, it is premature to reach definite conclusions with respect to the oncogenic mechanisms of the Cag type IV secretion system.

H. pylori plasmids and conjugation

About 50% of *H. pylori* clinical isolates carry plasmids of variable size (Penfold *et al.*, 1988), but with unknown function (cryptic plasmids). Despite the availability of three complete genomes of *H. pylori*, still little is known about the organization and gene content of plasmids. Many isolates carry small cryptic plasmids of 2–6 kb, some of which contain only the basic elements for replication. These include an origin of replication (*ori*), *repA* or *repB* genes, and, eventually one further short conserved open reading frame (*orf2*) with unknown function (minimal plasmids). Other *H. pylori* plasmids, such as pHel4, pHel5, pHPM186 or pHPM8 carry a number of additional genes of unknown function (extended plasmids) (Quiñones *et al.*, 2001; Hofreuter and Haas, 2002).

The first indication that conjugation of plasmids might exist in *H. pylori* originated from the sequence determination of plasmid pHel4 (Hofreuter and Haas, 2002). This plasmid carries a set of four open reading frames (*orf4C-orf4F*), encoding proteins with significant homologies to the MobA and MobC proteins of colicinogenic plasmids (Morlon *et al.*, 1988; Boyd *et al.*, 1989); *orf4C* encodes a putative relaxase of plasmid pHel4. Furthermore, the genes have a similar arrangement as the *mobA*, *mobB*, *mobC* and *mobD* genes of the colicinogenic *E. coli* plasmids pColA, pColE1 or pColD. Assuming that *orf4C* is involved in conjugation of pHel4, the question arises where the genes encoding the conjugation complex are located. Since there are obviously no such genes located on the plasmid, they might have been deleted from the plasmid, or alternatively, they could reside in the chromosome of *H. pylori*. Conjugation-like DNA exchange processes, which involved the transfer of chromosomally encoded antibiotic resistances, between *H. pylori* strains or between *H. pylori* and *Campylobacter jejuni* have been reported (Kuipers *et al.*, 1998; Oyarzabal *et al.*, 2007). Both DNase-sensitive (transformation) and DNase-resistant mechanisms involving cell-to-cell contact (conjugation) were assumed to contribute to DNA transfer. However, a possible conjugation machinery that would be involved in these processes was not identified.

Candidate type IV secretion systems providing a conjugation apparatus for mating pair formation could be the chromosomally encoded Cag, ComB or Tfs3 systems. Especially the Tfs3 system might be a promising candidate for such a function, but only a small percentage of *H. pylori* strains carry a complete *tfs3* locus (Kersulyte *et al.*, 2003), meaning that only such strains would be functional donor cells for conjugation of plasmids, such as pHel4. Two putative chromosomal genes encoding further putative relaxases have been identified in a plasticity region of *H. pylori* strain 26695 (Backert *et al.*, 1998). Whether they have a function in conjugative transfer of endogenous *H. pylori* plasmids or chromosomal regions has to be established. Using an *E. coli-H. pylori* shuttle plasmid (pHel) containing an *oriT* sequence of the conjugative IncPα plasmid RP4, but no mobilization (*mob*) genes, mobilization among clinical *H. pylori* strains could be shown (Backert *et al.*, 2005). For this plasmid transfer, both genes with homology to coupling proteins (*traG*) and to relaxases (*rlx*) on the *H. pylori* chromosome (*hp0996*, *hp1004*, *hp1006*) were important.

Acknowledgements

We are grateful to Stefan Kutter and Stefanie Rohrer for reading of the manuscript and helpful discussions. Limitations in space have prevented citation of some original or review papers, and we apologize to the authors whose reports were not referenced. Current research in our groups is supported by grants from the Deutsche Forschungsgemeinschaft (HA2697/9-1 to RH, Fi953/1-2 to W.F.) and the ERANET-Pathogenomics Network (HELDIVNET) from the Federal Ministry of Education and Research (to R.H.).

References

Aas, F.E., Lovold, C., and Koomey, M. (2002). An inhibitor of DNA binding and uptake events dictates the proficiency of genetic transformation in *Neisseria gonorrhoeae*: mechanism of action and links to Type IV pilus expression. Mol. Microbiol. *46*, 1441–1450.

Akopyants, N.S., Clifton, S.W., Kersulyte, D., Crabtree, J.E., Youree, B.E., Reece, C.A., Bukanov, N.O., Drazek, E.S., Roe, B.A., and Berg, D.E. (1998). Analyses of the *cag* pathogenicity island of *Helicobacter pylori*. Mol. Microbiol. *28*, 37–53.

Akopyanz, N., Bukanov, N.O., Westblom, T.U., and Berg, D.E. (1992a). PCR-based RFLP analysis of DNA sequence diversity in the gastric pathogen *Helicobacter pylori*. Nucleic Acids Res. *20*, 5137–5142.

Akopyanz, N., Bukanov, N.O., Westblom, T.U., Kresovich, S., and Berg, D.E. (1992b). DNA diversity among clinical isolates of *Helicobacter pylori* detected by PCR-based RAPD fingerprinting. Nucleic Acids Res. *20*, 5137–5142.

Alm, R.A., Ling, L.S., Moir, D.T., King, B.L., Brown, E.D., Doig, P.C., Smith, D.R., Noonan, B., Guild, B.C., deJonge, B.L., Carmel, G., Tummino, P.J., Caruso, A., Uria-Nickelsen, M., Mills, D.M., Ives, C., Gibson, R., Merberg, D., Mills, S.D., Jiang, Q., Taylor, D.E., Vovis, G.F., and Trust, T.J. (1999). Genomic-sequence comparison of two unrelated isolates of the human gastric pathogen *Helicobacter pylori*. Nature *397*, 176–180.

Amieva, M.R., Vogelmann, R., Covacci, A., Tompkins, L.S., Nelson, W.J., and Falkow, S. (2003). Disruption of the epithelial apical-junctional complex by *Helicobacter pylori* CagA. Science *300*, 1430–1434.

Ando, T., Israel, D.A., Kusugami, K., and Blaser, M.J. (1999). HP0333, a member of the dprA family, is involved in natural transformation in *Helicobacter pylori*. J. Bacteriol. *181*, 5572–5580.

Ando, T., Xu, Q., Torres, M., Kusugami, K., Israel, D.A., and Blaser, M.J. (2000). Restriction-modification system differences in *Helicobacter pylori* are a barrier to interstrain plasmid transfer. Mol. Microbiol. *37*, 1052–1065.

Andrzejewska, J., Lee, S.K., Olbermann, P., Lotzing, N., Katzowitsch, E., Linz, B., Achtman, M., Kado, C.I., Suerbaum, S., and Josenhans, C. (2006). Characterization of the pilin ortholog of the *Helicobacter pylori* type IV *cag* pathogenicity apparatus, a surface-associated protein expressed during infection. J. Bacteriol. *188*, 5865–5877.

Aras, R.A., Fischer, W., Perez-Perez, G.I., Crosatti, M., Ando, T., Haas, R., and Blaser, M.J. (2003). Plasticity of repetitive DNA sequences within a bacterial (Type IV) secretion system component. J. Exp. Med. *198*, 1349–1360.

Aras, R.A., Small, A.J., Ando, T., and Blaser, M.J. (2002). *Helicobacter pylori* interstrain restriction-modification diversity prevents genome subversion by chromosomal DNA from competing strains. Nucleic Acids Res. *30*, 5391–5397.

Argent, R.H., Kidd, M., Owen, R.J., Thomas, R.J., Limb, M.C., and Atherton, J.C. (2004). Determinants and consequences of different levels of CagA phosphorylation for clinical isolates of *Helicobacter pylori*. Gastroenterol. *127*, 514–523.

Asahi, M., Azuma, T., Ito, S., Ito, Y., Suto, H., Nagai, Y., Tsubokawa, M., Tohyama, Y., Maeda, S., Omata, M., Suzuki, T., and Sasakawa, C. (2000). *Helicobacter pylori* CagA protein can be tyrosine phosphorylated in gastric epithelial cells. J. Exp. Med. *191*, 593–602.

Atmakuri, K., Cascales, E., and Christie, P.J. (2004). Energetic components VirD4, VirB11 and VirB4 mediate early DNA transfer reactions required for bacterial type IV secretion. Mol. Microbiol. *54*, 1199–1211.

Azuma, T., Yamakawa, A., Yamazaki, S., Fukuta, K., Ohtani, M., Ito, Y., Dojo, M., Yamazaki, Y., and Kuriyama, M. (2002). Correlation between variation of the 3′ region of the *cagA* gene in *Helicobacter pylori* and disease outcome in Japan. J. Infect. Dis. *186*, 1621–1630.

Azuma, T., Yamakawa, A., Yamazaki, S., Ohtani, M., Ito, Y., Muramatsu, A., Suto, H., Yamazaki, Y., Keida, Y., Higashi, H., and Hatakeyama, M. (2004a). Distinct diversity of the *cag* pathogenicity island among *Helicobacter pylori* strains in Japan. J. Clin. Microbiol. *42*, 2508–2517.

Azuma, T., Yamazaki, S., Yamakawa, A., Ohtani, M., Muramatsu, A., Suto, H., Ito, Y., Dojo, M., Yamazaki, Y., Kuriyama, M., Keida, Y., Higashi, H., and Hatakeyama, M. (2004b). Association between diversity in the Src homology 2 domain--containing tyrosine phosphatase binding site of *Helicobacter pylori* CagA protein and gastric atrophy and cancer. J. Infect. Dis. *189*, 820–827.

Backert, S., Kwok, T., and König, W. (2005). Conjugative plasmid DNA transfer in *Helicobacter pylori* mediated by chromosomally encoded relaxase and TraG-like proteins. Microbiology *151*, 3493–3503.

Backert, S., Von Nickisch-Rosenegk, E., and Meyer, T.F. (1998). Potential role of two *Helicobacter pylori* relaxases in DNA transfer? [letter]. Mol. Microbiol. *30*, 673–674.

Backert, S., Ziska, E., Brinkmann, V., Zimny-Arndt, U., Fauconnier, A., Jungblut, P.R., Naumann, M., and Meyer, T.F. (2000). Translocation of the *Helicobacter pylori* CagA protein in gastric epithelial cells by a type IV secretion apparatus. Cell. Microbiol. *2*, 155–164.

Bacon, D.J., Alm, R.A., Burr, D.H., Hu, L., Kopecko, D.J., Ewing, C.P., Trust, T.J., and Guerry, P. (2000). Involvement of a plasmid in virulence of *Campylobacter jejuni* 81–176. Infect. Immun. *68*, 4384–4390.

Bagnoli, F., Buti, L., Tompkins, L., Covacci, A., and Amieva, M.R. (2005). *Helicobacter pylori* CagA induces a transition from polarized to invasive phenotypes in MDCK cells. Proc. Natl. Acad. Sci. USA *102*, 16339–16344.

Baltrus, D.A. and Guillemin, K. (2006). Multiple phases of competence occur during the *Helicobacter pylori* growth cycle. FEMS Microbiol. Lett. *255*, 148–155.

Barnard, F.M., Loughlin, M.F., Fainberg, H.P., Messenger, M.P., Ussery, D.W., Williams, P., and Jenks, P.J. (2004). Global regulation of virulence and the stress response by CsrA in the highly adapted human gastric pathogen *Helicobacter pylori*. Mol. Microbiol. *51*, 15–32.

Batchelor, R.A., Pearson, B.M., Friis, L.M., Guerry, P., and Wells, J.M. (2004). Nucleotide sequences and comparison of two large conjugative plasmids from different *Campylobacter* species. Microbiology *150*, 3507–3517.

Bayliss, R., Harris, R., Coutte, L., Monier, A., Fronzes, R., Christie, P.J., Driscoll, P.C., and Waksman, G. (2007). NMR structure of a complex between the VirB9/VirB7 interaction domains of the pKM101 type IV secretion system. Proc. Natl. Acad. Sci. USA *104*, 1673–1678.

Berge, M., Mortier-Barriere, I., Martin, B., and Claverys, J.P. (2003). Transformation of *Streptococcus pneumoniae* relies on DprA- and RecA-dependent protection of incoming DNA single strands. Mol. Microbiol. *50*, 527–536.

Berger, B.R. and Christie, P.J. (1994). Genetic complementation analysis of the *Agrobacterium tumefaciens virB* operon: *virB2* through *virB11* are essential virulence genes. J. Bacteriol. *176*, 3646–3660.

Beswick, E.J., Pinchuk, I.V., Suarez, G., Sierra, J.C., and Reyes, V.E. (2006). *Helicobacter pylori* CagA-dependent macrophage migration inhibitory factor produced by gastric epithelial cells binds to CD74 and stimulates procarcinogenic events. J. Immunol. *176*, 6794–6801.

Blaser, M.J. and Atherton, J.C. (2004). *Helicobacter pylori* persistence: biology and disease. J. Clin. Invest. *113*, 321–333.

Blomstergren, A., Lundin, A., Nilsson, C., Engstrand, L., and Lundeberg, J. (2004). Comparative analysis of the complete *cag* pathogenicity island sequence in four *Helicobacter pylori* isolates. Gene *328*, 85–93.

Bohne, J., Yim, A., and Binns, A.N. (1998). The Ti plasmid increases the efficiency of *Agrobacterium tumefaciens* as a recipient in *virB*-mediated conjugal transfer of an IncQ plasmid. Proc. Natl. Acad. Sci. USA 95, 7057–7062.

Boonjakuakul, J.K., Canfield, D.R., and Solnick, J.V. (2005). Comparison of *Helicobacter pylori* virulence gene expression *in vitro* and in the Rhesus macaque. Infect. Immun. *73*, 4895–4904.

Bourzac, K.M., Botham, C.M., and Guillemin, K. (2007). *Helicobacter pylori* CagA induces AGS cell elongation through a cell retraction defect that is independent of Cdc42, Rac1, and Arp2/3. Infect. Immun. *75*, 1203–1213.

Bourzac, K.M. and Guillemin, K. (2005). *Helicobacter pylori*-host cell interactions mediated by type IV secretion. Cell Microbiol. *7*, 911–919.

Bourzac, K.M., Satkamp, L.A., and Guillemin, K. (2006). The *Helicobacter pylori cag* pathogenicity island protein CagN is a bacterial membrane-associated protein that is processed at its C terminus. Infect. Immun. *74*, 2537–2543.

Boyd, A.C., Archer, J.A., and Sherratt, D.J. (1989). Characterization of the ColE1 mobilization region and its protein products. Mol. Gen. Genet. *217*, 488–498.

Brandt, S., Kwok, T., Hartig, R., König, W., and Backert, S. (2005). NF-κB activation and potentiation of proinflammatory responses by the *Helicobacter pylori* CagA protein. Proc. Natl. Acad. Sci. USA *102*, 9300–9305.

Buhrdorf, R., Förster, C., Haas, R., and Fischer, W. (2003). Topological analysis of a putative *virB8* homologue essential for the *cag* type IV secretion system in *Helicobacter pylori*. Int. J. Med. Microbiol. *293*, 213–217.

Bukanov, N.O. and Berg, D.E. (1994). Ordered cosmid library and high-resolution physical-genetic map of *Helicobacter pylori* strain NCTC11638. Mol. Microbiol. *11(3)*, 509–523.

Burns, D.L. (2003). Type IV transporters of pathogenic bacteria. Curr. Opin. Microbiol. *6*, 29–34.

Bury-Moné, S., Thiberge, J.M., Contreras, M., Maitournam, A., Labigne, A., and De, R.H. (2004). Responsiveness to acidity via metal ion regulators mediates virulence in the gastric pathogen *Helicobacter pylori*. Mol. Microbiol. *53*, 623–638.

Busler, V.J., Torres, V.J., McClain, M.S., Tirado, O., Friedman, D.B., and Cover, T.L. (2006). Protein-Protein Interactions among *Helicobacter pylori* Cag Proteins. J. Bacteriol. *188*, 4787–4800.

Cambronne, E.D. and Roy, C.R. (2006). Recognition and delivery of effector proteins into eukaryotic cells by bacterial secretion systems. Traffic. *7*, 929–939.

Cascales, E. and Christie, P.J. (2004). Definition of a bacterial type IV secretion pathway for a DNA substrate. Science *304*, 1170–1173.

Cendron, L., Seydel, A., Angelini, A., Battistutta, R., and Zanotti, G. (2004). Crystal structure of CagZ, a protein from the *Helicobacter pylori* pathogenicity island that encodes for a type IV secretion system. J. Mol. Biol. *340*, 881–889.

Censini, S., Lange, C., Xiang, Z., Crabtree, J.E., Ghiara, P., Borodovsky, M., Rappuoli, R., and Covacci, A. (1996). *cag*, a pathogenicity island of *Helicobacter pylori*, encodes type I-specific and disease-associated virulence factors. Proc. Natl. Acad. Sci. USA *93*, 14648–14653.

Chang, K.C., Yeh, Y.C., Lin, T.L., and Wang, J.T. (2001). Identification of genes associated with natural competence in *Helicobacter pylori* by transposon shuttle random mutagenesis. Biochem. Biophys. Res. Commun. *288*, 961–968.

Chang, Y.J., Wu, M.S., Lin, J.T., Pestell, R.G., Blaser, M.J., and Chen, C.C. (2006). Mechanisms for *Helicobacter pylori* CagA-induced cyclin D1 expression that affect cell cycle. Cell Microbiol. *8*, 1740–1752.

Chen, I. and Dubnau, D. (2004). DNA uptake during bacterial transformation. Nat. Rev. Microbiol. *2*, 241–249.

Christie, P.J., Atmakuri, K., Krishnamoorthy, V., Jakubowski, S., and Cascales, E. (2005). Biogenesis, architecture, and function of bacterial type IV secretion systems. Annu. Rev. Microbiol. *59*, 451–485.

Churin, Y., Al-Ghoul, L., Kepp, O., Meyer, T.F., Birchmeier, W., and Naumann, M. (2003). *Helicobacter pylori* CagA protein targets the c-Met receptor and enhances the motogenic response. J. Cell Biol. *161*, 249–255.

Couturier, M.R., Tasca, E., Montecucco, C., and Stein, M. (2006). Interaction with CagF is required for translocation of CagA into the host via the *Helicobacter pylori* type IV secretion system. Infect. Immun. *74*, 273–281.

Cox, J.M., Clayton, C.L., Tomita, T., Wallace, D.M., Robinson, P.A., and Crabtree, J.E. (2001). cDNA array analysis of *cag* pathogenicity island-associated *Helicobacter pylori* epithelial cell response genes. Infect. Immun. *69*, 6970–6980.

Crabtree, J.E., Covacci, A., Farmery, S.M., Xiang, Z., Tompkins, D.S., Perry, S., Lindley, I.J., and Rappuoli, R. (1995). *Helicobacter pylori*-induced interleukin-8 expression in gastric epithelial cells is associated with CagA positive phenotype. J. Clin. Pathol. *48*, 41–45.

Crawford, H.C., Krishna, U.S., Israel, D.A., Matrisian, L.M., Washington, M.K., and Peek, R.M., Jr. (2003). *Helicobacter pylori* strain-selective induction of matrix metalloproteinase-7 *in vitro* and within gastric mucosa. Gastroenterology *125*, 1125–1136.

Das, A., Anderson, L.B., and Xie, Y.H. (1997). Delineation of the interaction domains of *Agrobacterium tumefaciens* VirB7 and VirB9 by use of the yeast two-hybrid assay. J. Bacteriol. *179*, 3404–3409.

Das, A. and Xie, Y.H. (1998). Construction of transposon Tn*3phoA*: its application in defining the membrane topology of the *Agrobacterium tumefaciens* DNA transfer proteins. Mol. Microbiol. *27*, 405–414.

Davison, J. (1999). Genetic exchange between bacteria in the environment. Plasmid *42*, 73–91.

Donahue, J.P., Israel, D.A., Peek, R.M., Blaser, M.J., and Miller, G.G. (2000). Overcoming the restriction barrier to plasmid transformation of *Helicobacter pylori*. Mol. Microbiol. *37*, 1066–1074.

Draskovic, I. and Dubnau, D. (2005). Biogenesis of a putative channel protein, ComEC, required for DNA uptake: membrane topology, oligomerization and formation of disulphide bonds. Mol. Microbiol. *55*, 881–896.

Dubnau, D. (1999). DNA uptake in bacteria. Annu. Rev. Microbiol. *53*, 217–244.

Eisenbrandt, R., Kalkum, M., Lai, E.M., Lurz, R., Kado, C.I., and Lanka, E. (1999). Conjugative pili of IncP plasmids, and the Ti plasmid T pilus are composed of cyclic subunits. J. Biol. Chem. *274*, 22548–22555.

El Etr, S.H., Mueller, A., Tompkins, L.S., Falkow, S., and Merrell, D.S. (2004). Phosphorylation-independent effects of CagA during interaction between *Helicobacter pylori* and T84 polarized monolayers. J. Infect. Dis. *190*, 1516–1523.

Falush, D., Kraft, C., Taylor, N.S., Correa, P., Fox, J.G., Achtman, M., and Suerbaum, S. (2001). Recombination and mutation during long-term gastric colonization by *Helicobacter pylori*: estimates of clock rates, recombination size, and minimal age. Proc. Natl. Acad. Sci. USA *98*, 15056–15061.

Finkel, S.E. and Kolter, R. (2001). DNA as a nutrient: novel role for bacterial competence gene homologs. J. Bacteriol. *183*, 6288–6293.

Fischer, W., Püls, J., Buhrdorf, R., Gebert, B., Odenbreit, S., and Haas, R. (2001). Systematic mutagenesis of the *Helicobacter pylori cag* pathogenicity island: essential genes for CagA translocation in host cells and induction of interleukin-8. Mol. Microbiol. *42*, 1337–1348. Erratum in: Mol. Microbiol. 47, 1759 (2003).

Franco, A.T., Israel, D.A., Washington, M.K., Krishna, U., Fox, J.G., Rogers, A.B., Neish, A.S., Collier-Hyams, L., Perez-Perez, G.I., Hatakeyama, M., Whitehead, R., Gaus, K., O'brien, D.P., Romero-Gallo, J., and Peek, R.M., Jr. (2005). Activation of β-catenin by carcinogenic *Helicobacter pylori*. Proc. Natl. Acad. Sci. USA *102*, 10646–10651.

Fussenegger, M., Facius, D., Meier, J., and Meyer, T.F. (1996). A novel peptidoglycan-linked lipoprotein (ComL) that functions in natural transformation competence of *Neisseria gonorrhoeae*. Mol. Microbiol. *19*, 1095–1105.

Go, M.F., Kapur, V., Graham, D.Y., and Musser, J.M. (1996). Population genetic analysis of *Helicobacter pylori* by multilocus enzyme electrophoresis – extensive allelic diversity and recombinational population structure. J. Bacteriol. *178*, 3934–3938.

Gogarten, J.P. and Townsend, J.P. (2005). Horizontal gene transfer, genome innovation and evolution. Nat. Rev. Microbiol. *3*, 679–687.

Gomis-Rüth, F.X., Sola, M., De La Cruz, F., and Coll, M. (2004). Coupling factors in macromolecular type-IV secretion machineries. Curr. Pharm. Des. *10*, 1551–1565.

Goodman, S.D. and Scocca, J.J. (1988). Identification and arrangement of the DNA sequence recognized in specific transformation of *Neisseria gonorrhoeae*. Proc. Natl. Acad. Sci. USA *85*, 6982–6986.

Gorrell, R.J., Yang, J., Kusters, J.G., van Vliet, A.H., and Robins-Browne, R.M. (2005). Restriction of DNA encoding selectable markers decreases the transformation efficiency of *Helicobacter pylori*. FEMS Immunol. Med. Microbiol. *44*, 213–219.

Grahn, A.M., Haase, J., Bamford, D.H., and Lanka, E. (2000). Components of the RP4 conjugative transfer apparatus form an envelope structure bridging inner and outer membranes of donor cells: implications for related macromolecule transport systems. J. Bacteriol. *182*, 1564–1574.

Gressmann, H., Linz, B., Ghai, R., Pleissner, K.P., Schlapbach, R., Yamaoka, Y., Kraft, C., Suerbaum, S., Meyer, T.F., and Achtman, M. (2005). Gain and loss of multiple genes during the evolution of *Helicobacter pylori*. PLoS Genet. *1*, e43-(doi:10.1371/journal.pgen.0010043).

Guiral, S., Mitchell, T.J., Martin, B., and Claverys, J.P. (2005). Competence-programmed predation of non-competent cells in the human pathogen *Streptococcus pneumoniae*: genetic requirements. Proc. Natl. Acad. Sci. USA *102*, 8710–8715.

Hahn, J., Inamine, G., Kozlov, Y., and Dubnau, D. (1993). Characterization of *comE*, a late competence operon of *Bacillus subtilis* required for the binding and uptake of transforming DNA. Mol. Microbiol. *10*, 99–111.

Harris, R.L., Hombs, V., and Silverman, P.M. (2001). Evidence that F-plasmid proteins TraV, TraK and TraB assemble into an envelope-spanning structure in *Escherichia coli*. Mol. Microbiol. *42*, 757–766.

Hatakeyama, M. (2003). *Helicobacter pylori* CagA--a potential bacterial oncoprotein that functionally mimics the mammalian Gab family of adaptor proteins. Microbes Infect. *5*, 143–150.

Hatakeyama, M. (2006). *Helicobacter pylori* CagA -- a bacterial intruder conspiring gastric carcinogenesis. Int. J. Cancer *119*, 1217–1223.

Heintschel von Heinegg, E., Nalik, H.P., and Schmid, E.N. (1993). Characterisation of a *Helicobacter pylori* phage (HP1). J. Med. Microbiol. *38*, 245–249.

Higashi, H., Nakaya, A., Tsutsumi, R., Yokoyama, K., Fujii, Y., Ishikawa, S., Higuchi, M., Takahashi, A., Kurashima, Y., Teishikata, Y., Tanaka, S., Azuma, T., and Hatakeyama, M. (2004). *Helicobacter pylori* CagA induces Ras-independent morphogenetic response through SHP-2 recruitment and activation. J. Biol. Chem. *279*, 17205–17216.

Higashi, H., Tsutsumi, R., Fujita, A., Yamazaki, S., Asaka, M., Azuma, T., and Hatakeyama, M. (2002a). Biological activity of the *Helicobacter pylori* virulence factor CagA is determined by variation in the tyrosine phosphorylation sites. Proc. Natl. Acad. Sci. USA *99*, 14428–14433.

Higashi, H., Tsutsumi, R., Muto, S., Sugiyama, T., Azuma, T., Asaka, M., and Hatakeyama, M. (2002b). SHP-2 tyrosine phosphatase as an intracellular target of *Helicobacter pylori* CagA protein. Science *295*, 683–686.

Hofreuter, D. and Haas, R. (2002). Characterization of two cryptic *Helicobacter pylori* plasmids: a putative source for horizontal gene transfer and gene shuffling. J. Bacteriol. *184*, 2755–2766.

Hofreuter, D., Karnholz, A., and Haas, R. (2003). Topology and membrane interaction of *Helicobacter pylori* ComB proteins involved in natural transformation competence. Int. J. Med. Microbiol. *293*, 153–165.

Hofreuter, D., Odenbreit, S., and Haas, R. (2001). Natural transformation competence in *Helicobacter pylori* is mediated by the basic components of a type IV secretion system. Mol. Microbiol. *41*, 379–391.

Hohlfeld, S., Pattis, I., Püls, J., Plano, G.V., Haas, R., and Fischer, W. (2006). A C-terminal secretion signal is necessary, but not sufficient for type IV secretion of the *Helicobacter pylori* CagA protein. Mol. Microbiol. *59*, 1624–1637.

Höppner, C., Liu, Z., Domke, N., Binns, A.N., and Baron, C. (2004). VirB1 orthologs from *Brucella suis* and pKM101 complement defects of the lytic transglycosylase required for efficient type IV secretion from *Agrobacterium tumefaciens*. J. Bacteriol. *186*, 1415–1422.

Israel, D.A., Lou, A.S., and Blaser, M.J. (2000). Characteristics of *Helicobacter pylori* natural transformation. FEMS Microbiol. Lett. *186*, 275–280.

Jakubowski, S.J., Cascales, E., Krishnamoorthy, V., and Christie, P.J. (2005). *Agrobacterium tumefaciens* VirB9, an outer-membrane-associated component of a type IV secretion system, regulates substrate selection and T-pilus biogenesis. J. Bacteriol. *187*, 3486–3495.

Jakubowski, S.J., Krishnamoorthy, V., Cascales, E., and Christie, P.J. (2004). *Agrobacterium tumefaciens* VirB6 domains direct the ordered export of a DNA substrate through a type IV secretion System. J. Mol. Biol. *341*, 961–977.

Joyce, E.A., Gilbert, J.V., Eaton, K.A., Plaut, A., and Wright, A. (2001). Differential gene expression from two transcriptional units in the *cag* pathogenicity island of *Helicobacter pylori*. Infect. Immun. *69*, 4202–4209.

Judd, P.K., Kumar, R.B., and Das, A. (2005). The type IV secretion apparatus protein VirB6 of *Agrobacterium tumefaciens* localizes to a cell pole. Mol. Microbiol. *55*, 115–124.

Kalkum, M., Eisenbrandt, R., Lurz, R., and Lanka, E. (2002). Tying rings for sex. Trends Microbiol. *10*, 382–387.

Karita, M., Tummuru, M.K., Wirth, H.P., and Blaser, M.J. (1996). Effect of growth phase and acid shock on *Helicobacter pylori cagA* expression. Infect. Immun. *64*, 4501–4507.

Karnholz, A., Höfler, C., Odenbreit, S., Fischer, W., Hofreuter, D., and Haas, R. (2006). Functional and topological characterization of novel components of the *comB* DNA transformation competence system in *Helicobacter pylori*. J. Bacteriol. *188*, 882–893.

Kavermann, H., Burns, B.P., Angermüller, K., Odenbreit, S., Fischer, W., Melchers, K., and Haas, R. (2003). Identification and characterization of *Helicobacter pylori* genes essential for gastric colonization. J. Exp. Med. *197*, 813–822.

Kersulyte, D., Chalkauskas, H., and Berg, D.E. (1999). Emergence of recombinant strains of *Helicobacter pylori* during human infection. Mol. Microbiol. *31*, 31–43.

Kersulyte, D., Velapatiño, B., Mukhopadhyay, A.K., Cahuayme, L., Bussalleu, A., Combe, J., Gilman, R.H., and Berg, D.E. (2003). Cluster of type IV secretion genes in *Helicobacter pylori*'s plasticity zone. J. Bacteriol. *185*, 3764–3772.

Kim, S.Y., Lee, Y.C., Kim, H.K., and Blaser, M.J. (2006). *Helicobacter pylori* CagA transfection of gastric epithelial cells induces interleukin-8. Cell Microbiol. *8*, 97–106.

Koraimann, G. (2003). Lytic transglycosylases in macromolecular transport systems of Gram-negative bacteria. Cell Mol. Life Sci. *60*, 2371–2388.

Krause, S., Barcena, M., Pansegrau, W., Lurz, R., Carazo, J.M., and Lanka, E. (2000a). Sequence-related protein export NTPases encoded by the conjugative transfer region of RP4 and by the *cag* pathogenicity island of *Helicobacter pylori* share similar hexameric ring structures. Proc. Natl. Acad. Sci. USA. *97*, 3067–3072.

Krause, S., Pansegrau, W., Lurz, R., De La Cruz, F., and Lanka, E. (2000b). Enzymology of type IV macromolecule secretion systems: the conjugative transfer regions of plasmids RP4 and R388 and the *cag* pathogenicity island of *Helicobacter pylori* encode structurally and functionally related nucleoside triphosphate hydrolases. J. Bacteriol. *182*, 2761–2770.

Krueger, S., Hundertmark, T., Kalinski, T., Peitz, U., Wex, T., Malfertheiner, P., Naumann, M., and Roessner, A. (2006). *Helicobacter pylori* encoding the pathogenicity island activates matrix metalloproteinase 1 in gastric epithelial cells via JNK and ERK. J. Biol. Chem. *281*, 2868–2875.

Kuipers, E.J., Israel, D.A., Kusters, J.G., and Blaser, M.J. (1998). Evidence for a conjugation-like mechanism of DNA transfer in *Helicobacter pylori*. J. Bacteriol. *180*, 2901–2905.

Kundu, P., Mukhopadhyay, A.K., Patra, R., Banerjee, A., Berg, D.E., and Swarnakar, S. (2006). Cag pathogenicity island-independent upregulation of matrix metalloproteinases-9 and -2 secretion and expression in mice by *Helicobacter pylori* infection. J. Biol. Chem. *281*, 34651–34662.

Kutter, S., Buhrdorf, R., Haas, J., Schneider-Brachert, W., Haas, R., and Fisher, W. (2008). Protein subassemblies of the *Helicobacter pylori* Cag type IV secretion system revealed by localization and interaction studies. J. Bacteriol. *190* (in press).

Lai, E.M. and Kado, C.I. (2000). The T-pilus of *Agrobacterium tumefaciens*. Trends Microbiol. *8*, 361–369.

Lawley, T.D., Klimke, W.A., Gubbins, M.J., and Frost, L.S. (2003). F factor conjugation is a true type IV secretion system. FEMS Microbiol. Lett. *224*, 1–15.

Levine, S.M., Lin, E.A., Emara, W., Kang, J., Dibenedetto, M., Ando, T., Falush, D., and Blaser, M.J. (2007). Plastic cells and populations: DNA substrate characteristics in *Helicobacter pylori* transformation define a flexible but conservative system for genomic variation. FASEB J. doi: 10.1096/fj.07–8501com.

Lin, L.F., Posfai, J., Roberts, R.J., and Kong, H. (2001). Comparative genomics of the restriction-modification systems in *Helicobacter pylori*. Proc. Natl. Acad. Sci. USA. *98*, 2740–2745.

Liu, G., McDaniel, T.K., Falkow, S., and Karlin, S. (1999). Sequence anomalies in the *cag7* gene of the *Helicobacter pylori* pathogenicity island. Proc. Natl. Acad. Sci. USA *96*, 7011–7016.

Loh, J.T., Torres, V.J., and Cover, T.L. (2007). Regulation of *Helicobacter pylori cagA* expression in response to salt. Cancer Res. *67*, 4709–4715.

Marais, A., Mendz, G.L., Hazell, S.L., and Mégraud, F. (1999). Metabolism and Genetics of *Helicobacter pylori*: the Genome Era. Microbiol. Mol. Biol. Rev. *63*, 642–674.

Marshall, D.G., Dundon, W.G., Beesley, S.M., and Smyth, C.J. (1998). *Helicobacter pylori*--a conundrum of genetic diversity [In Process Citation]. Microbiology *144 (Pt 11)*, 2925–2939.

Mazodier, P. and Davies, J. (1991). Gene transfer between distantly related bacteria. Annu. Rev. Genet. *25*, 147–171.

Mégraud, F. (2004). Basis for the management of drug-resistant *Helicobacter pylori* infection. Drugs *64*, 1893–1904.

Merrell, D.S., Goodrich, M.L., Otto, G., Tompkins, L.S., and Falkow, S. (2003). pH-regulated gene expression of the gastric pathogen *Helicobacter pylori*. Infect. Immun. *71*, 3529–3539.

Mimuro, H., Suzuki, T., Tanaka, J., Asahi, M., Haas, R., and Sasakawa, C. (2002). Grb2 is a key mediator of *Helicobacter pylori* CagA protein activities. Mol. Cell *10*, 745–755.

Moese, S., Selbach, M., Brinkmann, V., Karlas, A., Haimovich, B., Backert, S., and Meyer, T.F. (2007). The *Helicobacter pylori* CagA protein disrupts matrix adhesion of gastric epithelial cells by dephosphorylation of vinculin. Cell Microbiol. *9*, 1148–1161.

Morlon, J., Chartier, M., Bidaud, M., and Lazdunski, C. (1988). The complete nucleotide sequence of the colicinogenic plasmid ColA. High extent of homology with ColE1. Mol. Gen. Genet. *211*, 231–243.

Moscoso, M. and Claverys, J.P. (2004). Release of DNA into the medium by competent *Streptococcus pneumoniae*: kinetics, mechanism and stability of the liberated DNA. Mol. Microbiol. *54*, 783–794.

Murata-Kamiya, N., Kurashima, Y., Teishikata, Y., Yamahashi, Y., Saito, Y., Higashi, H., Aburatani, H., Akiyama, T., Peek, R.M., Jr., Azuma, T., and Hatakeyama, M. (2007). *Helicobacter pylori* CagA interacts with E-cadherin and deregulates the β-catenin signal that promotes intestinal transdifferentiation in gastric epithelial cells. Oncogene *26*, 4617–4626.

Mushegian, A.R., Fullner, K.J., Koonin, E.V., and Nester, E.W. (1996). A family of lysozyme-like virulence factors in bacterial pathogens of plants and animals. Proc. Natl. Acad. Sci. USA *93*, 7321–7326.

Nagai, H., Cambronne, E.D., Kagan, J.C., Amor, J.C., Kahn, R.A., and Roy, C.R. (2005). A C-terminal translocation signal required for Dot/Icm-dependent delivery of the *Legionella* RalF protein to host cells. Proc. Natl. Acad. Sci. USA *102*, 826–831.

Naito, M., Yamazaki, T., Tsutsumi, R., Higashi, H., Onoe, K., Yamazaki, S., Azuma, T., and Hatakeyama, M. (2006). Influence of EPIYA-repeat polymorphism on the phosphorylation-dependent biological activity of *Helicobacter pylori* CagA. Gastroenterol. *130*, 1181–1190.

Naumann, M. (2005). Pathogenicity island-dependent effects of *Helicobacter pylori* on intracellular signal transduction in epithelial cells. Int. J. Med. Microbiol. *295*, 335–341.

Nedenskov-Sorensen, P., Bukholm, G., and Bovre, K. (1990). Natural competence for genetic transformation in *Campylobacter pylori*. J. Infect. Dis. *161*, 365–366.

O'Rourke, E.J., Chevalier, C., Pinto, A.V., Thiberge, J.M., Ielpi, L., Labigne, A., and Radicella, J.P. (2003). Pathogen DNA as target for host-generated oxidative stress: role for repair of bacterial DNA damage in *Helicobacter pylori* colonization. Proc. Natl. Acad. Sci. USA *100*, 2789–2794.

O'Rourke, E.J., Pinto, A.V., Petroni, E.A., Tolmasky, M.L., and Ielpi, L. (2004). Evidence for the active role of a novel nuclease from *Helicobacter pylori* in the horizontal transfer of genetic information. J. Bacteriol. *186*, 2586–2593.

Odenbreit, S., Gebert, B., Püls, J., Fischer, W., and Haas, R. (2001). Interaction of *Helicobacter pylori* with professional phagocytes: role of the *cag* pathogenicity island and translocation, phosphorylation and processing of CagA. Cell Microbiol. *3*, 21–31.

Odenbreit, S., Püls, J., Sedlmaier, B., Gerland, E., Fischer, W., and Haas, R. (2000). Translocation of *Helicobacter pylori* CagA into gastric epithelial cells by type IV secretion. Science *287*, 1497–1500.

Odenbreit, S., Till, M., and Haas, R. (1996). Optimized *blaM*-transposon shuttle mutagenesis of *Helicobacter pylori* allows the identification of novel genetic loci involved in bacterial virulence. Mol. Microbiol. *20*, 361–373.

Oh, J.D., Kling-Bäckhed, H., Giannakis, M., Xu, J., Fulton, R.S., Fulton, L.A., Cordum, H.S., Wang, C., Elliott, G., Edwards, J., Mardis, E.R., Engstrand, L.G., and Gordon, J.I. (2006). The complete genome sequence of a chronic atrophic gastritis *Helicobacter pylori* strain: evolution during disease progression. Proc. Natl. Acad. Sci. USA *103*, 9999–10004.

Oliveira, M.J., Costa, A.C., Costa, A.M., Henriques, L., Suriano, G., Atherton, J.C., Machado, J.C., Carneiro, F., Seruca, R., Mareel, M., Leroy, A., and Figueiredo, C. (2006). *Helicobacter pylori* induces gastric epithelial cell invasion in a c-Met and type IV secretion system-dependent manner. J. Biol. Chem. *281*, 34888–34896.

Oyarzabal, O.A., Rad, R., and Backert, S. (2007). Conjugative transfer of chromosomally encoded antibiotic resistance from *Helicobacter pylori* to *Campylobacter jejuni*. J. Clin. Microbiol. *45*, 402–408.

Pattis, I., Weiss, E., Laugks, R., Haas, R., and Fischer, W. (2007). The *Helicobacter pylori* CagF protein is a type IV secretion chaperone-like molecule that binds close to the C-terminal secretion signal of the CagA effector protein. Microbiology *153*, 2896–2909.

Peek, R.M.J. and Blaser, M.J. (2002). *Helicobacter pylori* and gastrointestinal tract adenocarcinomas. Nat. Rev. Cancer 2, 28–37.

Penfold, S.S., Lastovica, A.J., and Elisha, B.G. (1988). Demonstration of plasmids in *Campylobacter pylori*. J. Infect. Dis. *157*, 850–851.

Pflock, M., Finsterer, N., Joseph, B., Mollenkopf, H., Meyer, T.F., and Beier, D. (2006). Characterization of the ArsRS regulon of *Helicobacter pylori*, involved in acid adaptation. J. Bacteriol. *188*, 3449–3462.

Pillinger, M.H., Marjanovic, N., Kim, S.Y., Lee, Y.C., Scher, J.U., Roper, J., Abeles, A.M., Izmirly, P.I., Axelrod, M.J., Pillinger, M.Y., Tolani, S., Dinsell, V., Abramson, S.B., and Blaser, M.J. (2007). *Helicobacter pylori* stimulates gastric epithelial cell MMP-1 secretion via CagA-dependent and independent ERK activation. J. Biol. Chem. *282*, 18722–18731.

Pinto, A.V., Mathieu, A., Marsin, S., Veaute, X., Ielpi, L., Labigne, A., and Radicella, J.P. (2005). Suppression of homologous and homeologous recombination by the bacterial MutS2 protein. Mol. Cell *17*, 113–120.

Poppe, M., Feller, S.M., Römer, G., and Wessler, S. (2007). Phosphorylation of *Helicobacter pylori* CagA by c-Abl leads to cell motility. Oncogene *26*, 3462–3472.

Quiñones, M., Knesek, J.E., and McIntire, S.A. (2001). Sequence and gene expression analyses of plasmid pHPM8 from *Helicobacter pylori* reveal the presence of two operons with putative roles in plasmid replication and antibiotic activity. Plasmid *46*, 223–228.

Rabel, C., Grahn, A.M., Lurz, R., and Lanka, E. (2003). The VirB4 family of proposed traffic nucleoside triphosphatases: common motifs in plasmid RP4 TrbE are essential for conjugation and phage adsorption. J. Bacteriol. *185*, 1045–1058.

Rain, J.C., Selig, L., De Reuse, H., Battaglia, V., Reverdy, C., Simon, S., Lenzen, G., Petel, F., Wojcik, J., Schächter, V., Chemama, Y., Labigne, A., and Legrain, P. (2001). The protein-protein interaction map of *Helicobacter pylori*. Nature *409*, 211–215.

Ren, S., Higashi, H., Lu, H., Azuma, T., and Hatakeyama, M. (2006). Structural basis and functional consequence of *Helicobacter pylori* CagA multimerization in cells. J. Biol. Chem. *281*, 32344–32352.

Rieder, G., Merchant, J.L., and Haas, R. (2005). *Helicobacter pylori cag*-Type IV Secretion System Facilitates Corpus Colonization to Induce Precancerous Conditions in Mongolian Gerbils. Gastroenterol. *128*, 1229–1242.

Rohde, M., Püls, J., Buhrdorf, R., Fischer, W., and Haas, R. (2003). A novel sheathed surface organelle of the *Helicobacter pylori cag* type IV secretion system. Mol. Microbiol. *49*, 219–234.

Rosenstiel, P., Hellmig, S., Hampe, J., Ott, S., Till, A., Fischbach, W., Sahly, H., Lucius, R., Fölsch, U.R., Philpott, D., and Schreiber, S. (2006). Influence of polymorphisms in the NOD1/CARD4 and NOD2/CARD15 genes on the clinical outcome of *Helicobacter pylori* infection. Cell Microbiol. *8*, 1188–1198.

Saadat, I., Higashi, H., Obuse, C., Umeda, M., Murata-Kamiya, N., Saito, Y., Lu, H., Ohnishi, N., Azuma, T., Suzuki, A., Ohno, S., and Hatakeyama, M. (2007). *Helicobacter pylori* CagA targets PAR1/MARK kinase to disrupt epithelial cell polarity. Nature *447*, 330–333.

Savvides, S.N., Yeo, H.J., Beck, M.R., Blaesing, F., Lurz, R., Lanka, E., Buhrdorf, R., Fischer, W., Haas, R., and Waksman, G. (2003). VirB11 ATPases are dynamic hexameric assemblies: new insights into bacterial type IV secretion. EMBO J. *22*, 1969–1980.

Schilling, C.H., Covert, M.W., Famili, I., Church, G.M., Edwards, J.S., and Palsson, B.O. (2002). Genome-scale metabolic model of *Helicobacter pylori* 26695. J. Bacteriol. *184*, 4582–4593.

Schmid, E.N., von, R.G., and Ansorg, R. (1990). Bacteriophages in *Helicobacter* (*Campylobacter*) *pylori*. J. Med. Microbiol. *32*, 101–104.

Schmitt, W., Odenbreit, S., Heuermann, D., and Haas, R. (1995). Cloning of the *Helicobacter pylori recA* gene and functional characterization of its product. Mol. Gen. Genet. *248*, 563–572.

Schröder, G., Krause, S., Zechner, E.L., Traxler, B., Yeo, H.J., Lurz, R., Waksman, G., and Lanka, E. (2002). TraG-like proteins of DNA transfer systems and of the *Helicobacter pylori* type IV secretion system: inner membrane gate for exported substrates? J. Bacteriol. *184*, 2767–2779.

Schröder, G. and Lanka, E. (2005). The mating pair formation system of conjugative plasmids-A versatile secretion machinery for transfer of proteins and DNA. Plasmid *54*, 1–25.

Schulein, R., Guye, P., Rhomberg, T.A., Schmid, M.C., Schröder, G., Vergunst, A.C., Carena, I., and Dehio,

C. (2005). A bipartite signal mediates the transfer of type IV secretion substrates of *Bartonella henselae* into human cells. Proc. Natl. Acad. Sci. USA *102*, 856–861.

Scott, D.R., Marcus, E.A., Wen, Y., Oh, J., and Sachs, G. (2007). Gene expression *in vivo* shows that *Helicobacter pylori* colonizes an acidic niche on the gastric surface. Proc. Natl. Acad. Sci. USA *104*, 7235–7240.

Segal, E.D., Cha, J., Lo, J., Falkow, S., and Tompkins, L.S. (1999). Altered states: involvement of phosphorylated CagA in the induction of host cellular growth changes by *Helicobacter pylori*. Proc. Natl. Acad. Sci. USA *96*, 14559–14564.

Segal, E.D., Lange, C., Covacci, A., Tompkins, L.S., and Falkow, S. (1997). Induction of host signal transduction pathways by *Helicobacter pylori*. Proc. Natl. Acad. Sci. USA *94*, 7595–7599.

Selbach, M., Moese, S., Backert, S., Jungblut, P.R., and Meyer, T.F. (2004). The *Helicobacter pylori* CagA protein induces tyrosine dephosphorylation of ezrin. Proteomics *4*, 2961–2968.

Selbach, M., Moese, S., Hauck, C.R., Meyer, T.F., and Backert, S. (2002). Src is the kinase of the *Helicobacter pylori* CagA protein *in vitro* and *in vivo*. J. Biol. Chem. *277*, 6775–6778.

Selbach, M., Moese, S., Hurwitz, R., Hauck, C.R., Meyer, T.F., and Backert, S. (2003). The *Helicobacter pylori* CagA protein induces cortactin dephosphorylation and actin rearrangement by c-Src inactivation. EMBO J. *22*, 515–528.

Smeets, L.C., Arents, N.L., van Zwet, A.A., Vandenbroucke-Grauls, C.M., Verboom, T., Bitter, W., and Kusters, J.G. (2003). Molecular patchwork: Chromosomal recombination between two *Helicobacter pylori* strains during natural colonization. Infect. Immun. *71*, 2907–2910.

Smeets, L.C., Bijlsma, J.J., Boomkens, S.Y., Vandenbroucke-Grauls, C.M., and Kusters, J.G. (2000a). *comH*, a novel gene essential for natural transformation of *Helicobacter pylori*. J. Bacteriol. *182*, 3948–3954.

Smeets, L.C., Bijlsma, J.J., Kuipers, E.J., Vandenbroucke-Grauls, C.M., and Kusters, J.G. (2000b). The *dprA* gene is required for natural transformation of *Helicobacter pylori*. FEMS Immunol. Med. Microbiol. *27*, 99–102.

Smeets, L.C. and Kusters, J.G. (2002). Natural transformation in *Helicobacter pylori*: DNA transport in an unexpected way. Trends Microbiol. *10*, 159–162.

Solomon, J.M. and Grossman, A.D. (1996). Who's competent and when: regulation of natural genetic competence in bacteria. Trends. Genet. *12*, 150–155.

Spohn, G., Beier, D., Rappuoli, R., and Scarlato, V. (1997). Transcriptional analysis of the divergent *cagAB* genes encoded by the pathogenicity island of *Helicobacter pylori*. Mol. Microbiol. *26*, 361–372.

Stein, M., Bagnoli, F., Halenbeck, R., Rappuoli, R., Fantl, W.J., and Covacci, A. (2002). c-Src/Lyn kinases activate *Helicobacter pylori* CagA through tyrosine-phosphorylation of the EPIYA motifs. Mol. Microbiol. *43*, 971–980.

Stein, M., Rappuoli, R., and Covacci, A. (2000). Tyrosine phosphorylation of the *Helicobacter pylori* CagA antigen after *cag*-driven host cell translocation. Proc. Natl. Acad. Sci. USA *97*, 1263–1268.

Suerbaum, S. and Achtman, M. (1999). Evolution of *Helicobacter pylori*: the role of recombination [In Process Citation]. Trends. Microbiol. *7*, 182–184.

Suerbaum, S. and Josenhans, C. (2007). *Helicobacter pylori* evolution and phenotypic diversification in a changing host. Nat. Rev. Microbiol. *5*, 441–452.

Suzuki, M., Mimuro, H., Suzuki, T., Park, M., Yamamoto, T., and Sasakawa, C. (2005). Interaction of CagA with Crk plays an important role in *Helicobacter pylori*-induced loss of gastric epithelial cell adhesion. J. Exp. Med. *202*, 1235–1247.

Takata, T., Ando, T., Israel, D.A., Wassenaar, T.M., and Blaser, M.J. (2005). Role of *dprA* in transformation of *Campylobacter jejuni*. FEMS Microbiology Letters *252*, 161–168.

Tammer, I., Brandt, S., Hartig, R., König, W., and Backert, S. (2007). Activation of Abl by *Helicobacter pylori*: a novel kinase for CagA and crucial mediator of host cell scattering. Gastroenterol. *132*, 1309–1319.

Tanaka, J., Suzuki, T., Mimuro, H., and Sasakawa, C. (2003). Structural definition on the surface of *Helicobacter pylori* type IV secretion apparatus. Cell Microbiol. *5*, 395–404.

Terradot, L., Bayliss, R., Oomen, C., Leonard, G.A., Baron, C., and Waksman, G. (2005). Structures of two core subunits of the bacterial type IV secretion system, VirB8 from *Brucella suis* and ComB10 from *Helicobacter pylori*. Proc. Natl. Acad. Sci. USA *102*, 4596–4601.

Terradot, L., Durnell, N., Li, M., Li, M., Ory, J., Labigne, A., Legrain, P., Colland, F., and Waksman, G. (2004). Biochemical characterization of protein complexes from the *Helicobacter pylori* protein interaction map: strategies for complex formation and evidence for novel interactions within type IV secretion systems. Mol. Cell. Proteomics *3*, 809–819.

Thompson, L.J., Merrell, D.S., Neilan, B.A., Mitchell, H., Lee, A., and Falkow, S. (2003). Gene expression profiling of *Helicobacter pylori* reveals a growth-phase-dependent switch in virulence gene expression. Infect. Immun. *71*, 2643–2655.

Thompson, S. and Blaser, M.J. (1995). Isolation of the *Helicobacter pylori recA* gene and involvement of the *recA* region in resistance to low pH. Infect. Immun. *63*, 2185–2193.

Tomb, J.-F., White, O., Kerlavage, A.R., Clayton, R.A., Sutton, G.G., Fleischmann, R.D., Ketchum, K.A., Klenk, H.P., Gill, S., Dougherty, B.A., Nelson, K., Quackenbush, J., Zhou, L., Kirkness, E.F., Peterson, S., Loftus, B., Richardson, D., Dodson, R., Khalak, H.G., Glodek, A., McKenney, K., Fitzegerald, L.M., Lee, N., Adams, M.D., Hickey, E.K., Berg, D.E., Gocayne, J.D., Utterback, T.R., Peterson, J.D., Kelley, J.M., Cotton, M.D., Weidman, J.M., Fujii, C., Bowman, C., Watthey, L., Wallin, E., Hayes, W.S., Borodovsky, M., Karp, P.D., Smith, H.O., Fraser, C.M., and Venter, J.C. (1997). The complete genome sequence of the gastric pathogen *Helicobacter pylori*. Nature *388*, 539–547.

Tønjum, T., Håvarstein, L.S., Koomey, M., and Seeberg, E. (2004). Transformation and DNA repair: linkage by DNA recombination. Trends Microbiol. *12*, 1–4.

Torruellas-Garcia, J., Ferracci, F., Jackson, M.W., Joseph, S., Pattis, I., Plano, L.R., Fischer, W., and Plano, G.V. (2006). Measurement of effector protein injection by type III and type IV secretion systems using a 13-residue phosphorylatable GSK tag. Infect. Immun. *74*, 5645–5657.

Tsutsumi, R., Higashi, H., Higuchi, M., Okada, M., and Hatakeyama, M. (2003). Attenuation of *Helicobacter pylori* CagA-SHP-2 signaling by interaction between CagA and C-terminal Src kinase. J. Biol. Chem. *278*, 3664–3670.

Tsutsumi, R., Takahashi, A., Azuma, T., Higashi, H., and Hatakeyama, M. (2006). Focal adhesion kinase is a substrate and downstream effector of SHP-2 complexed with *Helicobacter pylori* CagA. Mol. Cell. Biol. *26*, 261–276.

Tummuru, M.K.R., Sharma, S.A., and Blaser, M.J. (1995). *Helicobacter pylori picB*, a homologue of the *Bordetella pertussis* toxin secretion protein, is required for induction of IL-8 in gastric epithelial cells. Mol. Microbiol. *18*, 867–876.

Umehara, S., Higashi, H., Ohnishi, N., Asaka, M., and Hatakeyama, M. (2003). Effects of *Helicobacter pylori* CagA protein on the growth and survival of B lymphocytes, the origin of MALT lymphoma. Oncogene *22*, 8337–8342.

van Doorn, L.J., Figueiredo, C., Sanna, R., Blaser, M.J., and Quint, W.G. (1999). Distinct variants of *Helicobacter pylori cagA* are associated with *vacA* subtypes. J. Clin. Microbiol. *37*, 2306–2311.

Vergunst, A.C., van Lier, M.C., den Dulk-Ras, A., Grosse Stüve, T.A., Ouwehand, A., and Hooykaas, P.J. (2005). Positive charge is an important feature of the C-terminal transport signal of the VirB/D4-translocated proteins of *Agrobacterium*. Proc. Natl. Acad. Sci. USA *102*, 832–837.

Viala, J., Chaput, C., Boneca, I.G., Cardona, A., Girardin, S.E., Moran, A.P., Athman, R., Memet, S., Huerre, M.R., Coyle, A.J., DiStefano, P.S., Sansonetti, P.J., Labigne, A., Bertin, J., Philpott, D.J., and Ferrero, R.L. (2004). Nod1 responds to peptidoglycan delivered by the *Helicobacter pylori cag* pathogenicity island. Nat. Immunol. *5*, 1166–1174.

Wen, Y., Marcus, E.A., Matrubutham, U., Gleeson, M.A., Scott, D.R., and Sachs, G. (2003). Acid-adaptive genes of *Helicobacter pylori*. Infect. Immun. *71*, 5921–5939.

Weydig, C., Starzinski-Powitz, A., Carra, G., Löwer, J., and Wessler, S. (2007). CagA-independent disruption of adherence junction complexes involves E-cadherin shedding and implies multiple steps in *Helicobacter pylori* pathogenicity. Exp. Cell Res. *313*, 3459–3471.

Xu, Q., Morgan, R.D., Roberts, R.J., and Blaser, M.J. (2000). Identification of type II restriction and modification systems in *Helicobacter pylori* reveals their substantial diversity among strains. Proc. Natl. Acad. Sci. USA *97*, 9671–9676.

Yamaoka, Y., El-Zimaity, H.M., Gutierrez, O., Figura, N., Kim, J.G., Kodama, T., Kashima, K., and Graham, D.Y. (1999). Relationship between the *cagA* 3′ repeat region of *Helicobacter pylori*, gastric histology, and susceptibility to low pH. Gastroenterol. *117*, 342–349.

Yamaoka, Y., Kodama, T., Kashima, K., Graham, D.Y., and Sepulveda, A.R. (1998). Variants of the 3′ region of the cagA gene in *Helicobacter pylori* isolates from patients with different H. pylori-associated diseases. J. Clin. Microbiol. *36*, 2258–2263.

Yamaoka, Y., Kudo, T., Lu, H., Casola, A., Brasier, A.R., and Graham, D.Y. (2004). Role of interferon-stimulated responsive element-like element in interleukin-8 promoter in *Helicobacter pylori* infection. Gastroenterol. *126*, 1030–1043.

Yamazaki, S., Yamakawa, A., Okuda, T., Ohtani, M., Suto, H., Ito, Y., Yamazaki, Y., Keida, Y., Higashi, H., Hatakeyama, M., and Azuma, T. (2005). Distinct diversity of *vacA*, *cagA*, and *cagE* genes of *Helicobacter pylori* associated with peptic ulcer in Japan. J. Clin. Microbiol. *43*, 3906–3916.

Yeh, Y.C., Lin, T.L., Chang, K.C., and Wang, J.T. (2003). Characterization of a ComE3 homologue essential for DNA transformation in *Helicobacter pylori*. Infect. Immun. *71*, 5427–5431.

Yeo, H.J., Savvides, S.N., Herr, A.B., Lanka, E., and Waksman, G. (2000). Crystal structure of the hexameric traffic ATPase of the *Helicobacter pylori* type IV secretion system. Mol. Cell *6*, 1461–1472.

Yeo, H.J., Yuan, Q., Beck, M.R., Baron, C., and Waksman, G. (2003). Structural and functional characterization of the VirB5 protein from the type IV secretion system encoded by the conjugative plasmid pKM101. Proc. Natl. Acad. Sci. USA *100*, 15947–15952.

Yokoyama, K., Higashi, H., Ishikawa, S., Fujii, Y., Kondo, S., Kato, H., Azuma, T., Wada, A., Hirayama, T., Aburatani, H., and Hatakeyama, M. (2005). Functional antagonism between *Helicobacter pylori* CagA and vacuolating toxin VacA in control of the NFAT signaling pathway in gastric epithelial cells. Proc. Natl. Acad. Sci. USA *102*, 9661–9666.

Yuan, Q., Carle, A., Gao, C., Sivanesan, D., Aly, K.A., Höppner, C., Krall, L., Domke, N., and Baron, C. (2005). Identification of the VirB4-VirB8-VirB5-VirB2 pilus assembly sequence of type IV secretion systems. J. Biol. Chem. *280*, 26349–26359.

Zahrl, D., Wagner, M., Bischof, K., Bayer, M., Zavecz, B., Beranek, A., Ruckenstuhl, C., Zarfel, G.E., and Koraimann, G. (2005). Peptidoglycan degradation by specialized lytic transglycosylases associated with type III and type IV secretion systems. Microbiology *151*, 3455–3467.

Zhu, Y., Wang, C., Huang, J., Ge, Z., Dong, Q., Zhong, X., Su, Y., and Zheng, S. (2007). The *Helicobacter pylori* virulence factor CagA promotes Erk1/2-mediated Bad phosphorylation in lymphocytes: a mechanism of CagA-inhibited lymphocyte apoptosis. Cell Microbiol. *9*, 952–961.

Zhu, Y., Zhong, X., Zheng, S., Du, Q., and Xu, W. (2005). Transformed immortalized gastric epithelial cells by virulence factor CagA of *Helicobacter pylori* through Erk mitogen-activated protein kinase pathway. Oncogene *24*, 3886–3895.

Gastric Biology of *Helicobacter pylori*

7

George Sachs, Yi Wen and David R. Scott

Abstract

Helicobacter pylori colonizes the highly acidic environment of the gastric mucosa. *H. pylori*, although a neutralophile, is able to flourish in its acidic gastric niche by buffering its periplasm to near neutrality using the mechanism of acid acclimation. This unique acid acclimation, in contrast to acid resistance or tolerance expressed by other neutralophiles, is dependent on urease activity, a proton gate urea channel, UreI, that enables rapid urea access to cytoplasmic urease, and a periplasmic membrane anchored α-carbonic anhydrase. The genes encoding these proteins increased transcription in acid *in vitro* and are members of the acid sensing ArsRS regulon that responds to a decline of periplasmic pH. Analysis of the *in vivo* transcriptome of *H. pylori* infecting the gerbil stomach also showed upregulation of 11 of these 12 genes as well as 23 other genes in the ArsRS regulon indicating that the site of infection is indeed acidic, lowering periplasmic pH to < 6.2. Other genes increased transcription *in vivo* including some for chemotaxis, motility, transport, cell envelope, adhesion, bioenergetics and metabolism providing information as to other requirements for colonization. Therefore although acid acclimation is necessary it may not be sufficient for gastric colonization.

Introduction

The normal gastric environment is highly acidic, with a median pH of 1.4 (Teyssen *et al.*, 1995). Many neutralophilic organisms, such as *V. cholera* or pathogenic *E. coli* can transit the stomach due to acid resistance or tolerance responses. These responses involve the uptake of amino acids such as glutamate, lysine or arginine followed by decarboxylation with consumption of a proton and export of the amine formed in exchange for the imported amino acid (Foster, 2004; Merrell and Camilli, 1999, 2000). These are able to buffer the cytoplasm to between pH 4.0 and 4.5 allowing survival in the gastric environment, but not colonization. Other systems such as the F_1F_0 ATPase can also be utilized for proton export (Fig. 7.1) (Suzuki *et al.*, 2003).

None of the acid resistance or tolerance response elements are able to buffer the periplasm of Gram-negative organisms to a relatively neutral pH in acidic media. Preservation of a near neutral periplasmic pH is essential for maintenance of an inner membrane potential that allows appropriate synthesis and folding of membrane proteins and electrogenic proton driven solute uptake and metabolite export (Padan *et al.*, 1981). The magnitude of the inward negative potential is determined by the proton motive force (pmf) equation that is applicable to both Gram-negative and Gram-positive bacteria (Mitchell, 1966) whereby the algebraic sum of the pH and potential gradient is maintained constant.

$$\text{pmf (mV)} = \Delta\mu_{H^+} = -RT/nf \ln[H^+]_{out}/[H^+]_{in} + \Delta\Psi = -61\Delta pH + PD$$

At a neutral pH of 7.0, the membrane potential of *E. coli* is about −180 mV (Kashket, 1981). When the periplasm pH falls, the membrane potential declines and at a medium pH ~4.0, the pmf is attributable entirely to the pH gradient with no inner membrane potential, until cytoplasmic pH falls.

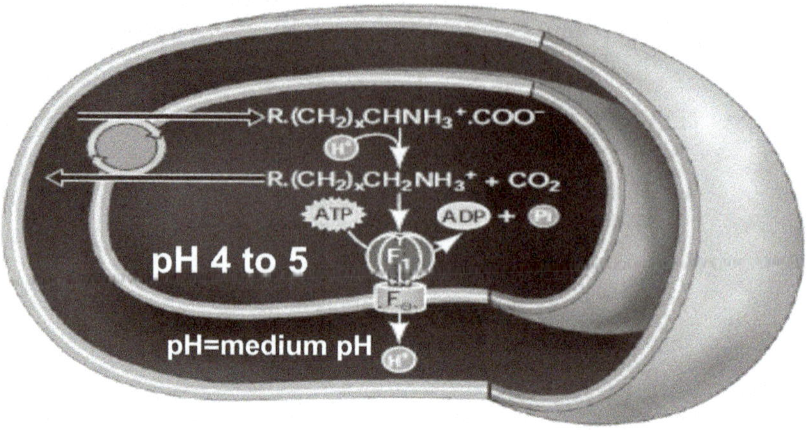

Figure 7.1 Control of pH homeostasis in bacteria. A schematic illustration of two of the established mechanisms of pH_i control by cytoplasmic or inner membrane proteins of neutralophiles in acidic media, involving the F_1F_0 ATPase and the amino acid decarboxylases accompanied by amino acid/decarboxylase product exchange. The F_1F_0 can extrude H^+ in the pumping direction rather than in the ATP synthase mode, and decarboxylation of the amino acid results in proton uptake at the methylene group that is decarboxylated and exchange of the amine for the amino acid at the inner membrane results in export of one H^+ per cycle. (From Sachs *et al.* (2005) with permission.)

The *Helicobacter* spp. that infect the stomach do not express any of the coupled decarboxylase/amine transporter family but do express the F_1F_0 ATPase; however, this alone would be insufficient to maintain colonization. Therefore, *H. pylori* has unique strategies to not only survive in acid but also to grow in an inimical acidic environment such as is likely to be found at their site of colonization in the stomach.

Acid acclimation

In this section, we focus on genes that encode proteins capable of altering pH either primarily or due to urease activity. Subsequently we shall discuss findings on the properties of some of these specific proteins involved in the acid response.

Thus far, among the neutralophiles, only gastric *Helicobacter* spp. are known to be able to buffer their periplasm in an acidic environment. This property is known as acid acclimation and enables colonization of the mammalian stomach. Periplasmic buffering by *H. pylori* relies largely on expression of a group of genes that we have termed acid acclimation genes.

The requirement for urease activity was known for some time, in that urease negative mutants were unable to colonize animal models such as the mouse (Karita *et al.*, 1995; Tsuda *et al.*, 1994a; Tsuda *et al.*, 1994b). *H. pylori* expresses the highest level of urease of any bacteria, perhaps representing as much as 10% of its protein (Bauerfeind *et al.*, 1997). Originally, it was thought that surface urease was the significant urease compartment, generating a neutralizing cloud of ammonia and the surface urease resulted from 'altruistic' lysis of neighbouring organisms (Dunn and Phadnis, 1998).

The urease gene cluster of *H. pylori* consists of a promoter followed by *ureA* and *ureB*, then another promoter followed by *ureI*, *ureE*, *ureF*, *ureG* and *ureH* (Akada *et al.*, 2000; Labigne *et al.*, 1991). Urease is a nickel-containing enzyme, assembled from UreA and UreB subunits to form the apoenzyme. Nickel is inserted by the action of the accessory urease proteins, UreE paired with UreG and UreF paired with UreH, similar to other bacterial ureases such as those expressed by *Klebsiella aerogenes* (Brayman and Hausinger, 1996; Colpas *et al.*, 1999; Moncrief and Hausinger, 1996, 1997; Mulrooney and Hausinger, 1990; Park *et al.*, 1994; Park and Hausinger, 1995; Voland *et al.*, 2003). All of these are soluble cytoplasmic proteins. The seventh gene, *UreI*, encodes a six-transmembrane proton gated urea channel. Initial experiments that showed the presence of a H^+-activated urea transporter in *H. pylori* measured the effect of pH on urease activity in intact and lysed bacteria (Scott *et al.*, 1998). The results are illustrated in Fig. 7.2. The data were obtained measuring $^{14}CO_2$ release from labelled urea, in contrast to previous experiments that had used pH measurements of weakly buffered media and hence did not define the effect of pH on intact bacteria because the pH change was too fast following urea addition.

UreI, being a putative membrane protein, was the obvious candidate for this acid-activated urea transporter. Deletion of *UreI* abolished the pH induced increase of urease activity seen in intact organisms but did not change the pH profile of urease activity from lysed *H. pylori* (Fig. 7.2). This provided evidence that UreI was likely a low pH-activated urea transporter or channel and enables urea access to intrabacterial urease in acidic medium. This was further substantiated by the demonstration that wild-type organisms survived down to at least pH 2.5 in the presence of physiological concentrations of urea (1–3 mM), but *ureI* deletion mutants required an external urea concentration of ~300 mmol/l for equivalent survival, a concentration required to achieve unfacilitated diffusion of urea across the inner membrane at a rate sufficient to counter intrabacterial acidification (Fig. 7.3). Moreover, whereas wild-type organisms were observed to elevate their periplasmic pH in acid medium with the addition of urea by visualizing BCECF free acid fluorescence by confocal microscopy, *UreI* deletion mutants did not buffer their periplasm (Fig. 7.4).

The concept that seemed likely was that acid activation of UreI allowed rapid access of urea to urease, with then the efflux of NH_3 allowing neutralization of the periplasm and the environment if it was weakly buffered. Hence, cytoplasmic urease and not surface urease was the important urease compartment essential for survival in acid in the presence of urea (Scott *et al.*, 2000).

After UreI was found to be necessary for survival in acid *in vitro*, it was shown that expression of UreI was essential for gastric colonization both for mouse and gerbil (Mollenhauer-Rektorschek *et al.*, 2002; Skouloubris *et al.*, 1998). In fact, *ureI* deletion mutants were only able to colonize the gerbil stomach if acid secretion was inhibited by inhibitors of the gastric H,K ATPase, but were eradicated if acid secretion was allowed to return (Mollenhauer-Rektorschek *et al.*, 2002).

Blue native gel electrophoresis showed that UreI was associated with UreA and UreB (Voland *et al.*, 2003) and then post-sectioning electron microscopy immuno-staining with antibodies against UreI and UreB showed that UreB localized to the membrane upon exposure of the organism to pH 5.5, a pH at which the channel was fully open in contrast to a pH of 7.0

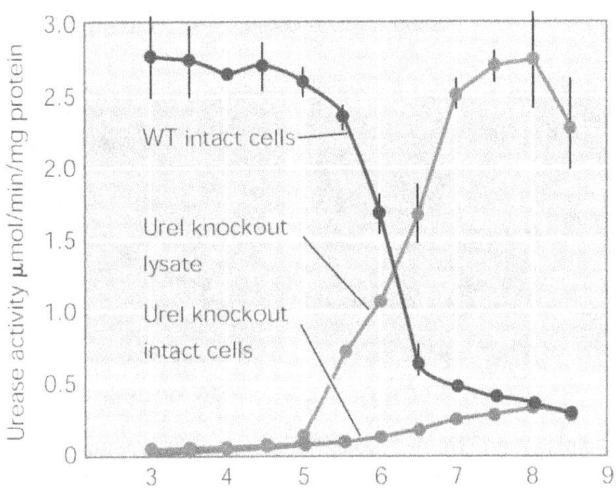

Figure 7.2 Urease activity in different settings. Urease activity of lysed wild-type or *ureI* deletion mutant bacteria show a pH optimum ~7.5 and no activity at pH 5.0 and below. Intact organisms show low activity at neutral pH and an increase to maximum when medium pH is ~5.0 and maintenance of activity to pH 3.0 (and even to pH 2.0). There is no urease activity below pH 5.0 in *ureI* deletion mutants. Note the absence of urease activity in intact *ureI* deletion mutants which have normal urease activity when lysed (From Sachs *et al.*, (2005) with permission.)

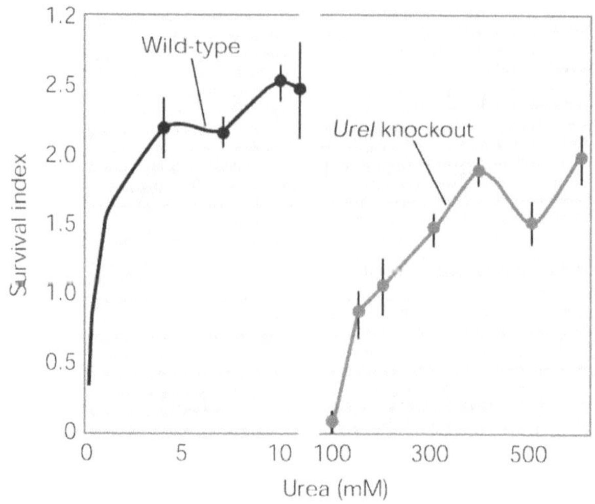

Figure 7.3 Survival of wild-type and *UreI* deletion mutants to 30 min exposure to fixed medium acidity as a function of different medium urea concentrations. A 300-fold increase in urea concentration is required for survival of the urel deletion mutants hence the increase in urea permeability due to UreI across the inner membrane is ~300-fold. The bypass of the *UreI* deletion mutant at high medium urea concentrations shows that UreI is indeed a urea transporter. (From Sachs *et al.* (2005) with permission.)

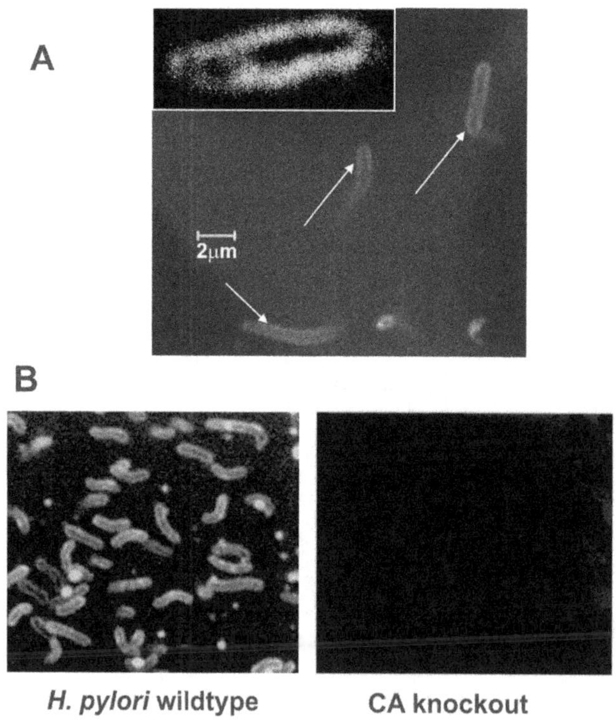

Figure 7.4 (A) Periplasmic pH control by intra-bacterial urease activity requires UreI: Using the pH sensitive fluorescent dye, BCECF to monitor changes in periplasmic pH, intrabacterial urease activity following urea addition increases pH of the periplasm (arrows). Increased fluorescence of the periplasm is further illustrated by the high magnification image (inset). Line indicates 2 μm. UreI deletion mutants show no periplasmic pH elevation. (B) Periplasmic pH control by *H. pylori* also requires periplasmic carbonic anhydrase: The effect of urea addition at pH 5.5 to wild type (left panel) and α-CA knockout organisms (right panel) again in the presence of BCECF free acid. In the absence of α-carbonic anhydrase, periplasmic buffering is abolished, demonstrating that periplasmic carbonic anhydrase activity is also essential for acid acclimation by *H. pylori*. Similar data are obtained in the presence of diamox added to wild-type organisms. (From Sachs *et al.* (2005) with permission.)

when the channel is fully closed (Fig. 7.5 (Hong *et al.*, 2003). This suggests that there is a complex formed between urease and UreI enabling a rapid response to medium acidity of the urease system with rapid exposure of urease to entering urea.

Up to this point, it seemed that the ability to infect the stomach by gastric *Helicobacter spp.*, that express UreI, (Scott *et al.*, 2000), was dependent entirely on cytoplasmic urease, the acid-gated urea channel, UreI, and efflux of NH_3. However, the NH_4^+/NH_3 couple has a pK_a of 9.2, and would not buffer at neutral pH.

Transcriptomal analysis was carried out in several laboratories in order to identify genes that were upregulated by acid exposure *in vitro* (Ang *et al.*, 2001; Bury-Mone *et al.*, 2004; Merrell *et al.*, 2003; Wen *et al.*, 2003). A survey of the genes that were progressively upregulated by exposure to pH 6.2, 5.5 and 4.5 in the absence and presence of urea found several genes that changed expression that could be thought of as either contributing buffering capacity such as periplasmic α-carbonic anhydrase, or generating urea such as arginase or ammonia such as amidase (*amiE*) or asparaginase. In addition there was upregulation of two hydrogenase accessory genes, *hypA* and *hypB* (Wen *et al.*, 2003). Expression of these genes that are involved in nickel regulation, is important for display of normal urease activity. Deletion of *hypA* or *hypB* reduced urease activity 40- and 200-fold, respectively (Mehta *et al.*, 2003; Olson *et al.*, 2001). In addition there was upregulation of *ureE* and *ureI*. The upregulation of this group of genes was blunted but not abolished by the presence of 5 mmol/l urea at pH 4.5 (Wen *et al.*, 2003). Hence, these genes were classified as the acid acclimation group expressed by *Helicobacter* spp. important for infection and colonization (Table 7.1).

The discovery that periplasmic α-carbonic anhydrase was upregulated in acid was amplified by the demonstration that deletion of *HP1186* or the inhibitor, acetazolamide, abolished the increase in periplasmic pH by the addition of urea in acidic media (Marcus *et al.*, 2005). The postulated mechanism of acid acclimation was then modified, wherein the bicarbonate generated in the periplasm from CO_2 diffusing out of the cytoplasm was responsible for buffering this compartment to about pH 6.3.

The acid acclimation genes were first identified by *in vitro* transcriptomal analysis. However, it is not clear if the gastric niche of *H. pylori* is relatively neutral or acidic pH. Therefore, a similar transcriptomal analysis was performed comparing organisms infecting the gerbil stom-

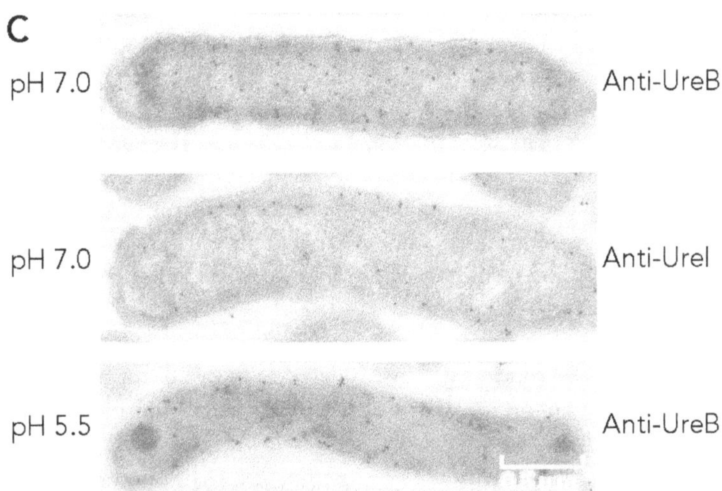

Figure 7.5 A post-section immuno-gold electron micrograph with (a) staining with anti-UreB at neutral pH showing mostly cytoplasmic location of UreB at a medium pH of 7.0 (b) staining with anti-UreI showing the expected membrane location of this protein and (c) staining with anti-UreB at a medium pH of 5.5 showing the large increase of UreB at the inner membrane, consistent with the association seen on Blue Native Gel electrophoresis(Voland *et al.*, 2003). (From (Sachs *et al.* (2005) with permission.)

Table 7.1 Increase of expression of acid acclimation genes in stomach compared with *in vitro* at pH 4.5

Accession no.	*In vivo* increase	Description	*In vitro* increase by reference
HP0072*	29.90 ± 8.23	Urease beta subunit (urea amidohydrolase) (ureB)	(Merrell *et al.*, 2003)
HP0073*	21.96 ± 4.38	Urease alpha subunit (ureA) (urea amidohydrolase)	(Bury-Mone *et al.*, 2004; Merrell *et al.*, 2003)
HP0071*†	5.04 ± 2.06	Urease accessory protein (ureI)	(Merrell *et al.*, 2003; Wen *et al.*, 2003)
HP0070*	1.79 ± 0.53	Urease accessory protein (ureE)	(Wen *et al.*, 2003)
HP0068*	3.64 ± 1.22	Urease accessory protein (ureG)	(Merrell *et al.*, 2003; Wen *et al.*, 2003)
HP0900	10.16 ± 1.39	Hydrogenase expression/formation protein (hypB)	(Wen *et al.*, 2003)
HP0869†	2.57 ± 0.19	Hydrogenase expression/formation protein (hypA)	(Bury-Mone *et al.*, 2004; Wen *et al.*, 2003)
HP1186†	2.68 ± 0.98	Carbonic anhydrase	(Wen *et al.*, 2003)
HP0294*	5.31 ± 1.78	Aliphatic amidase (amiE)	(Bury-Mone *et al.*, 2004; Merrell *et al.*, 2003; Wen *et al.*, 2003)
HP1238*	1.97 ± 0.25	Formamidase (amiF)	(Bury-Mone *et al.*, 2004; Merrell *et al.*, 2003)
HP0649	2.42 ± 0.60	Aspartase (aspA)	
HP1399*†	15.94 ± 2.90	Arginase (rocF)	(Bury-Mone *et al.*, 2004; Wen *et al.*, 2003)

Genes identified as members of the ArsRS regulon are in bold type.
*(Pflock *et al.*, 2006).
†(Wen *et al.*, 2006).

ach with organisms grown on plates at neutral pH and to gene expression at pH 4.5 *in vitro* (Scott *et al.*, 2007).

In addition to the acid acclimation genes already identified *in vitro* as responding to medium acidity, in the stomach there were some important additions to the data generated in our laboratory (Wen et al 2003), such as *ureA*, *ureB*, *ureG*, *amiF* and *aspA* that were upregulated only in the *in vivo* situation. Others had seen changes in the urease genes (Ang *et al.*, 2001; Merrell *et al.*, 2003). Since UreI is required for survival at a pH < 4.0 *in vitro*, it can be concluded that the pH of the site of colonization on the gastric surface is also < 4.0. Hence the environment of *H. pylori* in the stomach is acidic, not neutral.

Regulation of acid acclimation genes

H. pylori expresses four two-component systems (TCSs) composed of histidine protein kinase sensors and their cognate response regulators as well as two orphan response regulators. Of these, the HP0165/0166 TCS was upregulated in acid *in vitro* (Wen *et al.*, 2003). This system was then named the ArsRS regulon (Pflock *et al.*, 2004) to identify this two-component system as an acid response system. HP0165 is the histidine protein kinase sensor, ArsS, and HP0166 is the response regulator, ArsR. HP0165 was upregulated in mild acidity (pH 6.2), in the presence or absence of medium urea. This strongly suggested that this sensor, with several histidines in its periplasmic domain, is able to respond to a small decrease in pH of the periplasm from neutrality to 6.2 and does not require a large fall in cytoplasmic pH. This TCS has been investigated in some detail, by electrophoretic mobility shift assay and DNase foot printing and by investigation of the change of acid responsiveness of the expressed genes at pH 5.0 of a HP0165 deletion mutant (Wen *et al.*, 2006; Wen *et al.*, 2007, Pflock, 2006)

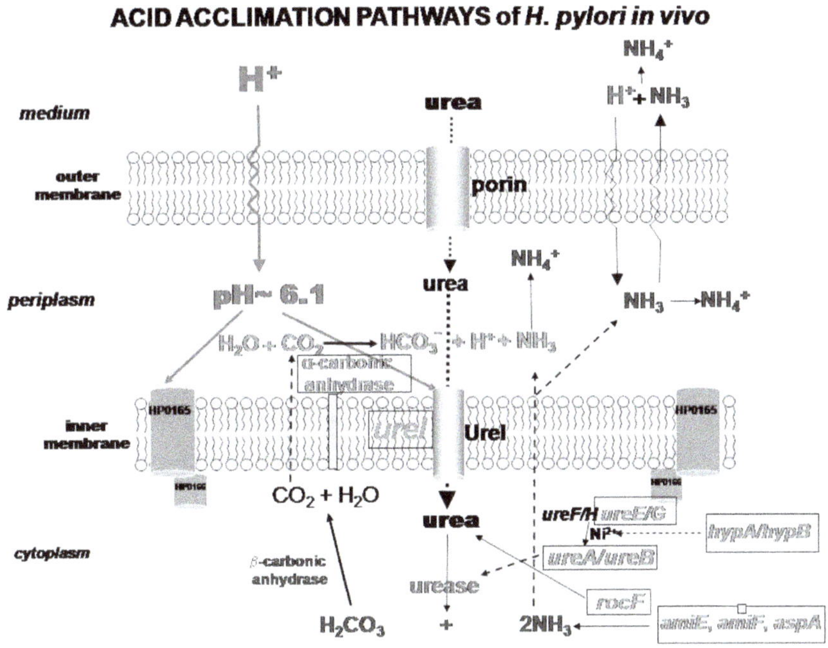

Figure 7.6 The mechanism of acid acclimation by *Helicobacter pylori in vivo*. Urea and protons enter the periplasm through outer membrane porins. Acid activation of UreI results in 300 fold acceleration of urea entry into the bacterial cytoplasm. At pH ~6.1 the ArsRS two-component system is also activated and results in the upregulation of the Ni^{2+} insertion genes, *ureE* and *ureG* that contribute to increase of urease activity by converting the apoenzyme into active urease. The potential ammonia producing enzymes, *amiE*, *amiF* and *aspA* as well as the urea producing *rocF* (arginase) are also upregulated by this two-component system as is the periplasmic α-carbonic anhydrase (HP1186) and *ureI*. Hydrolysis of urea in the cytoplasm results in the production of H_2CO_3 and $2NH_3$. The former is rapidly converted into CO_2 and H_2O by the cytoplasmic β-carbonic anhydrase and both the CO_2 and NH_3 diffuse rapidly through the inner membrane into the periplasm. There, the CO_2 is converted by α-carbonic anhydrase into HCO_3^- and H^+. The bicarbonate buffers the periplasm, maintaining periplasmic pH at 6.1 and the NH_3 neutralizes this proton liberated in this reaction as well as protons entering the periplasm. Ammonia is also able to diffuse through the outer membrane to neutralize medium acidity. Cytoplasmic NH_3 can also neutralize protons entering the cytoplasm as shown on the left.

As shown in Table 7.1, 10 of the 12 genes assigned to the acid acclimation group are regulated by ArsRS. Hence, this system plays an important role in the response of *H. pylori* to acidity both *in vitro* and *in vivo*. However there are many genes upregulated by acidity *in vitro* and gastric habitation that are not regulated by HP0166, and will be discussed below. The model of Fig. 7.6 shows our current concept of the response of this gene group where the *in vitro* and *in vivo* data are combined.

The acid acclimation genes present in *H. pylori* are just a fraction of the genes necessary for gastric colonization. It may be that the acid acclimation group is necessary but not sufficient for gastric colonization. We now discuss two proteins in the acid acclimation group that may be candidates for eradication of infection.

UreI and acid acclimation

The transport properties of UreI were studied using the *Xenopus laevis* heterologous protein expression system. Urea uptake is stimulated about 20-fold in *ureI* cRNA injected oocytes as the pH falls from 7.0 to 5.0, consistent with data from *H. pylori*. The pH–activation curve was identical to that of urease in intact bacteria with half-maximal activity at pH 5.9. The urea flux measured in UreI expressing oocyte was non-saturable, energy, temperature and voltage independent and symmetric (Weeks *et al.*, 2000). Hence, oocyte expression shows that UreI is a pH-gated urea channel.

Site-directed mutagenesis of UreI

Protonation/deprotonation of specific amino acid residues with a change in hydrogen bonding was deemed the likely basis for regulation of urea transport by UreI of *H. pylori/hepaticus*, particularly since there are no protonatable histidines in the periplasmic domain of the pH-insensitive *S. salivarius*. Although there is a decrease in cytoplasmic pH of *H. pylori* as external pH decreases, it is small and likely outside the range of protonation of almost any carboxylic amino acid in the presence of urea (Wen *et al.*, 2003). Therefore, the residues important for UreI gating must reside in the periplasmic domain. The first and second periplasmic loops of *H. pylori* each contain seven potential protonatable residues. The periplasmic C-terminus also contains a histidine residue that is potentially involved in proton gating. The absolute selectivity for urea as compared to thiourea, and the large number of tryptophan residues in the last three transmembrane segments suggest that the channel forms a fairly rigid structure and that only the gate for urea changes as a function of pH. The gate is presumably in the periplasmic domain. The amino acids involved were investigated by site-directed mutagenesis.

The effect of point mutations to the possible protonatable or charged residues in the periplasmic domains of *H. pylori* on urea uptake was measured using the *X. laevis* heterologous expression system (Weeks *et al.*, 2004). Alternatively, several histidine mutants were expressed in *H. pylori* and the rate of medium alkalinization determined at pH 2.2, 5.0 and 7.0 as an indirect measure of channel activity (Bury-Mone *et al.*, 2001). Mutations that affect the pH profile of urea entry can be considered as affecting proton gating, whereas mutations that abolish urea transport may change the protein conformation and thus may not reveal a site of proton gating and interpretation of the effect of these mutations as affecting only proton gating could be incorrect.

There are three histidine residues in the first periplasmic loop of *Hp* UreI, H54, H70 and H71. H54 is not present in all *Helicobacter ureI* gene sequences. H54 is present in the strain used for the oocyte uptake studies and replacing it with either arginine or glycine resulted in wild type channel activity indicating that this histidine residue is not involved in channel gating to increase open probability (Weeks *et al.*, 2004). Single replacement of H70 and H71 with glutamine had no effect on channel gating while replacement of both H70Q/H70Q increased urea uptake into oocytes at pH 7.5 by about 4-fold (5.33 ± 0.35 vs. 1.40 ± 0.13) over wild type but had no effect on channel gating, suggesting these mutations have no effect on channel opening but may be involved in channel closure at neutral pH. Replacement of either of these histidines with glycine resulted in complete inactivation of the channel when expressed in oocytes. Mutations of the carboxylates in the first periplasmic loop, E60D, E60Q, E63Q and D64N displayed wild-type channel activity. Thus in the first periplasmic loop, only the pair of histidines at position 70 and 71 play a role in pH regulation of *H. pylori* UreI activity (Fig. 7.7) and the nature of the mutation also has significant effects, perhaps due to a structural perturbation rather than an effect on acid gating (Weeks *et al.*, 2004). The structural perturbation may have different effects in the environment of the oocyte membrane as compared to the membrane of *H. pylori* perhaps reconciling some of the conflicting results obtained with mutants in the two systems (Bury-Mone *et al.*, 2001; Weeks *et al.*, 2004).

The second periplasmic loop and C-terminus contain three histidine residues (H123, H131, and H193) and five carboxylic amino acids (D126, D129, D130, E138, D140) and a charged lysine. Mutation of H123, H131 and H193 to arginine, asparagine, cysteine, glutamine or glycine strongly reduced or abolished urea transport at acidic pH without altering transport at neutral pH indicating these residues are important in acid gating of the channel. Mutations of the dicarboxylic amino acids of the second periplasmic loop showed that these amino acids in the second periplasmic loop are also involved in channel opening. For example, the D126N, D126E and D129E mutants transported urea at neutral pH just as in the wild type channel but at pH 5.0, transport was reduced by 30–50%. D130E and D130N displayed wild-type transport activity while E138D, E138Q and D140E and D140N were inactive at acid pH, showing that E138 and D140 were likely involved in acid activation or urea transport (Fig. 7.7).

To further explore the mechanism of pH gating of UreI, two other homologues of UreI

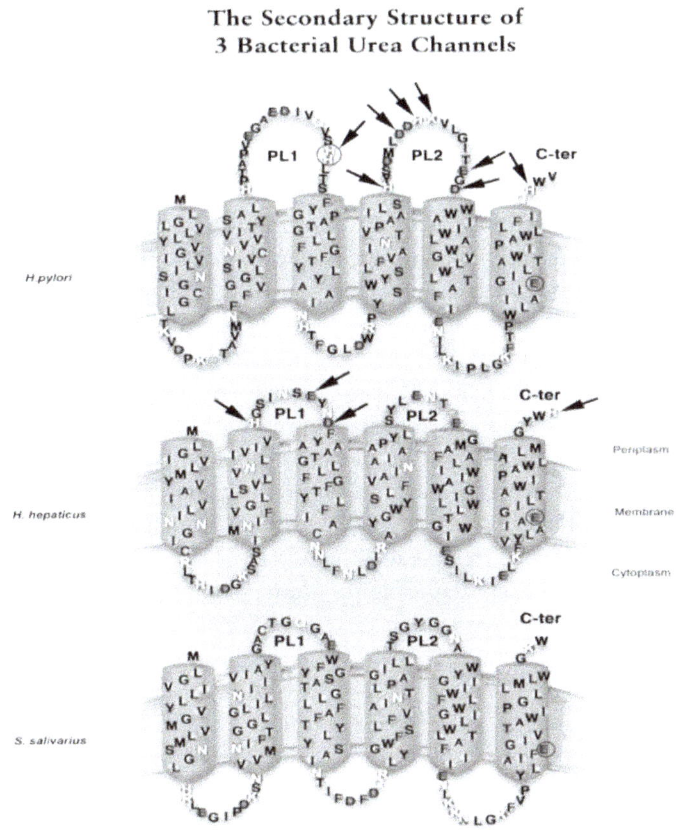

Figure 7.7 The predicted secondary structures of UreI from *H. pylori*, *H. hepaticus* and *S. salivarius* based on *in vitro* translation/insertion scanning and the TMpred algorithm (Weeks *et al.*, 2004). Arrows indicate mutations that affect the pH profile of urea uptake into oocytes with maintenance of urea transport in the first pair of UreI proteins. The circle in the sixth transmembrane domain indicates the single charged, conserved intramembranal glutamate which, when mutated to a neutral amino acid, did not affect urea transport. Black arrows show mutations preventing acid activation, circle over the HH residues in the first periplasmic loop of *H. pylori* indicates mutation that prevent channel closing at neutral pH in *H. pylori*. Black arrows also show proton-gating residues in *H. hepaticus*. There are no proton-gating residues in *S. salivarius*.

that showed different responses to changes in medium pH were investigated. Using site-directed mutagenesis the effect on the pH dependence of urea uptake into oocytes of UreI of *H. pylori* and *H. hepaticus* (proton-gated, higher pH of activation) and that of *Streptococcus salivarius* (pH-independent) were studied. Fig. 7.7 shows a model of the topology and amino acid composition of the UreI membrane proteins as determined by *in vitro* transcription/translation (Weeks *et al.*, 2000; Weeks *et al.*, 2004). These models provided the template for site-directed mutagenesis and construction of chimeras to elucidate the residues and regions important for transport and pH activation.

The UreI variant found in *H. hepaticus*, although acid activated, has a higher pH at which half-maximum transport is observed in oocytes, namely pH 6.8 (Weeks *et al.*, 2004). There is strong homology in the membrane domain, significant homology in the cytoplasmic domain and none in the periplasmic loops with the sequence of UreI of *H. pylori* although these also contain protonatable residues. All the protonatable amino acids of *H. hepaticus* could be mutated with retention of urea transport allowing direct analysis of the role of these amino acids without the caveat of full inactivation of the channel even at neutral pH as happens with mutation of the UreI of *H. pylori* (Weeks *et al.*, 2004). Whereas mainly mutations of the second, not the first, periplasmic loop of *H. pylori* affected the pH dependence of urea uptake, mutations mainly in the first periplasmic loop of *H. hepaticus* were those that affected the pH dependence of this UreI. These residues were H50, E56, and D59 in the first periplasmic loop. As with *Hp* mutation in the C terminus, H170, also affected pH dependence of urea uptake in oocytes (Fig. 7.7).

Streptococcus salivarius expresses a homologue of UreI whose expression along with the rest of the urease gene cluster is upregulated when the organism encounters an acidic environment. Also in contrast to *Helicobacter* spp. this is the first gene in the urease operon (Chen *et al.*, 2002; Chen *et al.*, 1998). *S. salivarius* UreI has about 82% similarity to the membrane domain of *H. pylori* UreI but none to the predicted periplasmic domains. Oocyte expression studies of *S. salivarius* UreI showed that it is also a urea channel, i.e. highly selective for urea, temperature independent, and non-saturable. Expression of this UreI in oocytes accelerated urea uptake equally at pH 5.0 and pH 7.5 indicating this urea channel was always open and did not show pH gating. Since this UreI is open independent of medium pH, it was possible to exploit this property to determine the periplasmic domains important in pH gating of UreI or *H. hepaticus* UreI by constructing chimeric UreI proteins by interchanging periplasmic loops between the two channels. The conserved glutamate in the sixth transmembrane domain of these three channels could be mutated to a neutral amino acid without effect on the pH dependence of urea uptake (Fig. 7.7).

Replacement of the first periplasmic loop of *Hp* UreI with that from *S. salivarius* UreI reduced urea uptake at pH 5.0 to 80% of wild-type *H. pylori* UreI activity while increasing uptake at pH 7.5 by three- to fourfold, results similar to the H70Q/H71Q double mutation. These mutational and the chimera data suggest that the first periplasmic loop of *H. pylori* is not involved in channel opening or urea transport in this variant of UreI, but may contribute to maintaining the channel in the closed state at neutral pH. A chimer between the second loop of *H. pylori* and the rest of *S. salivarius* was inactive. However, replacement of the complete periplasmic domain of *S. salivarius* with that of *H. hepaticus* resulted in pH independent urea uptake, hence the structure of the membrane domain of *S. salivarius* is dominant and unresponsive to the pH dependent conformational changes in the heterologous periplasmic domain of *H. hepaticus* (Weeks *et al.*, 2004).

Presumably the mechanism of proton gating, given the pH insensitivity of any chimeras containing the membrane domain of *S. salivarius* and the periplasmic domain of *H. hepaticus*, is dependent on a conformational change in the membrane domain of *H. pylori* or *hepaticus* induced by a change in the state of protonation of the histidines or carboxylates in the periplasmic domain that alter the conformation of the membrane domain in the acid gated UreI members probably due to changes in hydrogen bonding. A better understanding of the mechanism of proton gating and the path of urea transit will require a high-resolution crystal structure of UreI. The availability, known properties and lower complexity of the UreI of *H. hepaticus* and *S. salivarius* also make them candidates for crystallization studies.

A model was proposed for activation of UreI of *H. pylori* and *H. hepaticus* based on mutation, chimera and homologue data (Weeks *et al.*, 2004). A change in hydrogen bonding due to histidine protonation (pK_a ~6.0) with vicinal carboxylates or other H acceptors was thought to be responsible for the change in conformation of the periplasmic domain that was then transmitted to the membrane domain of *Helicobacter* to allow opening for the passage of urea. The change in interaction between the C terminus at TM6 and the loop between TM2 and interaction between TM2 and TM3 were thought particularly relevant. The UreI of *H. pylori* and *H. hepaticus* have a protonatable histidine in the C terminus, whereas that of the pH-insensitive *S. salivarius* has a lysine (K170), with a permanent positive charge. Permanent hydrogen bonding with this lysine and the single carboxylic acid (E59) in the first periplasmic loop would result in an open pH-independent channel.

α-Carbonic anhydrase and acid acclimation

The role of the periplasmic α-carbonic anhydrase (α-CA) (HP1186) in acid acclimation of *Helicobacter pylori* has been investigated (Marcus *et al.*, 2005). Since α-CA catalyses the conversion of CO_2 to HCO_3^-, the role of the other product of urea hydrolysis, CO_2, in periplasmic buffering was studied using α-CA deletion mutants and the carbonic anhydrase inhibitor acetazolamide. Western analysis confirmed that α-CA was bound to the inner membrane. Immunoblots and PCR confirmed the absence of the enzyme and the gene in the α-CA knock-out. In the mutants or in the presence of acetazolamide, there was a ~3 log decrease in acid survival. In acid, absence of α-CA activity decreased membrane integrity as observed using membrane permeant and impermeant fluorescent DNA dyes. The increase in membrane potential and cytoplasmic buffering following urea addition to wild-type organisms in acid was absent in the α-CA knock-out mutants or in the presence of acetazolamide although UreI and urease remained fully functional. At acidic pH, the elevation of cytoplasmic and periplasmic pH with urea was abolished in the absence of α-CA activity. Hence, buffering of the periplasm to a pH consistent with viability depends not only on NH_3 efflux from the cytoplasm but also on the conversion of the CO_2, produced by urease to HCO_3^- by the periplasmic α-CA.

The *HP1186* gene, encoding the periplasmic α-carbonic anhydrase of *H. pylori*, is transcribed by the vegetative sigma factor σ^{80} and regulated by the HP0165–0166 two-component system (ArsRS) in response to low-pH conditions (Wen *et al.*, 2007; Wen *et al.*, 2003).

Evidence demonstrating that ArsRS acts as a transcriptional regulator of α-carbonic anhydrase in response to acidic pH comes from the observation that the mRNA levels of *HP1186* gene were increased about four-fold in wild-type *H. pylori* strain when treated at pH4.5, whereas no such upregulation of α-carbonic anhydrase mRNA was detected in the $HP0165^-$ background. EMSA and footprint experiments with purified HP0166-His_6 (Wen *et al.*, 2006) have revealed that the HP0166 response regulator protein binds directly to the promoter of *HP1186* gene. When a mutated response regulator HP0166-D52N which replaced the aspartic acid residue at position 52 with asparagine was used for EMSA, no binding was observed to the promoter of *HP1186* (Wen *et al.*, 2006), demonstrating that the phosphorylated from of HP0166 is required to efficiently bind the promoter of *HP1186*. This strongly supports the conclusion that the acid-induced activation of *HP1186* promoter is regulated by the ArsRS two-component system.

To investigate the mechanism that regulates the expression of *HP1186* in response to low-pH and the role of the ArsRS two-component system in this acid-inducible regulation, Northern blot analysis was performed with RNAs isolated from two different wild-type *H. pylori* strains (26695 and 43504) and mutants with ArsS deletions, after exposure to either neutral pH or low pH (pH 4.5). ArsS-dependent upregulation of *HP1186* α-carbonic anhydrase in response to low pH was found in both wild-type strains. Western blot analysis of *H. pylori* membrane proteins confirmed the regulatory role of ArsS in HP1186 expression in response to low pH. Analysis of the promoter region of *HP1186* shows two possible transcription start points (TSP_1 and TSP_2) located 43 and 11 bp 5′ of ATG start codon, respectively, suggesting that there are two promoters transcribing the *HP1186* gene. Quantitative primer extension analysis showed that the promoter from TSP_1 (43 bp 5′ of ATG start codon) is a pH-dependent promoter and is able to be regulated by ArsRS in environmental acidity, whereas the promoter from TSP_2 may be responsible for control of the basal transcription of *HP1186* α-carbonic anhydrase.

These data were obtained mostly *in vitro*. To determine whether acid acclimation or other processes could be found occurring in the organisms inhabiting the gerbil stomach, transcriptomal analysis was performed and compared to *in vitro* gene expression. This would not only reveal the pH at the gastric surface where the organism resides, but also show what genes other than the acid acclimation group changed expression, giving some information as to the other requirements for colonization.

Acidity responsive genes

About 200 genes are upregulated by acid exposure for 30 min *in vitro* and about 100 downregulated. After infection of the gerbil stomach the 261 genes of known function that were upregulated more than twofold in the stomach are shown in the pie chart of Fig. 7.8. Since only about 40 of these are regulated by ArsRS, there are other regulatory mechanisms for the gene expression in the stomach.

A major category of genes belongs to those involved in translation, suggesting that the

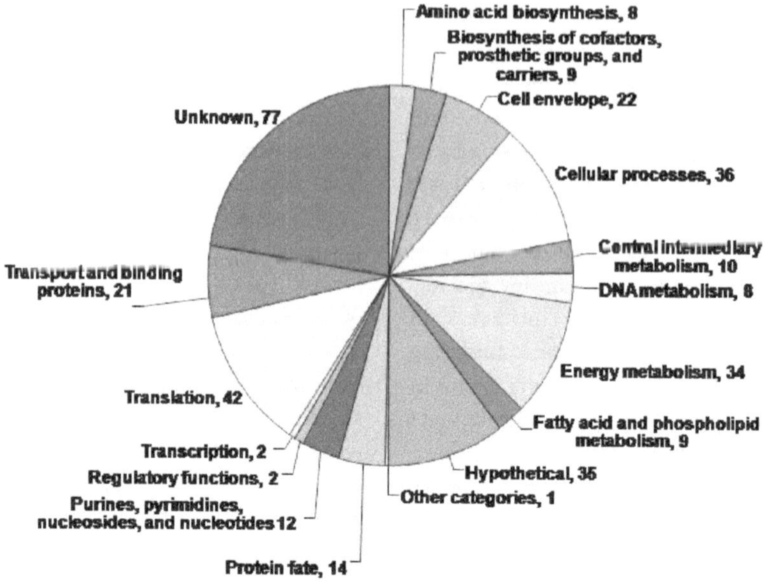

Figure 7.8 Pie chart showing the different genes upregulated in the stomach relative to the transcriptome determined at pH 4.5 *in vitro* classified as to possible function as well as 77 genes of unknown function.

mRNA isolated from gastric dwelling bacteria is derived from organisms in growth phase as compared to those being grown on plates. There was also a large change in genes involved in energy metabolism dependent in part on activation of glucose metabolism and an increase of bioenergetic functions. Also, shown in Table 7.2 is the distribution of genes regulated *in vivo* as a function of the size of the gene family. The last column shows the % of genes in a given family that are upregulated *in vivo* as compared with pH 4.5 *in vitro*.

These results illustrate the breadth of response that *H. pylori* has to gastric colonization. Although the majority of the acid acclimation genes are upregulated in the stomach, there was upregulation of many members of other groups, such as those involved in motility and chemotaxis, oxidative stress, ATP synthesis and pathogenicity.

The genes found to be upregulated in the stomach that are known to be regulated by HP0166 in addition to the acid acclimation genes are listed in Table 7.3. As discussed above and presented in more detail below, many other gene groups increase expression with gastric infection, showing that there are other pathways of regulation of gene expression utilized by *H. pylori*.

Other regulatory genes involved in acid resistance

The urease system is the major component of acid acclimation and acid resistance in *H. pylori*, however urease independent mechanisms are also present. In addition to the regulation by ArsRS of the ammonia providing amidases, *amiE* and *amiF*, the ferric uptake regulator (*fur*) is also involved in their regulation. Homologues of *fur* in many bacteria regulate the response to iron limitation (Escolar *et al.*, 1999). *In H. pylori* under iron restricted conditions, Fur binds the *pfr* promoter repressing the expression of iron-storage proteins thus enhancing iron uptake capability (Bereswill *et al.*, 2000). Conversely, in the presence of iron, fur binds iron and binds to promoters of iron-uptake genes repressing iron-uptake (Hantke, 2001). In *S. enteric* serovar *Typhimurium* Fur has been shown to be important in the regulation of acid resistance, independent of its role in iron regulation (Foster, 1991; Foster and Hall, 1992).

Fur mutants of *H. pylori* were not affected in their ability to survive acid shock down to pH 3.5 (Bijlsma *et al.*, 2002). However growth of these mutants was impacted by low pH. Fur deletion mutants developed smaller colonies when grown on acidic plates and also growth in broth culture was impaired at acidic pH when grown in the

Table 7.2 Groups of genes of known function that are upregulated more than twofold in the stomach

Functional group	Number of genes per group*	Number of genes upregulated *in vivo*	Per cent of genes upregulated more than twofold *in vivo*
Acid acclimation	12	12	100
Motility, chemotaxis	47	9	19.15
Pathogenicity, anaerobic metabolism, regulation	66	9	13.64
Oxidative stress	9	3	33.33
Ion and Iron, ABC, glucose transporters	72	3	4.17
ATP synthase, Krebs cycle, glycolysis	76	9	11.84
Amino acid, cofactor, purine, pyrimidine biosynthesis	107	6	5.61
Fatty acid and phospholipid metabolism	25	7	28.00
Adhesion, cell surface	91	3	3.30
Protein synthesis, turnover stabilization, DNA metabolism	221	6	2.71

*Grouped according to TIGR annotation for *Helicobacter pylori* 26695.

Table 7.3 ArsRS regulated genes with increased transcription more than twofold *in vivo.*

Accession no.	*In vivo* mean ± SEM	Description
HP0380*	3.78 ± 1.18	Glutamate dehydrogenase (gdhA)
HP0306*	6.15 ± 2.39	Glutamate-1-semialdehyde 2,1-aminomutase (hemL)
HP1456*†	3.97 ± 0.91	Membrane-associated lipoprotein (lpp20)
HP0389*	3.12 ± 1.19	Superoxide dismutase (sodB)
HP1192*	5.38 ± 1.66	Secreted protein involved in flagellar motility
HP0243*	7.11 ± 0.82	Neutrophil-activating protein (napA) (bacterioferritin)
HP0875*†	14.02 ± 3.65	Catalase
HP1563*	16.03 ± 5.27	Alkyl hydroperoxide reductase (tsaA)
HP0924*	3.45 ± 0.48	4-Oxalocrotonate tautomerase (dmpl)
HP0824*	13.35 ± 4.99	Thioredoxin (trxA)
HP0757*	2.05 ± 0.57	Beta-alanine synthetase homologue
HP0825*	10.54 ± 4.29	Thioredoxin reductase (trxB)
HP1340*	2.94 ± 0.36	Biopolymer transport protein (exbD)
HP1341*	3.35 ± 0.56	Siderophore-mediated iron transport protein (tonB)
HP1432*	5.87 ± 0.72	Histidine- and glutamine-rich protein
HP1427*	13.24 ± 7.03	Histidine-rich, metal-binding polypeptide (hpn)
HP1098*	4.29 ± 0.36	Conserved hypothetical secreted protein
HP0891*	4.86 ± 1.29	Conserved hypothetical protein
HP1286*	6.06 ± 2.10	Conserved hypothetical secreted protein
HP0120*	2.21 ± 0.68	*H. pylori* predicted coding region HP0120
HP1022*	2.25 ± 0.80	*H. pylori* predicted coding region HP1022
HP0305*	2.57 ± 0.65	*H. pylori* predicted coding region HP0305
HP0492*	4.81 ± 1.80	*H. pylori* predicted coding region HP0492

*(Pflock *et al.*, 2006).
†(Wen *et al.*, 2006).

absence of urea. However, in the presence of 5 mmol/l urea, growth of the *fur* mutant in acid was identical to the wild-type strain.

Fur regulates the expression of amidase (*amiE*) and formamidase (*amiF*) activity in a pH-dependent manner (van Vliet *et al.*, 2003). At pH 5.5, amidase and formamidase activity and transcription both increased significantly. *fur* mutant amidase and formamidase activities were constitutively higher than the wildtype organism and decreased pH failed to increase the activity of either enzyme (Bijlsma *et al.*, 2002). Therefore, the acidic growth deficiency of the *fur* mutant in the absence of urea may be due to the decrease in amidase and formamidase activity and *fur* represses their expression.

Acidification results in increased expression of the *nikR* gene encoding the NikR regulator which mediates the induction of urease at supra-physiological levels of Ni^{2+}. *nikR* represses *fur* transcription, therefore *nikR* mutants have increased amidase and formamidase activities similar to the *fur* knockout mutants. Taken together, these data suggest a regulatory cascade of the transcriptional induction of *amiE* and *amiF* in which protons directly interact with NikR, de-repressing *fur* which in turn de-represses *amiE* and *amiF*, increasing the concentrations of these enzymes with a subsequent increase in NH_3 raising the pH of the cytoplasm in the presence of external acidity. Therefore, the Fur and NikR regulators augment ArsRS regulation of acid acclimation to high acidity by increasing cytoplasmic pH and survival at times when urea concentrations are limiting. However, these regulators do not influence periplasmic pH, the key parameter for gastric colonization.

Genes upregulated in the stomach relative to pH 4.5 *in vitro*

Here, we shall discuss some of the various genes that increase in the stomach as compared to pH 4.5 *in vitro* and may be related to colonization or to adaptation to the habitat and not necessarily reflect requirement for colonization capability.

Motility and chemotaxis

A number of genes involved in motility and chemotaxis increase expression in the organisms inhabiting the stomach (Table 7.4, Table 7.2, second row). Of these, 14 genes are involved in flagellar construction and four are chemotaxis-related proteins. Many of these have been shown to be essential for infection (McGee *et al.*, 2005). Only three of these genes are regulated by ArsRS, hence other regulation is present. The chemotaxis genes *cheY* are histidine kinase dependent regulators and *cheA* is the kinase sensor (Foynes *et al.*, 2000). The organisms move away from acidic pH *in vitro* and the methyl accepting chemotaxis *tlpB* gene was shown essential for infection of the mouse stomach (Foynes *et al.*, 2000). The increase in motility that should reflect upregulation of this group of genes may enable the organism to find its favoured niche of gastric habitation.

Transporters

The gastric environment is a unique habitat, not only for its acidity but also in terms of the various nutrients and limited ionic environment. It is therefore expected that adaptation to the gastric environment involves regulation of both organic and inorganic transporter encoding genes. Twenty-six genes encoding dipeptide or amino acid transporters and cation transporters were upregulated in the stomach (Table 7.5). The dipeptide transporters are widely distributed in bacteria and presumably allow uptake of dipeptides as a source of amino acids. They are proton gradient driven and predicted to have 12 transmembrane segments (Hagting *et al.*, 1997).

The glucose/galactose transporter (HP1174) was upregulated both *in vivo* and *in vitro* acidic conditions as would be expected since glucose appears to be the major energy source for the organism in the stomach and *in vitro* (Mendz *et al.*, 1993). The genes encoding two glutamine transporters were also enhanced in expression as might be expected since glutamine can act as an energy source *in vitro* (Mendz and Hazell, 1995).

Genes involved in iron, copper (*copP*, HP1073), cadmium (*cadA*, HP0791), and cobalt (*corA* HP1344) transport were upregulated in the stomach. Maintenance of iron homeostasis has been shown to be crucial for survival of *H. pylori* in acid *in vitro* (Gancz *et al.*, 2006). The *fur* gene was unchanged even though it has been

Table 7.4 Gastric expression of motility and chemotaxis genes

Accession no.	*In vivo* mean ± SEM	*In vitro* pH 4.5 mean ± SEM	Description
HP0601	20.69 ± 14.06	2.03 ± 0.05	Flagellin A (flaA)
HP0115	2.19 ± 1.51	2.45 ± 0.16	Flagellin B (flaB)
HP0870	4.53 ± 0.79	1.18 ± 0.21	Flagellar hook (flgE)
HP0908	3.07 ± 0.39	0.88 ± 0.21	Flagellar hook (flgE)
HP1585*	3.93 ± 0.44	0.65 ± 0.16	Flagellar basal body rod protein (flgG)
HP1559†	2.98 ± 0.77	2.13 ± 0.12	Flagellar basal body rod protein (flgB)
HP1557	4.45 ± 0.95	4.81 ± 1.33	Flagellar basal-body protein (fliE)
HP0325	2.27 ± 0.50	3.23 ± 0.09	Flagellar basal-body L-ring protein (flgH)
HP0907	3.56 ± 0.68	1.62 ± 0.26	Hook assembly protein, flagella (flgD)
HP1233	2.50 ± 0.54	0.62 ± 0.07	*H. pylori* predicted coding region HP1233 flgJ-like
HP1192*	5.38 ± 1.66	14.12 ± 6.74	Secreted protein involved in flagellar motility
HP0232	4.17 ± 1.37	1.00 ± 0.01	Secreted protein involved in flagellar motility
HP1462	2.22 ± 0.61	0.75 ± 0.05	Secreted protein involved in flagellar motility
HP1067*	2.34 ± 0.66	2.31 ± 0.29	Chemotaxis protein (cheY)
HP0393	4.38 ± 0.72	2.24 ± 0.10	Chemotaxis protein (cheV)
HP0019	2.79 ± 0.75	2.80 ± 0.20	Chemotaxis protein (cheV)
HP0103	2.87 ± 1.76	0.55 ± 0.11	Methyl-accepting chemotaxis protein (tlpB)
HP0599	10.17 ± 0.24	1.36 ± 0.05	Methyl-accepting chemotaxis protein MCP-like protein)

Genes identified as members of the ArsRS regulon are in bold type.
*(Pflock *et al.*, 2006).
†(Wen *et al.*, 2006).

considered as important for infection (van Vliet *et al.*, 2002). Also in this group are ion binding proteins, for example, two genes encoding histidine rich proteins, *hpn* (HP1427) and HP1432 have been implicated in nickel storage. Recent work indicates that HP1427 is a reservoir for nickel even protecting against excess nickel when expressed in *E. coli* (Ge *et al.*, 2006). There are several genes implicated in copper export including *copB* (Ge and Taylor, 1996). *cadA* is known to encode an important cadmium resistance factor (Herrmann *et al.*, 1999). It has also been shown that magnesium uptake by *corA* is important for viability of the organism (Pfeiffer *et al.*, 2002). Two members of the resistance-nodulation-division family of antibiotic efflux systems increased expression *in vivo*. The *hefABC* operon encodes an antibiotic efflux system consisting of HefA, an outer membrane protein, HefB, a membrane fusion protein and HefC, the efflux pump protein found in the inner membrane (Bina *et al.*, 2000). *hefA* (HP0605) and *hefB* (HP606) both increased expression *in vivo* but not in the *in vitro* acid induced transcriptome, indicating that low pH is not the stimulus for the increase in this export system.

Perhaps in the gerbil stomach, protection against excess copper, cadmium and cobalt is a requirement for habitation. However, it is interesting that only the periplasmic and outer membrane members of this operon, which are exposed to the extreme gastric environment, increased expression, possibly due to increased degradation *in vivo*.

The increased expression in the stomach of genes encoding organic substrate transporters indicates a requirement for an increase of these transporters for infection of the stomach as compared with growth the *in vitro* situation. Moreover, increased expression of the ion transporter genes also indicates that some of the alkaline earth metals are present in sufficient

Table 7.5 Gastric expression of transport genes

Accession no.	*In vivo* mean ± SEM	*In vitro* pH 4.5 mean ± SEM	Description
HP1174*	8.77 ± 3.89	6.07 ± 0.58	Glucose/galactose transporter (gluP)
HP0055	2.48 ± 0.76	1.06 ± 0.18	Proline permease (putP)
HP1169	7.82 ± 2.80	1.02 ± 0.13	Glutamine ABC transporter, permease protein (glnP)
HP1172	3.89 ± 1.02	2.64 ± 0.27	Glutamine ABC transporter, glutamine-binding protein
HP0302	6.77 ± 1.70	2.84 ± 0.56	Dipeptide ABC transporter, ATP-binding protein (dppF)
HP0301	5.15 ± 0.83	4.04 ± 0.36	Dipeptide ABC transporter, ATP-binding protein (dppD)
HP0853	3.50 ± 0.53	0.87 ± 0.13	ABC transporter, ATP-binding protein (yheS)
HP0299	2.07 ± 0.39	3.78 ± 0.87	Dipeptide ABC transporter, permease protein (dppB)
HP1251	2.94 ± 0.95	0.83 ± 0.05	Oligopeptide ABC transporter, permease protein (oppB)
HP1562	3.04 ± 0.18	0.63 ± 0.10	Iron (III) ABC transporter (ceuE)
HP1400	2.39 ± 0.27	0.96 ± 0.04	Iron (III) dicitrate transport protein (fecA)
HP0686*	2.30 ± 0.99	0.98 ± 0.11	Iron (III) dicitrate transport protein (fecA)
HP1561	2.25 ± 0.12	1.49 ± 0.07	Iron (III) ABC transporter (ceuE)
HP1512	4.03 ± 0.98	0.45 ± 0.04	Iron-regulated outer membrane protein (frpB)
HP1341*†	3.35 ± 0.56	0.82 ± 0.13	Siderophore-mediated iron transport protein (tonB)
HP0791	2.08 ± 0.42	1.74 ± 0.09	Cadmium-transporting ATPase, P-type (cadA)
HP1073	3.00 ± 0.62	0.60 ± 0.08	Copper ion-binding protein (copP)
HP1344	3.75 ± 0.87	1.58 ± 0.12	magnesium and cobalt transport protein (corA)
HP1427*	13.24 ± 7.03	1.71 ± 0.19	histidine-rich, metal binding polypeptide (hpn)
HP1432*	5.87 ± 0.72	3.02 ± 1.46	Histidine- and glutamine-rich protein
HP1340*	2.94 ± 0.36	0.71 ± 0.17	Biopolymer transport protein (exbD)
HP1445	2.16 ± 0.65	0.42 ± 0.05	Biopolymer transport protein (exbB)
HP1378	3.17 ± 0.67	0.96 ± 0.32	Competence lipoprotein (comL)
HP0605	4.92 ± 0.77	0.66 ± 0.46	Membrane fusion protein, hefABC efflux system (hefA)
HP0606	2.77 ± 0.92	0.76 ± 0.04	Membrane fusion protein hefABC efflux system (hefB)
HP0818	3.55 ± 0.09	0.80 ± 0.25	Osmoprotection protein (proWX)

Genes identified as members of the ArsRS regulon are in bold type.
*(Pflock *et al.*, 2006).
†(Wen *et al.*, 2006).

concentration so as to require upregulation of ion export systems. Six of the 26 genes in this cluster were upregulated *in vitro* and all the genes increased more *in vivo* than *in vitro*. Evidently environmental factors, other than pH, impact expression of these genes, although six have been shown to be in the HP0166 regulon (Pflock *et al.*, 2006; Wen *et al.*, 2006) and *tonB* was increased after attachment to AGS cells (Kim *et al.*, 2002).

Oxidative stress

Helicobacter spp. are microaerophiles, requiring 5% oxygen and normal atmospheric 20% oxygen is lethal. The organism increases metabolism in the stomach as compared to 30 min exposure to pH 4.5 in broth and associated with this, there is a requirement for increased expression of genes encoding protective proteins such as superoxide dismutase, catalase and alkyl hydroperoxide reductase. Nine genes encoding oxidative stress proteins were also upregulated *in vivo* as well as *in vitro*, indicating that this is an important element in the gastric survival of the organism (Table 7.6). The increase in oxidative metabolism, such as by complex I, may be responsible for the increased expression of the oxidative stress genes. The oxygen tension in the stomach is likely to undergo wide variations since, with stimulation of secretion, there is blood perfusion of the gastric surface (Guth, 1992), which does not occur in the resting stomach. Hence, higher oxygen tension at the gastric surface may occur in regions when

Table 7.6 Gastric expression of oxidative stress and anaerobic genes

Accession no.	*In vivo* mean ± SEM	*In vitro* pH 4.5 mean ± SEM	Description
HP0389*	3.12 ± 1.19	1.82 ± 0.32	Superoxide dismutase (*sodB*)
HP0875*†	14.02 ± 3.65	3.75 ± 0.32	Catalase
HP1563*	16.03 ± 5.27	1.64 ± 0.15	Alkyl hydroperoxide reductase (*tsaA*)
HP0224	8.19 ± 2.54	1.70 ± 0.30	Peptide methionine sulfoxide reductase (*msrA*)
HP0825*	10.54 ± 4.29	1.67 ± 0.32	Thioredoxin reductase (*trxB*)
HP0824*	13.35 ± 4.99	2.49 ± 0.29	Thioredoxin (*trxA*)
HP1458	4.64 ± 1.01	1.07 ± 0.11	Thioredoxin
HP1496	4.02 ± 0.69	1.05 ± 0.18	General stress protein (*ctc*)
HP0243*	7.11 ± 0.82	2.39 ± 0.12	Neutrophil-activating protein (*napA*)

Genes identified as members of the ArsRS regulon are in bold type.
*(Pflock *et al.*, 2006).
†(Wen *et al.*, 2006).

acid secretion is stimulated. In addition there is gastric upregulation of thioredoxin reductase and peptide methionine sulphoxide reductase genes involved in reversal of disulfide bond formation and reversal of methionine oxidation (Holmgren, 1985; Lowther *et al.*, 2000). The thioredoxins are also involved as antioxidants suggesting increased oxygen stress in the stomach (Wang *et al.*, 2006). The multiplicity of oxidative stress genes upregulated *in vitro* and even more *in vivo* emphasizes the need for protection against oxidative stress in the organism. Further, six of these genes interact with HP0166 suggesting that in part the oxidative stress response is also induced by acidity. *napA* (HP0243) has a positively charged surface at acidic pH due to histidine protonation that may allow binding to DNA and regulation of transcription (Cooksley *et al.*, 2003; Tonello *et al.*, 1999). Therefore, we have classified this as an acid stress related gene. All nine genes in this cluster are increased more *in vivo* and three are activated *in vitro*.

Cell envelope, surface and adhesion

In vivo there was an increase of expression of 30 genes involved in cell envelope regulation and outer-membrane proteins including 14 members of the *hop* family of porins. In this large group of *in vivo* upregulated genes, only *hopD* (HP0025), *omp11* (HP0472), *lpp20* (HP1456) and *envA* (HP1052) increased expression *in vitro* when exposed to pH 4.5 in the absence of urea. Expression of *omp11* and *envA*, along with *flaA* (see above), is regulated by the alternative σ^{28} factor.

Several genes regulating cell surface composition were also upregulated. Many of these are involved in regulation of cell surface polysaccharides and may respond to conditions at the surface of the epithelial cells to which *H. pylori* adheres. Further, a large number of these interact with ArsR, suggesting that part of their upregulation is dependent on the acidity of the environment (Pflock *et al.*, 2006; Wen *et al.*, 2003).

Adherence to gastric surface cells offers a distinct advantage for colonization over free swimming bacteria, which are rapidly shed when confronted by rapid secretion of mucus, peristalsis, and stomach empting. Furthermore, delivery of CagA and other bacterial proteins and induction of pathology via the T4SS system is dependent on adhesion (Hessey *et al.*, 1990).

Adhesion requires a bacterial membrane adhesin protein and a host cell receptor. One of these, BabA, binds the fucosylated blood group antigens Lewis b (Le^b) and H1 but 28% to 62% of *H. pylori* strains do not produce BabA, like the strain 69A used here (Heuermann and Haas, 1998) suggesting the presence of other adhesins (Odenbreit, 2005). There was an 8 fold increase of expression of the *babA* gene *in vivo*, showing that gene regulatory machinery is intact. Two other OMPs, AlpA and AlpB (adherence associated lipoprotein), have been identified as adhesins, although the host cell receptors are unknown. In the gerbil stomach, both of these adhesin genes increased expression independently of pH by at least fivefold.

Studies of *H. pylori* gene regulation following attachment to AGS cells found an increase in expression of *omp6* and *omp19* but there was no increase of expression of these genes in the *in vivo* transcriptome (Kim *et al.*, 2004). Five of these 30 outer membrane protein or cell surface genes were identified as interacting with HP0166 in the comparison of the HP0165 deletion mutant at pH 5.0 and may therefore respond to acidity *in vivo* but, except for *lpp20*, not *in vitro* (Table 7.7). Of the 30 genes in this group, only three were increased at pH 4.5 *in vitro* but six interact with HP0166. HP0196 was secreted after AGS cell attachment (Kim *et al.*, 2002). Since only 2 of the 25 genes showed a response to acidity *in vitro* this group therefore is classified as largely gastric responsive.

Bioenergetics

In contrast to the *in vitro* acid exposed transcriptome, six components of the cytoplasmic and membrane domains of the F_1F_0 ATPase were upregulated in the gerbil stomach (Table 7.8). This complex can function either as an ATP synthase, using energy from the inward proton motive force or export protons as an ATPase. Subunits of both the cytoplasmic and membrane domains of the complex were upregulated in the stomach. The beta subunit of F_1 (HP1132) showed particularly high levels of expression *in vivo*. Therefore in the gastric environment there is an increase of oxidative ATP biosynthesis as compared to exposure to acidity *in vitro*. None of the genes in this group are part of the ArsRS regulon and are not directly dependent on the acidity of the environment since their expression is unchanged in acidic *in vitro* conditions. The increased level could be the result of increased energy demand in the stomach requiring higher levels of ATP synthesis or utilization of the F_1F_0 ATPase to export protons.

Expression of metabolism genes

Twenty-eight genes thought to play a role in oxidative and other metabolic pathways increased in the transcriptome of the gastric-dwelling bacteria (Table 7.9). HP0642 encoding NAD(P)H-flavin oxidoreductase increased expression *in vivo* and in culture at pH 4.5 in a microaerobic atmosphere (Wen *et al.*, 2003). For the other genes implicated in electron transport and transmembrane proton motive force generation, their upregulation was greater in the gastric environment. Several genes concerned with aerobic metabolism, namely those encoding NAD(P)H ubiquinone oxido-reductase (HP1266, HP1267, HP1269) were upregulated *in vivo*. The ferredoxin oxido-reductase genes, HP0589, HP0590 and HP0591 involved in the generation of NADH or NADPH for oxidation by complex I, the first step in the respiratory chain (Friedrich *et al.*, 1995), were also upregulated in the *in vitro* acid exposed transcriptome, although not to the same extent as in the stomach, while the pyruvate ferredoxin oxido-reductase genes, HP1108, HP1110 and HP1111, that catalyse the oxidation of pyruvate to acetyl CoA and CO_2 (Hughes *et al.*, 1995) only increased expression in the gerbil stomach. Flavodoxin is the likely electron acceptor for the pyruvate oxidation complex and appears essential for bacterial survival (Cremades *et al.*, 2005). Cytochrome c553 is part of the succinate dehydrogenase complex (Chen *et al.*, 1999). It seems that NADPH is more rapidly oxidized than NADH, implying that this substrate is the major reducing agent in *H. pylori* (Chang *et al.*, 1995).

Several glycolytic, two Entner–Doudoroff (HP1099 and HP1100) and substrate specific genes were upregulated in the stomach environment, showing that glucose utilization is important in gastric habitation by *H. pylori*, as well as *in vitro* acidic conditions (Mendz *et al.*, 1994). Except for the genes encoding glucose-6-phospate isomerase, phosphoglycerate kinase and lactic dehydrogenase, the glycolytic genes increased expression to a greater extent *in vivo* than under acidic conditions *in vitro*. It appears that the Embden–Meyerhof pathway and the Krebs cycle are dominant in the organism when inhabiting the stomach. This could be due to the higher acidity in the stomach or due to another aspect of the gastric environment.

The upregulation of the gene encoding 4-oxalocrotonate tautomerase may imply that *H. pylori* can utilize aromatic hydrocarbons as an energy source but it is not certain that this is the function of the enzyme in proteobacteria. This is the only gene in this group for which there is evidence for regulation by HP0166 (Pflock *et al.*,

Table 7.7 Gastric expression of cell envelope and surface genes

Accession no.	*In vivo* mean ± SEM	*in vitro* pH 4.5 mean ± SEM	Description
HP0025	4.02 ± 0.96	2.72 ± 0.10	Outer membrane protein (omp2)
HP0127	8.20 ± 2.40	1.05 ± 0.20	Outer membrane protein (omp4)
HP0252	5.53 ± 1.19	0.56 ± 1.93	Outer membrane protein (omp7)
HP0317*	7.04 ± 1.84	0.98 ± 0.05	Outer membrane protein (omp9)
HP0324	3.59 ± 0.47	1.05 ± 0.18	Outer membrane protein (omp10)
HP0472*†	7.12 ± 4.80	5.92 ± 0.29	Outer membrane protein (omp11)
HP0638	2.04 ± 0.64	1.67 ± 0.60	Outer membrane protein (omp13)
HP0671*	3.52 ± 0.43	0.99 ± 0.03	Outer membrane protein (omp14)
HP0706	3.89 ± 0.44	1.29 ± 0.07	Outer membrane protein (omp15)
HP0912	6.22 ± 2.21	0.70 ± 0.91	Outer membrane protein (omp20)
HP0913	5.26 ± 1.52	0.87 ± 0.16	Outer membrane protein (omp21)
HP1107	2.35 ± 0.43	1.81 ± 0.04	Outer membrane protein (omp23)
HP1157	5.14 ± 2.19	1.21 ± 0.06	Outer membrane protein (omp26)
HP1177	5.03 ± 1.57	0.29 ± 0.21	Outer membrane protein (omp27)
HP1243*	8.66 ± 1.00	0.68 ± 0.12	Outer membrane protein (omp28)
HP1469	2.88 ± 1.04	0.85 ± 0.17	Outer membrane protein (omp31)
HP1564*	6.89 ± 2.33	1.27 ± 0.14	Outer membrane protein (putative)
HP0003	4.49 ± 1.02	1.44 ± 0.11	3-Deoxy-D-manno-octulosonic acid 8-phosphate synthetase (kdsA)
HP0160	3.10 ± 0.89	0.97 ± 0.08	Cysteine-rich protein D (hcpD, pbp4)
HP0196	5.73 ± 1.79	0.77 ± 0.08	UDP-3–O-(3-hydroxymyristoyl) glucosamine N-acyltransferase D
HP0679	2.64 ± 0.90	1.39 ± 0.21	Lipopolysaccharide biosynthesis protein (wbpB)
HP0957	2.24 ± 0.70	0.27 ± 0.03	3-Deoxy-D-manno-octulosonic-acid transferase (kdtA)
HP1052	9.11 ± 3.01	1.90 ± 0.15	UDP-3–O-acyl N-acetylglcosamine deacetylase (envA)
HP1373	2.01 ± 0.63	0.57 ± 0.07	Rod shape-determining protein (mreB)
HP1375	3.68 ± 1.27	0.60 ± 0.04	UDP-N-acetylglucosamine acyltransferase (lpxA)
HP1543	2.08 ± 0.46	0.64 ± 0.06	Toxr-activated gene (tagE)
HP0655	3.30 ± 0.7	0.86 ± 0.15	Protective surface antigen D15
HP0410	8.70 ± 3.22	1.29 ± 0.09	Putative neuraminyllactose-binding hemagglutinin homologue hpaA)
HP1456*†	3.97 ± 0.91	3.32 ± 0.66	Membrane-associated lipoprotein (lpp20)
HP0045	3.50 ± 0.98	0.92 ± 0.16	GDP-fucose synthetase (fcl, wbcJ)

Genes identified as members of the ArsRS regulon are in bold type.
*(Pflock *et al.*, 2006).
†(Wen *et al.*, 2006).

2006). 3-Oxoadipate CoA transferase may also be involved in aromatic catabolism and has been suggested to be in part responsible for clarithromycin resistance of *H. pylori* (Hsieh *et al.*, 1998). Glycolate oxidase, *glcd* (HP0509), is one of the five subunits of the system in *E. coli* that forms glyoxylate, which then can be converted either to malate or glycerate phosphate entering either glycolysis or the tricarboxylic acid cycle (Pellicer *et al.*, 1996). The presence of this homologue in *H.*

Table 7.8 Gastric expression of bioenergetic genes

Accession no.	*In vivo* mean ± SEM	*In vitro* pH 4.5 mean ± SEM	Description
HP0828	2.86 ± 0.13	0.45 ± 0.12	ATP synthase F_0, subunit a (*atpB*)
HP1136	3.05 ± 0.98	0.80 ± 0.18	ATP synthase F_0, subunit b (*atpF*)
HP1212	3.39 ± 0.48	0.60 ± 0.19	ATP synthase F_0, subunit c (*atpE*)
HP1134	2.83 ± 0.82	0.44 ± 0.01	ATP synthase F_1, subunit alpha (*atpA*)
HP1132	10.99 ± 6.04	0.96 ± 0.03	ATP synthase F_1, subunit beta (*atpD*)
HP1133	5.76 ± 1.05	0.85 ± 0.10	ATP synthase F_1, subunit gamma (*atpG*)

Table 7.9 Gastric expression of metabolic genes

Accession no.	*In vivo* mean ± SEM	*In vitro* pH 4.5 mean ± SEM	Description
HP1108	3.47 ± 0.61	1.39 ± 0.39	Pyruvate ferredoxin oxidoreductase, gamma subunit
HP1110	2.25 ± 0.29	1.08 ± 0.21	Pyruvate ferredoxin oxidoreductase, alpha subunit
HP1111	6.34 ± 2.95	1.54 ± 0.27	Pyruvate ferredoxin oxidoreductase, beta subunit
HP1161	6.18 ± 2.40	1.23 ± 0.09	Flavodoxin (fldA)
HP1540	2.41 ± 0.47	0.66 ± 0.19	Ubiquinol cytochrome c oxidoreductase, (fbcF)
HP0642	4.99 ± 1.10	6.74 ± 2.15	NAD(P)H-flavin oxidoreductase
HP1266	3.59 ± 0.79	0.92 ± 0.19	NADH-ubiquinone oxidoreductase, NQO3 subunit (NQO3)
HP1267	2.20 ± 0.43	0.77 ± 0.03	NADH-ubiquinone oxidoreductase, NQO8 subunit (NQO8)
HP1269	2.12 ± 0.22	0.90 ± 0.05	NADH-ubiquinone oxidoreductase, NQO10 subunit (NQO10)
HP1227	5.34 ± 0.73	0.90 ± 0.20	Cytochrome c553 fur
HP0779	6.46 ± 2.50	0.83 ± 0.26	Aconitase B (acnB)
HP1325	3.45 ± 0.27	0.65 ± 0.04	Fumarase (fumC)
HP0191	2.74 ± 0.73	0.69 ± 0.27	Fumarate reductase, iron–sulphur subunit (frdB)
HP1099	5.42 ± 0.91	3.00 ± 0.54	2-Keto-3-deoxy-6-phosphogluconate aldolase (eda)
HP1100	5.93 ± 1.91	3.11 ± 0.58	6-Phosphogluconate dehydratase
HP1103	2.49 ± 0.61	1.45 ± 0.33	Glucokinase (glk)
HP1166	2.28 ± 1.17	9.17 ± 2.99	Glucose-6-phosphate isomerase (pgi)
HP0176	2.45 ± 0.38	1.01 ± 0.09	Fructose-bisphosphate aldolase (tsr)
HP1385	2.95 ± 0.50	1.57 ± 0.16	Fructose-1,6-bisphosphatase
HP1346	2.03 ± 0.39	0.95 ± 0.23	Glyceraldehyde-3-phosphate dehydrogenase (GAP)
HP1345	2.70 ± 0.40	2.38 ± 0.26	Phosphoglycerate kinase
HP0961	2.32 ± 0.50	0.93 ± 0.05	glycerol-3-phosphate dehydrogenase, $NAD(P)^+$ dependent
HP0154	5.76 ± 0.79	1.00 ± 0.10	Enolase (eno)
HP1222	2.94 ± 0.68	3.60 ± 0.88	D-Lactate dehydrogenase (dld)
HP0924*	3.45 ± 0.48	2.01 ± 0.54	4-Oxalocrotonate tautomerase (dmpl)
HP0692	8.66 ± 4.05	4.84 ± 1.51	3-Oxoadipate coA-transferase subunit B (yxjE)
HP0509	2.14 ± 0.31	0.70 ± 0.02	Glycolate oxidase subunit (glcD)
HP0620	4.13 ± 1.16	0.59 ± 0.10	Inorganic pyrophosphatase (ppa)

Genes identified as members of the ArsRS regulon are in bold type.
*(Pflock *et al.*, 2006).

pylori suggests similar metabolic capabilities of this organism. Pyrophosphatase (HP0620) gene expression is increased *in vivo* (Wood, 1977). This avoids the accumulation of pyrophosphate and provides phosphate for metabolic purposes.

There have been several enzymes of the Krebs cycle identified in *H. pylori* both by gene homology and directly by measurement of metabolism. Of these, aconitase, fumarase and fumarate reductase were upregulated *in vivo* but not in the *in vitro* acid induced transcriptome. The enzymes identified by measurement of metabolism include the different subunits of pyruvate ferredoxin oxidoreductase, citrate synthase, isocitrate dehydrogenase and succinate and malate dehydrogenase (Pitson *et al.*, 1999). Fumarate reductase that catalyses the formation of succinate is essential for infection of the mouse stomach and one of the three subunits of this enzyme, *frdB* (HP0191) is upregulated under gastric conditions (Ge, 2002). The combined data implicate the tricarboxylic acid cycle as an energy source for the organism in the stomach. The isocitrate dehydrogenase generates NADP from NAD^+, providing the major electron donor for complex I in the organism.

Exposure of *H. pylori* to the gastric environment is associated with increased expression of oxidative metabolism genes as well as those involved in glycolysis and in the Krebs cycle. The increased expression of these genes is largely independent of acidity and probably reflects the metabolism necessary to colonize in the environment of the stomach. Of the 28 genes in this group, only eight increased *in vitro* at pH 4.5; hence we categorized them as those primarily regulated *in vivo*. Since only HP0924 gene expression is regulated by ArsRS, most of these genes are probably not periplasmic pH regulated.

Anaerobic metabolism

Seven genes involved in anaerobic metabolism increased in the stomach. Thus, hydrogenase gene expression increased *in vivo*. *H. pylori* can oxidize H_2 and H_2 is available in the stomach (Brown *et al.*, 1987; Levitt, 1969; Olson and Maier, 2002). Deletion of one of the hydrogenase subunits impaired colonization of the mouse stomach (Olson and Maier, 2002). The presence of H_2 in the stomach probably accounts for their upregulation. The ferredoxin like protein is implicated in reduction of various substrates such as metronidazole, contributing to resistance of *H. pylori* to this antibiotic (Jenks and Edwards, 2002).

The three subunits of ferredoxin oxidoreductase genes increased expression in the gastric environment and also *in vitro* showing that the organism is capable of anaerobic metabolism in both conditions perhaps using aromatic compounds as electron acceptors (Buckel and Golding, 2006) (Table 7.10). Again, given the absence of HP0166 interaction except for HP0588, a ferredoxin like protein, the increased expression of these genes is likely induced by factors in the gastric environment other than the ArsRS. It is of interest that expression of all of these genes is regulated by the *fur* gene, emphasizing the importance of iron homeostasis for this group of genes (Ernst *et al.*, 2005).

Pathogenicity

Ten gastric upregulated genes fall into this group. Many *H. pylori* strains contain the Cag pathogenicity island, whose genes encode for many components of a type IV secretory system (T4SS) (Akopyants *et al.*, 1998; Odenbreit *et al.*, 2000). Of these, *cag26* (HP0547), *cag 25* (HP0546) and *cag20* (HP0541) are upregulated greater than fourfold in the gastric environment. *cag26* encodes the protein, CagA which has been associated with virulence of infection and is exported and injected into the host cell (Odenbreit *et al.*, 2000). CagA association with the inner membrane of *H. pylori* increases in acidic pH medium *in vitro* (Wu *et al.*, 2005). Also included in this cluster is the *virB11/ptlH* (HP0525) homologue that is found in *cagII* part of the *cag* PAI and is therefore common to all strains of *H. pylori* and has about 30% homology to the *A. tumefaciens virB11* gene involved in DNA transfer and to the *B. pertussis* gene *ptlH* which constitutes part of the machinery required for delivery of pertussis toxin to host cells (Akopyants *et al.*, 1998).

Given the increased expression of *cag25* and *cag26* and *cag20* in response to acid stress *in vivo*, the elevated expression of the other genes in this pathogenicity cluster indicate an increased

Table 7.10 Gastric expression of genes involved with anaerobic metabolism

Accession no.	In vivo mean ± SEM	*In vitro* pH 4.5 mean ± SEM	Description
HP0631	2.64 ± 0.61	1.55 ± 0.06	Quinone-reactive Ni/Fe hydrogenase, small subunit (hydA)
HP0632	3.30 ± 1.39	1.45 ± 0.09	Quinone-reactive Ni/Fe hydrogenase, large subunit (hydB)
HP0633	2.14 ± 0.51	1.26 ± 0.08	Quinone-reactive Ni/Fe hydrogenase, cytochrome b subunit hydC)
HP0588*	7.60 ± 2.06	2.95 ± 0.17	Ferrodoxin-like protein
HP0589	12.31 ± 3.39	2.59 ± 0.32	Ferredoxin oxidoreductase, alpha subunit
HP0590	5.39 ± 2.41	2.63 ± 0.45	Ferredoxin oxidoreductase, beta subunit
HP0591	4.08 ± 0.51	2.58 ± 0.35	Ferredoxin oxidoreductase, gamma subunit

Genes identified as members of the ArsRS regulon are in bold type.
*(Wen *et al.*, 2006).

role in the stomach whether due to acid or cell adhesion (Table 7.11). Five of the nine genes are upregulated at pH 4.5 *in vitro*. None has been found to respond to HP0166.

Regulatory genes

Two genes suggested to be regulatory factors increased with gastric infection (Table 7.12). The gene, HP1043 was upregulated fourfold in the stomach but not in the *in vitro* acid-induced transcriptome, suggestive of an important role in gastric colonization. *tenA* was also upregulated both *in vitro* and *in vivo*. HP0166 was upregulated in the *in vitro* acid induced transcriptome but not *in vivo* and is part of the acid response two-component system ArsRS (Pflock *et al.*, 2006; Wen *et al.*, 2003). Many genes known to be regulated by HP0166 were upregulated in the stomach as discussed above.

Expression of the cytoplasmic response regulator, HP1043, is essential for the organism and is regulated post-transcriptionally (Beier and Frank, 2000; Muller *et al.*, 2007). The receiver domain of this protein is atypical and may not require phosphorylation for function (Schar *et al.*, 2005). HP1043 is classified as an orphan response regulator since its cognate kinase remains unknown (Muller *et al.*, 2007). It is one of the three known response regulators in *H. pylori*. There are also two orphan response regulators (Beier and Frank, 2000). The *tenA* homologues in *B. subtilis* are apparently a thiaminase and not a transcriptional regulator (Toms *et al.*, 2005). Alternatively, it could function as an activator that increases production of extracellular proteases as shown for *Pyrococcus* (Itou *et al.*, 2004). The encoded protein is predicted to contain a single transmembrane segment, with a large portion of

Table 7.11 Gastric expression of pathogenicity island (PAI) genes

Accession no.	*In vivo* mean ± SEM	*In vitro* pH 4.5 mean ± SEM	Description
HP0547	21.10 ± 5.06	1.37 ± 0.41	cag pathogenicity island protein (*cag26*) *cagA*
HP0546	5.90 ± 2.71	1.03 ± 0.04	cag pathogenicity island protein (*cag25*)
HP0543	3.06 ± 0.15	2.02 ± 0.15	cag pathogenicity island protein (*cag22*)
HP0542	2.66 ± 0.49	3.40 ± 1.05	cag pathogenicity island protein (*cag21*)
HP0541	4.22 ± 0.59	2.45 ± 0.45	cag pathogenicity island protein (*cag20*)
HP0532	3.46 ± 1.37	2.06 ± 0.04	cag pathogenicity island protein (*cag12*)
HP0531	3.04 ± 0.86	1.07 ± 0.03	cag pathogenicity island protein (*cag11*)
HP0520	2.26 ± 0.51	2.47 ± 0.18	cag pathogenicity island protein (*cag1*)
HP0525	2.47 ± 0.50	0.90 ± 0.20	virb11/ptlH homologue

Table 7.12 Gastric expression of regulatory genes

Accession no.	*In vivo* mean ± SEM	*In vitro* pH 4.5 mean ± SEM	Description
HP1043	3.96 ± 1.52	1.57 ± 0.11	Response regulator
HP1287	2.73 ± 0.21	4.97 ± 0.52	Transcriptional regulator (tenA)

the N terminus exposed to the cytoplasm and the C terminus periplasmic. There are two histidines in this domain that could respond to periplasmic pH. *tenA* responds to acid *in vitro* and hence is probably regulated by the acid environment in the stomach, although not by the HP0165/0166 two-component system.

Other genes

Gene expression in other families was also upregulated and these are summarized in Table 7.13.

A large number of protein related and cell division genes increased in the bacteria colonizing the stomach relative to the pH 4.5 *in vitro* conditions. This likely reflects that the bacteria were in growth phase on the gastric surface when isolated. Amino acid and cofactor biosynthetic message was also increased probably due to the differences in the environment in the stomach as compared to the plate or broth culture conditions. The same is probably correct for the finding of upregulation of some purine and pyrimidine biosynthesis genes. The increase in genes encoding phospholipid and fatty acid metabolism reflects perhaps increased membrane biosynthesis. The one hundred five genes encoding proteins of unknown function are also of obvious interest but without assigned function their role in the gastric biology of *H. pylori* cannot be ascertained at present.

The gene expression changes seen with acid exposure *in vitro* and with gastric colonization reveal that *H. pylori* in the gerbil stomach is indeed exposed to acidic pH. From the data presented here, it is not possible to be precise about the pH to which the organism is chronically exposed, but, given the acid survival of *ureI* deletion mutants, the pH is < 4.0. Almost certainly the increased transcription of many genes in the acid acclimation group suggests an even lower pH. Critical to the survival of this gastric pathogen is its ability to regulate its periplasmic pH in an acidic environment, depended on the expression of the acid acclimation group of genes. The other gene families that are regulated in the stomach provides some insight as to the environmental conditions facing *H. pylori* in this animal model. The increased synthesis of flagellar and chemotactic proteins is consistent with a vital role for motility in infection. Probably there is continuous motion of the organism to optimize its environment be it pH, urea or nutrient conditions. The increase of various transporters is a response to concentrations of divalent cations and organic compounds in its gastric environment. The variation in its metabolism and oxygen concentrations in the stomach likely accounts for the changes in the oxidative stress genes. The changes in

Table 7.13 Other genes increased in the stomach

Functional group	Number of genes upregulated more than twofold *in vivo*
DNA metabolism	9
Protein synthesis and cell division	46
Protein degradation and stabilization	19
Amino acid and cofactor biosynthesis	25
Purine and pyrimidine biosynthesis	9
Fatty acid and phospholipid metabolism	11
Unknown and hypothetical	105

envelope and surface genes again may result from their proximity to the gastric surface cells in the gerbil stomach. The increase in some of the F_1F_0 genes could argue for increased metabolism and ATP requirement or an increase of proton export in acid. The changes seen in metabolically related genes must derive from stimulation of specific metabolic pathways involving, for example, glucose utilization. The increase in genes encoding genes involved in anaerobiosis must reflect transient conditions at the gastric surface. The increase in expression of the PAI genes and *cag26* is most simply interpreted as a consequence of adhesion to the gastric surface. The increase of HP1043 and *tenA* suggests that *H. pylori* also undergoes some degree of cytoplasmic acidification beyond what happens *in vitro* at pH 4.5. It will be intriguing to define which of all these genes is essential for infection and therefore a viable target for eradication.

References

Akada, J.K., Shirai, M., Takeuchi, H., Tsuda, M., and Nakazawa, T. (2000). Identification of the urease operon in *Helicobacter pylori* and its control by mRNA decay in response to pH. Mol. Microbiol. *36*, 1071–1084.

Akopyants, N.S., Clifton, S.W., Kersulyte, D., Crabtree, J.E., Youree, B.E., Reece, C.A., Bukanov, N.O., Drazek, E.S., Roe, B.A., and Berg, D.E. (1998). Analyses of the cag pathogenicity island of *Helicobacter pylori*. Mol. Microbiol. *28*, 37–53.

Ang, S., Lee, C.Z., Peck, K., Sindici, M., Matrubutham, U., Gleeson, M.A., and Wang, J.T. (2001). Acid-induced gene expression in *Helicobacter pylori*: study in genomic scale by microarray. Infect. Immun. *69*, 1679–1686.

Bauerfeind, P., Garner, R., Dunn, B.E., and Mobley, H.L. (1997). Synthesis and activity of *Helicobacter pylori* urease and catalase at low pH. Gut *40*, 25–30.

Beier, D., and Frank, R. (2000). Molecular characterization of two-component systems of *Helicobacter pylori*. J. Bacteriol. *182*, 2068–2076.

Bereswill, S., Greiner, S., van Vliet, A.H., Waidner, B., Fassbinder, F., Schiltz, E., Kusters, J.G., and Kist, M. (2000). Regulation of ferritin-mediated cytoplasmic iron storage by the ferric uptake regulator homolog (Fur) of *Helicobacter pylori*. J. Bacteriol. *182*, 5948–5953.

Bijlsma, J.J., Waidner, B., Vliet, A.H., Hughes, N.J., Hag, S., Bereswill, S., Kelly, D.J., Vandenbroucke-Grauls, C.M., Kist, M., and Kusters, J.G. (2002). The *Helicobacter pylori* homologue of the ferric uptake regulator is involved in acid resistance. Infect. Immun. *70*, 606–611.

Bina, J.E., Alm, R.A., Uria-Nickelsen, M., Thomas, S.R., Trust, T.J., and Hancock, R.E. (2000). *Helicobacter pylori* uptake and efflux: basis for intrinsic susceptibility to antibiotics in vitro. Antimicrob. Agents Chemother. *44*, 248–254.

Brayman, T.G., and Hausinger, R.P. (1996). Purification, characterization, and functional analysis of a truncated *Klebsiella aerogenes* UreE urease accessory protein lacking the histidine-rich carboxyl terminus. J. Bacteriol. *178*, 5410–5416.

Brown, N.J., Rumsey, R.D., and Read, N.W. (1987). Adaptation of hydrogen analysis to measure stomach to caecum transit time in the rat. Gut *28*, 849–854.

Buckel, W., and Golding, B.T. (2006). Radical enzymes in anaerobes. Annu. Rev. Microbiol. *60*, 27–49.

Bury-Mone, S., Skouloubris, S., Labigne, A., and De Reuse, H. (2001). The *Helicobacter pylori* UreI protein: role in adaptation to acidity and identification of residues essential for its activity and for acid activation. Mol. Microbiol. *42*, 1021–1034.

Bury-Mone, S., Thiberge, J.M., Contreras, M., Maitournam, A., Labigne, A., and De Reuse, H. (2004). Responsiveness to acidity via metal ion regulators mediates virulence in the gastric pathogen *Helicobacter pylori*. Mol. Microbiol. *53*, 623–638.

Chang, H.T., Marcelli, S.W., Davison, A.A., Chalk, P.A., Poole, R.K., and Miles, R.J. (1995). Kinetics of substrate oxidation by whole cells and cell membranes of *Helicobacter pylori*. FEMS Microbiol. Lett. *129*, 33–38.

Chen, M., Andersen, L.P., Zhai, L., and Kharazmi, A. (1999). Characterization of the respiratory chain of *Helicobacter pylori*. FEMS Immunol. Med. Microbiol. *24*, 169–174.

Chen, Y.Y., Betzenhauser, M.J., and Burne, R.A. (2002). cis-Acting elements that regulate the low-pH-inducible urease operon of *Streptococcus salivarius*. Microbiology *148*, 3599–3608.

Chen, Y.Y., Hall, T.H., and Burne, R.A. (1998). *Streptococcus salivarius* urease expression: involvement of the phosphoenolpyruvate:sugar phosphotransferase system. FEMS Microbiol. Lett. *165*, 117–122.

Colpas, G.J., Brayman, T.G., Ming, L.J., and Hausinger, R.P. (1999). Identification of metal-binding residues in the *Klebsiella aerogenes* urease nickel metallochaperone, UreE. Biochemistry *38*, 4078–4088.

Cooksley, C., Jenks, P.J., Green, A., Cockayne, A., Logan, R.P., and Hardie, K.R. (2003). NapA protects *Helicobacter pylori* from oxidative stress damage, and its production is influenced by the ferric uptake regulator. J. Med. Microbiol. *52*, 461–469.

Cremades, N., Bueno, M., Toja, M., and Sancho, J. (2005). Towards a new therapeutic target: *Helicobacter pylori* flavodoxin. Biophys Chem. *115*, 267–276.

Dunn, B.E., and Phadnis, S.H. (1998). Structure, function and localization of *Helicobacter pylori* urease. Yale J. Biol. Med. *71*, 63–73.

Ernst, F.D., Bereswill, S., Waidner, B., Stoof, J., Mader, U., Kusters, J.G., Kuipers, E.J., Kist, M., van Vliet, A.H., and Homuth, G. (2005). Transcriptional profiling of *Helicobacter pylori* Fur- and iron-regulated gene expression. Microbiology *151*, 533–546.

Escolar, L., Perez-Martin, J., and de Lorenzo, V. (1999). Opening the iron box: transcriptional metalloregulation by the Fur protein. J. Bacteriol. *181*, 6223–6229.

Foster, J.W. (1991). *Salmonella* acid shock proteins are required for the adaptive acid tolerance response. J. Bacteriol. *173*, 6896–6902.

Foster, J.W. (2004). *Escherichia coli* acid resistance: tales of an amateur acidophile. Nat. Rev. Microbiol. *2*, 898–907.

Foster, J.W., and Hall, H.K. (1992). Effect of *Salmonella typhimurium* ferric uptake regulator (*fur*) mutations on iron- and pH-regulated protein synthesis. J. Bacteriol. *174*, 4317–4323.

Foynes, S., Dorrell, N., Ward, S.J., Stabler, R.A., McColm, A.A., Rycroft, A.N., and Wren, B.W. (2000). *Helicobacter pylori* possesses two CheY response regulators and a histidine kinase sensor, CheA, which are essential for chemotaxis and colonization of the gastric mucosa. Infect. Immun. *68*, 2016–2023.

Friedrich, T., Steinmuller, K., and Weiss, H. (1995). The proton-pumping respiratory complex I of bacteria and mitochondria and its homologue in chloroplasts. FEBS Lett. *367*, 107–111.

Gancz, H., Censini, S., and Merrell, D.S. (2006). Iron and pH homeostasis intersect at the level of Fur regulation in the gastric pathogen *Helicobacter pylori*. Infect. Immun. *74*, 602–614.

Ge, R., Watt, R.M., Sun, X., Tanner, J.A., He, Q.Y., Huang, J.D., and Sun, H. (2006). Expression and characterization of a histidine-rich protein, Hpn: potential for Ni2+ storage in *Helicobacter pylori*. Biochem. J. *393*, 285–293.

Ge, Z. (2002). Potential of fumarate reductase as a novel therapeutic target in *Helicobacter pylori* infection. Expert Opin. Ther Targets *6*, 135–146.

Ge, Z., and Taylor, D.E. (1996). *Helicobacter pylori* genes *hpcopA* and *hpcopP* constitute a cop operon involved in copper export. FEMS Microbiol. Lett. *145*, 181–188.

Guth, P.H. (1992). Current concepts in gastric microcirculatory pathophysiology. Yale J. Biol. Med. *65*, 677–688.

Hagting, A., vd Velde, J., Poolman, B., and Konings, W.N. (1997). Membrane topology of the di- and tripeptide transport protein of *Lactococcus lactis*. Biochemistry *36*, 6777–6785.

Hantke, K. (2001). Iron and metal regulation in bacteria. Curr. Opin. Microbiol. *4*, 172–177.

Herrmann, L., Schwan, D., Garner, R., Mobley, H.L., Haas, R., Schafer, K.P., and Melchers, K. (1999). *Helicobacter pylori cadA* encodes an essential Cd(II)-Zn(II)-Co(II) resistance factor influencing urease activity. Mol. Microbiol. *33*, 524–536.

Hessey, S.J., Spencer, J., Wyatt, J.I., Sobala, G., Rathbone, B.J., Axon, A.T., and Dixon, M.F. (1990). Bacterial adhesion and disease activity in *Helicobacter* associated chronic gastritis. Gut *31*, 134–138.

Heuermann, D., and Haas, R. (1998). A stable shuttle vector system for efficient genetic complementation of *Helicobacter pylori* strains by transformation and conjugation. Mol. Gen. Genet. *257*, 519–528.

Holmgren, A. (1985). Thioredoxin. Annu. Rev. Biochem. *54*, 237–271.

Hong, W., Sano, K., Morimatsu, S., Scott, D.R., Weeks, D.L., Sachs, G., Goto, T., Mohan, S., Harada, F., Nakajima, N., *et al.* (2003). Medium pH-dependent redistribution of the urease of *Helicobacter pylori*. J. Med. Microbiol. *52*, 211–216.

Hsieh, P.F., Yang, J.C., Lin, J.T., and Wang, J.T. (1998). Molecular mechanisms of clarithromycin resistance in *Helicobacter pylori*. J. Formos Med. Assoc. *97*, 445–452.

Hughes, N.J., Chalk, P.A., Clayton, C.L., and Kelly, D.J. (1995). Identification of carboxylation enzymes and characterization of a novel four-subunit pyruvate:flavodoxin oxidoreductase from *Helicobacter pylori*. J. Bacteriol. *177*, 3953–3959.

Itou, H., Yao, M., Watanabe, N., and Tanaka, I. (2004). Structure analysis of PH1161 protein, a transcriptional activator TenA homologue from the hyperthermophilic archaeon *Pyrococcus horikoshii*. Acta Crystallogr. D Biol. Crystallogr. *60*, 1094–1100.

Jenks, P.J., and Edwards, D.I. (2002). Metronidazole resistance in *Helicobacter pylori*. Int. J. Antimicrob Agents *19*, 1–7.

Karita, M., Tsuda, M., and Nakazawa, T. (1995). Essential role of urease *in vitro* and in vivo *Helicobacter pylori* colonization study using a wild-type and isogenic urease mutant strain. J. Clin. Gastroenterol. *21 Suppl 1*, S160–163.

Kashket, E.R. (1981). Effects of aerobiosis and nitrogen source on the proton motive force in growing *Escherichia coli* and *Klebsiella pneumoniae* cells. J. Bacteriol. *146*, 377–384.

Kim, N., Marcus, E.A., Wen, Y., Weeks, D.L., Scott, D.R., Jung, H.C., Song, I.S., and Sachs, G. (2004). Genes of *Helicobacter pylori* regulated by attachment to AGS cells. Infect. Immun. *72*, 2358–2368.

Kim, N., Weeks, D.L., Shin, J.M., Scott, D.R., Young, M.K., and Sachs, G. (2002). Proteins released by *Helicobacter pylori* in vitro. J. Bacteriol. *184*, 6155–6162.

Labigne, A., Cussac, V., and Courcoux, P. (1991). Shuttle cloning and nucleotide sequences of *Helicobacter pylori* genes responsible for urease activity. J. Bacteriol. *173*, 1920–1931.

Levitt, M.D. (1969). Production and excretion of hydrogen gas in man. N. Engl. J. Med. *281*, 122–127.

Lowther, W.T., Brot, N., Weissbach, H., Honek, J.F., and Matthews, B.W. (2000). Thiol-disulfide exchange is involved in the catalytic mechanism of peptide methionine sulfoxide reductase. Proc. Natl. Acad. Sci. USA *97*, 6463–6468.

Marcus, E.A., Moshfegh, A.P., Sachs, G., and Scott, D.R. (2005). The periplasmic alpha-carbonic anhydrase activity of *Helicobacter pylori* is essential for acid acclimation. J. Bacteriol. *187*, 729–738.

McGee, D.J., Langford, M.L., Watson, E.L., Carter, J.E., Chen, Y.T., and Ottemann, K.M. (2005). Colonization and inflammation deficiencies in Mongolian gerbils infected by *Helicobacter pylori* chemotaxis mutants. Infect. Immun. *73*, 1820–1827.

Mehta, N., Olson, J.W., and Maier, R.J. (2003). Characterization of *Helicobacter pylori* nickel metabolism accessory proteins needed for maturation of both urease and hydrogenase. J. Bacteriol. *185*, 726–734.

Mendz, G.L., and Hazell, S.L. (1995). Aminoacid utilization by *Helicobacter pylori*. Int. J. Biochem. Cell Biol. *27*, 1085–1093.

Mendz, G.L., Hazell, S.L., and Burns, B.P. (1993). Glucose utilization and lactate production by *Helicobacter pylori*. J. Gen. Microbiol. *139*, 3023–3028.

Mendz, G.L., Hazell, S.L., and Burns, B.P. (1994). The Entner–Doudoroff pathway in *Helicobacter pylori*. Arch. Biochem. Biophys. *312*, 349–356.

Merrell, D.S., and Camilli, A. (1999). The *cadA* gene of *Vibrio cholerae* is induced during infection and plays a role in acid tolerance. Mol. Microbiol. *34*, 836–849.

Merrell, D.S., and Camilli, A. (2000). Regulation of *Vibrio cholerae* genes required for acid tolerance by a member of the 'ToxR-like' family of transcriptional regulators. J. Bacteriol. *182*, 5342–5350.

Merrell, D.S., Goodrich, M.L., Otto, G., Tompkins, L.S., and Falkow, S. (2003). pH-regulated gene expression of the gastric pathogen *Helicobacter pylori*. Infect. Immun. *71*, 3529–3539.

Mitchell, P. (1966). Chemiosmotic coupling in oxidative and photosynthetic phosphorylation. Biol. Rev. Camb Phil. Soc. *41*, 445–502.

Mollenhauer-Rektorschek, M., Hanauer, G., Sachs, G., and Melchers, K. (2002). Expression of UreI is required for intragastric transit and colonization of gerbil gastric mucosa by *Helicobacter pylori*. Res. Microbiol. *153*, 659–666.

Moncrief, M.B., and Hausinger, R.P. (1996). Purification and activation properties of UreD–UreF–urease apoprotein complexes. J. Bacteriol. *178*, 5417–5421.

Moncrief, M.B., and Hausinger, R.P. (1997). Characterization of UreG, identification of a UreD-UreF-UreG complex, and evidence suggesting that a nucleotide-binding site in UreG is required for in vivo metallocenter assembly of *Klebsiella aerogenes* urease. J. Bacteriol. *179*, 4081–4086.

Muller, S., Pflock, M., Schar, J., Kennard, S., and Beier, D. (2007). Regulation of expression of atypical orphan response regulators of *Helicobacter pylori*. Microbiol. Res. *162*, 1–14.

Mulrooney, S.B., and Hausinger, R.P. (1990). Sequence of the *Klebsiella aerogenes* urease genes and evidence for accessory proteins facilitating nickel incorporation. J. Bacteriol. *172*, 5837–5843.

Odenbreit, S. (2005). Adherence properties of *Helicobacter pylori*: impact on pathogenesis and adaptation to the host. Int. J. Med. Microbiol. *295*, 317–324.

Odenbreit, S., Puls, J., Sedlmaier, B., Gerland, E., Fischer, W., and Haas, R. (2000). Translocation of *Helicobacter pylori* CagA into gastric epithelial cells by type IV secretion. Science *287*, 1497–1500.

Olson, J.W., and Maier, R.J. (2002). Molecular hydrogen as an energy source for *Helicobacter pylori*. Science *298*, 1788–1790.

Olson, J.W., Mehta, N.S., and Maier, R.J. (2001). Requirement of nickel metabolism proteins HypA and HypB for full activity of both hydrogenase and urease in *Helicobacter pylori*, pp. 176–182.

Padan, E., Zilberstein, D., and Schuldiner, S. (1981). pH homeostasis in bacteria. Biochim. Biophys. Acta *650*, 151–166.

Park, I.S., Carr, M.B., and Hausinger, R.P. (1994). *In vitro* activation of urease apoprotein and role of UreD as a chaperone required for nickel metallocenter assembly. Proc. Natl. Acad. Sci. USA *91*, 3233–3237.

Park, I.S., and Hausinger, R.P. (1995). Evidence for the presence of urease apoprotein complexes containing UreD, UreF, and UreG in cells that are competent for in vivo enzyme activation. J. Bacteriol. *177*, 1947–1951.

Pellicer, M.T., Badia, J., Aguilar, J., and Baldoma, L. (1996). glc locus of *Escherichia coli*: characterization of genes encoding the subunits of glycolate oxidase and the glc regulator protein. J. Bacteriol. *178*, 2051–2059.

Pfeiffer, J., Guhl, J., Waidner, B., Kist, M., and Bereswill, S. (2002). Magnesium uptake by CorA is essential for viability of the gastric pathogen *Helicobacter pylori*. Infect. Immun. *70*, 3930–3934.

Pflock, M., Dietz, P., Schar, J., and Beier, D. (2004). Genetic evidence for histidine kinase HP165 being an acid sensor of *Helicobacter pylori*. FEMS Microbiol. Lett. *234*, 51–61.

Pflock, M., Finsterer, N., Joseph, B., Mollenkopf, H., Meyer, T.F., and Beier, D. (2006). Characterization of the ArsRS regulon of *Helicobacter pylori*, involved in acid adaptation. J. Bacteriol. *188*, 3449–3462.

Pitson, S.M., Mendz, G.L., Srinivasan, S., and Hazell, S.L. (1999). The tricarboxylic acid cycle of *Helicobacter pylori*. Eur. J. Biochem. *260*, 258–267.

Sachs, G., Weeks, D.L., Wen, Y., Marcus, E.A., Scott, D.R., and Melchers, K. (2005). Acid acclimation by *Helicobacter pylori*. Physiology (Bethesda) *20*, 429–438.

Schar, J., Sickmann, A., and Beier, D. (2005). Phosphorylation-independent activity of atypical response regulators of *Helicobacter pylori*, pp. 3100–3109.

Scott, D.R., Marcus, E.A., Weeks, D.L., Lee, A., Melchers, K., and Sachs, G. (2000). Expression of the *Helicobacter pylori ureI* gene is required for acidic pH activation of cytoplasmic urease. Infect. Immun. *68*, 470–477.

Scott, D.R., Marcus, E.A., Wen, Y., Oh, J., and Sachs, G. (2007). Gene expression *in vivo* shows that *Helicobacter pylori* colonizes an acidic niche on the gastric surface. Proc. Natl. Acad. Sci. USA *104*, 7235–7240.

Scott, D.R., Weeks, D., Hong, C., Postius, S., Melchers, K., and Sachs, G. (1998). The role of internal urease in acid resistance of *Helicobacter pylori*. Gastroenterology *114*, 58–70.

Skouloubris, S., Thiberge, J.M., Labigne, A., and De Reuse, H. (1998). The *Helicobacter pylori* UreI protein is not involved in urease activity but is essential for bacterial survival in vivo. Infect. Immun. *66*, 4517–4521.

Suzuki, T., Murakami, T., Iino, R., Suzuki, J., Ono, S., Shirakihara, Y., and Yoshida, M. (2003). F0F1-ATPase/synthase is geared to the synthesis mode by conformational rearrangement of epsilon subunit in response to proton motive force and ADP/ATP balance. J. Biol. Chem. *278*, 46840–46846.

Teyssen, S., Chari, S.T., Scheid, J., and Singer, M.V. (1995). Effect of repeated boluses of intravenous omeprazole and primed infusions of ranitidine on

24-hour intragastric pH in healthy human subjects. Dig. Dis. Sci. *40*, 247–255.

Toms, A.V., Haas, A.L., Park, J.H., Begley, T.P., and Ealick, S.E. (2005). Structural characterization of the regulatory proteins TenA and TenI from *Bacillus subtilis* and identification of TenA as a thiaminase II. Biochemistry *44*, 2319–2329.

Tonello, F., Dundon, W.G., Satin, B., Molinari, M., Tognon, G., Grandi, G., Del Giudice, G., Rappuoli, R., and Montecucco, C. (1999). The *Helicobacter pylori* neutrophil-activating protein is an iron-binding protein with dodecameric structure. Mol. Microbiol. *34*, 238–246.

Tsuda, M., Karita, M., Mizote, T., Morshed, M.G., Okita, K., and Nakazawa, T. (1994a). Essential role of *Helicobacter pylori* urease in gastric colonization: definite proof using a urease-negative mutant constructed by gene replacement. Eur. J. Gastroenterol. Hepatol. 6 Suppl. 1, S49–52.

Tsuda, M., Karita, M., Morshed, M.G., Okita, K., and Nakazawa, T. (1994b). A urease-negative mutant of *Helicobacter pylori* constructed by allelic exchange mutagenesis lacks the ability to colonize the nude mouse stomach. Infect. Immun. *62*, 3586–3589.

van Vliet, A.H., Ketley, J.M., Park, S.F., and Penn, C.W. (2002). The role of iron in *Campylobacter* gene regulation, metabolism and oxidative stress defense. FEMS Microbiol. Rev. *26*, 173–186.

van Vliet, A.H., Stoof, J., Poppelaars, S.W., Bereswill, S., Homuth, G., Kist, M., Kuipers, E.J., and Kusters, J.G. (2003). Differential regulation of amidase- and formamidase-mediated ammonia production by the *Helicobacter pylori* fur repressor. J. Biol. Chem. *278*, 9052–9057.

Voland, P., Weeks, D.L., Marcus, E.A., Prinz, C., Sachs, G., and Scott, D. (2003). Interactions among the seven *Helicobacter pylori* proteins encoded by the urease gene cluster. Am. J. Physiol. Gastrointest. Liver Physiol. *284*, G96-G106.

Wang, G., Alamuri, P., and Maier, R.J. (2006). The diverse antioxidant systems of *Helicobacter pylori*. Mol. Microbiol. *61*, 847–860.

Weeks, D.L., Eskandari, S., Scott, D.R., and Sachs, G. (2000). A H^+-gated urea channel: the link between *Helicobacter pylori* urease and gastric colonization. Science *287*, 482–485.

Weeks, D.L., Gushansky, G., Scott, D.R., and Sachs, G. (2004). Mechanism of proton gating of a urea channel. J. Biol. Chem. *279*, 9944–9950.

Wen, Y., Feng, J., Scott, D.R., Marcus, E.A., and Sachs, G. (2006). Involvement of the HP0165-HP0166 two-component system in expression of some acidic-pH-upregulated genes of *Helicobacter pylori*. J. Bacteriol. *188*, 1750–1761.

Wen, Y., Feng, J., Scott, D.R., Marcus, E.A., and Sachs, G. (2007). The HP0165-HP0166 Two-Component System (ArsRS) Regulates Acid-Induced Expression of HP1186 {alpha}-Carbonic Anhydrase in *Helicobacter pylori* by Activating the pH-Dependent Promoter, pp. 2426–2434.

Wen, Y., Marcus, E.A., Matrubutham, U., Gleeson, M.A., Scott, D.R., and Sachs, G. (2003). Acid-adaptive genes of *Helicobacter pylori*. Infect. Immun. *71*, 5921–5939.

Wood, H.G. (1977). Some reactions in which inorganic pyrophosphate replaces ATP and serves as a source of energy. Fed Proc. *36*, 2197–2206.

Wu, H., Nakano, T., Daikoku, E., Morita, C., Kohno, T., Lian, H.H., and Sano, K. (2005). Intrabacterial proton-dependent CagA transport system in *Helicobacter pylori*. J. Med. Microbiol. *54*, 1117–1125.

Metal Metabolism and Transport in *Helicobacter pylori*

8

Jeroen Stoof, Clara Belzer and Arnoud H.M. van Vliet

Abstract

The members of the genus *Helicobacter* are among the most successful colonizers of the mammalian gastrointestinal and hepatobiliary tract, and are thought to have co-evolved during the development of the mammalian digestive system. The best-studied *Helicobacter* species is the human pathogen *H. pylori*, which colonizes the gastric mucosa, and this colonization results in chronic gastritis, which may develop into peptic ulcer disease or gastric adenocarcinoma. The longevity of gastric colonization suggests that *H. pylori* must be able to cope with the different stresses encountered in the gastric mucosa, such as low pH and restriction of essential metals such as iron and nickel. In this chapter we will review the current knowledge on the role of metal metabolism in the physiology of *H. pylori*, and discuss the contribution of the regulatory mechanisms involved in controlling metal metabolism in the adaptation of *H. pylori* to its gastric niche.

Introduction

Transition metals play a vital role in the metabolism of all organisms, and this is reflected by the wide variety of chemical reactions in which they are involved. Metals can be cofactors of enzymes, catalysing basic functions such as electron transport, redox reactions and energy metabolism, and are essential for maintaining the osmotic pressure of the cell (van Vliet *et al.*, 2001a; Belzer *et al.*, 2007a; Whitmire *et al.*, 2007). However, the intracellular metal concentration needs to be carefully balanced, since both metal limitation and metal overload delay growth and can cause cell death. For example, imbalanced iron metabolism can lead to the formation of reactive oxygen species via the Fenton and Haber–Weiss reactions ($Fe^{2+} + O_2 \rightarrow Fe^{3+} + O_2^-$ and $Fe^{2+} + H_2O_2 \rightarrow Fe^{3+} + OH^- + OH^\bullet$). Therefore metal homeostasis is considered to be of critical importance to all living organisms, and is maintained by balancing import, efflux, metabolism and storage of metals.

Helicobacter species are microaerophilic bacteria, which colonize mucosal surfaces of the gastrointestinal and hepatobiliary tract of mammals, including humans. Colonization is usually lifelong, and often leads to the development of a wide range of inflammatory diseases. The best-studied *Helicobacter* species is the human pathogen *Helicobacter pylori*, which colonizes the mucosal surface overlaying the epithelial cells in the stomach of humans and some non-human primates (Kusters *et al.*, 2006).

The mucosal surfaces of the mammalian gastrointestinal and hepatobiliary tracts are niches with continuous changes in environmental conditions (van Vliet *et al.*, 2001a). An example is the pH in the gastric mucosa where *H. pylori* resides. Many studies assume there is a pH gradient ranging from pH 4 at the luminal side to pH 6.5 at the epithelial surface, but acid shocks have been predicted to occur (Stingl *et al.*, 2002; Schreiber *et al.*, 2004; Schreiber *et al.*, 2005). However, there are also studies which implicate that the gastric mucosa is much more acidic, and that *H. pylori* may be subjected to low pH throughout its growth cycle (Scott *et al.*, 2007). The environmental pH influences the bioavailability of metals like iron and nickel, and metal concentrations may also fluctuate due to release of nutrients from the damages epithelium and diffusion of metals from food. The bioavailability of metals in the gastric mucosa has not been determined so far, and the only information available is from determinations of metal concentrations in non-cancerous and cancerous stomach tissues. Nickel, copper and iron concentrations being significantly higher in cancerous tissues, while calcium concentrations were lowered (Yaman *et al.*, 2007). In contrast to the gastric situation where bacterial colonization is usually limited, the intestinal tract contains many bacterial species and thus enterohepatic *Helicobacter* species have to compete for metals and other nutrients. This implies that *Helicobacter* species in both the gastric and enteric environments probably have to cope with both conditions of metal-restriction and metal-overload.

The chronicity of infection with *H. pylori* suggests that this bacterium is well adapted to the constraints imposed by its host environment. Transition metals like nickel and iron play an essential role in this adaptation, and in this chapter we will discuss the wealth of knowledge on these subjects in *H. pylori*, including metal transport, storage and regulation of genes involved in metal metabolism.

Role of metals in *Helicobacter* physiology

Metals play essential roles in many cellular processes, by functioning as cofactor of enzymes, as electron donor or acceptor in redox reactions, or to maintain cell turgor or a proton motive force. These diverse functions are carried out by several different metals. A wealth of information is available for the role of metals like iron and nickel in *Helicobacter*, since these have a key role in virulence and metabolism of *Helicobacter* species. However, for many other metals there is little or no experimental information available for *Helicobacter*, and thus only bioinformatic predictions are available to suggest a role in *Helicobacter* physiology or metabolism. Thus, we will discuss only briefly the role of iron and nickel in *Helicobacter* physiology below.

Nickel

The transition metal nickel (atomic number 28, atomic weight 58.69) is the co-factor of the enzyme urease, one of the most abundant enzymes of *H. pylori* (Hu *et al.*, 1992). Urease converts urea in two ammonia and one bicarbonate molecule, and both the reaction products contribute heavily to acid resistance of *H. pylori* (Sachs *et al.*, 2003; Marcus *et al.*, 2005). It has been estimated that up to 10% of the soluble protein of *H. pylori* can consist of the urease subunits UreA and UreB (Hu *et al.*, 1992), and expression of urease is coupled to nickel availability (van Vliet *et al.*, 2001b). In standard growth conditions *H. pylori* produces both activated urease holoenzyme and inactive urease apo-enzyme, and is able to rapidly activate the urease apo-enzyme by inserting the nickel cofactor (van Vliet *et al.*, 2002a; Stingl and De Reuse, 2005; Seshadri *et al.*, 2007). The second nickel-containing enzyme of *H. pylori* is the Ni/Fe hydrogenase, which functions in the bidirectional conversion of hydrogen to two protons and two electrons (Olson *et al.*, 2001). Hydrogen is produced by colonic bacteria, but is not used by the mammal host, while *H. pylori* can use it for energy production (Olson and Maier, 2002). Inactivation of either urease or hydrogenase results in an inability of *H. pylori* to colonize the gastric environment in a variety of animal models (Tsuda *et al.*, 1994; Olson and Maier, 2002), thus highlighting the importance of both nickel-cofactored enzymes in pathogenesis of *H. pylori* infection.

Iron

The transition metal iron (atomic number 26, atomic weight 55.85) is the co-factor of many essential enzymes, either in its elemental state or complexed in haem (van Vliet *et al.*, 2001a; Andrews *et al.*, 2003). Iron is a versatile metal, which in nature mostly exists in either the ferrous (2+) or ferric (3+) state. The ferrous state is stable in anaerobic, acidic conditions, but rapidly converts to the insoluble ferric form in aerobic conditions at neutral or alkaline pH (Andrews *et al.*, 2003). The conversion of ferrous to ferric iron in the presence of oxygen can also result in the formation of reactive oxygen species as mentioned previously, and thus intracellular iron metabolism is carefully controlled by storing excess iron in an inactive but reusable form in the Pfr ferritin (Bereswill *et al.*, 1998b; Waidner *et al.*, 2002a). The link between iron and oxidative stress defence is also apparent through the iron-responsive control of the iron-containing superoxide dismutase SodB in *H. pylori* (Ernst *et al.*, 2005b).

Metal transport

The evolution of membranes has allowed organisms to maintain constant intracellular conditions mostly independent of the conditions outside of the cell. The lack of free transport over membranes has necessitated the development of specific transporters for solutes, nutrients and metals (Wooldridge and Williams, 1993). Gram-negative bacteria such as *Helicobacter* species also have an outer membrane, and the size of some metal complexes has resulted in the evolution of specific transporter proteins in both the outer membrane and the cytoplasmic membrane, dependent on the metal cofactor involved (van Vliet *et al.*, 2001a). Next to systems importing essential metal ions, there are also efflux systems actively pumping out specific metal ions when their concentrations exceed certain levels. The joint action of import and efflux systems, coupled with storage systems available for iron and nickel, allow the cell to maintain the concentrations of the different ions between the permissive values (van Vliet *et al.*, 2001a).

The availability of the complete genome sequence of *H. pylori*, combined with studies using insertional mutagenesis, has allowed for the study of different metal transport systems in *H. pylori*, and we will discuss the current knowledge on nickel and iron transporters, and include magnesium, copper, potassium and sodium transporters and the metal efflux systems. The different metal transporters of *H. pylori* which have been identified and for which experimental data are available are displayed in Fig. 8.1.

Iron

At mucosal surfaces, availability of free iron is far below the concentration required to support bacterial growth (Ratledge and Dover, 2000), since most iron is complexed into haem compounds or ferritin, or is chelated by lactoferrin or transferrin (Andrews *et al.*, 2003). However, iron may be present as ferrous iron in the acidic, microaerobic gastric environment, and this form of iron is thought to be able to pass through the porins of the outer membrane, and thus only requires transport over the cytoplasmic membrane. In *H. pylori* this ferrous iron transport is mediated

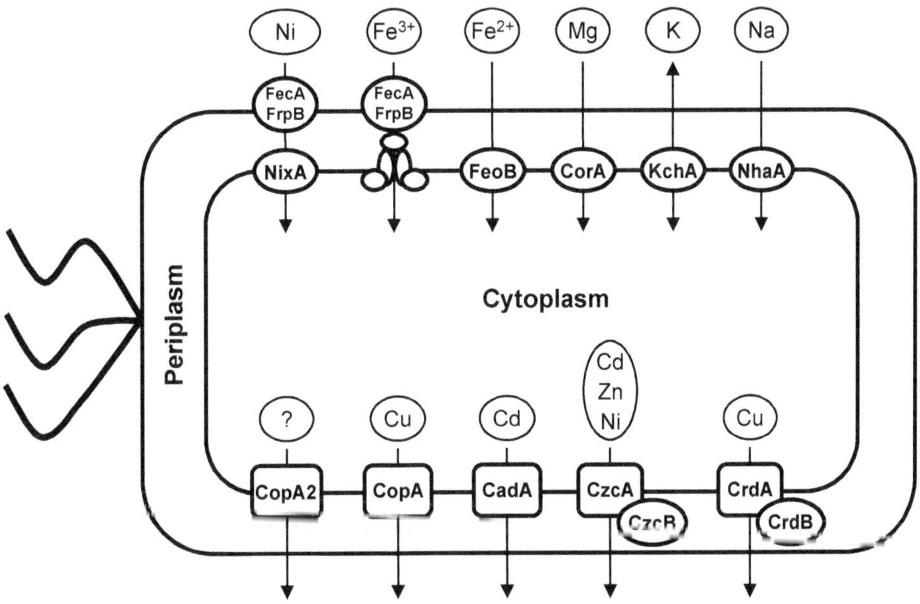

Figure 8.1 Overview of the different systems involved in mediating metal transport and metal homeostasis in *Helicobacter pylori*. Arrows denote the direction of transport.

by the FeoB protein (Velayudhan *et al.*, 2000). FeoB-mediated iron acquisition is of major importance to *H. pylori*, as *feoB* mutants are unable to colonize the gastric mucosa (Velayudhan *et al.*, 2000). Iron is often also present as insoluble ferric iron complexes, which are too large to pass through the bacterial outer membrane. Therefore transport of ferric iron requires an outer membrane receptor to transport the iron over the outer membrane, as well as a ABC transporter which transports the iron from the periplasm to the cytoplasm (van Vliet *et al.*, 2001a; Andrews *et al.*, 2003). Iron sources available in the gastric mucosa are lactoferrin, haem compounds, and iron derived from pepsin-degraded food. Iron is predicted to display increased solubility in the acidic conditions in the gastric mucosa, and eukaryotic iron-complexing proteins display lowered binding affinity at these conditions. *H. pylori* has been reported to be able to utilize haem compounds and ferric citrate as sole iron source (Dhaenens *et al.*, 1999; Velayudhan *et al.*, 2000), and it may be able to also use human lactoferrin, although this is still under debate (Husson *et al.*, 1993; van Vliet *et al.*, 2001a). The *H. pylori* genome sequence contains three copies of the putative ferric citrate outer membrane receptor FecA, and three genes encoding proteins with relatively weak homology to the enterochelin outer membrane receptor FrpB of *Neisseria* species (Carson *et al.*, 1999), for which in *H. pylori* the substrate is still unknown. There are two copies of the periplasmic binding protein CeuE, and finally a single inner membrane permease (FecD) and ATP-binding protein (FecE) (Fig. 8.1). Mutants have been described in two of the *fecA* genes (Velayudhan *et al.*, 2000; Schauer *et al.*, 2007), and in the HP1512 *frpB* gene (Davis *et al.*, 2006; Schauer *et al.*, 2007). Rather surprisingly, most of these mutations did not affect iron transport, suggesting that the other copies of the FecA and FrpB proteins may be able to complement the mutation in a single copy, or may have acquired a different function, which is also reflected by the differential regulation of the *H. pylori* FecA and FrpB orthologues (van Vliet *et al.*, 2002b; Davis *et al.*, 2006; Ernst *et al.*, 2006; Schauer *et al.*, 2007). Thus, the contribution of these outer membrane proteins in ferric iron uptake of *H. pylori* remains to be elucidated.

Nickel

Although the exact source and bioavailability of nickel to *H. pylori* are not known, it is well known that the bacterium is able to cater for the high level expression of its urease enzyme, and as such is predicted to be able to scavenge sufficient nickel. The presence of higher nickel concentrations in gastric cancer tissues (a disease associated with the presence of *H. pylori*) suggests that *H. pylori* is able to acquire nickel even at nickel-restricted conditions occurring in the human stomach (Yaman *et al.*, 2007). Nickel availability in human serum is low (2–11 nmol/l) (Christensen *et al.*, 1999), and the nickel concentration in ingested food varies significantly depending on the diet and on food sources, with vegetarian food sources containing higher concentrations of nickel (Sunderman, 1993). It is therefore predicted that *H. pylori* uses efficient high-affinity nickel transporters to cater for the requirement of the urease and hydrogenase enzymes.

Uncomplexed nickel is thought to be able to cross the outer membrane via porins, and thus only requires transport over the cytoplasmic membrane. In *H. pylori* this transport is mediated via the NixA protein, a 37 kDa protein of the HoxN-type of nickel/cobalt transporters (Mobley *et al.*, 1995). Mutation of the *nixA* gene resulted in decreased levels of nickel transport, which was coupled to reduced urease activity (Bauerfeind *et al.*, 1996), and mutation of *nixA* complemented nickel-sensitivity of a *H. pylori nikR* mutant (Ernst *et al.*, 2005c). Absence of *nixA* also reduced gastric colonization levels in a mouse model of *H. pylori* infection, which is likely to be due to the reduced urease activity in a *nixA* mutant (Nolan *et al.*, 2002). Other cytoplasmic membrane nickel transporters have been proposed, like the *abcCD* and the *dpp* loci, but neither of these systems is currently considered to function in nickel uptake (Hendricks and Mobley, 1997; Davis and Mobley, 2005).

There is recent evidence that nickel uptake by *H. pylori* may also require pH-dependent transport over the outer membrane, a feature not previously described in Gram-negative bacteria. Mutation of the gene encoding the FrpB4 (HP1512) outer membrane protein, and absence of the associated ExbB-ExbD-TonB system both resulted in a decrease of nickel transport at acidic pH (Schauer *et al.*, 2007), and in a separate study absence of HP1512 induced cytoplasmic nickel-deficient conditions in *H. pylori* (Davis *et al.*, 2006). This was corroborated by the finding that expression of the OM transporters HP1512 and FecA3 (HP1400) as well as the ExbB-ExbD-TonB system is controlled by the NikR nickel-regulatory protein in response to nickel (Davis *et al.*, 2006; Ernst *et al.*, 2006). However, the nickel complex transported by this system is not yet known, and thus this interesting adaptation of nickel transport by *H. pylori* requires further investigation.

Copper

The transition metal copper is a cofactor for several proteins involved in electron transport, such as oxidases and hydroxylases, but may also contribute to the formation of reactive oxygen species (Bayle *et al.*, 1998). Therefore, the cell is required to maintain homeostatic control of the copper concentration, as with iron and nickel. *H. pylori* expresses several proteins which are either involved in copper transport, or may act as copper chaperones. It is currently unclear how copper is transported into *H. pylori*, since a specific copper importer is absent (Herrmann *et al.*, 1999; Mobley *et al.*, 1999). However, *H. pylori* expresses several proteins thought to be involved in copper export, including the CopA and CopA2 P-type ATPases, and the Crd copper resistance determinant (Melchers *et al.*, 1998; Waidner *et al.*, 2002b). Furthermore, *H. pylori* also expresses a small protein, CopP, that may function as copper chaperone, shuttling copper between processes (Beier *et al.*, 1997; Bayle *et al.*, 1998).

Potassium

Potassium is the most abundant ion in all living cells with concentrations ranging up to 0.5 M even when extracellular levels are very low, and potassium is essential for maintaining osmolarity and turgor pressure of the cell (van Vliet *et al.*, 2001a; Stingl *et al.*, 2007). The first response of bacteria when adapting to high osmolarity is the increase in uptake of potassium. Adaptation to low osmolarity leads to the efflux of potassium and other osmotic solutes, through non-specific channel proteins. The *H. pylori* genome lacks the constitutive low affinity potassium importers and inducible high-affinity potassium importers and exporters known to mediate potassium homeostasis in other bacteria (Alm *et al.*, 1999; Tomb *et al.*, 1997), and only seems to contain a single potassium channel required for K^+ uptake (HpKchA, HP0490) (Stingl *et al.*, 2007) and a putative KefB orthologue (HP0471).

Sodium

The role of sodium in bacteria is predominantly in transport activities for different nutrients, since Na^+ functions as symporter (Na^+ is cotransported with the solute) and antiporter (Na^+ is transported in the reverse direction to the solute). Sodium also functions in maintaining the proton gradient necessary for the proton motive force, and thus in energy production (van Vliet *et al.*, 2001a). The intracellular sodium concentration of bacteria is usually lower than that of the environment. This is mostly achieved by the export of sodium ions by sodium-proton antiporters. These assist in maintaining intracellular pH, osmolarity and salt concentration. *H. pylori* contains a gene encoding a sodium–proton antiporter homologue, the *nhaA* gene (HP1552). The *H. pylori nhaA* gene was able to complement an *E. coli* mutant lacking all sodium–proton antiporter activity (Inoue *et al.*, 2001). An interesting difference between *E. coli* and *H. pylori* NhaA was found by comparing their respective activities at different pH values. *H. pylori* NhaA showed high activity at pH values between 6.0 and 8.5, while *E. coli* NhaA was only active above pH 8 (Inoue *et al.*, 2001). The activity of *H. pylori* NhaA at higher proton concentrations is likely to reflect an adaptation to the acidic conditions in the gastric mucosa. It was recently postulated that a stretch of 40 extra amino acids, that is only present in *H. pylori* NhaA, might mediate its altered pH sensitivity (Inoue *et al.*, 1999). *H. pylori* contains a second gene encoding a putative sodium–proton antiporter homologue (HP1183), as well as four homologues of transport systems that probably use sodium as symport or counterport ion, namely the sodium–glutamate (HP1506), sodium–proline (HP0055) and sodium–alanine (HP0942) symporters, and L-serine permease (HP0133) (Tomb *et al.*, 1997; Alm *et al.*, 1999).

Magnesium

Magnesium is the most abundant divalent cation in living cells, and is often found bound to cellular polyanions such as nucleic acids. Furthermore it regularly functions as cofactor in conjunction with ATP in many enzymatic reactions, like those catalysing modification, replication and transcription of nucleic acids. Because it is continuously complexed, magnesium does not contribute much to osmotic processes (Smith and Maguire, 1998). Many *H. pylori* enzymes probably use magnesium as cofactor, like its restriction-modifications systems, but this has not been investigated experimentally. Bacteria require micromolar concentrations of magnesium for growth, and thus have the need for active magnesium acquisition. Magnesium uptake has been extensively studied in *Salmonella typhimurium*, which contains four magnesium transport systems, the CorA, MgtA, MgtB and MgtE transporters (Smith and Maguire, 1998). *H. pylori* contains only a CorA homologue (HP1344), which is thought to mediate both import and export of magnesium, cobalt and nickel (Smith and Maguire, 1998; Pfeiffer *et al.*, 2002). A *H. pylori* CorA mutant was only capable of growing in medium supplemented with Mg^{2+} concentrations exceeding 20 mM (Pfeiffer *et al.*, 2002), suggesting that there are no other high-affinity Mg^{2+} transporters in *H. pylori*.

Metal efflux systems

Although many metal efflux systems display selectivity for their target metal, it can often not be avoided that toxic concentrations of metal build up in the cytoplasm, or that toxic metals enter the cell. This problem is mostly solved by either specific or non-specific efflux. Metal efflux is of great importance to *H. pylori*, since its urease enzyme is sensitive to several transition metals (Perez-Perez *et al.*, 1994; Herrmann *et al.*, 1999), and this may result in acid-sensitivity. Two types of metal efflux systems have been identified in *H. pylori*: the P-type ATPases, which are single component transporters which require ATP for their transport function, and the multi-component RND-like efflux pumps. Two of the three P-type ATPases are suggested to be CopA orthologues, which function in copper metabolism, whereas the third P-type ATPase in *H. pylori* is the CadA protein, which is homologous to bacterial cadmium and copper P-type ATPases. The *H. pylori* P-type ATPases contain eight transmembrane domains (Melchers *et al.*, 1996), and mutational studies demonstrated that CadA is involved in resistance to cadmium, zinc and cobalt, but not to copper or nickel (Herrmann *et al.*, 1999). This indicates that the CadA exporter is not specific for a single metal.

The other type of metal-efflux systems in *H. pylori* consists of the RND-type cation-proton antiporters, and usually contain an cytoplasmic membrane transporter, a periplasm-spanning protein and an OM protein thought to be involved in transport over the OM. The *H. pylori* genome contains two Czc-like systems, the Crd copper resistance determinant (Waidner *et al.*, 2002b) and a cobalt–zinc–nickel (Czn) resistance determinant (Stähler *et al.*, 2006). The latter system is also implicated in acid resistance (possibly via nickel efflux) (Stähler *et al.*, 2006) and efflux of antibiotics (van Amsterdam *et al.*, 2005). The interaction or cooperation between the different metal efflux systems deserves further investigation.

Metal storage

Both the import and efflux of metals are energy-consuming processes, but while the cell avoids unnecessary export of vital nutrients, it will also need to ensure that higher concentrations of metals cannot cause cellular damage. Therefore, cells have developed mechanisms to store essential metals in an inactive, safe but reusable form. Two types of such proteins have been identified in *H. pylori*, and those are involved in iron and nickel metabolism.

Iron

Two iron storage proteins have been identified in *H. pylori*, and these represent the two known classes of bacterial ferritins: the non-haem ferritin Pfr and the bacterioferritin HP-NAP (Dps) (Bereswill *et al.*, 1998b; Tonello *et al.*, 1999). The Pfr ferritin serves as an intracellular iron deposit, storing ferric iron in an inactive form. The Pfr protein protects *H. pylori* against intracellular iron toxicity and free iron-mediated oxidative stress (Bereswill *et al.*, 1998b), and iron stored in Pfr can be released and reused by the cell to support growth under iron-limited conditions (Waidner *et al.*, 2002a). In contrast, the HP-NAP protein is both homologous to bacterioferritins as well as to the DNA-binding proteins of the Dps family (Tonello *et al.*, 1999). Mutation of HP-NAP did not affect iron metabolism of *H. pylori*, but mutations affecting expression of HP-NAP or mutation of the gene encoding HP-NAP result in increased oxidative damage of genomic DNA, suggestive of a Dps-like role for HP-NAP (Barnard *et al.*, 2004; Wang *et al.*, 2006a; Wang *et al.*, 2006b). A role of HP-NAP in *H. pylori* iron storage, although suggested, is yet to be demonstrated.

Nickel

Several urease- and hydrogenase-associated systems have adapted to the central role of nickel (Maier *et al.*, 2007). The accessory proteins involved in hydrogenase biogenesis also affect urease activity (Benoit *et al.*, 2007), and the HspA (GroES) chaperone contains a nickel-binding motif (Kansau *et al.*, 1996). Next to the nickel transporters and urease/hydrogenase accessory proteins, *H. pylori* also expresses one or two small, very histidine-rich proteins (Hpn) which show strong binding to nickel (Gilbert *et al.*, 1995; Ge *et al.*, 2006; Maier *et al.*, 2007). Mutation of the *hpn* gene rendered *H. pylori* sensitive to nickel (Mobley *et al.*, 1999), and it has subsequently been shown to compete with urease for available intracellular nickel at neutral pH (Seshadri *et al.*, 2007). At acidic pH nickel is released from Hpn, thus resulting in activation of urease at times most needed, and prevention of aberrant urease activity at neutral pH (Seshadri *et al.*, 2007).

Metal-responsive regulatory systems

The potential toxicity of transition metals necessitates tight control of their intracellular concentration, and has been shown to represent an excellent model for the study of gene regulation in many bacteria, including *H. pylori*. Rapid responses to changes in environmental conditions are often mediated via changes in transcription of sets of genes that encode some factor involved in the dealing with these stresses. In the case of metal metabolism, the regulatory system controls the expression of the importers and exporters, the storage and often also the utilization of the specific metal. Due to the close connection between metal-availability and environmental pH, metal-responsive regulatory systems are often involved in acid resistance, e.g. in *Salmonella* species and also in *H. pylori* (Hall and Foster, 1996; Bijlsma *et al.*, 2002).

Metal-responsive regulatory systems usually consist of a regulatory protein that contains domains which sense metal, and interact with their target promoters (Escolar *et al.*, 1999). The metal-binding domain usually allows multimerization upon binding of the target metal, and this activation of the protein results in binding of the regulator to its target promoters. Binding of the regulator subsequently results in induction or repression of transcription of the genes encoding the corresponding uptake, efflux and/or storage systems (Andrews *et al.*, 2003). Three metal-responsive systems have to date been identified in *H. pylori*: the nickel-responsive regulator NikR (van Vliet *et al.*, 2002a), the iron-responsive regulator Fur (Bereswill *et al.*, 1998a), and the copper-responsive two-component regulatory system CrdRS (Waidner *et al.*, 2005). We will also briefly discuss the ArsRS two-component regulatory system, since this is thought to be the major effector protein mediating acid-responsive regulation (Pflock *et al.*, 2006a; Wen *et al.*, 2006).

NikR

The NikR (HP1338) regulator belongs to the class of ribbon–helix–helix (RHH) proteins, and was originally identified in *E. coli* as the repressor of the nickel uptake operon *nikABCDE* (De Pina *et al.*, 1999). Binding of nickel-cofactored NikR to the *nikA* promoter region results in repression of transcription, thus allowing NikR to block nickel transport in conditions where free nickel is available in the cytoplasm (Chivers and Sauer, 2000). The *H. pylori nikR* gene was first identified as being involved in nickel-responsive induction of the urease system (van Vliet *et al.*, 2001b; van Vliet *et al.*, 2002a), and has since then been shown to encode the regulatory protein controlling nickel transport and nickel metabolism in *H. pylori* by regulating the expression of nickel transporter genes (*nixA*, *fecA3* and *frpB4*

(Ernst *et al.*, 2005c; Davis *et al.*, 2006; Ernst *et al.*, 2006)), nickel storage (*hpn* (Contreras *et al.*, 2003)) and nickel utilization (*ureA* (Delany *et al.*, 2005; Ernst *et al.*, 2005c; Abraham *et al.*, 2006)). It also indirectly affects iron metabolism by influencing expression of the *fur* gene (Bury-Mone *et al.*, 2004; van Vliet *et al.*, 2004b; Delany *et al.*, 2005). The crystal structure of the *H. pylori* NikR protein has been determined (Dian *et al.*, 2006), and it was shown that NikR acts as a nickel-activated tetrameric protein, similar to the *E. coli* NikR protein (Delany *et al.*, 2005; Dian *et al.*, 2006). The *H. pylori* NikR system clearly differs from the *E. coli* counterpart, since it does not require a low-affinity nickel-binding site, but can use other cations as well, and contains a nine-amino-acid N-terminal extension which influences recognition of target sequences (Benanti and Chivers, 2007).

Fur

The ferric uptake regulator (Fur) protein (HP1027) is probably the best-studied metal-responsive regulatory protein, and was first isolated as the repressor of iron uptake in *E. coli* (Hantke, 2001). In *E. coli*, Fur plays a central role in metabolism of the bacterium, since it controls iron uptake, iron storage, oxidative stress resistance and species of small RNA molecules that influence genes outside the classical iron homeostasis systems (Masse and Gottesman, 2002; Andrews *et al.*, 2003). In *H. pylori* Fur mediates regulation of outer membrane proteins thought to be involved in iron acquisition (Delany *et al.*, 2001a; van Vliet *et al.*, 2002b; Ernst *et al.*, 2005a; Gancz *et al.*, 2006), regulates Pfr-mediated iron storage (Bereswill *et al.*, 2000; Delany *et al.*, 2001b), regulates expression of the iron-cofactored superoxide dismutase SodB (Ernst *et al.*, 2005a; Ernst *et al.*, 2005b), but is also involved in regulation of nitrogen metabolism in *H. pylori* through the AmiE aliphatic amidase and AmiF formamidase (van Vliet *et al.*, 2003). The *H. pylori* Fur protein is a bifunctional regulatory protein when compared to its *E. coli* counterpart, as the iron-free (apo)form of *H. pylori* Fur represses the expression of Pfr and SodB (Bereswill *et al.*, 2000; Delany *et al.*, 2001b; Ernst *et al.*, 2005b). This form of regulation may compensate for the absence in *H. pylori* of small RNAs like RyhB (Masse and Gottesman, 2002). Expression of *H. pylori* Fur is autoregulated in response to iron availability (Delany *et al.*, 2002), but also controlled by the NikR regulator (see above). Absence of Fur results in a decrease of acid-resistance of *H. pylori* (Bijlsma *et al.*, 2002), and although the exact mechanism underlying this phenomenon is not yet known, it may well be mediated through the role of Fur in expression of the ArsRS system (Bury-Mone *et al.*, 2004; Lee *et al.*, 2004).

ArsRS

The ArsRS (HP0166-HP0165) proteins constitutes a two-component regulatory system (Bijlsma and Groisman, 2003) that is required for the transcriptional responses to changes in pH (Wen *et al.*, 2003; Bury-Mone *et al.*, 2004; Pflock *et al.*, 2005; Pflock *et al.*, 2006b; Scott *et al.*, 2007; Wen *et al.*, 2007). The ArsS protein is a histidine kinase protein, which is thought to autophosphorylate once the periplasmic pH drops, and subsequently transfers the phosphor group to the cytoplasmic response regulator ArsR (Schar *et al.*, 2005). Phosphorylation activates the ArsR protein, which then regulates the expression of a large number of genes involved in the adaptation to acidity, e.g. urease, the amidases, arginase and Hpn-mediated nickel storage (Bury-Mone *et al.*, 2004; Pflock *et al.*, 2006a; Wen *et al.*, 2006). Transcription of a part of the ArsR-regulon is also pH-responsive in the absence of the ArsS histidine kinase, and ArsR does not require phosphorylation for its function, suggesting that ArsR may also directly sense changes in cytoplasmic pH through a yet unknown mechanism (Pflock *et al.*, 2006a; Wen *et al.*, 2006). The ArsRS system may not directly be involved in metal metabolism of *H. pylori*, but indirectly affects metal homeostasis through its control of urease transcription which is shared with NikR (van Vliet *et al.*, 2004a; Ernst *et al.*, 2005c; Pflock *et al.*, 2005).

CrdRS

The *H. pylori* CrdRS (HP1365-HP1364) proteins constitute a two-component regulatory system that is required for the copper-responsive induction of the copper-resistance determinant CrdAB–CzcBA (Waidner *et al.*, 2002b; Waidner *et al.*, 2005). Its exact role in the metabolism of other metals is not known, nor is it known through which mechanism senses copper, but absence of either CrdR or CrdS disturbs the regulation of the *crdAB* operon (Waidner *et al.*, 2005), and other candidates for copper-regulatory systems are not apparent in the *H. pylori* genome sequences available to date. The CrdRS system plays an important role in gastric colonization in a mouse infection model for *H. pylori* (Panthel *et al.*, 2003), but its direct role in acid resistance and acid-responsive gene regulation of *H. pylori* is still unclear and may reflect differences between strains (Loh and Cover, 2006; Pflock *et al.*, 2007).

Conclusions

Metal metabolism plays an important role in the physiology of *H. pylori*, since despite its small genome, it contains several systems contributing to metal homeostasis. These systems also display significant differences when compared with their enterobacterial counterparts, or even with systems present in other enteric *Helicobacter* species (Dhaenens *et al.*, 1999; Belzer *et al.*, 2005; Belzer *et al.*, 2007b), and this is likely to represent the evolutionary adaptation to the conditions occurring in the gastric mucosa. There are still many questions to be answered about metal metabolism and transport in *H. pylori* and other *Helicobacter* species, and these include but are not limited to elucidation of the role of outer membrane receptors in nickel uptake, the role of metals in acid-adaptation, and the way that *H. pylori* uses overlapping regulons and regulatory cascades in adaptation to changes in the conditions in the gastric mucosa.

Acknowledgements

We thank the present and past members of the Erasmus MC *Helicobacter* research group, and our collaborators for their support and input over the years.

References

Abraham, L.O., Li, Y. and Zamble, D.B. (2006). The metal and DNA-binding activities of *Helicobacter pylori* NikR. J. Inorg Biochem. *100*, 1005–1014.

Alm, R.A., Ling, L.S., Moir, D.T., King, B.L., Brown, E.D., Doig, P.C., Smith, D.R., Noonan, B., Guild, B.C., deJonge, B.L., Carmel, G., Tummino, P.J., Caruso, A., Uria-Nickelsen, M., Mills, D.M., Ives, C., Gibson, R., Merberg, D., Mills, S.D., Jiang, Q., Taylor, D.E., Vovis, G.F. and Trust, T.J. (1999). Genomic-sequence comparison of two unrelated isolates of the human gastric pathogen *Helicobacter pylori*. Nature *397*, 176–180.

Andrews, S.C., Robinson, A.K. and Rodriguez-Quinones, F. (2003). Bacterial iron homeostasis. FEMS Microbiol. Rev. *27*, 215–237.

Barnard, F.M., Loughlin, M.F., Fainberg, H.P., Messenger, M.P., Ussery, D.W., Williams, P. and Jenks, P.J. (2004). Global regulation of virulence and the stress response by CsrA in the highly adapted human gastric pathogen *Helicobacter pylori*. Mol. Microbiol. *51*, 15–32.

Bauerfeind, P., Garner, R.M. and Mobley, H.L.T. (1996). Allelic exchange mutagenesis of *nixA* in *Helicobacter pylori* results in reduced nickel transport and urease activity. Infect. Immun. *64*, 2877–2880.

Bayle, D., Wangler, S., Weitzenegger, T., Steinhilber, W., Volz, J., Przybylski, M., Schafer, K.P., Sachs, G. and Melchers, K. (1998). Properties of the P-type ATPases encoded by the *copAP* operons of *Helicobacter pylori* and *Helicobacter felis*. J. Bacteriol. *180*, 317–329.

Beier, D., Spohn, G., Rappuoli, R. and Scarlato, V. (1997). Identification and characterization of an operon of *Helicobacter pylori* that is involved in motility and stress adaptation. J. Bacteriol. *179*, 4676–4683.

Belzer, C., Stoof, J., Beckwith, C.S., Kuipers, E.J., Kusters, J.G. and van Vliet, A.H.M. (2005). Differential regulation of urease activity in *Helicobacter hepaticus* and *Helicobacter pylori*. Microbiology *151*, 3989–3995.

Belzer, C., Stoof, J. and van Vliet, A.H.M. (2007a). Metal-responsive gene regulation and metal transport in *Helicobacter* species. Biometals *20*, 417–429.

Belzer, C., van Schendel, B.A., Kuipers, E.J., Kusters, J.G. and van Vliet, A.H. (2007b). Iron-responsive repression of urease expression in *Helicobacter hepaticus* is mediated by the transcriptional regulator Fur. Infect. Immun. *75*, 745–752.

Benanti, E.L. and Chivers, P.T. (2007). The N-terminal arm of the *Helicobacter pylori* Ni^{2+}-dependent transcription factor NikR is required for specific DNA-binding. J. Biol. Chem *282*, 20365–20375.

Benoit, S.L., Mehta, N., Weinberg, M.V., Maier, C. and Maier, R.J. (2007). Interaction between the *Helicobacter pylori* accessory proteins HypA and UreE is needed for urease maturation. Microbiology *153*, 1474–1482.

Bereswill, S., Lichte, F., Vey, T., Fassbinder, F. and Kist, M. (1998a). Cloning and characterization of the *fur* gene from *Helicobacter pylori*. FEMS Microbiol. Lett. *159*, 193–200.

Bereswill, S., Waidner, U., Odenbreit, S., Lichte, F., Fassbinder, F., Bode, G. and Kist, M. (1998b). Structural, functional and mutational analysis of the *pfr* gene encoding a ferritin from *Helicobacter pylori*. Microbiology *144*, 2505–2516.

Bereswill, S., Greiner, U., van Vliet, A.H.M., Waidner, B., Fassbinder, F., Schiltz, E., Kusters, J.G. and Kist, M. (2000). Regulation of ferritin-mediated cytoplasmic iron storage by the Ferric Uptake Regulator homolog (Fur) of *Helicobacter pylori*. J. Bacteriol. *182*, 5948–5953.

Bijlsma, J.J. and Groisman, E.A. (2003). Making informed decisions, regulatory interactions between two-component systems. Trends Microbiol. *11*, 359–366.

Bijlsma, J.J.E., Waidner, B., van Vliet, A.H.M., Hughes, N.J., Hag, S., Bereswill, S., Kelly, D.J., Vandenbroucke-Grauls, C.M.J.E., Kist, M. and Kusters, J.G. (2002). The Ferric Uptake Regulator (Fur) homologue of *Helicobacter pylori* is involved in acid resistance. Infect. Immun. *70*, 606–611.

Bury-Moné, S., Thiberge, J.M., Contreras, M., Maitournam, A., Labigne, A. and De Reuse, H. (2004). Responsiveness to acidity via metal ion regulators mediates virulence in the gastric pathogen *Helicobacter pylori*. Mol. Microbiol. *53*, 623–638.

Carson, S.D., Klebba, P.E., Newton, S.M. and Sparling, P.F. (1999). Ferric enterobactin binding and utilization by *Neisseria gonorrhoeae*. J. Bacteriol. *181*, 2895–2901.

Chivers, P.T. and Sauer, R.T. (2000). Regulation of high-affinity nickel uptake in bacteria, Ni^{2+}- dependent interaction of NikR with wild-type and mutant operator sites. J. Biol. Chem. *275*, 19735–19741.

Christensen, J.M., Kristiansen, J., Nielsen, N.H., Menne, T. and Byrialsen, K. (1999). Nickel concentrations in serum and urine of patients with nickel eczema. Toxicol. Lett. *108*, 185–189.

Contreras, M., Thiberge, J.M., Mandrand-Berthelot, M.A. and Labigne, A. (2003). Characterization of the roles of NikR, a nickel-responsive pleiotropic autoregulator of *Helicobacter pylori*. Mol. Microbiol. *49*, 947–963.

Davis, G.S. and Mobley, H.L. (2005). Contribution of *dppA* to urease activity in *Helicobacter pylori* 26695. *Helicobacter* *10*, 416–423.

Davis, G.S., Flannery, E.L. and Mobley, H.L. (2006). *Helicobacter pylori* HP1512 is a nickel-responsive NikR-regulated outer membrane protein. Infect. Immun. *74*, 6811–6820.

De Pina, K., Desjardin, V., Mandrand-Berthelot, M.A., Giordano, G. and Wu, L.F. (1999). Isolation and characterization of the *nikR* gene encoding a nickel-responsive regulator in *Escherichia coli*. J. Bacteriol. *181*, 670–674.

Delany, I., Pacheco, A.B.F., Spohn, G., Rappuoli, R. and Scarlato, V. (2001a). Iron-dependent transcription of the *frpB* gene of *Helicobacter pylori* is controlled by the Fur protein. J. Bacteriol. *183*, 4932–4937.

Delany, I., Spohn, G., Rappuoli, R. and Scarlato, V. (2001b). The Fur repressor controls transcription of iron-activated and -repressed genes in *Helicobacter pylori*. Mol. Microbiol. *42*, 1297–1309.

Delany, I., Spohn, G., Pacheco, A.B., Ieva, R., Alaimo, C., Rappuoli, R. and Scarlato, V. (2002). Autoregulation of *Helicobacter pylori* Fur revealed by functional analysis of the iron-binding site. Mol. Microbiol. *46*, 1107–1122.

Delany, I., Ieva, R., Soragni, A., Hilleringmann, M., Rappuoli, R. and Scarlato, V. (2005). *In vitro* analysis of protein-operator interactions of the NikR and Fur metal-responsive regulators of coregulated genes in *Helicobacter pylori*. J. Bacteriol. *187*, 7703–7715.

Dhaenens, L., Szczebara, F., Van Nieuwenhuyse, S. and Husson, M.O. (1999). Comparison of iron uptake in different *Helicobacter* species. Res. Microbiol. *150*, 475–481.

Dian, C., Schauer, K., Kapp, U., McSweeney, S.M., Labigne, A. and Terradot, L. (2006). Structural basis of the nickel response in *Helicobacter pylori*, crystal structures of *Hp*NikR in apo and nickel-bound states. J. Mol. Biol. *361*, 715–730.

Ernst, F.D., Bereswill, S., Waidner, B., Stoof, J., Mader, U., Kusters, J.G., Kuipers, E.J., Kist, M., van Vliet, A.H.M. and Homuth, G. (2005a). Transcriptional profiling of *Helicobacter pylori* Fur- and iron-regulated gene expression. Microbiology *151*, 533–546.

Ernst, F.D., Homuth, G., Stoof, J., Mader, U., Waidner, B., Kuipers, E.J., Kist, M., Kusters, J.G., Bereswill, S. and van Vliet, A.H.M. (2005b). Iron-responsive regulation of the *Helicobacter pylori* iron-cofactored superoxide dismutase SodB is mediated by Fur. J. Bacteriol. *187*, 3687–3692.

Ernst, F.D., Kuipers, E.J., Heijens, A., Sarwari, R., Stoof, J., Penn, C.W., Kusters, J.G. and van Vliet, A.H.M. (2005c). The nickel-responsive regulator NikR controls activation and repression of gene transcription in *Helicobacter pylori*. Infect. Immun. *73*, 7252–7258.

Ernst, F.D., Stoof, J., Horrevoets, W.M., Kuipers, E.J., Kusters, J.G. and van Vliet, A.H.M. (2006). NikR mediates nickel-responsive transcriptional repression of the *Helicobacter pylori* outer membrane proteins FecA3 (HP1400) and FrpB4 (HP1512). Infect. Immun. *74*, 6821–6828.

Escolar, L., Perez-Martin, J. and de Lorenzo, V. (1999). Opening the iron-box, transcriptional metalloregulation by the Fur protein. J. Bacteriol. *181*, 6223–6229.

Gancz, H., Censini, S. and Merrell, D.S. (2006). Iron and pH homeostasis intersect at the level of Fur regulation in the gastric pathogen *Helicobacter pylori*. Infect. Immun. *74*, 602–614.

Ge, R., Watt, R.M., Sun, X., Tanner, J.A., He, Q.Y., Huang, J.D. and Sun, H. (2006). Expression and characterization of a histidine-rich protein, Hpn, potential for Ni^{2+} storage in *Helicobacter pylori*. Biochem. J. *393*, 285–293.

Gilbert, J.V., Ramakrishna, J., Sunderman, F.W., Jr., Wright, A. and Plaut, A.G. (1995). Protein Hpn, cloning and characterization of a histidine-rich metal-binding polypeptide in *Helicobacter pylori* and *Helicobacter mustelae*. Infect. Immun. *63*, 2682–2688.

Hall, H.K. and Foster, J.W. (1996). The role of *fur* in the acid tolerance response of *Salmonella typhimurium* is physiologically and genetically separable from its role in iron acquisition. J. Bacteriol. *178*, 5683–5691.

Hantke, K. (2001). Iron and metal regulation in bacteria. Curr. Opin. Microbiol. *4*, 172–177.

Hendricks, J.K. and Mobley, H.L. (1997). *Helicobacter pylori* ABC transporter, effect of allelic exchange mutagenesis on urease activity. J. Bacteriol. *179*, 5892–5902.

Herrmann, L., Schwan, D., Garner, R., Mobley, H.L., Haas, R., Schafer, K.P. and Melchers, K. (1999). *Helicobacter pylori cadA* encodes an essential Cd(II)-Zn(II)-Co(II) resistance factor influencing urease activity. Mol. Microbiol. *33*, 524–536.

Hu, L.T., Foxall, P.A., Russell, R. and Mobley, H.L. (1992). Purification of recombinant *Helicobacter pylori* urease apoenzyme encoded by *ureA* and *ureB*. Infect. Immun. *60*, 2657–2666.

Husson, M.O., Legrand, D., Spik, G. and Leclerc, H. (1993). Iron acquisition by *Helicobacter pylori*, importance of human lactoferrin. Infect. Immun. *61*, 2694–2697.

Inoue, H., Sakurai, T., Ujike, S., Tsuchiya, T., Murakami, H. and Kanazawa, H. (1999). Expression of functional Na^+/H^+ antiporters of *Helicobacter pylori* in antiporter-deficient *Escherichia coli* mutants. FEBS Lett. *443*, 11–16.

Inoue, H., Tsuboi, Y. and Kanazawa, H. (2001). Chimeric Na(+)/H(+) antiporters constructed from NhaA of *Helicobacter pylori* and *Escherichia coli*, implications for domains of NhaA for pH sensing. J. Biochem. (Tokyo) *129*, 569–576.

Kansau, I., Guillain, F., Thiberge, J.M. and Labigne, A. (1996). Nickel binding and immunological properties of the C-terminal domain of the *Helicobacter pylori* GroES homologue (HspA). Mol. Microbiol. *22*, 1013–1023.

Kusters, J.G., van Vliet, A.H.M. and Kuipers, E.J. (2006). Pathogenesis of *Helicobacter pylori* infection. Clin. Microbiol. Rev. *19*, 449–490.

Lee, H.W., Choe, Y.H., Kim, D.K., Jung, S.Y. and Lee, N.G. (2004). Proteomic analysis of a ferric uptake regulator mutant of *Helicobacter pylori*, regulation of *Helicobacter pylori* gene expression by ferric uptake regulator and iron. Proteomics *4*, 2014–2027.

Loh, J.T. and Cover, T.L. (2006). Requirement of histidine kinases HP0165 and HP1364 for acid resistance in *Helicobacter pylori*. Infect. Immun. *74*, 3052–3059.

Maier, R.J., Benoit, S.L. and Seshadri, S. (2007). Nickel-binding and accessory proteins facilitating Ni-enzyme maturation in *Helicobacter pylori*. Biometals *20*, 655–664.

Marcus, E.A., Moshfegh, A.P., Sachs, G. and Scott, D.R. (2005). The periplasmic alpha-carbonic anhydrase activity of *Helicobacter pylori* is essential for acid acclimation. J. Bacteriol. *187*, 729–738.

Masse, E. and Gottesman, S. (2002). A small RNA regulates the expression of genes involved in iron metabolism in *Escherichia coli*. Proc. Natl. Acad. Sci. USA *99*, 4620–4625.

Melchers, K., Weitzenegger, T., Buhmann, A., Steinhilber, W., Sachs, G. and Schafer, K.P. (1996). Cloning and membrane topology of a P type ATPase from *Helicobacter pylori*. J. Biol. Chem. *271*, 446–457.

Melchers, K., Herrmann, L., Mauch, F., Bayle, D., Heuermann, D., Weitzenegger, T., Schuhmacher, A., Sachs, G., Haas, R., Bode, G., Bensch, K. and Schafer, K.P. (1998). Properties and function of the P type ion pumps cloned from *Helicobacter pylori*. Acta Physiol. Scand. Suppl. *643*, 123–135.

Mobley, H.L., Garner, R.M. and Bauerfeind, P. (1995). *Helicobacter pylori* nickel-transport gene *nixA*, synthesis of catalytically active urease in *Escherichia coli* independent of growth conditions. Mol. Microbiol. *16*, 97–109.

Mobley, H.L., Garner, R.M., Chippendale, G.R., Gilbert, J.V., Kane, A.V. and Plaut, A.G. (1999). Role of Hpn and NixA of *Helicobacter pylori* in susceptibility and resistance to bismuth and other metal ions. Helicobacter *4*, 162–169.

Nolan, K.J., McGee, D.J., Mitchell, H.M., Kolesnikow, T., Harro, J.M., O'Rourke, J., Wilson, J.E., Danon, S.J., Moss, N.D., Mobley, H.L. and Lee, A. (2002). In vivo behavior of a *Helicobacter pylori* SS1 *nixA* mutant with reduced urease activity. Infect. Immun. *70*, 685–691.

Olson, J.W., Mehta, N.S. and Maier, R.J. (2001). Requirement of nickel metabolism proteins HypA and HypB for full activity of both hydrogenase and urease in *Helicobacter pylori*. Mol. Microbiol. *39*, 176–182.

Olson, J.W. and Maier, R.J. (2002). Molecular hydrogen as an energy source for *Helicobacter pylori*. Science *298*, 1788–1790.

Panthel, K., Dietz, P., Haas, R. and Beier, D. (2003). Two-component systems of *Helicobacter pylori* contribute to virulence in a mouse infection model. Infect. Immun. *71*, 5381–5385.

Perez-Perez, G.I., Gower, C.B. and Blaser, M.J. (1994). Effects of cations on *Helicobacter pylori* urease activity, release, and stability. Infect. Immun. *62*, 299–302.

Pfeiffer, J., Guhl, J., Waidner, B., Kist, M. and Bereswill, S. (2002). Magnesium uptake by CorA is essential for viability of the gastric pathogen *Helicobacter pylori*. Infect. Immun. *70*, 3930–3934.

Pflock, M., Kennard, S., Delany, I., Scarlato, V. and Beier, D. (2005). Acid-induced activation of the urease promoters is mediated directly by the ArsRS two-component system of *Helicobacter pylori*. Infect. Immun. *73*, 6437–6445.

Pflock, M., Finsterer, N., Joseph, B., Mollenkopf, H., Meyer, T.F. and Beier, D. (2006a). Characterization of the ArsRS regulon of *Helicobacter pylori*, involved in acid adaptation. J. Bacteriol. *188*, 3449–3462.

Pflock, M., Kennard, S., Finsterer, N. and Beier, D. (2006b). Acid-responsive gene regulation in the human pathogen *Helicobacter pylori*. J. Biotechnol. *126*, 52–60.

Pflock, M., Muller, S. and Beier, D. (2007). The CrdRS (HP1365-HP1364) two-component system is not involved in pH-responsive gene regulation in the *Helicobacter pylori* strains 26695 and G27. Curr. Microbiol. *54*, 320–324.

Ratledge, C. and Dover, L.G. (2000). Iron metabolism in pathogenic bacteria. Annu. Rev. Microbiol. *54*, 881–941.

Sachs, G., Weeks, D.L., Melchers, K. and Scott, D.R. (2003). The gastric biology of *Helicobacter pylori*. Annu. Rev. Physiol. *65*, 349–369.

Schar, J., Sickmann, A. and Beier, D. (2005). Phosphorylation-independent activity of atypical response regulators of *Helicobacter pylori*. J. Bacteriol. *187*, 3100–3109.

Schauer, K., Gouget, B., Carriere, M., Labigne, A. and de Reuse, H. (2007). Novel nickel transport mechanism across the bacterial outer membrane energized by the TonB/ExbB/ExbD machinery. Mol. Microbiol. *63*, 1054–1068.

Schreiber, S., Konradt, M., Groll, C., Scheid, P., Hanauer, G., Werling, H.O., Josenhans, C. and Suerbaum, S.

(2004). The spatial orientation of *Helicobacter pylori* in the gastric mucus. Proc. Natl. Acad. Sci. USA. *101*, 5024–5029.

Schreiber, S., Bucker, R., Groll, C., Azevedo-Vethacke, M., Garten, D., Scheid, P., Friedrich, S., Gatermann, S., Josenhans, C. and Suerbaum, S. (2005). Rapid loss of motility of *Helicobacter pylori* in the gastric lumen in vivo. Infect. Immun. *73*, 1584–1589.

Scott, D.R., Marcus, E.A., Wen, Y., Oh, J. and Sachs, G. (2007). Gene expression in vivo shows that *Helicobacter pylori* colonizes an acidic niche on the gastric surface. Proc. Natl. Acad. Sci. USA. *104*, 7235–7240.

Seshadri, S., Benoit, S.L. and Maier, R.J. (2007). Roles of His-rich Hpn and Hpn-like proteins in *Helicobacter pylori* Nickel Physiology. J. Bacteriol. *189*, 4120–4126.

Smith, R.L. and Maguire, M.E. (1998). Microbial magnesium transport, unusual transporters searching for identity. Mol. Microbiol. *28*, 217–226.

Stähler, F.N., Odenbreit, S., Haas, R., Wilrich, J., van Vliet, A.H.M., Kusters, J.G., Kist, M. and Bereswill, S. (2006). The novel *Helicobacter pylori* CznABC metal efflux pump is required for cadmium, zinc and nickel resistance, for urease modulation and for gastric colonization. Infect. Immun. *74*, 3845–3852.

Stingl, K., Altendorf, K. and Bakker, E.P. (2002). Acid survival of *Helicobacter pylori*, how does urease activity trigger cytoplasmic pH homeostasis? Trends Microbiol. *10*, 70–74.

Stingl, K. and De Reuse, H. (2005). Staying alive overdosed, how does *Helicobacter pylori* control urease activity? Int. J. Med. Microbiol. *295*, 307–315.

Stingl, K., Brandt, S., Uhlemann, E.M., Schmid, R., Altendorf, K., Zeilinger, C., Ecobichon, C., Labigne, A., Bakker, E.P. and de Reuse, H. (2007). Channel-mediated potassium uptake in *Helicobacter pylori* is essential for gastric colonization. EMBO J. *26*, 232–241.

Sunderman, F.W. (1993). Biological monitoring of nickel in humans. Scand. J. Work Environm. Health *19*, 34–38.

Tomb, J.F., White, O., Kerlavage, A.R., Clayton, R.A., Sutton, G.G., Fleischmann, R.D., Ketchum, K.A., Klenk, H.P., Gill, S., Dougherty, B.A., Nelson, K., Quackenbush, J., Zhou, L., Kirkness, E.F., Peterson, S., Loftus, B., Richardson, D., Dodson, R., Khalak, H.G., Glodek, A., McKenney, K., Fitzegerald, L.M., Lee, N., Adams, M.D., Hickey, E.K., Berg, D.E., Gocayne, J.D., Utterback, T.R., Peterson, J.D., Kelley, J.M., Cotton, M.D., Weidman, J.M., Fujii, C., Bowman, C., Watthey, L., Wallin, E., Hayes, W.S., Borodovsky, M., Karpk, P.D., Smith, H.O., Fraser, C.M. and Venter, J.C. (1997). The complete genome sequence of the gastric pathogen *Helicobacter pylori*. Nature 388, 539–547.

Tonello, F., Dundon, W.G., Satin, B., Molinari, M., Tognon, G., Grandi, G., del Guicide, G., Rappuoli, R. and Montecucco, C. (1999). The *Helicobacter pylori* neutrophil-activating protein is an iron-binding protein with dodecameric structure. Mol. Microbiol. *34*, 238–246.

Tsuda, M., Karita, M., Morshed, M.G., Okita, K. and Nakazawa, T. (1994). A urease-negative mutant of *Helicobacter pylori* constructed by allelic exchange mutagenesis lacks the ability to colonize the nude mouse stomach. Infect. Immun. *62*, 3586–3589.

van Amsterdam, K., Bart, A. and van der Ende, A. (2005). A *Helicobacter pylori* TolC efflux pump confers resistance to metronidazole. Antimicrob. Agents Chemother. *49*, 1477–1482.

van Vliet, A.H.M., Bereswill, S. and Kusters, J.G. (2001a). Ion Metabolism and Transport. In, *Helicobacter pylori*, Physiology and Genetics. H.L.T. Mobley, G.L. Mendz, and S.L. Hazell, ed. ASM Press, Washington, D.C. p. 193–206.

van Vliet, A.H.M., Kuipers, E.J., Waidner, B., Davies, B.J., de Vries, N., Penn, C.W., Vandenbroucke-Grauls, C.M.J.E., Kist, M., Bereswill, S. and Kusters, J.G. (2001b). Nickel-responsive induction of urease expression in *Helicobacter pylori* is mediated at the transcriptional level. Infect. Immun. *69*, 4891–4897.

van Vliet, A.H.M., Poppelaars, S.W., Davies, B.J., Stoof, J., Bereswill, S., Kist, M., Kuipers, E.J., Penn, C.W. and Kusters, J.G. (2002a). NikR mediates nickel-responsive transcriptional induction of urease expression in *Helicobacter pylori*. Infect. Immun. *70*, 2846–2852.

van Vliet, A.H.M., Stoof, J., Vlasblom, R., Wainwright, S.A., Hughes, N.J., Kelly, D.J., Bereswill, S., Bijlsma, J.J.E., Hoogenboezem, T., Vandenbroucke-Grauls, C.M.J.E., Kist, M., Kuipers, E.J. and Kusters, J.G. (2002b). The role of the Ferric Uptake Regulator (Fur) in regulation of *Helicobacter pylori* iron uptake. *Helicobacter 7*, 237–244.

van Vliet, A.H.M., Stoof, J., Poppelaars, S.W., Bereswill, S., Homuth, G., Kist, M., Kuipers, E.J. and Kusters, J.G. (2003). Differential regulation of amidase- and formamidase-mediated ammonia production by the *Helicobacter pylori* Fur repressor. J. Biol. Chem. *278*, 9052–9057.

van Vliet, A.H.M., Ernst, F.D. and Kusters, J.G. (2004a). NikR-mediated regulation of *Helicobacter pylori* acid adaptation. Trends Microbiol. *12*, 489–494.

van Vliet, A.H.M., Kuipers, E.J., Stoof, J., Poppelaars, S.W. and Kusters, J.G. (2004b). Acid-responsive gene induction of ammonia-producing enzymes in *Helicobacter pylori* is mediated via a metal-responsive repressor cascade. Infect. Immun. *72*, 766–773.

Velayudhan, J., Hughes, N.J., McColm, A.A., Bagshaw, J., Clayton, C.L., Andrews, S.C. and Kelly, D.J. (2000). Iron acquisition and virulence in *Helicobacter pylori*, a major role for FeoB, a high-affinity ferrous iron transporter. Mol. Microbiol. *37*, 274–286.

Waidner, B., Greiner, S., Odenbreit, S., Kavermann, H., Velayudhan, J., Stähler, F., Bisse, E., van Vliet, A.H.M., Andrews, S.C., Kusters, J.G., Kelly, D.J., Haas, R., Kist, M. and Bereswill, S. (2002a). Essential role of ferritin Pfr in *Helicobacter pylori* iron metabolism and gastric colonization. Infect. Immun. *70*, 3923–3929.

Waidner, B., Melchers, K., Ivanov, I., Loferer, H., Bensch, K.W., Kist, M. and Bereswill, S. (2002b). Identification by RNA profiling and mutational analysis of the novel copper resistance determinants CrdA

(HP1326), CrdB (HP1327), and CzcB (HP1328) in *Helicobacter pylori*. J. Bacteriol. *184*, 6700–6708.

Waidner, B., Melchers, K., Stahler, F.N., Kist, M. and Bereswill, S. (2005). The *Helicobacter pylori* CrdRS two-component regulation system (HP1364/HP1365) is required for copper-mediated induction of the copper resistance determinant CrdA. J. Bacteriol. *187*, 4683–4688.

Wang, G., Alamuri, P. and Maier, R.J. (2006a). The diverse antioxidant systems of *Helicobacter pylori*. Mol. Microbiol. *61*, 847–860.

Wang, G., Hong, Y., Olczak, A., Maier, S.E. and Maier, R.J. (2006b). Dual roles of *Helicobacter pylori* NapA in inducing and combating oxidative stress. Infect. Immun. *74*, 6839–6846.

Wen, Y., Marcus, E.A., Matrubutham, U., Gleeson, M.A., Scott, D.R. and Sachs, G. (2003). Acid-adaptive genes of *Helicobacter pylori*. Infect. Immun. *71*, 5921–5939.

Wen, Y., Feng, J., Scott, D.R., Marcus, E.A. and Sachs, G. (2006). Involvement of the HP0165-HP0166 two-component system in expression of some acidic-pH-upregulated genes of *Helicobacter pylori*. J. Bacteriol. *188*, 1750–1761.

Wen, Y., Feng, J., Scott, D.R., Marcus, E.A. and Sachs, G. (2007). The HP0165-HP0166 two-component system (ArsRS) regulates acid-induced expression of HP1186 alpha-carbonic anhydrase in *Helicobacter pylori* by activating the pH-dependent promoter. J. Bacteriol. *189*, 2426–2434.

Whitmire, J.M., Gancz, H. and Merrell, D.S. (2007). Balancing the double-edged sword, metal ion homeostasis and the ulcer bug. Curr. Med. Chem. *14*, 469–478.

Wooldridge, K.G. and Williams, P.H. (1993). Iron uptake mechanisms of pathogenic bacteria. FEMS Microbiol. Rev. *12*, 325–348.

Yaman, M., Kaya, G. and Yekeler, H. (2007). Distribution of trace metal concentrations in paired cancerous and non-cancerous human stomach tissues. World J. Gastroenterol. *13*, 612–618.

Replication, Partitioning, Segregation and Cell Division in *Helicobacter pylori*

9

Teruko Nakazawa and Hiroaki Takeuchi

Abstract

Chromosome replication and cell division are essential steps for bacterial proliferation. Recent studies in *Escherichia coli* and some other bacteria have revealed important insights into the whole process. The complete genome sequences of three strains of *Helicobacter pylori* indicated that homologues of most genes involved in replication and segregation are present, suggesting that the basic mechanisms are common. However, some orthologues for cell division are missing in *H. pylori* possibly because of its unique niche in gastric mucus as well as the lifestyle. Analysis of the replication machinery of *H. pylori* has started very recently. The replication origin *oriC*, the initiator protein DnaA, and the replication helicase DnaB was identified. In addition, the ParABS system for chromosome partitioning was identified. FtsK-*dif* system for chromosome segregation and FtsZ-ring system for cell division might be functioning in *H. pylori* like in many other bacteria although further studies are required to elucidate the process.

Introduction

Helicobacter pylori, a spiral microaerophilic bacterium, is associated with chronic gastritis, a process that increases risk for developing peptic ulcer, non-cardial adenocarcinoma, and gastric mucosa-associated lymphoid tissue lymphoma. The infection occurs early in childhood when the gastric acidity, the major antimicrobial defence of the stomach, is immature. Upon ingestion, *H. pylori* migrates through gastric mucous layers and attaches to gastric epithelial cell surface with adhesins to access nutrients from the host. Urease activity and motility with flagella are important factors for survival in the stomach. A virulent type bacterium injects CagA into the host epithelial cell for setting a proliferation platform. A vacuolating cytotoxin, VacA, may also contribute for colonization, although its precise role is still obscure. When the energy supply for bacterial biosynthetic activities reaches to a steady state, the bacterium is ready to initiate replication. Under unfavourable conditions such as oxygen depletion or presence of antimicrobial agents, *H. pylori* changes the shape from a spiral to a coccoid form, a dormant or resting state, that appears to be excreted into faeces. When ingested, the metabolic activities of the coccoid form may be regained by using stored polyphosphate as an urgent energy.

Bacterial proliferation starts by chromosome replication followed by division of a parental cell. DNA replication and cell division are tightly coupled, so that the entire genetic material is copied exactly once before each cell division and segregate exactly into two daughter cells.

The genome of *H. pylori*, an ε-Proteobacterium consists of a single circular chromosome, and the complete genome sequences of three strains have so far been reported; 26695 obtained from a patient with superficial gastritis (Tomb *et al.*, 1997), J99 from a patient with duodenal ulcer disease (Alm *et al.*, 1999), and HPAG1 from a patient with chronic atrophic gastritis (Oh *et al.*, 2006). The size, annotated numbers of ORF, and numbers of strain-specific ORF are summarized in Table 9.1. Each genome

Table 9.1 Comparison of the sequenced *H. pylori* 26695, J99 and HPAG1 genomes

Genome features	*H. pylori* strain		
	26695	J99	HPAG1
Size (base pairs)	1,667,867	1,643,831	1,596,366
G+C content (%)	39	39	39
ORF[a]	1,590	1,496	1,536
Strain-specific ORF[b]	132 (90)	58 (37)	43 (28)

[a]These numbers are obtained from the original annotations of the 26695 (Tomb *et al.*, 1997), J99 (Alm *et al.*, 1999), and HPAG1 (Oh *et al.*, 2006) genomes.
[b]Determined by Oh *et al.* (2006). Numbers of genes for hypothetical proteins are presented in parentheses.

has 3–8% strain-specific genes, and most of these genes do not have assigned functions and are located in the plasticity zones (Oh *et al.*, 2006).

Replication

Detailed analysis of the replication process in bacteria is largely focused on *E. coli*, a representative γ-Proteobacterium, whereas studies on *H. pylori* started only recently. Homologues of the genes involved in replication of *E. coli* are present in the *H. pylori* genomes (Table 9.2), suggesting that the basic mechanism of replication is similar. However, orthologues of certain *E. coli* genes are apparently missing from the *H. pylori* genome, suggesting some differences in the proliferation machinery. Such differences may be attributed to smaller genome size of *H. pylori* and remote subdivisions of the Proteobacteria. The differences may also be associated with the lifestyle, i.e. *H. pylori* is a mucus-associated oligotroph proliferating at a fixed temperature with slow doubling time, whereas *E. coli* is a free-living prototroph proliferating under various conditions at variable temperatures with faster doubling time.

Origin of replication

In bacteria, replication of circular chromosome starts from a replication origin *oriC* and proceeds bi-directionally until the replication forks reach the termination site. Eubacterial *oriC* contains DnaA boxes which are the binding site for the initiator protein DnaA for replication. By using a skewed oligomer method a putative *oriC* of *H. pylori* strain 26695 was reported to be in between 1,566,000 and 1,640,000 (Salzberg *et al.*, 1998).

More recently, the *oriC* region of strain J99 was determined by detrended [G-C] DNA walks on the whole genome (Zawilak *et al.*, 2001). A switch of G/C DNA asymmetry was located at position 1,556,599 that is only 1.2 kb away from the 3′ end of *dnaA*. A similar switch was found on the 26695 genome at position 1 606 201 that is also very close to *dnaA*, 1.4 kb away from its 3′ end.

By searching the neighbouring sequences, Zawilak *et al.* (2001) found typical DnaA boxes in the promoter region of *dnaA*, and the consensus sequence for DnaA box, 5′-T(T/C) ATTCACA-3′, was deduced (Fig. 9.1). In contrast to *E. coli* DnaA boxes that are separated by considerably long spaces, five DnaA boxes of *H. pylori* are arranged in a compact space, only 150 bp in between the 5′ ends of *dnaA* and *punB*. AT-rich sequences for putative unwinding site for initiation are also found in this region (Fig. 9.1).

The biological activity of the DnaA boxes was shown by specific binding of the purified C-terminal DNA binding domain of *H. pylori* DnaA by using DNase I footprinting and gel retardation analysis (Zawilak *et al.*, 2001). Among the DnaA boxes, two pairs of closely spaced DnaA boxes, box 2–3 and box 4–5 have the highest affinity towards the *H. pylori* DnaA protein (Zawilak-Pawlik *et al.*, 2005).

The region of *dnaA* and its surrounding sequences of the three genomes are well conserved among the three *H. pylori* genomes except that the region of strainHPAG1 is inverted compared with the 26695 and J99 genomes (Fig. 9.2) (Oh *et al.*, 2006). The DnaA boxes of the three strains are located at the same position with the same orientation. It can be concluded that the region depicted in Fig. 9.2 is *oriC* of *H. pylori*. A circular

Table 9.2 Proteins involved in chromosome replication of *E. coli*[a] and their homologues recognized in *H. pylori*

Function	Protein (gene)	ORF in *H. pylori* strain[b]		
		26695	J99	HPAG1
Initiation				
Initiator	DnaA (*dnaA*)	HP1529	Jhp1417	HPAG1_1387
Replicative helicase	DnaB (*dnaB*)	HP1362	jhp1280	HPAG1_1309
	DnaC (*dnaC*)	□[c]	□	□
RNA primase	DnaG (*dnaG*)	HP0012	jhp0010	HPAG1_0012
Single-strand DNA binding protein	SSB	HP1245	jhp1166	HPAG1_1188
Repressor	SeqA (*seqA*)	□	□	□
DNA synthesis				
DNA polymerase III				
Sliding clamp				
β subunit	DnaN (*dnaN*)	HP0500	jhp0452	HPAG1_0476
Clamp loader				
γ/τ subunit	DnaX (*dnaX*)	HP0717	jhp0655	HPAG1_0702
δ-subunit	HolA (*holA*)	□	□	HPAG1_1190
δ′ subunit	HolB (*holB*)	HP1231	jhp1152	HPAG1_1173
Core				
α subunit	DnaE (*dnaE*)	HP1460	jhp1353	HPAG1_1455
ε subunit	DnaQ (*dnaQ*)	HP1387	jhp1438	HPAG1_1481
DNA ligation				
DNA ligase	Lig (*lig*)	HP0615	jhp0558	HPAG1_0596

[a]Functions and proteins in *E. coli* are adopted from a review (O'Donnell, 2006).
[b]ORFs are annotate in strain 26695 (Tomb *et al.*, 1997), J99 (Alm *et al.*, 1999) and HPAG1 (Oh *et al.*, 2006).
[c]Homologues are not assigned.

A

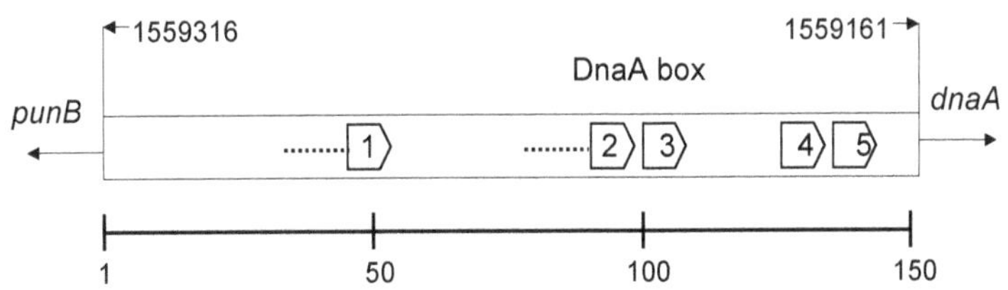

B

H. pylori T T/C A T T C A C A

E. coli T T A T N C A C A

Figure 9.1 A. Location of five DnaA boxes in between *dnaA* and *punB* (purine nucleoside phosphorylase) in the *H. pylori* J99 genome (Zawilak *et al.*, 2001). Ten-bp AT-rich sequences are represented by dashed lines. (B) The consensus sequence for DnaA boxes of *H. pylori* J99 and *E. coli*.

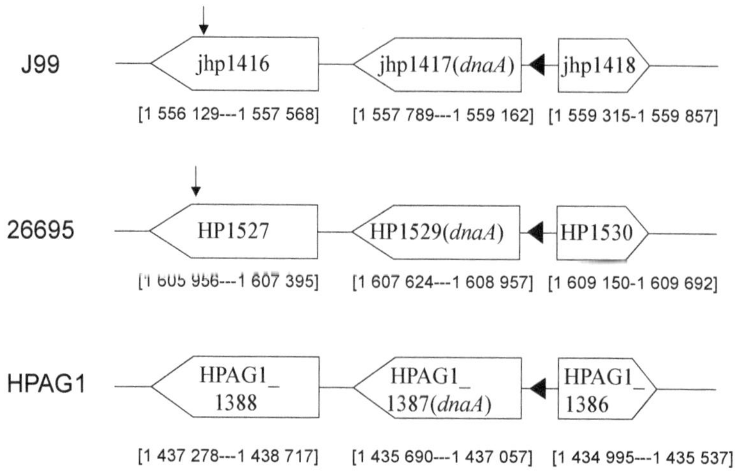

Figure 9.2 The *oriC* region in the genome of *H. pylori* J99, 26695, and HPAG1. The region of strain HPAG1 is inverted compared with those of 26695 and J99. Arrowhead represents location and orientation of DnaA boxes, vertical arrows represent the position of a switch of G/C DNA asymmetry located at 1,556,599 and 1,606,201 in the genome of J99 and 26695, respectively (Zawilak *et al.*, 2001). The location of each ORF in the genome is indicated in parentheses. The lengths of the space between *dnaA* and *punB* in J99, 26695, and HPAG1 are 150 bp, 193 bp, and 153 bp, respectively. The genes jhp1418, HP1530, and HPAG_1386 are for *punB* and jhp1416, HP1527, and HPAG1_1388 are for conserved hypothetical proteins.

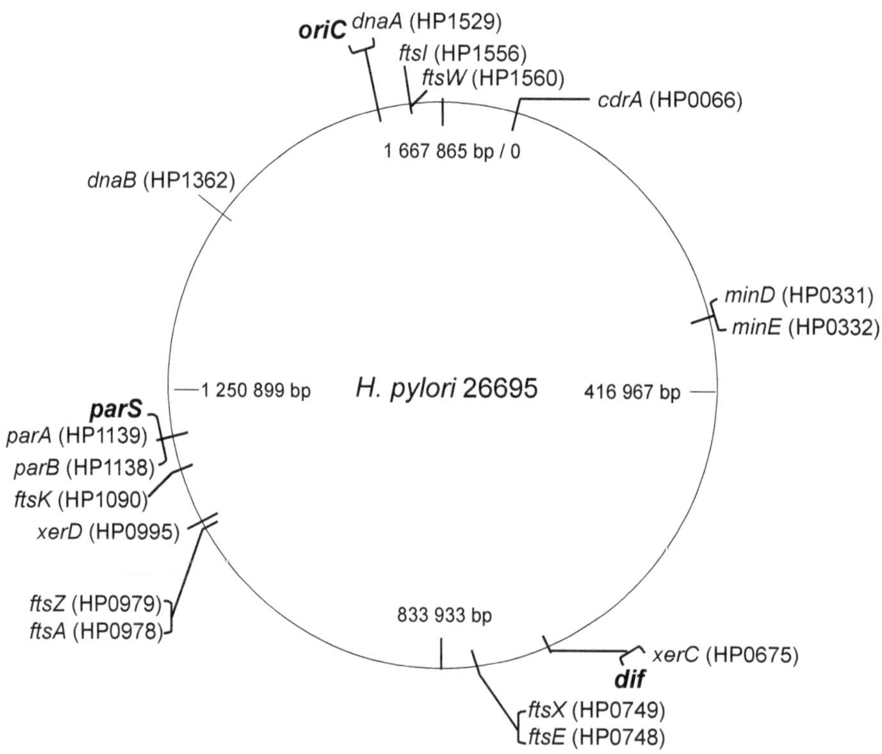

Figure 9.3 Circular map of *H. pylori* 26695 representing *oriC*, *parS,* and *dif* sites involved in replication, partition and segregation, respectively, together with the associated genes. The *dif* site is tentatively assigned next to *xerC* recombinase (Hendrickson and Lowrence, 2006). Genes possibly involved in the cell division are also indicated. See details in the text.

map of the whole genome of *H. pylori* 26695 representing the *oriC* region is shown (Fig. 9.3).

Initiator DnaA

The homology amongst the initiator protein DnaA from various bacteria are striking, although there is no similarity between bacterial DnaA and Archaebacterial or eukaryotic ORC complex subunits.

A comparison of the amino acid sequences of the DnaA proteins revealed that certain regions of the protein are more highly conserved than others, and on this basis the protein was divided into four domains. Various mutational and biochemical studies with the *E. coli* DnaA have since assigned distinct roles to those domains (Magdalena *et al.*, 2005; Kaguni, 2006). *H. pylori* DnaA appears to have similar domain structures to *E. coli* DnaA. Domain I, a short N-terminal sequence conserved among most DnaA proteins may mediate interactions between DnaA protein molecules, which are necessary for initiation. Domain II may be a site of interaction with replicative helicase DnaB. The highest degree of conservation is found in domain III which contains a Walker motif for ATP binding and hydrolysis. Domain IV is involved in DNA binding and also well conserved. A region has been identified at the starting region of Domain IV that is involved in interactions between the DnaA protein and the cell membrane phospholipid.

The *H. pylori* DnaA protein has been purified from *E. coli* cells harbouring the *dnaA* gene of *H. pylori* using polyclonal antibodies raised against domain IV of *H. pylori* DnaA (Zawilak *et al.*, 2003). The size of the protein is 52 kDa as predicted, and a large part of the protein in crude extract of *H. pylori* was found in the inner membrane fraction, suggesting that an active initiation complex of ATP-DnaA is formed in the inner membrane similar to *E. coli* DnaA. In addition, the amount of DnaA was abundant, estimated to be approximately 3,000 moles per cell. Gel retardation analysis revealed that the purified *H. pylori* DnaA binds to the *oriC* region of *H. pylori* DNA in the presence of ATP resulting to form high molecular nucleoprotein complexes. *H. pylori* DnaA prefers the *H. pylori oriC* fragment with two tandem DnaA boxes (box 2–3 and Box 4–5 in Fig. 9.1) over non-box DNA fragments. The binding affinity of *H. pylori* DnaA to its *oriC* region was estimated to be 4nM that is lower than that of *E. coli* DnaA, 1nM, to the cognate *oriC* region (Zawilak *et al.*, 2003).

DnaB helicase

Binding of ATP-DnaA to the DnaA box in the *oriC* region is followed by unwinding of the adjacent AT-rich sequences in the *oriC* region. Subsequently DnaB, the replicative DNA helicase is recruited at the *oriC* region with the help of DnaA. The *dnaB* gene of three *H. pylori* genomes locates close and approximately 200 kb apart from *oriC* (Fig. 9.3). The *dnaB* gene of strain 26695 was cloned and characterized (Soni *et al.*, 2003). The DnaB sequence is 32% identical and 57% similar to *E. coli* DnaB that was shown to complement a temperature-sensitive mutant of *E. coli dnaB* at the non-permissive temperature, confirming that the cloned *H. pylori* DnaB is the replicative helicase. Interestingly, *H. pylori* does not have an obvious DnaC homologue which is essential in *E. coli* for loading DnaB. Using two different temperature-sensitive mutants of *E. coli dnaC*, it was shown that function of *E. coli* DnaC is complemented solely by *H. pylori* DnaB, suggesting that *H. pylori* DnaB can bypass *E. coli* DnaC activity (Soni *et al.*, 2005).

DnaB forms a hexameric ring that moves in a 5′ to 3′ direction resulting in the separation of double strands of the helix to make available single strands that are able to act as templates for daughter DNA synthesis (O'Donnell, 2006, Herrick and Sclavi, 2007). Separate helicase hexamers seems to bind to each strand, and the two helicase molecules move apart in opposite directions. This creates two replication forks that will also move in opposite directions around the chromosome to replicate the DNA bidirectionally. *E. coli* DnaB interacts with DnaG primase that acts at the initiation of replication and elongation of the lagging strand to prime Okazaki fragment synthesis. *H. pylori* has a DnaG primase homologue (Table 9.2) that has 36 % identity and 56 % similarity to *E. coli* DnaG.

H. pylori also has a homologue of the single-strand DNA binding protein of *E. coli* that may stabilize unwound DNA and play an essential role in replication. The *E. coli* single-strand

DNA binding protein is a homotetramer of 18.8 kb subunits that binds co-operatively to single-stranded DNA in a sequence non-specific fashion. Its role in replication is to help DnaB for opening and unwinding of the DNA duplex and probably to protect single-stranded regions from nucleolytic attack. The strands of the double helix, once separated, are maintained in the separated state by the single-strand binding protein until two complete replicating complexes, poised to move in opposite directions, are established at the origin. The single-strand-binding proteins are eventually replaced by DNA polymerase III (O'Donnell, 2006).

DNA synthesis

Chromosomal DNA synthesis is catalysed by DNA polymerase III. The polymerase III holoenzyme of *E. coli* contains 10 different proteins assorting into three major functional units, the enzyme core, the β sliding clamp, and the γ/τ -δδ' complex clamp loader. The polymerase III holoenzyme, hexameric DnaB helicase, DnaG primase, and single-strand-binding protein interact in a constant state of change for functioning for replication of both leading and lagging strands (O'Donnell, 2006). *H. pylori* has homologues of β subunit DnaN, γ/τ subunits DnaX, δ subunit HolA, and δ' subunit HolB, as well as the core α subunit DnaE and the core ε subunit DnaQ (Table 9.2).

The replication complexes may remain tethered to the DNA throughout replication, and DnaG primase, whose continuing role is to synthesize RNA primers as required for the discontinuously replicated lagging strand, associates at the fork when required, and disassociates from it as each lagging-strand fragment is completed. Then the DNA fragments are joined by the nick-sealing DNA ligase whose homologue is also found in *H. pylori* (Table 9.2). Eventually, the divergently moving forks meet on the far side of the chromosome.

Regulation of replication

DnaA binds with ATP at Walker motif in domain III to become an active form that is autocatalytically hydrolysed to ADP-DnaA in a process termed regulatory inactivation of DnaA (Kaguni, 2006). Only a newly synthesized ATP-DnaA can bind to the DnaA box in *oriC* region to initiate replication. Completion of replication is not a prerequisite for new initiation. Under optimal nutrient conditions *E. coli* cells re-initiate replication before completion of the previous round and accelerate the cell cycle. In such a case sequestration of the origin by SeqA-dam methylase system is required; SeqA binding to hemi-methylated GATC sequences near the origin of replication is crucial to set the synchrony of the secondary replication initiation before completion of the previous round of replication. No SeqA homologues are found in *H. pylori*.

The regulatory mechanisms that modulate the replication will differ and are likely to reflect the bacterial lifestyle. Bacteria capable of multifork replication like *E. coli* possess certain requirements to prevent mis-initiations like the SeqA-dam methylase system. On the other hand, slow-growing oligotrophs such as *H. pylori* might not need such repressing systems probably because the energy metabolism as well as the biosynthesis of precursor molecules for proliferation including replication proceeds in a constant rate under fixed temperature. The origin of *H. pylori* genome is located in the putative promoter site of *dnaA*, suggesting that repression of DnaA on its own gene expression is the major regulatory system of replication in *H. pylori*.

Partitioning and segregation

The *E. coli* chromosome replicates bidirectionally, and each half of the chromosome is replicated by a different replication fork. To determine the arrangement of different replication forks during replication in *E. coli*, two specific chromosome sites are differentially labelled and their locations are traced (Nielsen *et al.*, 2006). The results reported are summarized in Table 9.3.

At the pre-replication stage, the origin is at the cell centre, the terminus is near the new pole. At the stage of mid-replication, the forks have dissociated from the cell centre, and newly replicated DNA products segregate to opposite sides of the cell centre. As replication progresses, the new origins attach to the cell quarter positions and the two arms of each nascent chromosome are sorted out so that each arm lies on either side

Table 9.3 Chromosome arrangement during replication[a]

Stage	Localization		
	Origin	Replication forks and replicated arms	Terminus
Pre-replication	Centre		Pole
Mid-replication		Either side of the cell centre	Centre
Pre-termination	Quarter positions	Sorted to either side of the quarters	Centre
Termination	Quarter positions	Forks at the centre Arms at either side of the quarters	Centre

[a]The results of Nielsen *et al.* (2006) are summarized.

of the cell quarters. At the termination stage, the origins are at the quarter positions, the terminus and forks are at the cell centre, and the arms of the two new chromosomes are arranged with respect to the cell quarters.

Chromosome replication, partitioning, and segregation require spatial as well as temporal regulatory systems. Two systems, the ParABS system and the FtsK-*dif* system have been extensively studied, and homologous genes involved in these systems are assigned in *H. pylori* (Table 9.4).

ParABS system for partitioning

ParABS system consists of ParA/ParB family of proteins together with the centromere-like DNA sequence *parS* that is involved in chromosome and plasmid partitioning during bacterial cell division. Almost all bacteria have the ParABS system for chromosome partitioning with the exception of certain γ-Proteobacteria including *E. coli*. The ParA and ParB homologues of *Bacillus subtilis* are Soj and Spo0J, respectively, and several centromere-like *parS* sequences are located in the origin-proximal region of the bacterial chromosome. The ParA and ParB proteins are not only important in chromosome partitioning but also serve a checkpoint function in linking chromosome segregation to cell division and/or DNA replication (Ghosh *et al.*, 2006).

ParA is an ATPase that may oscillate between ATP-bound and ADP-bound states in order to drive the partitioning cycle. ParB is a DNA-binding protein that interacts with one or more *parS* sequences and also with ParA. The ParA/ParB complex contributes to accurate chromosome partitioning, separation of replicated sister origins, and regulation of replication.

In *B. subtilis* the Spo0J protein (ParB homologue) is required for accurate chromosome segregation and regulates the developmental switch from vegetative growth to sporulation. Spo0J recognizes at least eight *parS* sites located near the origin of replication. Spo0J forms discrete foci associated with the *oriC* region of the chromosome throughout the cell cycle. It was also demonstrated that Spo0J associates with several kilobases of DNA flanking *parS* through a *parS*-dependent nucleation event that promotes lateral spreading of Spo0J along the chromosome (Murray *et al.*, 2006). Thus Spo0J could form as kinetochore-like scaffolds near *oriC* and interact with motor proteins including Soj (ParA homologue) to either push or pull chromosomes apart.

In the *H. pylori* genome *parA* (*soj*) and *parB* (*spo0J*) homologues are located next to each other possibly forming an operon *parBA* (Table 9.4). The N-terminal region of *H. pylori* ParA are most conserved that contains a domain for interaction with ParB as well as ATP-binding motifs that are representative of a superfamily of ATPases involved in partitioning of bacterial plasmids and chromosomes. The C-terminus of ParB responsible for DNA-binding and dimerization is relatively non-conserved. By cloning and expressing *parA* and *parB* of *H. pylori* 26695 in *E. coli*, the ATPase activity of ParA (65 kDa homodimer) and DNA-binding activity of ParB (30 kDa) were demonstrated (Lee *et al.*, 2006). In addition, two putative *parS* sequences, 17 bp

Hp-parS1, and 15 bp *Hp-parS2*, were identified by searching *parS*-homologous sequences of *B. subtilis*. Both the *parS* sequences are located in a hypothetical locus, HP1142, two genes apart from of *parA*. The HP1142 homologues were also found in J99 and HPAG1 at similar locations, although the latter sequence was less conserved. The specific binding of *H pylori* ParB to both *Hp-parS1* and *Hp-parS2* was demonstrated by gel retardation analysis. The ParABS system of *H. pylori* 26695 is approximately 400 kbp apart from *oriC* (Fig. 9.3) that is within the origin-proximal 20–30% of the circular chromosome as observed in *B. subtilis* and other species of bacteria.

Subcellular localization of ParA and ParB in growing *H. pylori* was determined by immunofluorescence microscopy with polyclonal antibodies raised against the purified proteins (Lee *et al.*, 2006). The results are interpreted in light of the chromosome arrangement during replication in *E. coli* (Table 9.3). ParB assembled into discrete foci, localizing near the centre in normal size cells (pre-replication stage), separated into two spots in elongated cells (replicating stage), and finally localizing at quarter positions in dividing cells (post-replication stage). Co-localization of ParB with nucleoid was visualized by DAPI staining of DNA. In contrast to ParB, ParA distributed irregularly within the cells. It is plausible that ParA ATPase acts as a motor to drive ParB on the *parS* sequences so that newly replicated chromosome arms are accurately arranged. Interestingly, the intracellular levels of both ParA and ParB reduced significantly in the coccoid cells, suggesting that the Par proteins are expressed only in actively dividing cells (Lee *et al.*, 2006).

FtsK-dif system for segregation

Bacteria with circular chromosomes have evolved systems that ensure dimeric chromosomes to be resolved to monomers prior to cell division. In *E. coli* and many other bacteria the resolution process of chromosome dimer is mediated by a site-specific recombinase, XerC/XerD complex, that acts at the terminus site known as *dif*. Deletion of *dif* or mutations in *xerC* or *xerD* resulted in the development of a subpopulation of filamentous cells containing abnormally partitioned nucleoids. The 28-bp *dif* sequences of *E. coli* and *B. subtilis* have been reported (Sciochetti *et al.*, 2001). *H. pylori* has homologues of XerC and XerD (Table 9.4), but their functions remain to be characterized.

To direct DNA translocation into the appropriate daughter cell, the replication terminus must be at the cell centre prior to cell division (Table 9.3). A new bioinformatical method to identify the terminus of replication in bacterial genomes has been developed (Hendrickson and Lawrence, 2006), and oligonucleotides with the following properties were searched in 45 completely sequenced genomes: the over-abundant G-rich octamers that are skewed on leading strands and enriched near the terminus of replication. Such sequences were found in 38 of the 45 bacterial species including *H. pylori*. It was concluded that these sequences function as <u>a</u>rchitecture <u>im</u>parting <u>s</u>equences (AIMS) and that these must have roles in terminal replication. In *H. pylori* J99, AIMS was identified as AGTAGGG and the terminus was located at 685 000 within the sequence of a hypothetical protein jhp0616 next to *xerC* (jhp0617). It should be noted that representative AIMSs of *E. coli* are identical to the FtsK-orienting polar sequence (KOPS) as discussed below.

Hendrickson *et al.* (Hendrickson et al., 2007) further analysed *E. coli* K12 and other γ-Proteobacteria as well as Firmicutes and concluded that most replication forks terminate at the *dif* site rather than at the previously suggested Ter sites that are more distantly situated away from the newly identified replication terminus. These results highlight the central role of the *dif* site in integrating chromosome mobilization, recombination and replication termination, and suggests that the terminus identified above (Hendrickson and Lawrence, 2006) must be the *dif* site. As for the *dif* site in *H. pylori* 26695, we tentatively located it next to *xerC* (HP0675) (Fig. 9.3), since the *xerC* and its surrounding regions of J99 and 26695 are well conserved.

The *dif* sequence only works with FtsK, a recently identified DNA translocation protein widely distributed among bacteria. The *E. coli* DNA translocase FtsK is an oligomeric ATP-fuelled motor that coordinates chromosome segregation and cell division. FtsK assembles at

Table 9.4 Proteins involved in chromosome segregation and cell division of *E. coli* and their homologues recognized in *H. pylori*

Process and function	Protein (gene)	ORF in *H. pylori* strain[b]		
		26695	J99	HPAG1
Segregation				
Partition	ParA (*parA*)	HP1139	jhp1067	HPAG1_1077
	ParB (*parB*)	HP1138	jhp1066	HPAG1_1076
Recombination	XerC (*xerC*)	HP0675	jhp0617	HPAG1_0658
	XerD (*xerD*)	HP0995	jhp0941, jhp0951	□[c]
DNA translocation	FtsK (*ftsK*)	HP1090	jhp0335	HPAG1_0357
Cell division				
Ring formation				
Proto-ring	FtsZ (*ftsZ*)	HP0979	jhp0913	HPAG1_0959
	FtsA (*ftsA*)	HP0978	jhp0912	HPAG1_0960
	ZipA (*zipA*)	□	□	□
Periplasmic connector	FtsQ (*ftsQ*)	□	□	□
	FtsB (*ftsB*)	□	□	□
	FtsL (*ftsL*)	□	□	□
	FtsW (*ftsW*)	□	□	□
Septal peptidoglycan factory	FtsI (*ftsI*)	HP1560	jhp1468	HPAG1_1509
	FtsN (*ftsN*)	HP1556	jhp1464	HPAG1_1505
Ring inhibitor	MinC (*minC*)	HP1053	jhp0372	HPAG1_0394
	MinD (*minD*)	HP0331	jhp0314	HPGA1_0360
	MinE (*minD*)	HP0332	jhp0315	HPAG1_0361
Required in low-salt medium	FtsE (*ftsE*)	HP0748	jhp0685	HPAG1_0733
	FtsX (*ftsX*)	HP0749	jhp0686	HPAG1_0734
Division inhibitor	CdrA (*cdrA*)[d]	HP0066	jhp0061	HPAG1_0066
				HPAG1_0067

[a]Functions and proteins in *E. coli* are adopted from reviews (Michie and Lowe, 2006; Lutkenhaus, 2007).
[b]ORFs are annotated in strain 26695 (Tomb *et al.*, 1997), J99 (Alm *et al.*, 1999), and HPAG1 (Oh *et al.*, 2006).
[c]Homologues are not assigned.
[d]Identified in *H. pylori* HPK5 (Takeuchi *et al.*, 1998). See details in the text.

the division septum and translocates the DNA to bring the two *dif* sites in the dimer into close proximity. Upon formation of the XerCD-*dif* synapse, FtsK stimulates recombination by direct interaction with XerD (Higgins, 2007).

The FtsK monomer of *E. coli* consists of three domains: a 217-residue N-terminal transmembrane domain, a poorly conserved and putatively unstructured 600-residue linker and a 512-residue motor domain ($FtsK_C$) that has been classified as a member of the AAA+ family of ATPases and shown to translocate duplex DNA. $FtsK_C$ is further divided into three domains, α, β, and γ. The α (818 to 943 residue) and β (944 to 1258 residue) domains form a functional motor capable of translocating along DNA (Ptacin *et al.*, 2006). The domains have a ring-like assembly with a central cavity to accommodate double-stranded DNA (Sivanathan *et al.*, 2006).

The γ domain (1261 to 1329 residue) interacts with 8-bp 5′-GGGNAGGG-3′ consensus DNA sequences termed KOPS (FtsK-orienting polar sequences) or FRS (FtsK-recognition sequence). KOPSs are located predominantly on the leading-strand for replication on each replichore, where they form a pattern of oriented repeated sequences with their density increasing as terminus is approached. The γ domain forms a winged helix and recognizes and binds with KOPS for DNA translocation depending on ATP hydrolysis. KOPSs have been found to control translocation directionality. When the KOPS approaches FtsK fixed at the septum from the 3′ end, the DNA translocation stops (non-permissive orientation). On the other hand, when KOPS is approaching from the 5′ end (permissive orientation), DNA translocation is not obstructed. As the KPOSs are over-represented on the chromosome's leading strand and switch strand at *dif*, they can guide FtsK towards *dif* from any location on the chromosome (Bigot *et al.*, 2006: Sivanathan, *et al.*, 2006).

KOPS-guided and FtsK promoted DNA translocation brings newly replicated *dif* sites to the septum. Then FtsK stimulates decatenation of links between sister chromosomes and activates the site-specific XerC/D recombinase to form daughter chromosomes from dimers (Yates, *et al.*, 2006). Since skewed sequence motifs similar to KPOS are assigned in *H. pylori* as described above (Hendrickson and Lawrence, 2006), and homologues of FtsK are present in *H. pylori* (Table 9.4), similar KOPS-guided translocation of DNA towards *dif* may be performed by FtsK for chromosome segregation.

Cell division

The final step of bacterial proliferation is cell division. The main cytoskeletal element involved in cell division is the Z ring, a membrane-associated structure consisting of polymers of tubulin-related protein FtsZ, one of the *fts* gene products that were termed because of its defective phenotype to form filamentous cells at high temperature. FtsZ is the first protein known to localize to the mid-cell position prior to septum invagination. During cell division, more than 10 accessory proteins are believed to localize to the cell centre that are involved in septum formation and separation of the parental cell to produce daughter cells (Michie and Lowe, 2006). Several Fts and related protein homologues are assigned in *H. pylori* (Table 9.4).

FtsZ for Z-ring formation

FtsZ is a highly conserved protein found in virtually all Eubacteria and Archaea. The centrally located Z-ring is formed by bi-directional polymerization from a single nucleation site. It is assumed that the Z-ring reduces in diameter during cell division, drawing the cytoplasmic membrane inwards. GTP-bound FtsZ forms polymers in the presence of calcium that assemble laterally resulting to form bundles or sheets. It was demonstrated that a specific conformational change in the guanine moiety is associated with bundling (Marrington *et al.*, 2004). The 'flipping' of the guanine moiety in FtsZ may explain the link between reduced GTPase activity, increased polymer stability, and polymer bundling.

The assembly and positioning of the Z ring is regulated by the Min system consisting of MinC, MinD, and MinE (Lutkenhaus, J., 2007). These proteins cooperate to position the Z ring at mid-cell, the centre of the proliferating cell, by inhibiting the formation of the Z ring away from the cell centre. MinC is the effector of the MinCDE system responsible for repressing cell division by antagonizing FtsZ polymerization. MinD, a membrane-bound ATPase, undergoes a remarkable oscillation. At one time, MinD-ATP molecules are accumulating along the inner membrane at a polar zone and extending towards mid-cell. During this process MinC associating with MinD inhibits the Z-ring formation. Then MinD starts to shrink toward the pole by the stimulatory activity of MinE on MinD-ATPase, resulting to release MinD from the inner membrane. As the polar zone disappears, a new polar zone at the other pole is established, and the cycle is restarted. Such oscillation is remarkably rapid, approximately 20 s at 20°C. Homologues of MinC, MinD and MinE are present in *H. pylori* (Table 9.4), although the homology of MinC is very low.

Assembly of cell division complex

During cell division in *E. coli*, FtsZ is tethered by FtsA or ZipA and remained as a ring at the leading edge of the constricting division septum. An actin-related FtsA is an ATP-binding protein associated with junctions of inner and outer membranes with the C-terminal sequence (Michie and Lowe, 2006). FtsA is conserved among many bacteria including *H. pylori* (Table 9.4), whereas ZipA is present only in *E. coli* and the closely related bacteria.

A concerted mode of assembly of several components of the divisome in *E. coli* is proposed (Vincente and Rico, 2006) that three multimolecular complexes are formed; the proto-ring, the periplasmic connector, and the peptidoglycan factory. The proto-ring at the mid cell consists of FtsZ and FtsA or ZipA, where membrane-bound FtsA or ZipA stabilizes the FtsZ ring and might signal the cell to initiate constriction via peptidoglycan synthesis. FtsK, a DNA translocase discussed in the previous section, is next to be recruited to the proto-ring and fixed to the septum by the N-terminal transmembrane domain. Three proteins, FtsQ, FtsB and FtsL each containing cytoplasmic, membrane and periplasmic domains, could connect the cytoplasmic ring (formed by the proto-ring and FtsK) and the septal peptidoglycan factory consisting of FtsW, a putative transporter of the precursors, and FtsI, which is a septation-specific transpeptidase. FtsN is a late assembling protein, although its function is unknown.

In addition to proto-ring proteins, FtsA and FtsZ, homologues of proteins in the peptidoglycan factory, FtsW and FtsI, are found in *H. pylori* (Table 9.4). On the other hand, homologues of periplasmic proteins that connect proto-ring and peptidoglycan factory are missing, possibly because structures of these proteins are divergent.

Homologues of *E. coli* FtsE–FtsX complex, a putative ABC transporter required for growth in low-salt media are also present in *H. pylori* (Table 9.4).

Cell-division related gene, *cdrA*

A cell division-related gene *cdrA* locating next to the urease gene cluster of *H. pylori* HPK5 was cloned and characterized (Takeuchi *et al.*, 1998). The putative gene product, CdrA, is homologous to the C-terminal half of a putative ATP binding protein HP0066, jhp0061, and HPAG1_0066 of strain 26695, J99, and HPAG1, respectively, locating at the equivalent position (Table 9.4). However, the size of HPK5 CdrA (367-residue polypeptide) is much smaller than those of the homologues (approximately 831-residue polypeptide). *E. coli* seems to lack an obvious homologue, although some similarity was found in the β-domain of FtsK including an ATP-binding motif.

The *cdrA* gene of *H. pylori* HPK5 is dispensable and unique in having a repressive role on cell division (Takeuchi *et al.*, 2006). Under high salt conditions, cell division of the wild type strain is impaired forming multinuclear filaments in contrast to the apparently normal cell division of a *cdrA*-disrupted mutant. Intriguingly, the filaments of the wild type appear to have septum consisting only with inner membrane without cell wall structure as observed by transmission electron microscopy. These results raise a possibility that the septum with cytoplasmic membrane is formed before synthesis of septal peptidoglycan, and the *cdrA* product suppresses the septal peptidoglycan synthesis. Incidentally, *H. pylori* lacks the homologues of connector proteins that link cytoplasmic ring and the septal peptidoglycan factory, further suggesting that this bacterium might have a unique mechanism of septum formation. In the late stationary phase of culture, the wild type cells tend to become ghosts and aggregates, whereas the mutant cells are spiral and viable. In addition, the mutant cells are more resistant to ampicillin killing.

In spite of the size difference, the *cdrA* gene homologues in *H. pylori* 26695, and J99, seem also to have a repressive role on cell division (Takeuchi *et al.*, 2006), and microarray analyses of a strain of *H. pylori* revealed that the *cdrA* gene was lost in colonies recovered from infected mouse (Israel *et al.*, 2001a), and isolates from the source of the patient of *H. pylori* J99 after 6 years had a deletion of the *cdrA* homologue jhp0067 (Israel *et al.*, 2001b). As the secretion of proinflammatory cytokines induced by *cdrA*-positive strains tends to be higher than that by *cdrA*-negative strains, deletion of *cdrA* might be an adaptation to the microenvironment in the stomach for persistent colonization (Takeuchi *et al.*, unpublished data).

Concluding remarks and future perspectives

Bacteria grow and divide by a process referred to as binary fission, a process in which a cell increases in size and splits into two daughter cells. Recent structural and molecular studies of *E. coli* and some other bacteria have significantly advanced our understanding of how the processes of replication (O'Donnell, 2006), partitioning (Ghosh *et al.*, 2006), segregation (Higgins, 2007), and cell division (Lutkenhaus, 2007) are performed. By searching available *H. pylori* genome sequences, we could identify most of the homologous genes involved in the process (Table 9.2 and Table 9.4), indicating that the fundamental process for bacterial replication and cell division is common.

The replication origin *oriC* of *H. pylori* was identified. The DnaA boxes are located in the narrow promoter region of *dnaA*, suggesting an autonomous regulation and tight coupling of *dnaA* expression and replication initiation. DnaB helicase consisting of the replication machinery and the ParABS system for chromosome partitioning have also been characterized. The FtsK-*dif* system and the FtsZ-ring system for chromosome segregation and cell division, respectively, might be similar to those in *E. coli*, although their characterization awaits future studies. Among the cell-division orthologues, periplasmic connector proteins to connect proto-ring and peptidoglycan factory seem to be unique. These unique gene products might be good targets for developing new anti-*H. pylori* drugs, since these are accessible from the bacterial cell surface.

H. pylori is a genetically diverse pathogen with several strain-specific flexible genes including genes unique to *H. pylori* such as those for restriction-modification, transposase, and cell surface proteins. The *cdrA* gene involved in suppression of cell division may be one of such genes. It can be hypothesized that CdrA is associated with acute inflammation at the onset of *H. pylori* infection and is inactivated or deleted after prolonged infection as a result of bacterial adaptation. It should be clarified whether CdrA inhibits septal peptidoglycan synthesis, whether its inhibition is occurring *in vivo*, and what kind of pathophysiology is associated with CdrA. Future studies on this line may help our understanding of how *H. pylori* persists in the ecological niche.

References

Alm, R.A., Ling, L.S., Moir, D.T., King, B.L., Brown, E.D., Doig, P.C., Smith, D.R., Noonan, B., Guild, B.C., deJonge, B.L., Carmel, G., Tummino, P.J., Caruso, A., Uria-Nickelsen, M., Mills, D.M., Ives, C., Gibson, R., Merberg, D., Mills, S.D., Jiang, Q., Taylor, D.E., Vovis, G.F., and Trust, T.J. (1999). Genomic-sequence comparison of two unrelated isolates of the human gastric pathogen *Helicobacter pylori*. *Nature* 397, 176–180.

Bigot, S., Saleh, O.A., Cornet, F., Allemand, J.F., Barre, F.X. (2006). Oriented loading of FtsK on KOPS. Nat. Str. Mol. Biol. *13*, 1026–1028.

Ghosh, S.K., Hajra, S., Paek, A., and Jayaram, M. (2006). Mechanisms for chromosome and plasmid segregation. Annu. Rev. Biochem. 75, 11–41.

Hendrickson, H., and Lawrence, J.G. (2006). Selection for chromosome architecture in bacteria. J. Mol. Evol. 62, 615–629.

Hendrickson, H., and Jeffrey G. Lawrence J.G. (2007). Mutational bias suggests that replication termination occurs near the dif site, not at Ter sites. Mol. Microbiol. *64*, 42–56.

Herrick, J., and Sclavi, B. (2007). Ribonucleotide reductase and the regulation of DNA replication: an old story and an ancient heritage. Mol. Microbiol. *63*, 22–34.

Higgins, N.P. (2007). What's the Dif? Mol. Microbiol. *64*, 1–4.

Israel, D.A., Salama, N., Arnold, C.N., Moss, S.F., Ando, T., Wirth, H.P., Tham, K.T., Camorlinga, M., Blaser, M.J., Falkow, S., and Peek, R.M. (2001a). *Helicobacter pylori* strain-specific differences in genetic content, identified by microarray, influence host inflammatory responses. J. Clin. Invest. *107*, 611–620.

Israel, D.A., Salama, N., Krishna, U., Rieger, U.M., Atherton, J.C., Falkow, S., and Peek, R.M. (2001b). *Helicobacter pylori* genetic diversity within the genetic niche of a single human host. Proc. Natl. Acad. Sci. U.S.A. *98*, 14625–14630.

Kaguni, J.M. (2006). DnaA: controlling the initiation of bacterial DNA replication and more. Annu. Rev. Microbiol. *60*, 351–375.

Lee MJ, Liu CH, Wang SY, Huang CT, Huang H. (2006). Characterization of the Soj/Spo0J chromosome segregation proteins and identification of putative *parS* sequences in *Helicobacter pylori*. Biochem. Biophys. Res. Commun. *342*, 744–750.

Lutkenhaus, J. (2007). Assembly and dynamics of the bacterial MinCDE system and spatial regulation of the Z ring. Annu. Rev. Biochem. *76*, (in press).

Magdalena, M.F., Simmons, L.A., and Kaguni, J.M. (2005). An essential tryptophan of *Escherichia coli* DnaA protein functions in oligomerization at the *E. coli* replication origin. J. Biol. Chem. *280*, 24627–24633.

Marrington, R., Small, E., Rodger, A., Dafforn, T.R., and Addinall, S.G. (2004). FtsZ fiber bundling is triggered by a conformational change in bound GTP. J. Biol. Chem. *279*, 48821–48829.

Michie, K.A., and Lowe, J. (2006). Dynamic filaments of the bacterial cytoskeleton. Annu. Rev. Biochem. *75*, 467–92.

Murray, H., Ferreira, H., and Jeff Errington, J. (2006). The bacterial chromosome segregation protein Spo0J spreads along DNA from *parS* nucleation sites. Mol. Microbiol. *61*, 1352–1361.

Nielsen, H.J., Ottesen, J.R., Youngren, B., Austin, S.J., and Hansen, F.G. (2006). The *Escherichia coli* chromosome is organized with the left and right chromosome arms in separate cell halves. Mol. Microbiol. *62*, 331–338.

O'Donnell, M. (2006). Replisome architecture and dynamics in *Escherichia coli*. J. Biol. Chem. *281*, 10653–10656.

Oh, J.D., Kling-Backhed, H., Giannakis, M., Xu, J., Fulton, R.S., Fulton, L.A., Cordum, H.S., Wang, C., Elliott, G., Edwards, J., Mardis, E.R., Engstrand, L.G., and Gordon, J.I. (2006). The complete genome sequence of a chronic atrophic gastritis *Helicobacter pylori* strain: Evolution during disease progression. Proc. Natl. Acad. Sci. USA *103*, 9999–10004.

Ptacin, J.L., Nöllmann, M., Bustamante, C., and Nicholas R Cozzarelli, N.R. (2006). Identification of the FtsK sequence-recognition domain. Nat. Str. Mol. Biol. *13*, 1023–1025.

Salzberg, S.L., Salzberg, A.J., Kerlavage, A.R., and Tomb, J.F. (1998). Skewed oligomers and origin of replication. Gene *217*, 57–67.

Sciochetti, S.A., Piggot, P.J., and Blakely, G.W. (2001). Identification and characterization of the *dif* site from *Bacillus subtilis*. J. Bacteriol. *183*, 1058–1068.

Sivanathan, V., Allen, M.D., de Bekker, C., Baker, R., Arciszewska, L.K., Freund, S.M., Bycroft, M., Lowe, J., and Sherratt, D.J. (2006). The FtsK gamma domain directs oriented DNA translocation by interacting with KOPS. Nat. Struct. Mol. Biol. *13*, 965–72.

Soni, R.K., Mehra, P., Choudhury, N.R., Mukhopadhyay, G., and Dhar, S.K. (2003). Functional characterization of *Helicobacter pylori* DnaB helicase. Nucleic Acids Res. *31*, 6828–6840.

Soni, R.K., Mehra, P., Mukhopadhyay, G., and Dhar, S.K. (2005). *Helicobacter pylori* DnaB helicase can bypass *Escherichia coli* DnaC function *in vivo*. Biochem. J. *389*, 541–548.

Takeuchi, H., Nakazawa, T., Okamoto, T., Shirai, M., Kimoto, M., Nishioka, M., Con, S.A., Morimoto, N., and Sugiura, T. (2006). Cell elongation and cell death of *Helicobacter pylori* is modulated by the disruption of *cdrA* (cell division-related gene A). Microbiol. Immunol. *50*, 487–497.

Takeuchi, H., Shirai, M., Akada, J.K., Tsuda, M., and Nakazawa, T. (1998). Nucleotide Sequence and Characterization of *cdrA*, a Cell Division-Related Gene of *Helicobacter pylori*. J. Bacteriol. *180*, 5263–5268.

Tomb, J.F., White, O., Kerlavage, A.R., Clayton, R.A., Sutton, G.G., Fleischmann, R.D., Ketchum, K.A., Klenk, H.P., Gill, S., Dougherty, B.A., Nelson, K., Quackenbush, J., Zhou, L., Kirkness, E.F., Peterson, S., Loftus, B., Richardson, D, , Dodson, R., Khalak, H.G., Glodek, A., McKenney, K., Fitzegerald, L.M., Lee, N., Adams, M.D., Hickey, E.K., Berg, D.E., Gocayne, J.D., Utterback, T.R., Peterson, J.D., Kelley, J.M., Cotton, M.D., Weidman, J.M., Fujii, C., Bowman, C., Watthey, L., Wallin, E., Hayes, W.S., Borodovsky, M., Karp, P.D., Smith, H.O., Fraser, C.M., and Venter, J.C. (1997). The complete genome sequence of the gastric pathogen *Helicobacter pylori*. *Nature* **388,** 539–547.

Vicente, M., and Rico, A.I. (2006). The order of the ring: assembly of *Escherichia coli* cell division components. Mol. Microbiol. *61*, 5–8.

Yates, J., Zhekov, I., Baker, R., Eklund, B., Sherratt, D.J., and Arciszewska, L.K. (2006). Dissection of a functional interaction between the DNA translocase, FtsK, and the XerD recombinase. Mol. Microbiol. *59*, 1754–1766.

Zawilak-Pawlik, A., Agnieszka Kois, A., Majka, J., Jakimowicz, D., Smulczyk-Krawczyszyn, A., Messer, W., and Zakrzewska-Czerwińska, J. (2005). Architecture of bacterial replication initiation complexes: orisomes from four unrelated bacteria. Biochem. J. *389*, 471–481.

Zawilak, A., Cebrat, S., Mackiewicz, P., Król-Hulewicz, A., Jakimowicz, D., Messer, W., Gosciniak, G., and Zakrzewska-Czerwinska, J. (2001). Identification of a putative chromosomal replication origin from *Helicobacter pylori* and its interaction with the initiator protein DnaA. Nucleic Acids Res. *29*, 2251–2259.

Zawilak, A., Durrant, M.C., Jakimowicz, P., Backert, S., and Zakrzewska-Czerwiska, J. (2003). DNA binding specificity of the replication Initiator protein DnaA from *Helicobacter pylori*. J. Mol. Biol. *334*, 933–947.

Molecular Mechanisms of Host Adaptation in *Helicobacter*

10

Stephan C. Schuster, Nicola E. Wittekindt and Bodo Linz

Abstract

Helicobacter pylori has lived in close association with its human host for hundreds of thousands of years. This has resulted in the highly adapted strains, on which today's clinical and molecular research is based. Through the host jump of a *Helicobacter* from an early human to large felines, it became possible to study the adaptation process in a genomic comparative analysis. In this chapter we describe the molecular mechanisms that have driven the genomic evolution of *Helicobacter* in its host and have enabled the enormous plasticity of the bacterium's genome. Based on its unusually high mutation and recombination rate, we propose a six-step model for host adaptation that also includes *Helicobacter*'s ability for horizontal gene transfer in combination with its natural competence for DNA uptake.

Introduction

After *Helicobacter* was recognized as a ubiquitous human pathogen that infects more than 50% of today's human population, it became a logical target for genome analysis. In this regard, Tomb *et al.* published the first complete genome of any *Helicobacter* strain in 1997, strain 26695, isolated from a gastritis patient in the United Kingdom (Tomb *et al.*, 1997). Two years later, strain J99, isolated from a patient with duodenal ulcer, was published by Alm and colleagues (Alm *et al.*, 1999). Despite the small size of the genomes with little more than 1.6 million nucleotides and no described plasmids it took more than 7 years for publication of a third *Helicobacter pylori* genome isolated from a patient with chronic atrophic gastritis (ChAG) (Oh *et al.*, 2006). Given the relevance of *Helicobacter* in human pathogenicity and the severity of the clinical outcome, the small number of completed genomes comes at a surprise particularly when compared to other bacterial genera of human pathogens. In case of the genera *Staphylococcus* (2.5–2.9 Mb), *Pseudomonas* (4.5–7.1 Mb) and *Escherichia* (4.6–5.6 Mb), 14, 12 and 8 genomes are available, respectively, despite the fact that all of these genomes are much larger than that of *Helicobacter* (~1.6 Mb). Currently only *H. pylori* is recognized as a human specific *Helicobacter* whereas there have been species-specific *Helicobacter* described for almost all known vertebrates (Solnik and Schauer, 2001). In most cases the phylogenetic relationships between these *Helicobacter* strains reflect the evolutionary history and radiation of their hosts. While there are several *Helicobacter* genome projects under way of non-human hosts, currently only the two species *Helicobacter hepaticus* and *Helicobacter acinonychis* have been published (Suerbaum *et al.*, 2003; Eppinger *et al.*, 2006). Comparative genomic analysis of these genomes raised the question whether *H. hepaticus* is a true *Helicobacter* because it is not a gastric but an enterohepatic bacterium as it infects the liver of mice. Moreover, phylogenetic analyses of the 16S rRNA separates the different *Helicobacter* isolates into two major phylogenetic groupings (Solnick and Schauer, 2001), that correspond to the gastric and enterohepatic species (Fig. 10.1). There are only two exceptions, the gastric species *Helicobacter mustelae* (from ferret) and *Helicobacter suncus* (from house musk shrew) whose

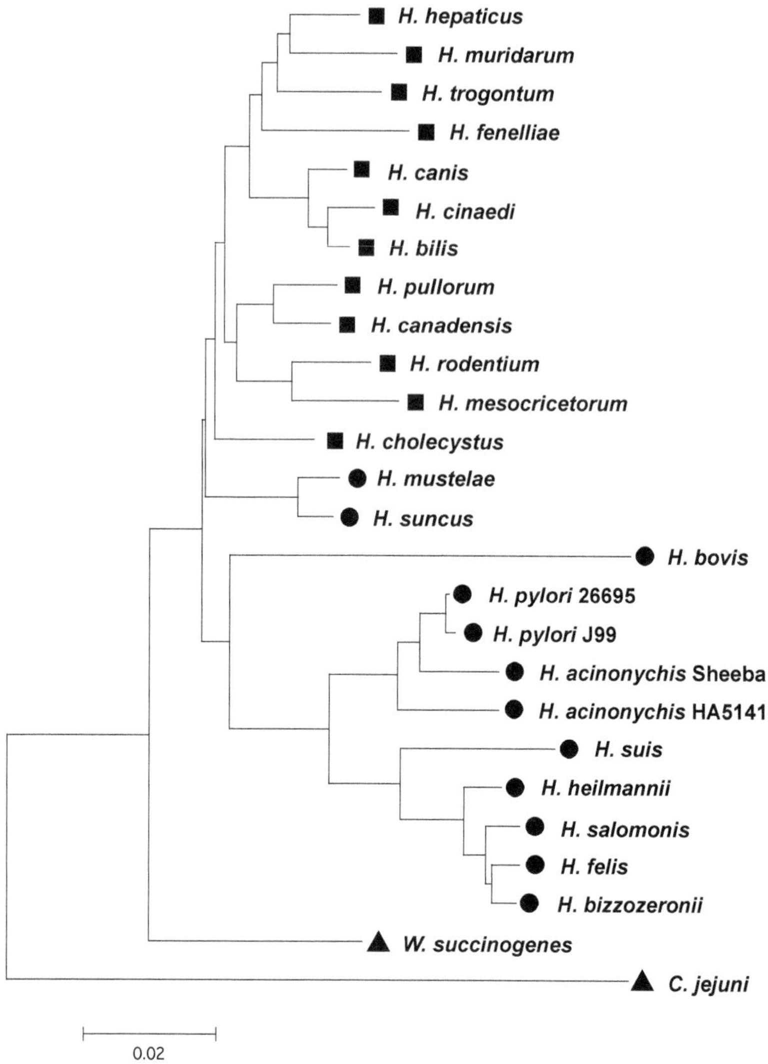

Figure 10.1 Phylogenetic tree of the 16S rRNA sequences from different ε-proteobacteria. Gastric *Helicobacter* are indicated by circles, enterohepatic *Helicobacter* are labelled with rectangles. Non-*Helicobacter* ε-proteobacteria are depicted by triangles.

16S rRNA sequences group within the enterohepatic species cluster. However, the 16S rRNA phylogeny is often not congruent with that of other genes in the chromosome. For example, the two *H. acinonychis* strains HA5141 and Sheeba share almost identical sequences in seven housekeeping genes that are scattered around the chromosome (Gressmann *et al.*, 2005) but their 16S rRNA sequences (Eaton *et al.*, 1993; Eppinger *et al.*, 2006) are very different (Fig. 10.1). It is therefore currently discussed which phylogenetic and other typing approaches should be used for the differentiation of gastric versus related *Helicobacter* strains (Linz and Schuster, 2007). The classification of *H. acinonychis* may serve as an example for this endeavour. Originally isolated in the course of stomach biopsies performed on captive large felines with symptoms of gastritis, *H. acinonychis* was expected to classify as a gastric strain specific to carnivores of the pantheris family. Initial typing carried out by 16S rRNA sequencing revealed an unexpected phylogenetic proximity to the human *H. pylori* strains (Eaton *et al.*, 1993; Dailidiene *et al.*, 2004). This suggested for the first time that Helicobacters might not strictly radiate vertically with their hosts but might also be capable of horizontal transmission between mammalian species. This scenario of a host jump set the stage for analysing the direction of transmission between the human and the animal host. Comparative genomic analysis revealed the molecular mechanisms that are needed for the adaptation process in a new host (see below).

Genome plasticity in Helicobacter

With the availability of the first two *Helicobacter* genomes of strains 26695 and J99, it became possible to assess the syntenic order between the two. In this regard, Alm and coworkers (Alm *et al.*, 1999) showed that ten genomic rearrangements have occurred between both genomes. One major rearrangement shows a symmetrical positioning between the origin and terminus of replication, indicating a recent recombinatorial event. Increased rates of recombination appear to be a hallmark of *Helicobacter* genome dynamics, particularly when compared to other bacterial phyla (Fig. 10.2A). This observation was confirmed with the third *Helicobacter* sequence of strain HPAG1, which was shown to harbour the same order of intra-chromosomal recombinatorial events. All three strains are believed to share a recent evolutionary history of less than 100,000 years, as strain 26695 was isolated from a European patient and was assigned to the bacterial population hpEurope, while strain J99 possesses an African genotype (Falush *et al.*, 2003). As European and African human populations have separated approximately 60,000 years ago (Liu *et al.*, 2006), the same is true for the *Helicobacter* strains that migrated together with their hosts (Linz *et al.*, 2007). This intimate host pathogen relationship allowed for the typing of the human host populations with greater resolution than would have been possible by just studying human genetic markers. The higher resolution is due to a faster ticking clock of *Helicobacter* (Björkholm *et al.*, 2001; Eppinger *et al.*, 2006; Linz and Schuster, 2007) than the one of its host, as well as by use of molecular typing mechanisms, such as multi-locus sequence typing (MLST) (Achtman *et al.*, 1999). MLST, in contrast to 16S analysis, targets 7 housekeeping genes present in all genomes, and therefore provides not only more sequence information, but also analyses sequence intervals which may evolve faster than the more functionally constrained 16S and 23S rRNA genes. Thus, the phylogenetic resolution of the MLST-typing allowed for the identification of not only defined migrating groups, but also enabled the differentiation between sociological groupings by using *H. pylori* sequences as molecular markers (Falush *et al.*, 2003, Linz *et al.*, 2007, Wirth *et al.*, 2004).

Mutation and recombination shape microbial genomes

Two processes drive the evolution of an organism, namely mutation and recombination (for a review see Eisen, 2007). Both these mechanisms also come to play in the process, which helps a bacterium adapting to its host. Mutation has three principle causes, (i) copying parental DNA, (ii) errors in chromosome segregation, and (iii) modification of DNA in between replication cycles. These modifications can either occur spontaneously or be induced through harmful chemicals or ionizing radiation. As spontaneous mutations occur naturally very infrequently, the above means are used in laboratory environments to alter replication accuracy or to induce chemical aberration of the nucleotides. The resulting point mutations can be categorized as either transitions or transversions. Transitions, e.g. the exchange of a purine base for another purine (G to A) or likewise a pyrimidine for a pyrimidine (C to T), are much more frequently found than transversions (exchange of a purine for a pyrimidine or vice versa). Furthermore, C to T transitions are found to occur naturally more frequently than any other point mutation, as they result of a simple deamination reaction. In addition to single nucleotide exchanges, insertions and deletions are the second major group of contributors which widely affect the transcriptional and translational processes in the cell. This is because any change altering the nucleotide number within an existing open reading frame by a number that is not dividable by three results in a so-called frameshift mutation. As a consequence, the amino acid sequence resulting from the translational process is altered beginning with the position of the insertion or deletion of nucleotide sequence, as now a different reading frame is being read. Insertions in general occur in a random pattern and the inserted sequence

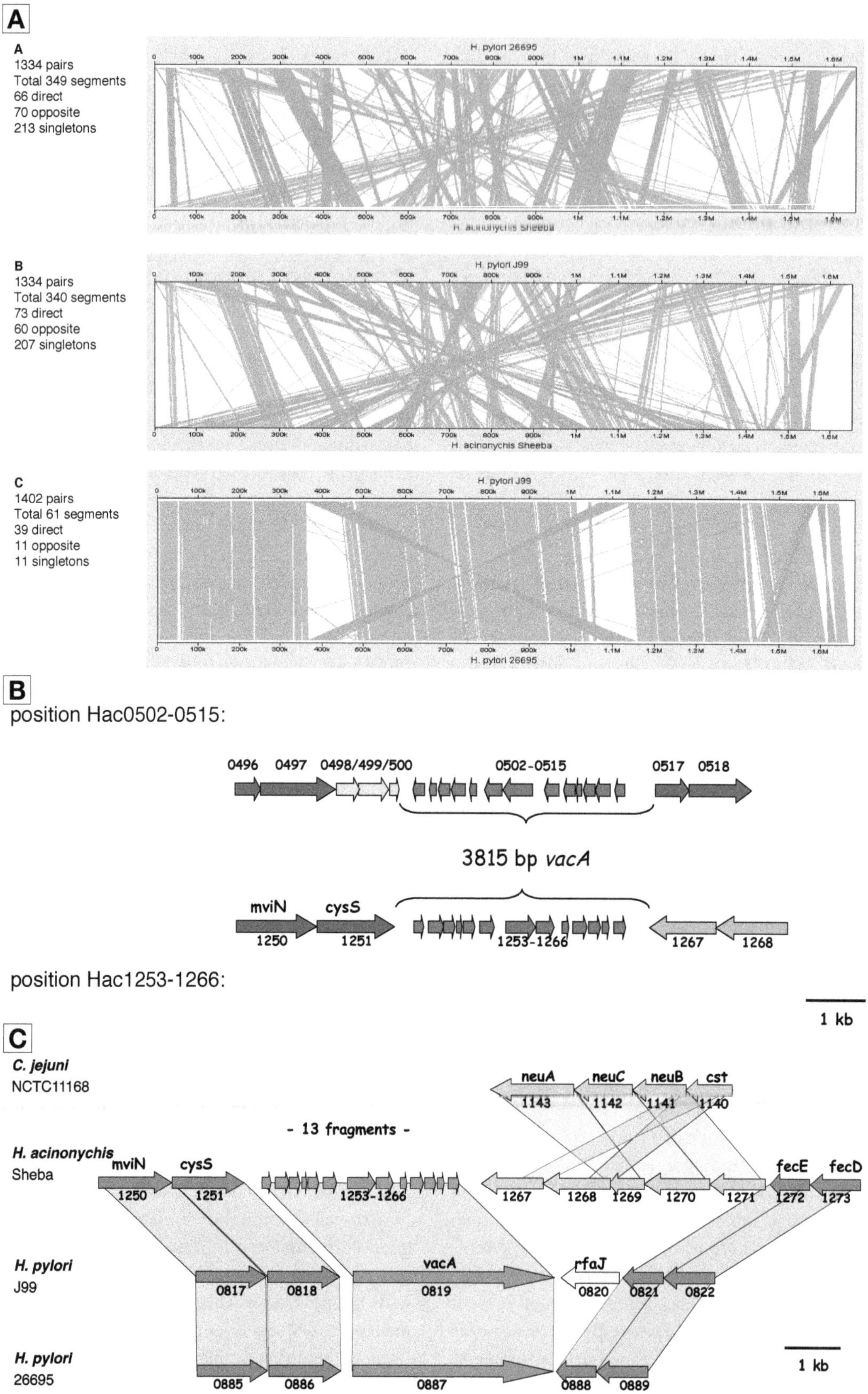
A
A
1334 pairs
Total 349 segments
66 direct
70 opposite
213 singletons
H. pylori 26695
H. acinonychis Sheeba
B
1334 pairs
Total 340 segments
73 direct
60 opposite
207 singletons
H. pylori J99
H. acinonychis Sheeba
C
1402 pairs
Total 61 segments
39 direct
11 opposite
11 singletons
H. pylori J99
H. pylori 26695
B
position Hac0502-0515:
0496
0497
0498/499/500
0502-0515
0517
0518
3815 bp vacA
mviN
cysS
1250
1251
1253-1266
1267
1268
position Hac1253-1266:
1 kb
C
C. jejuni
NCTC11168
neuA
neuC
neuB
cst
1143
1142
1141
1140
- 13 fragments -
H. acinonychis
Sheeba
mviN
cysS
1250
1251
1253-1266
1267
1268
1269
1270
1271
1272
1273
fecE
fecD
H. pylori
J99
0817
0818
vacA
0819
rfaJ
0820
0821
0822
H. pylori
26695
0885
0886
0887
0888
0889
1 kb

is frequently copied from an existing location elsewhere on the chromosome. This process therefore results in the duplication of sets of genes. One interesting example for this process is the duplication of an already inactivated *vacA* gene in the *H. acinonychis* genome (Eppinger *et al.*, 2006). This gene, despite multiple disruptions of its open reading frame, was shown to be duplicated and modified by only one nucleotide exchange since its duplication, independent of the fact that both copies do not provide any selection advantage to their carrier (Fig. 10.2B). Alternatively, the source of the to be inserted DNA fragment may be extrachromosomal, e.g. mobile genetic elements, phages, or plasmids. A relevant additional source of DNA in *Helicobacter* is found through the ability to import extracellular, naked DNA from the environment in a process that is called natural transformation. All of the above mechanisms can contribute to a phenomenon termed horizontal gene transfer, the genetic exchange not within a lineage but in between species. The two processes of deletions and insertions are key to maintaining a steady flow of incoming genetic information without inflating the genome size of a microbial organism indefinitely. Again the genome of *H. acinonychis* may serve as the best example among the helicobacters for this dynamic process. In Eppinger *et al.* 2006 it was shown that *H. acinonychis* not only has lost at least ten but probably many more genes that were previously found in all 56 *H. pylori* isolates of a globally representative strain collection (Gressmann *et al.*, 2005), but also has incorporated a new, non-*Helicobacter* specific genetic repertoire. The most striking example for such an insertion of genetic material is the *neuACB* and *cst* gene cluster which was shown to increase the pathogenicity in *Bacillus* and *Campylobacter* strains, but has not been previously observed in any other *Helicobacter* genome. These genes allow *H. acinonychis* to attach sialic acid to its cell surface carbohydrates and thereby evade detection by the immune system.

Inversions and translocations

A further mechanism to reorganize an organism's genetic material without adding additional information content is found in form of inversions and translocations. Inversions excise stretches of DNA and reintegrate them in opposite orientation at either the same location or in a symmetrical fashion in regard to the chromosome's origin and terminus of replication. In contrast to the symmetric event, translocations move a sequence interval from one position on the genome to another without reversing the transcriptional orientation. Both effects are likely to be the most powerful force that is shaping not only the genomes of the genus of Helicobacters, but also those of related epsilon-proteobacteria. As the relative position of a gene on the chromosome strongly influences its transcriptional activity, relocation on the chromosome can have dramatic effects on gene expression. In those cases, however, where inversions and translocations generate breakpoints within existing genes, they may result in a complete disruption of gene function. Examples for mutation through recombination are found in abundant numbers in all gastric and hepatic Helicobacters and have been studied in detail for the *H. acinonychis* genome. Twenty fragmented genes were identified among the 182 genes that are flanking 91 chromosomal rearrangements in the *H. acinonychis* genome. Six of these functionally inactivated genes coded for outer membrance proteins (OMPs), three for fucosyl-transferases and four for restriction and modification enzymes, whereas the proposed function of the remaining seven genes was unknown (Eppinger *et al.*, 2006).

Alterations on the nucleotide level are, however, most likely to occur in the process of DNA replication. Mechanistically they may result

Figure 10.2 (opposite) Genomic changes accompanying the host jump. (A) Recombination is vastly increased in *Helicobacter* species that have undergone a host jump (panel A: synteny between *H. pylori* 26695 and *H. acinonychis* strain Sheeba; panel B: between *H. pylori* J99 and *H. acinonychis;* panel C: between *H. pylori* 26695 and *H. pylori* J99). Recombination frequency of strains having remained within the same host is found to be lesser and therefore results in smaller genomic variation. (B) Gene duplication of the two *vacA* gene loci in the *H. acinonychis* genome. Only a single nucleotide difference was found to differentiate both copies. The gene duplication must therefore have occurred after gene inactivation through fragmentation. (C) Examples for insertion, deletion, gene fragmentation, and horizontal gene transfer all within a short sequence interval (*10kb*) of the H. acinonychis genome.

from the violation of the base-pairing during the extension of the replicated strand. These violations may be due to an increased error rate of the polymerase enzyme or through tautomerization reactions of the keto- or amino-forms of the four nucleotides or, more frequently, through chemical modifications of nucleotides on the template strand. In this regard it has been observed that *Helicobacter* does show a higher rate of mutation than has been observed for other bacterial genera (Björkholm *et al.*, 2001; LeClerc *et al.*, 1996; Oliver *et al.*, 2001; Velicer *et al.* 2006). As the process of single nucleotide mutations is counteracted through gene conversion, e.g. the replacement of a mutated gene copy through the wild type one, it is not possible to deconvolute the contributing effects through comparative genomics. Future studies are needed to test the activity of purified DNA modifying enzymes of *Helicobacter in vitro* in comparison to enzymes of other bacterial model systems.

Lastly, during replication, slip strand mispairing (SSM) can introduce deviating numbers of nucleotides between template and replicated strand, thereby resulting in a temporary mispairing between the two. SSM is generally corrected by error checking mechanisms that remove any loop structures, which may have been generated during replication. Nevertheless, SSM gives rise to a large number of contingency genes, also known as phase variable genes, and phase variation is a common mechanism among many pathogenic bacteria (Moxon *et al.*, 1994). While comparative whole genome analyses revealed more than 100 putative phase variable genes in three genomes of the pathogenic *Neisseria* spp. (genome size 2.2 MB) (Snyder *et al.*, 2001), only 46 putative phase variable genes were identified in the 1.6 MB genomes of the two *H. pylori* genomes 26695 and J99. A systematic PCR and sequencing analysis for the presence of these genes in a collection of 23 *H. pylori* strains representative for the global genetic diversity confirmed phase variability for 30 genes (Salaun *et al.*, 2004). Phase variation is an important mechanism in the process of host adaptation and, not surprisingly, it is often associated with virulence determinants. Contingency genes are characterized by homopolymeric tracts and dinucleotide repeats, low-complexity zones that facilitate SSM (Moxon *et al.*, 1994). As a consequence, erroneously copied contingency genes show a higher frequency of frameshift mutations, an effect that may be deleterious for many gene products but conveys biological fitness in case of many OMPs that need to be altered in their overall composition in order to evade immunodetection. In addition to genes encoding OMPs, which are the most frequent class of contingency genes in *H. pylori*, phase variation has been detected in genes of the LPS biosynthesis, fucosyltransferases, glycosyltransferases and an acetylgalactoamyl transferase (Saunders *et al.*, 1998; Salaun *et al.*, 2004), the variable expression of which will also change immunogenic bacterial cell surface components. Besides the modification of antigenic properties and thus the evasion of immune response, phase variation has also been observed for genes coding for restriction-modification (R-M) systems, the benefit of which to the bacterium is less obvious (Salaun *et al.*, 2004). However, *H. pylori* isolates exhibit significant inter-strain variation in R-M system activity, that in turn affects the strains transformability with foreign DNA (Aras *et al.*, 2002). A concurrent SSM shutdown of several genes of the R-M system might facilitate the acquisition and incorporation of foreign DNA into the genome, which would then add new and potentially useful features like the specific sialylation gene cluster that has been imported into *H. acinonychis* (Eppinger *et al.*, 2006).

Lack of DNA repair mechanisms

Missing DNA repair contributes to an increased mutation rate in *Helicobacter*. Eisen and co-workers suggested in 1997 (Eisen *et al.*, 1997) that *Helicobacter*, despite the fact that its genome harbours homologues of the MutS and UvrD genes, does not possess appropriate DNA mismatch repair. This was concluded through the observation that both genes are paralogues but not orthologues of the genes that serve these functions in other microbial systems and, accordingly, that the *mutS* gene from *H. pylori* belongs to the *mutS*2 cluster instead of *mutS*1 cluster of the *mutS* genes (Eisen, 1998). Indeed, later work provided experimental evidence that MutS dependent mismatch repair was absent from *H. pylori* because inactivation of the *mutS* gene in a variety of clinical isolates had no substantial effect on the mutation frequency (Björkholm *et al.*, 2001). In fact, the spontaneous mutation rate in the individual *H. pylori* strains varied between 10^{-5} and 10^{-7} (Björkholm *et al.*, 2001) which was comparable to that in mutator strains of *Escherichia coli* (LeClerc *et al.*, 1996) and *Pseudomonas aeruginosa* (Oliver et al., 2001) but which was orders of magnitude higher than the average *E. coli* mutation rate. The mutation rate in *H. pylori* is so high that it should be deleterious, but it is obviously not. A likely reason is gene conversion or so far unrecognized DNA repair systems.

An important prerequisite for effective gene conversion is an efficient DNA uptake system. The DNA of some naturally transformable pathogenic bacteria, for example Neisseria and Haemophilus, contains abundant specific sequence motifs that enhance the uptake of within-species DNA during transformation (Smith *et al.*, 1999). These uptake sequences, that facilitate the intra-specific horizontal gene transfer, are often located 3′ of ORFs where they are believed to be involved in transcription termination. However, although *H. pylori* DNA does not possess any specific DNA uptake signals (Saunders *et al.*, 1999), it is nevertheless highly efficient in natural transformation. The DNA uptake involves a type IV secretion apparatus that is encoded by the *comB* locus (Hofreuter *et al.*, 1998; Hofreuter *et al.*, 2001), but is apparently also facilitated by a number of other genes (Smeets *et al.*, 2002; Yeh *et al.*, 2003). Natural transformation and recombination in *H. pylori* is very frequent, in fact its recombination rate is by far exceeding that of other bacteria (Suerbaum *et al.*, 1998). A mathematical model based on Bayesian statistics estimated a recombination rate per nucleotide of 6.9×10^{-5} (Falush *et al.*, 2001), a value that is similar to *H. pylori*'s mutation rate in antibiotic resistance genes (Björkholm *et al.*, 2001). However, frequent recombination is not restricted to *H. pylori* as increased rates of recombination are also observed in other ε-proteobacteria. Intraspecific recombination is also frequent in *C. jejuni* (Suerbaum *et al.*, 2001) where it has generated abundant allelic diversity from a small number of polymorphic nucleotides.

Putting things to a test: the *H. acinonychis* genome

The closest relative of *H. pylori* is *H. acinonychis*, a bacterium infecting the stomach of large felines. It was named after the host from which it was first isolated, the cheetah *Acinonyx jubatus* (Eaton *et al.*, 1993), and has since been isolated from lions and tigers suffering from gastritis (Schroder *et al.*, 1998; Dailidiene *et al.*, 2004). However, all currently available *H. acinonychis* strains were isolated from captive animals, and attempts to identify this bacterium among free-living carnivores have failed (Terio *et al.*, 2005; Linz *et al.*, unpublished). The close relation between *H. pylori* and *H. acinonychis* is surprising, because hosts more closely related to large felines, namely house cats and dogs, harbour the more distantly related *Helicobacter* species *Helicobacter felis*, *Helicobacter salomonis* and *Helicobacter bizzozeronii* (Fig. 10.1; Solnick and Schauer, 2001). This implies a jump between host species followed by adaptation to a new host. It has been suggested that *H. pylori* evolved from a recent zoonosis from a domesticated animal because isolates from Peruvians with primarily Amerindian ancestry showed higher sequence similarity to Spanish as opposed to East Asian strains (Kersulyte *et al.*, 2000), which was interpreted as the introduction of *H. pylori* by European conquerors and hence the absence of *H. pylori* from the Pre-Colombian Americas. However, the subsequent identification of sequences of East Asian ancestry among strains from Native Americans suggested that *H. pylori* was already associated with humans when Asian peoples crossed the Bering strait >13,000 years ago (Ghose *et al.*, 2002; Yamaoka *et al.*, 2002; Falush *et al.*, 2003). This argues against a zoonotic origin from a domesticated animal, because domestication of animals started ca. 10,000 years ago, perhaps with the exception of dogs, whose domestication from wolves possibly happened somewhat earlier (Diamond, 1997).

However, the above analyses were based on sequences of housekeeping genes, which do not provide evidence for the directionality of a host jump. Thus the question remained: Who ate whom? Did a carnivore get infected by eating a human who was colonized with *H. pylori* or did a human infect himself by eating a large feline carrying *H. acinonychis*? In order to answer this question, the complete genome sequence was determined from the *H. acinonychis* strain Sheeba (Eppinger *et al.*, 2006).

The Sheeba genome is very similar to those of *H. pylori* 26695 and J99, and all three genomes share numerous coding sequences with high mean genetic similarity between orthologous CDSs. However, it is not orthologous CDSs but rather genomic differences that may be important for the host jump. Genes that become redundant or even deleterious in the new host will be inactivated and/or subsequently lost. Indeed, the frequency of fragmented genes was particularly high in the Sheeba genome, with 255 CDSs derived from 92 fragmented genes scattered around the chromosome, which were the result of numerous frameshifts and stop codons. The most frequent class of fragmented genes encodes hypothetical proteins (45 fragmented genes), and their relevance to the host jump is unclear. Other frequent gene classes encode eight transport systems and six transposases. Fragmented genes that potentially played a role in the host jump encode 12 OMPs, 11 R-M enzymes, three fucosyl transferases and two VacA vacuolating cytotoxins. The two *vacA* orthologues probably arose via a duplication event after fragmentation into 13 CDSs, because their sequences differed by only one nucleotide (Fig. 10.2B). Even among the 1306 genes that are common to the genomes of Sheeba and the two *H. pylori*, 45 are fragmented in Sheeba but intact in 26695 and J99, whereas only one gene is fragmented in both 26695 and J99 but intact in Sheeba. Furthermore, over ten times as many genes in the Sheeba genome are fragmented as in J99, and almost three times as many as in 26695. Thus, the frequency of fragmented genes strongly implies that the host jump occurred from humans to large felines. Sequencing of ten fragmented genes, including seven coding for OMPs, two hypothetical proteins and one VacA orthologue, from three additional *H. acinonychis* isolates confirmed that fragmentation is found throughout the species and thus was probably a crucial event for the host jump (Eppinger *et al.*, 2006).

In addition to functional inactivation of deleterious genes, the host jump may have been facilitated by the acquisition of novel genes by horizontal gene transfer (HGT). Among the 37 genes that were uniformly present in all seven *H. acinonychis* and absent from all 23 representative *H. pylori* strains we identified a five-gene cluster, consisting of *neuACB* and two *cst* genes, which are syntenic with those of the *Bacillus cereus* virulence plasmid pBC218 and the chromosome of *C. jejuni* NCTC11168. These genes encode proteins involved in sialic acid synthesis (*neuACB*) and transfer (*cst*) which alter bacterial cell surface carbohydrates.

When did it happen? Ds, the synonymous genetic distance, calculated from 612 conserved orthologues from all three genomes, was twice as high between Sheeba and *H. pylori* as Ds was between 26695 and J99. Strain 26695 belongs to the population hpEurope and J99 to hpAfrica1 (Falush *et al.*, 2003). The split between 26695 and J99 was set to 100,000 years ago, which was justified by the intimate association between humans and *H. pylori* and the assumed split between African and non-African humans approx. 100,000 years ago. This resulted in a suggested date for the host jump approx. 200,000 years ago (Eppinger *et al.*, 2006). However, this divergence time from the last common ancestor between *H. pylori* and *H. acinonychis* probably constitutes an overestimate by at least twofold. Firstly, later work (Linz *et al.*, 2007) estimated an African origin for *H. pylori* populations other than hpAfrica2 to 60,000 years ago, which in turn suggests that the last common ancestor of *H. pylori* and *H. acinonychis* still existed 120,000 years ago. Secondly, the adaptation to a new host via simultaneous functional inactivation of so many genes requires a mutational burst (Fig. 10.3), which also results in an elevated number of the synonymous differences that were used for the age calculation. Thus, a more realistic estimate for the date of the host jump is probably 100,000 years. Accordingly, the synonymous molecular clock rate for *H. pylori*, which was estimated to be $6.2–9.2\times10^{-7}$ should be adjusted to $1–1.5\times10^{-6}$, a clock rate three orders of magnitude faster than the molecular clock rate for the endosymbiotic bacteria from aphids, *Buchnera acidicola* (2×10^{-9}) (Moran *et al.*, 1993). Such a high mutation rate is close to being deleterious, and all clinical *H. pylori* strains may therefore be considered mutator strains (Wang *et al.*, 1999),

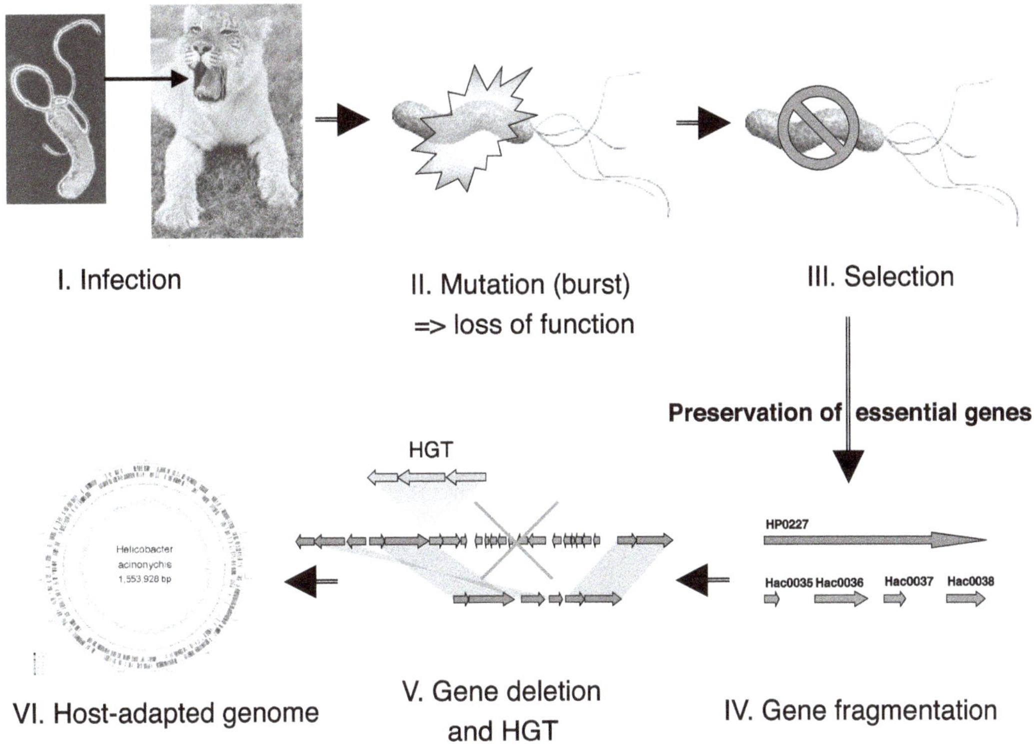

Figure 10.3 Model for host jump. (HGT: Horizontal Gene Transfer)

unlike *E. coli* where mutants in mismatch repair are rare among natural isolates. The high frequency of recombination in *H. pylori* (Suerbaum *et al.*, 1998; Falush *et al.*, 2001) may counteract this mutation rate through the replacement of deleterious mutations by imported DNA of neighbouring *H. pylori* cells, a mechanism that is also known as gene conversion. Remaining deleterious or disadvantageous nucleotide exchanges will be subsequently removed from the bacterial population by purifying selection.

A model for host adaptation

The *H. acinonychis* host jump was the first documented example for a bacterial transmission between host species on the genomic level. This gives the unique opportunity to study the mechanisms and possible causes that drive host adaptation at the molecular level within the infecting agent (see Fig. 10.2C). We therefore like to explore the series of events that are likely to occur after the transmission to a new host and propose that many of these events also come to play during the initial transmission of *H. pylori* upon inter-person contact. In case of *H. acinonychis* the transmission is likely to have occurred through a carnivore devouring a human prey, thereby ingesting gastric tissue (Fig. 10.3I). This would allow for a direct propagation of the human *H. pylori* into the gastric mucus of the large feline. Inter human transmission, however, may occur through close personal contact such as breast feeding, kissing or through smear infection. Once the incoming *Helicobacter* has reached its potential niche in the stomach, it needs to establish persistence. In the case of *Helicobacter* this involves first and most importantly to overcome immediate excretion through the flow of the mucus and the intestinal peristaltic. *Helicobacter* therefore has to adhere to the stomach epithelium by making contact to the stomach walls. This is obtained by attaching outer cell wall components to the surfaces of host cells. Involved components may include outer membrane components or the flagellar filaments. Once this has been achieved, gastric *Helicobacter* start to modify their immediate environment, which is highly acidic, through the production of ammonia by degradation of urea, using the enzyme urease. All these effects of the bacterial agent are likely to no longer remain un-noted by the host's defence system, which continuously monitors for non-self biological material. This marks the onset of an on-going struggle between host and pathogenic microorganisms, where both sides have evolved fascinating strategies that ensure their survival. In case of a host jump however there is no time to be lost for the incoming bacterium to make itself invisible to the host innate immuno-response system, as this likely will clear any detected non-commensal bacteria. Being naive about the host, bacteria may, however, opt for computing the best possible version of themselves through their sheer population numbers in combination with all the above described mechanisms that may contribute to increasing the offspring's variability. However, mechanisms to increase the dynamics of the genome, its gene content, as well as the coding variation are probably not actively controlled and employed by the microorganism, but rather result from measures taken by the host in order to destroy the infecting agent.

In this regard, part of the immune response induces severe oxidative stress that causes an increase in the mutation rate, a mutational burst (Fig. 10.3II), which results in frequent synonymous and non-synonymous nucleotide changes including premature stop codons. Furthermore, nucleotide insertions and/or deletions generate break points on one or both strands of the chromosome, thereby resulting in shifts of reading frames. This may lead to altered protein sequences together with more frequent stop codons, and in consequence to fragmented, inactivated genes. These mutagenizing events are clearly not targeted to particular genes or regions and therefore occur throughout the entire genome. If the disadvantageous mutations are not repaired in important genes, its carrier will be subjected to purifying selection, e.g. the individual will be removed from the population pool (Fig. 10.3III). Alternatively, these inferior alleles may be replaced by gene conversion, i.e. the import of released DNA from lysed neighbouring bacterial cells. This genomic repair mechanism must be of high relevance for the biological fitness of *Helicobacter*, which has maintained an unusual high rate of DNA uptake and natural competence that was not observed in other epsilon-proteobacteria.

Selection is strongly directed towards the functional inactivation of genes encoding surface epitopes that are recognized by the host's immune system, particularly OMPs, and towards changes in carbohydrate compounds of the bacteria's cell envelope through fragmentation of glycosyltransferases and fucosyltransferases from the LPS biosynthesis pathway (Fig. 10.3IV). Besides an increased mutation rate, the stress also enhances the frequency of intra-chromosomal rearrangements (Linz and Schuster, 2007), which may further contribute to the functional inactivation of those genes that are flanking the chromosomal breakpoints. Moreover, flanking genes that encode OMPs may be reorganized by the shuffling of mosaic blocks, which also varies their surface epitopes (Eppinger *et al.*, 2006).

The inactivation of genes that encode restriction and modification enzymes leads to a decreased defence against import of foreign DNA and thus facilitates horizontal gene transfer (HGT) (Fig. 10.3V). Horizontal import of genes from unrelated organisms that are already adapted to this host may play a crucial role during adaptation if genes are imported whose products alter surface components, for example the cluster of sialic acid synthesis (*neuACB*) and transfer (*cst*) genes in *H. acinonychis* that modify bacterial cell surface carbohydrates by sialylation (Eppinger *et al.*, 2006). However, a lowered defence against the import of foreign DNA may also result in the import of undesired DNA. The *H. acinonychis* Sheeba genome harbours two prophages, both of which have been imported by HGT as they were not found among the 23 representative *H. pylori* isolates analysed.

Thus, the functional inactivation by fragmentation and subsequent deletion of a large number of genes coding for OMPs as well as the modification of bacterial cell surface carbohydrates through changes in the LPS biosynthesis genes and horizontal import of new gene clusters reflect the evasion of deleterious responses from the immune system, which finally results in a genome that is adapted to the new host (Fig. 10.3VI).

References

Achtman, M., Azuma, T., Berg, D.E., Ito, Y., Morelli, G., Pan, Z.-J., Suerbaum, S., Thompson, S., van der Ende, A., and van Doorn, L.J. (1999). Recombination and clonal groupings within *Helicobacter pylori* from different geographical regions. Mol. Microbiol. *32*, 459–470.

Alm, R.A., Ling, L.-S.L., Moir, D.T., King, B.L., Brown, E.D., Doig, P.C., Smith, D.R., Noonan, B., Guild, B.C., deJonge, B.L., *et al.* (1999). Genomic-sequence comparison of two unrelated isolates of the human gastric pathogen *Helicobacter pylori.* Nature *397*, 176–180.

Aras, R.A., Small, A.J., Ando T., and Blaser, M.J. (2002). *Helicobacter pylori* interstrain restriction-modification diversity prevents genome subversion by chromosomal DNA from competing strains. Nucleic Acids Res. *30*, 5391–5397.

Björkholm, B., Sjölund, M., Falk, P.G., Berg, O.G., Engstrand, L., and Andersson, D.I. (2001). Mutation frequency and biological cost of antibiotic resistance in *Helicobacter pylori.* Proc. Natl. Acad. Sci. USA. *98*, 14607–14612.

Dailidiene D., Dailide G., Ogura K., Zhang M., Mukhopadhyay A.K., Eaton K.A., Cattoli G., Kusters J.G., and Berg D.E. (2004). *Helicobacter acinonychis*: genetic and rodent infection studies of a *Helicobacter pylori*-like gastric pathogen of cheetahs and other big cats. J. Bacteriol. *186*, 356–365.

Diamond, J. (1997). Guns, germs, and steel (W.W. Norton & Co., New York).

Eaton K.A., Radin M.J., Kramer L, Wack R, Sherding R, Krakowka S, Fox J.G., and Morgan D.R. (1993). Epizootic gastritis associated with gastric spiral bacilli in cheetahs (*Acinonyx jubatus*). Vet. Pathol. *30*, 55–63.

Eisen J.A., Kaiser D., and Myers R.M. (1997). Gastrogenomic delights: a movable feast. Nat. Med. *10*, 1076–1078.

Eisen, J.A. (1998). A phylogenomic study of the MutS family of proteins. Nucl. Acids Res. *26(18)*, 4291–300.

Eisen, J.A. (2007). Generation of variation through mutation and recombination. In Evolution, N.H. Barton, D.E.G. Briggs, J.A. Eisen, D.B. Goldstein and N.H. Patel, ed. (Cold Spring Harbor Laboratory Press).

Eppinger, M., Baar, C., Linz, B., Raddatz, G., Lanz, C., Keller, H., Morelli, G., Gressmann, H., Achtman, M., and Schuster, S.C. (2006). Who ate whom? Adaptive *Helicobacter* genomic changes that accompanied a host jump from early humans to large felines. PLoS Genet. *2*, e120.

Falush, D., Kraft, C., Taylor, N.S., Correa, P., Fox, J.G., Achtman, M., and Suerbaum, S. (2001). Recombination and mutation during long-term gastric colonization by *Helicobacter pylori*: estimates of clock rates, recombination size, and minimal age. Proc. Natl. Acad. Sci. USA 98, 15056–15061.

Falush, D., Wirth, T., Linz, B., Pritchard, J. K., Stephens, M., Kidd, M., Blaser, M.J., Graham, D.Y., Vacher, S., Perez-Perez, G.I., *et al.* (2003). Traces of human migrations in *Helicobacter pylori* populations. Science *299*, 1582–1585.

Ghose, C., Perez-Perez, G.I., Dominguez-Bello, M.G., Pride, D.T., Bravi, C.M., and Blaser, M.J. (2002). East Asian genotypes of *Helicobacter pylori* strains in Amerindians provide evidence for its ancient human carriage. Proc. Natl. Acad. Sci. USA. *99*, 15107–15111.

Gressmann, H., Linz, B., Ghai, R., Pleissner, K.-P., Schlapbach, R., Yamaoka, Y., Kraft, C., Suerbaum, S., Meyer, T.F., and Achtman, M. (2005). Gain and loss of multiple genes during the evolution of *Helicobacter pylori*. PLoS Genet. *1*, e43.

Hofreuter, D., Odenbreit, S., Henke, G., and Haas, R. (1998). Natural competence for DNA transformation in *Helicobacter pylori*: identification and genetic characterization of the comB locus. Mol. Microbiol. *28*, 1027–1038.

Hofreuter, D., Odenbreit, S., and Haas, R. (2001). Natural transformation competence in *Helicobacter pylori* is mediated by the basic components of a type IV secretion system. Mol. Microbiol. 41, 379–391.

Kersulyte, D., Mukhopadhyay, A.K., Velapatino, B., Su, W., Pan, Z., Garcia, C., Hernandez, V., Valdez, Y., Mistry, R.S., Gilman, R.H., *et al.* (2000). Differences in genotypes of *Helicobacter pylori* from different human populations. J. Bacteriol. *182*, 3210–3218.

LeClerc, J.E., Li, B., Payne, W.L, and Cebula, T.A (1996). High mutation frequencies among *Escherichia coli* and Salmonella pathogens. Science *274*, 1208–1211.

Linz, B. and Schuster, S.C. (2007). Genomic diversity in *Helicobacter* and related organisms. Res. Microbiol. *158*, 737–744.

Linz, B., Balloux, F., Moodley, Y., Manica, A., Liu, H., Roumagnac, P., Falush, D., Stamer, C., Prugnolle, F., Van Der Merwe, S.W., *et al.* (2007). An African origin for the intimate association between humans and *Helicobacter pylori.* Nature *445*, 915–918.

Liu, H., Prugnolle, F., Manica, A., and Balloux, F. (2006). A geographically explicit genetic model of worldwide human-settlement history. Am. J. Hum. Genet. *79*, 230–237.

Moran, N.A., Munson, M.A., Baumann, P., and Ishikawa, H. (1993). A molecular clock in endosymbiotic bacteria is calibrated using the insect hosts. Proc. R. Soc. Lond. B *253*, 167–171.

Moxon, E.R., Rainey, P.B., Nowak, M.A., and Lenski, R.E. (1994). Adaptive evolution of highly mutable loci in pathogenic bacteria. Curr. Biol. *4*, 24–33.

Oh, J.D., Kling-Backhed, H., Giannakis, M., Xu, J., Fulton, R.S., Fulton, L.A., Cordum, H.S., Wang, C., Elliott, G., Edwards, J., *et al.* (2006). The complete genome sequence of a chronic atrophic gastritis *Helicobacter pylori* strain: evolution during disease progression. Proc. Natl. Acad. Sci. USA *103*, 9999–10004.

Oliver, A., Cantón, R., Campo, P., Baquero, F., and Blázquez, J. (2001). High frequency of hypermutable *Pseudomonas aeruginosa* in cystic fibrosis lung infection. Science *288*, 1251–1254.

Salaun, L., Linz, B., Suerbaum, S., and Saunders, N.J. (2004). The diversity within an expanded and redefined repertoire of phase-variable genes in *Helicobacter pylori*. Microbiology *150*, 817–830.

Saunders, N.J., Peden, J.F., Hood, D.W., and Moxon, E.R. (1998). Simple sequence repeats in the *Helicobacter pylori* genome. Mol. Microbiol. 27, 1291–1298.

Saunders, N.J., Peden, J.F., and Moxon, E.R. (1999). Absence in *Helicobacter pylori* of an uptake sequence for enhancing uptake of homospecific DNA during transformation. Microbiology *145*, 3523–3528.

Schröder, H.D., Ludwig, C., Jakob, W., Reischl, U., Stolte, M., and Lehn, N. (1998). Chronic gastritis in tigers associated with *Helicobacter acinonyx*. J. Comp Pathol. *119*, 67–73.

Smeets, L.C., and Kusters, J.G. (2002). Natural transformation in *Helicobacter pylori*: DNA transport in an unexpected way. Trends Microbiol. *10*, 159–162.

Smith H.O., Gwinn M.L., and Salzberg S.L. (1999). DNA uptake signal sequences in naturally transformable bacteria. Res. Microbiol. *150*, 603–616.

Snyder, L.A., Butcher, S.A., and Saunders, N.J. (2001). Comparative whole-genome analyses reveal over 100 putative phase-variable genes in the pathogenic *Neisseria* spp. Microbiology *147*, 2321–2332.

Solnick, J.V., and Schauer, D.B. (2001). Emergence of diverse *Helicobacter* species in the pathogenesis of gastric and enterohepatic diseases. Clin. Microbiol. Rev. *14*, 59–97.

Suerbaum, S., Maynard Smith, J., Bapumia, K., Morelli, G., Smith, N.H., Kunstmann, E., Dyrek, I., and Achtman, M. (1998). Free recombination within *Helicobacter pylori*. Proc. Natl. Acad. Sci. USA *95*, 12619–12624.

Suerbaum, S., Lohrengel, M., Sonnevend, A., Ruberg, F., and Kist, M. (2001). Allelic diversity and recombination in *Campylobacter jejuni*. J. Bacteriol.. *183*, 2553–2559.

Suerbaum, S., Josenhans, C., Sterzenbach, T., Drescher, B., Brandt, P., Bell, M., Droge, M., Fartmann, B., Fischer H.P., Ge, Z., *et al.* (2003). The complete genome sequence of the carcinogenic bacterium *Helicobacter hepaticus*. Proc. Natl. Acad. Sci. USA *100*, 7901–7906.

Terio, K.A., Munson, L., Marker, L., Aldridge, B.M., and Solnick, J.V. (2005). Comparison of *Helicobacter* spp. in cheetahs (*Acinonyx jubatus*) with and without gastritis. J. Clin. Microbiol. *43*, 229–234.

Tomb, J.F., White, O., Kerlavage, A.R., Clayton, R.A., Sutton, G.G., Fleischmann, R.D., Ketchum, K.A., Klenk, H.P., Gill, S., Dougherty, B.A. *et al.* (1997). The complete genome sequence of the gastric pathogen *Helicobacter pylori*. Nature *388*, 539–547.

Velicer, G.J., Raddatz, G., Keller, H., Deiss, S., Lanz, C., Dinkelacker, I., and Schuster S.C. (2006). Comprehensive mutation identification in an evolved bacterial cooperator and its cheating ancestor. Proc. Natl. Acad. Sci. USA *103*, 8107–8112.

Wang, G., Humayun, M.Z., and Taylor, D.E. (1999). Mutation as an origin of genetic variability in *Helicobacter pylori*. Trends Microbiol. *7*, 488–493.

Wirth, T., Wang, X., Linz, B., Novick, R.P., Lum, J.K., Blaser, M.J., Morelli, G., Falush, D., and Achtman, M. (2004). Distinguishing human ethnic groups by means of sequences from *Helicobacter pylori*: Lessons from Ladakh. Proc. Natl. Acad. Sci. USA. *101*, 4746–4751.

Yamaoka, Y., Orito, E., Mizokami, M., Gutierrez, O., Saitou, N., Kodama, T., Osato, M.S., Kim, J.G., Ramirez, F.C., Mahachai, V., and Graham, D.Y. (2002). *Helicobacter pylori* in North and South America before Columbus. FEBS Lett. *517*, 180–184.

Yeh, Y.C., Lin, T.L., Chang, K.C., and Wang, J.T. (2003). Characterization of a ComE3 homologue essential for DNA transformation in *Helicobacter pylori*. Infect. Immun. *71*, 5427–5431.

Genomotyping of *Helicobacter pylori* and its Host: Microarray-Based Insights on Gene Variation, Expression and Function

11

Olivier Humbert, Delia M. Pinto-Santini and Nina R. Salama

Abstract

Whole genome microarray analysis or genomotyping has been used to query many aspects of *Helicobacter pylori* biology. Comparative genomic hybridization has yielded new insights into the genome wide extent of genetic diversity both between *H. pylori* strains from different geographic regions as well as within the bacterial population of single infected hosts. Transcriptional profiling of bacterial RNA has shown that *H. pylori* has a complex transcriptional response to a variety of environmental stresses encountered in the stomach in spite of having few annotated transcription factors in its genome. Transcriptional profiling of natural (human) and experimental hosts has revealed molecular signatures of disease progression and successful vaccination. Additional microarray based technologies for genetic screens for infectivity, localization of transcription factor binding sites and whole genome mutational analysis are beginning to provide molecular details of the mechanisms by which *H. pylori* establishes a persistent infection in the human stomach.

Introduction

Microarray technology came into use a little over ten years ago (Schena *et al.*, 1995). The basic method involves high-density printing of probes specific for all or a subset of genes in the genome of an organism on glass, nylon or another solid substrate. Designing probes to minimize cross hybridization allows simultaneous measurements of most or all the genes in the genome in a single hybridization using a very small volume (e.g. 10–20 μl under a coverslip). We refer the reader to several reviews on this technology for a more complete discussion (Dalma-Weiszhausz *et al.*, 2006; Heller, 2002; Wolber *et al.*, 2006), but will briefly highlight some features of this experimental method as it relates to the studies reviewed in this chapter.

There are several array platforms including cDNA-based arrays, long oligo-based arrays (50–70mer), and GeneChip type arrays which use short oligos and a combination of perfect and mismatch oligos to determine specific hybridization. Similarly, tiling arrays use short overlapping probes to query all bases in a particular region of the genome. The long oligonucleotide and cDNA arrays generally employ a co-hybridization strategy where an experimental and control sample are each labelled with different fluorescent dyes and the ratio of signal intensity in each dye emission channel determines the relative level of each product in the different samples. In order to compare more than two samples with this scheme, a common reference sample must be employed. The GeneChip technology hybridizes a single sample at a time and a variety of standards are used to calibrate the signal from each chip such that comparison between chips can be made. While each platform uses different hybridization protocols and data extraction methods they all provide a relative measure of each gene across samples.

All microarrays rely on the complementarity of DNA for their specificity. This allows a wide variety of different experiments to be analysed on microarrays. The most common application of microarrays is to monitor RNA transcripts levels in gene expression or transcriptional profiling studies. In this case RNA purified from each experimental condition is reverse transcribed to incorporate the label and make a complementary copy that will hybridize to the sense strand printed on the microarray. Transcriptional profiling has been used extensively to monitor changes in *H. pylori* gene expression in response to various environmental conditions, during infection of a host and in the context of putative transcriptional regulator mutants. In addition host cell transcriptional profiling has been used in a variety of studies to better understand host responses to bacterial infection.

Another common use of microarrays is to determine gene content and copy number differences in different cell populations. These studies, referred to as comparative genomic hybridization (array CGH), use genomic DNA as the template for labelling. Such studies have enhanced our understanding of *H. pylori* strain variation among clinical isolates. Tiling microarrays have been used to identify single nucleotide polymorphisms (SNP) among different clones, again increasing our understanding of *H. pylori* strain variation and adaptation to selection conditions. Recently microarrays have been used to map the chromosomal location of DNA binding proteins in the cell. For this type of analysis, called Chip on Chip, proteins are cross-linked to DNA using formaldehyde. The DNA is then fragmented and the protein of interest purified by immunoprecipitation with a specific antibody. The DNA associated with the protein can then be eluted, amplified and hybridized to the array. While this method has been applied extensively in eukaryotic systems, *H. pylori* transcription factors are just starting to be analysed in this manner. Microarrays have also been used to track the behaviour of transposon mutants both *in vitro* and during infection. In this application, the DNA sequences next to the transposon insertion site are amplified and labelled for microarray hybridization allowing localization of the insertion sites. Clearly, microarrays are powerful and rapidly evolving experimental tools that will continue to be applied to understanding *H. pylori*–host interactions.

Data analysis represents a formidable challenge in microarray experiments. Because so many measurements are performed simultaneously and each has an associated error rate, determining differences that are truly significant can be challenging. Early studies used arbitrary cut-offs such as changes of twofold or more to select genes for further analysis. Increasingly, statistical methods that incorporate analyses of variance or permutation testing are used (Dudoit *et al.*, 2003; Mills *et al.*, 2001; Reimers and Carey, 2006; Tusher *et al.*, 2001). These methods provide *P*-values, allowing detection of even subtle differences with confidence of a low false discovery rate. A consequence of the increasingly more sophisticated data analysis methods is that early papers provide smaller gene lists than later papers making cross comparisons difficult. Luckily, most researchers deposit their data in public databases, allowing the possibility of re-analysis. The programs Cluster and Treeview have been extremely useful in organizing and visualizing microarray data (Eisen *et al.*, 1998). Cluster uses hierarchical clustering to group genes with similar patterns of change across samples together. These patterns can then be visualized as a heat map graphic where intensity of the colour corresponds to magnitude of expression differences. Most microarray experiments output fairly long lists of genes with known, partially known or poorly known functions. Increasingly researchers have turned to programs that mine public annotation databases to understand whether particular functional classes or cellular process are affected in their experiments using programs such as GoMiner (Zeeberg *et al.*, 2005), DAVID (Huang *et al.*, 2007)} and GOurmet (Doherty *et al.*, 2006). This type of analysis has been particularly useful in understanding the complexity of the host response to *H. pylori* infection. As the *H. pylori* genome annotation becomes more complete through functional experiments and bioinformatic approaches, these programs will be applied to *H. pylori* microarray data sets as well.

Microarray-based genotyping of *H. pylori* strains

Before whole genome sequences and microarray techniques were available, bacterial variants were characterized by phenotypic studies including serology, biochemical tests and multilocus enzyme electrophoresis (MLEE), which generated valuable data on the diversity of bacterial populations. Pre-genome techniques for molecular typing were developed which allowed a better, although still qualitative, estimate of nucleotide level variation, such as restriction length polymorphisms (RFLP) and arbitrary primed PCR (AP-PCR). Since variations in these patterns can correspond to silent or synonymous base substitutions, without dramatic proteome changes, these techniques are good for typing but have limited utility for correlation with other phenotypic characteristics. Additionally, the identity and genome location of the genetic differences revealed by these techniques cannot be assessed. High throughput sequencing sparked the development of multilocus sequence typing (MLST), in which fragments of approximately 500 nucleotides typically from seven housekeeping genes are sequenced from several isolates. MLST is more quantitative because it reveals changes in nucleotide sequences that can be used to estimate the relatedness among isolates but genome-wide assessment of variation is not possible. MLST has also been used to gain insights into *H. pylori* sequence diversity worldwide (Achtman *et al.*, 1999; Falush *et al.*, 2003; Linz *et al.*, 2007) and for classification purposes.

Several *H. pylori* strains have been sequenced, including two *cag* pathogenicity island (PAI)+ clinical isolates, 26695 (from a patient with superficial gastritis) (Tomb *et al.*, 1997) and J99 (from a patient with duodenal ulcer disease) (Alm *et al.*, 1999). Most array CGH experiments utilize arrays containing PCR products representing genes encoded by these two strains. Recently, a third strain, HPAG1, isolated from a patient with chronic atrophic gastritis (ChAG) was sequenced (Oh *et al.*, 2006). Whole genome sequences and DNA microarrays techniques provide a global and quantitative understanding of genomic diversity and an opportunity to assess gene content variation in an unprecedented way. Techniques such as RFLP and AP-PCR are often used for an initial survey of genetic variants in the populations under study, followed by array CGH for a global assessment. There are several outcomes of *H. pylori* infection, ranging from chronic active gastritis and peptic ulcers to gastric cancer and mucosa-associated lymphoid tissue (MALT) lymphoma. These diverse outcomes have led to the hypothesis that different diseases are associated with bacterial genetic diversity. Indeed there is a high degree of heterogeneity at the DNA level among *H. pylori* strains and several studies have pointed to high rates of recombination, high mutation rates and to the ability to take up DNA as key forces generating such diversity (Falush *et al.*, 2003; Suerbaum *et al.*, 1998). Analysis of the genetic composition of the *H. pylori* population, by cataloguing the genes and allelic variants and correlating them with different phenotypes, may help determine how different strains can contribute to different kinds of disease. In the following sections we discuss what has been learned about *H. pylori* genetic variation using microarray-based whole genome comparisons.

The extent of *H. pylori* genetic variation

The extent of allelic diversity in *H. pylori* populations has been described for several virulence factors, including *cagA*, *vacA* and various adhesion genes. In each case some alleles correlate with the pathogenic potential of the strain. Correlation of certain allelic variants with geographic regions has also been described. Our group first explored the extent of *H. pylori* gene content diversity using DNA microarray by examining 15 clinical isolates on 26695-J99-based arrays (Salama *et al.*, 2000). This study identified 362 variable genes shared by only a subset of the 15 strains tested and extended the findings from comparisons of the 26695 and J99 genome sequences (Alm *et al.*, 1999), where 105 and 73 unique genes were detected, respectively. The core genome, defined by genes present in all strains studied, consisted of 1281 genes. The strain variable genes represented 22% of all *H. pylori* genes. Many of the strain-variable genes had no predicted function and 58% of those had no sequence-based database homology. When looking at the covariance of genes among strains, the *cag* PAI, *babA* and two restriction modification systems were shown to vary together (reviewed by (Joyce *et al.*, 2002). The nature of this covariance is not understood, but could reflect selection for this combination of genes. No correlation was found with a particular disease outcome (Joyce *et al.*, 2002) although this observation was not conclusive since the collection of strains examined was not from a structured epidemiological study and did not include strains from asymptomatic patients.

A second array CGH study analysed a collection of strains previously characterized by MLST that represent geographically diverse hosts (54 total) (Gressmann *et al.*, 2005). Array CGH of these strains further reduced the core *H. pylori* genome to 1150 genes shared by all the strains in this collection. The functional categories of the variable genes included restriction modification genes, insertion sequences, transposons and outer membrane proteins. Several studies have confirmed those categories as the most variable (Israel *et al.*, 2001; Kivi *et al.*, 2007; Kraft *et al.*, 2006; Oh *et al.*, 2006; Salama *et al.*, 2000). Many genes in this flexible pool also encode *H. pylori*-specific proteins or conserved proteins of unknown functions. As observed previously (Salama *et al.*, 2000), many of these genes are located in the plasticity zones and *cag* PAI, but the authors also identified other variable regions scattered around the genome called rA to rX. Altogether, these observations highlight the dynamic nature of the *H. pylori* genome and provide many avenues for future investigation.

The core genome is present in all strains and up to 25% of genes vary between strains. Characterization of the genes in the 'core' pool will define the essential components of *H. pylori's* lifestyle. Questions regarding the 'flexible' pool include: How are the variable genes gained or lost? What are the adaptive capabilities conferred by this variation? Phenotypic analysis of these genes will provide clues about the adaptive capabilities of this bacterium and identify the source of this gene pool.

All array CGH-based studies provide information about coding sequences, allowing detection of presence or absence of genes that are only in the reference strains. Variation in intergenic regions has not been analysed because non-coding sequences are not represented on the arrays that have been used. Identification of a gene as absent may result from the complete absence of the coding sequence, the presence of divergent sequences or the insertion of a completely new sequence derived from another source. A program called GACK has been developed for microarray-based genomotyping analysis, which estimates the level of hybridization of a particular gene to its counterpart in the array giving a probability for absence or sequence divergence of the gene (Kim *et al.*, 2002). Most studies have not used this kind of analysis so the true reason for lack of detection by arrays remains undetermined. Sequencing from the flanking regions of the predicted absent gene is needed to understand the nature of this variation. The arrays platforms used in the studies described here cannot detect other genetic changes like point mutations and frame shifts, which can also have a significant effect at the transcriptome and proteome level. The implementation of microarray technology like 'comparative genome sequencing', CGS (Albert *et al.*, 2005) in the study of genetic variation of *H. pylori* is promising in this regard. This method uses tiling arrays for a sequenced bacterial genome to map mutations approximately in a first step and then locate and identify the precise mutation by using custom-designed resequencing arrays in a second step. This approach was used for the detailed characterization of the genetic changes associated with the acquisition of metronidazole resistance by *H. pylori* (Albert *et al.*, 2005).

How and when is the genetic diversity generated

Although many studies have shown that *H. pylori* gene content is variable, we do not know how and when this diversity is generated or the implication of bacterial genome variation on disease outcome. *H. pylori* strains isolated from human hosts with different diseases, at different times post-infection, in the context of disease progression or within families have been analysed to address this issue (Israel *et al.*, 2001; Kivi *et al.*, 2007; Kraft *et al.*, 2006; Oh *et al.*, 2006; Salama, 2007). Studies using animal hosts have also been undertaken to look at the evolution of strains during transmission to new hosts as well as to correlate gene content with pathologic outcome (Bjorkholm *et al.*, 2001; Bjorkholm *et al.*, 2002; Solnick *et al.*, 2004; Thompson *et al.*, 2004)

Human studies

In order to study genomic changes during chronic infection, Israel, *et al.* evaluated the genomic changes of strains infecting the patient from which strain J99 was obtained after 6 years of evolution in the gastric mucosa (Israel *et al.*,

2001). Thirteen additional isolates were obtained from this patient, which differed in gene content from the J99 original strain but were highly related to it. In addition to gene loss, gain of genes was observed. Where did these new genes come from? These results prompted the hypothesis that a transiently infecting strain contributed these genes.

Kraft *et al.* extended the previous study of only one patient by analysing 21 pairs of isolates from patients at two different times during infection, separated by three months to four years (Kraft *et al.*, 2006). The analysis used genomic microarrays to assess genetic changes in *H. pylori* strains during chronic infection, but did not correlate changes with progression of disease. Of the 21 pairs only nine were found to have potential genomic changes at one or more loci as suggested by differential hybridization on the array. Upon sequencing changes from five of the pairs were confirmed and included both gain and loss of sequences in six regions, as well as the presence of divergent sequences most probably due to homologous recombination events in two other regions. Changing loci included the *cag* PAI, plasticity zone, insertion sequence elements and restriction modification systems as well as housekeeping genes. This collection of isolates had been previously analysed by MLST to determine the rate of recombination and mutation allowing the authors to estimate that gene content changes were 650 times less frequent than homologous replacement events. These results provide more evidence that diversity is generated by recombination between co-colonizing strains either transiently or persistently living in the stomach. Alternatively, intragenomic recombination between members of gene families could also give rise to variation of this type. To distinguish between these possibilities, more isolates need to be analysed at each time point to determine whether multiple strains did colonize these hosts. This study does, however, confirm the importance of recombination as a force generating *H. pylori* diversity.

The above mentioned studies used populations with a low prevalence of *H. pylori* infection, where the probability of mixed infection is lower than in high prevalence populations. Our group recently analysed the genetic composition of isolates at different times post-infection in a Mexican population, where prevalence of infection is 80% (Salama *et al.*, 2007). Multiple colonies isolated from 12 patients were analysed by RAPD and AFLP. Of the 12 patients, 11 were colonized by only one strain population. Array CGH showed more than 95% of the gene loci were conserved in isolates from the same patient with only 24–67 genes varying between single colony isolates. In the single patient that had a mixed infection, multiple colonies of each strain type were observed. Using the array results as a guide, over 6 kb were tested by both sequencing and PCR fragment length polymorphisms to detect possible recombination events. Only one such event was observed among 16 single-colony isolates analysed, suggesting that even during mixed infection, genetic exchange among distinct strains is rare.

If transmission represents a bottleneck that reduces diversity by strongly selecting colonization-competent clones that then diversify over time through mutation and recombination, isolates taken from children (most recent infections) may show genetic homogeneity, as compared with isolates taken from adults (longer time to diversify). However, in the Mexican population, both children and adults were infected with multiple but related strain variants (Salama *et al.*, 2007), suggesting that either variation is selected very close to the time of transmission, or that during transmission a population of strain variants are passed to the new host. Analyses from a human challenge study suggested that transmission to a new human host does not represent a strict bottleneck because no gene loss or divergence could be detected up to three months after transmission of a single strain to four unrelated hosts (Salama *et al.*, 2007).

In order to obtain evidence of transmission of a population of strains during natural infection, analysis of strain populations within families is required. Kivi, *et al.*, analysed variation of strains within a single family by collecting isolates from one mother and her three children (Kivi *et al.*, 2007). Isolates were typed by PCR-based typing, sequencing of five loci, and array CGH. The mother was infected by two different strains and the children were all infected with only one strain closely related to one of the strains infecting the mother. Isolates from the same individual differed by 0.1% and related isolates from different individuals by 0.2%. The two different strains identified in the family differed by 8% and by 18% relative to the sequenced strains 26695 and J99. This study does not support the hypothesis of induced variation during transmission. Analysis of more families will be required to further test this idea. It is also possible that other kinds of changes not detected by microarray approaches (e.g. point mutations or epigenetic changes) are responsible for adaptation during transmission to new hosts.

To gain insights into the evolution of strains during disease progression, Oh *et al.* characterized strains colonizing the gastric environment over time in individuals with different (evolving) gastric pathologies (Oh *et al.*, 2006). They genotyped isolates from patients with chronic atrophic gastritis (ChAG), a phenotype associated with higher risk of developing gastric cancer, who were enrolled in a case–control epidemiological study of gastric cancer. They also analysed isolates collected during progression from ChAG to gastric adenocarcinoma within a 4-year period. One strain from a ChAG patient (HPAG1) was selected for whole genome sequencing. HPAG1 could establish residency in gastric epithelial stem cells expanded in a gnotobiotic transgenic mouse model. This model has an expanded gastric epithelial stem cell population because of loss of acid producing parietal cells, a condition also seen in humans with ChAG and gastric adenocarcinoma. Genomic arrays constructed based on the HPAG1 sequence allowed comparison with strains from other patients in the same study with the same disease diagnosis. Analysis of the HPAG1 genome sequence, identified 1536 genes, of which 43 were unique to the strain, as compared to J99 and 26695. Comparisons with seven other ChAG strains showed 1025 common genes, while 12 genes were unique to HPAG1. Of the 1025 genes shared between the ChAG strains, 121 were absent from the 1150 genes in the so-called pan-genome. The authors proposed that these strain variable genes represent possible 'ChAG-associated' genes. Whether this set of genes was selected during progression to this pathology or if they have a direct impact on disease progression remains to be explored. The authors looked for changes in gene content after inoculation of HPAG1 into two mouse backgrounds with different genetic makeup but found no gene changes, even one year post-infection. This observation could reflect lack of selective pressures for diversification in the mouse stomach or indicate that these strains are at a dead-end point for the ability to diversify. Comparison of strains taken from a patient with chronic gastritis which progressed to gastric carcinoma revealed loss of DNA repair-related genes, suggesting that genetic instability could have been a selected trait during disease progression, a period where the changing gastric environment poses pressure on the bacterial population for adaptation to a new niche.

Animal studies

Animal models have been used to explore *H. pylori* adaptation to new hosts and to test whether certain genotypes are associated with specific disease outcomes. Bjorkholm, *et al.* (Bjorkholm *et al.*, 2001) took two isolates from the stomach of a 90-year-old patient with gastric ulcer, and characterized them by array CGH against 26695-J99 coding sequences. Comparison of the two isolates revealed: one lacking the 27 genes of the *cag* PAI plus 10 additional genes (37 total) and the other lacking 29 genes but containing the *cag* PAI. Comparisons of the isolates with another 14 strains using hierarchical cluster analysis showed that the isolates were more closely related to each other than to any other of the genotyped strains. Nevertheless, the two isolates differed in their ability to infect mice. The *cag* PAI+ clone was able to infect both conventionally raised mice and germ-free mice but the *cag* PAI– clone colonized only germ-free mice. In order to determine what changes occurred during adaptation to mice, the *cag* PAI+ clone was compared with post-infection isolates from both conventionally raised mice and germ-free mice. In both scenarios the isolates post infection remained cagA+, maintaining all the *cag* PAI genes. Whether the PAI remained functional (able to translocate CagA or induce IL-8) is unknown. Sequencing of three genes (*cagA*, *recA* and *16srRNA*) showed no nucleotide substitutions, even when passed 50 times *in vitro*. These results suggest that substantial genomic changes are not required to adapt to the mouse stomach. Other possible explanations include lower adaptive selective pressures in mice as compared to humans or simply that more time is required for changes to occur.

Thompson *et al.* also studied the evolution of gene content during adaptation of human clinical isolates to the murine gastric environment (Thompson *et al.*, 2004). They compared two mouse-adapted strains: SS200 (which they mouse adapted in this study) and SS1 (a previously mouse adapted strain). When comparing the original isolates to the mouse-adapted isolates, a few changes in gene content were detected during adaptation. These results contrast with observations by Bjorkholm *et al.* (2001), who could detect no changes during adaptation to mice and additionally suggested that changes might be strain-specific. Interestingly, Thompson *et al.* identified the loss of the adhesion gene *babA* in the pre-mouse isolates as well as the mouse-adapted ones.

Solnick *et al.* observed loss of *babA* during adaptation of the human *H. pylori* isolate J166 to a rhesus macaque host by array CGH (Solnick *et al.*, 2004). By sequencing the flanking regions where the array detected the change, they discovered that loss of *babA* signal on the array was due to new sequences inserted on the *babA* locus. In one case the original *babA* was replaced with *babB* sequences by gene conversion. They also detected a case where *babA* was not expressed due to the generation of a frameshift mutation caused by slipped-strand mispairing in a 5′ poly(CT) tract of the gene. The phenomenon of gene conversion occurs *in vitro* but at a very low frequency, suggesting that adaptation to the new host selected the replacement of *babA* with *babB*. These changes might have a selective advantage during persistence and indicate a requirement for *H. pylori* to remodel its cell surface during adaptation to a new host. These results also illustrate the need to devise ways to detect other changes than just gain or loss of genes on a global scale.

In an effort to identify genomic changes associated with a more pronounced immune response another study by Bjorkholm *et al.* (2002) genotyped clinical isolates from a case–control study for gastric cancer. In previous work they had identified an association between the ability to bind to gastric mucosa and a more pronounced host inflammatory response. Based on these results the clinical isolates were classified by their ability to bind gastric tissue sections taken from two different mouse backgrounds, one expressing Lewis b-containing receptors, relevant to human gastric tissue, the other expressing NeuAcα2,3Galβ1,4-glycan receptors, which are expressed during gastric atrophy. Although some of the strains that bind these receptors *in vitro* could colonize mice expressing the human receptors, no correlation was found between levels of expression of selected biomarkers induced during mice infection, such as Hsp70, pIgR and IL-1β and ability to bind these receptors *in vitro*. However, there were strain differences associated with tissue pathology. The authors identified several strains, designated 'robust response' strains, which induced more severe inflammation in both mouse backgrounds. To establish a genetic signature of this response they selected genes that were present in most of the strains causing a robust response in mice and absent in most of the strains unable to colonize mice. Several genes satisfied this criterion in both mouse backgrounds including genes belonging to the restriction-modification category. Since the presence of this category of genes correlated with a more robust response in both mouse backgrounds, these genes may regulate some aspect of gene expression, perhaps through epigenetic regulation. The challenge will be to determine how (if) restriction-modification genes impact bacterial gene expression and disease outcome in humans.

Conclusions from *H. pylori* gene content studies

The use of genome-wide microarray analysis of *H. pylori* strains has allowed an exquisite description of bacterial gene content across different patient populations and experiments have also begun to explore when and how diversity is generated as well as its possible impact on disease outcome. Interestingly, two levels of variation are detected, one reflecting the long-term evolutionary history of *H. pylori* with their human hosts as revealed by the diversity of gene content when comparing strains across global populations. The second level was revealed by single-colony isolates from the same host to show that they differ both at the single nucleotide level and in gene content. While these isolates from the same host do not have identical genotypes, they are much more related to each other than to isolates from unrelated hosts. This has led to the idea that people are infected with a population of related strain variants perhaps reflecting short-term evolution within an individual human host during disease progression and/or transmission. More experiments are required to explore whether the short-term genetic variation does affect disease progression and to characterize intrafamilial transmission. Sequencing of strains from patients with more severe disease (gastric cancer or MALT lymphoma) might reveal bacterial genetic changes not detected in strains from patients with gastritis and ulcers. Several studies have confirmed that the variable genes are enriched with genes of the *cag* PAI and plasticity zones, and include various functional categories such as transposases, restriction-modification enzymes and outer membrane proteins. Although some outer membrane proteins have been implicated with more severe disease outcomes, or adaptation to new hosts, the role of the other functional categories is still unclear.

Bacterial transcriptome studies

Using DNA microarrays to elucidate regulons and stimulons in bacteria

The understanding of bacterial pathogenesis relies on the investigation of the underlying genetic and biochemical mechanisms used by microbes to modify the host environment provided. Transcriptional regulation constitutes a primary and rapid response used by microbial organisms to respond to changes in environmental conditions, such as nutritional status and environmental stresses. Due to the central role that this type of regulation plays in bacteria, the elucidation of regulatory networks has become a key but challenging task in the field of microbiology.

Profiling of gene expression using microarrays has provided a powerful approach to study *regulons*, a set of transcriptional units controlled by a single regulatory protein, and *stimulons*, a group of transcriptional units that are differentially expressed in response to environmental perturbation. To identify a regulon, the standard protocol is to compare the expression profile of strains carrying an active form of the regulator with strains carrying a less active or inactive form under inducing conditions. The study of stimulons typically relies on the identification of genes that are differentially expressed in a wild-type strain after perturbation of the environment. These transcriptional responses can be large and complex when they encompass multiple regulons.

Characterization of the physiological state of *H. pylori* during infection

One would ultimately like to use microarrays to study pathogens during infection to visualize gene expression in complex multicellular environments. For *H. pylori* this effort has been particularly challenging because of the restricted number of bacteria that can be recovered from infection models. The methods currently available limit the yield of bacterial RNA obtained after purification from host debris. Despite these difficulties, some studies have attempted to profile the physiological state of *H. pylori* during infection, either by examining bacterial gene transcription in coculture with a gastric cell line or after isolation from the gerbil stomach.

Although fundamental differences exist between cancer-derived gastric cell lines and the epithelium of the stomach, Kim *et al.* analysed the transcription profile of the *H. pylori* strain 69a in coculture with AGS cells (Kim *et al.*, 2004). Without purifying the bacterial RNA from the eukaryotic RNA and using a radiolabelling approach, the authors identified 22 upregulated and 21 downregulated genes under the conditions examined as compared with *H. pylori* grown *in vitro* under standard conditions. The upregulated genes included some in the pathogenicity island and genes involved in cell growth such as the termination factor Rho and the ATP-dependent RNA helicase DEAD-box family. Surprisingly, adhesion of *H. pylori* to AGS cells resulted in upregulation of only one of the 27 *cag* PAI genes (*cag3*). Flagellar genes and outer membrane proteins were both up- and downregulated, suggesting major changes in the composition of the cell envelope perhaps to optimize *H. pylori* adhesion to epithelial cells. Extrapolation from this study to the stomach environment is limited because high multiplicity of infection (1,000:1), a relatively short incubation time (4 hours), and because some of the microarray data could not be validated by qPCR or RT-PCR. The authors speculate that this discrepancy is attributed to low levels of gene expression since no signal contamination from the AGS cells RNA was found.

To more accurately assess *H. pylori* gene transcription in the stomach, the same group isolated bacterial RNA from infected gerbil stomachs following 10 days of infection (Scott *et al.*, 2007). *H. pylori* RNA was isolated from the fundic and antral mucosa, enriched using bacterial mRNA enrichment kits and compared with *H. pylori* RNA from *in vitro* cultures. Although the authors were specifically interested in differentially regulated genes involved in acid acclimation, they identified a large number of upregulated genes that could be divided into 16 functional groups. Many of the genes identified are under the control of the acid-sensing two-component system ArsS/ArsR (such as *rocF*, *urea A/B*, *urea E/G*, *urea I*, amiE/F), which underscores that gastric acidity is a major stress encountered by *H. pylori* during infection. The transcription of several motility and chemotaxis genes was also increased more than twofold, as were nine genes of the *cag* PAI (including *cagA*), many of which are also regulated by low pH *in vitro*. While these studies represent a starting point for understanding the transcriptional program *in vivo*, the low sensitivity remains a major issue.

Characterization of H. pylori *stimulons in* vitro

To circumvent the technical difficulties associated with the isolation of sufficient bacterial RNA from a complex multicellular environment, an alternative approach has been to replicate certain aspects of the *in vivo* environment in broth culture. For the study of *H. pylori*, these include acidic pH, iron starvation and stationary phase of growth. Due to its unique niche, many of the survival strategies used by *H. pylori* are not conserved in other well-studied bacterial species. Microarray technology, because it allows an unbiased look at all genes, constitutes an ideal tool to elucidate how *H. pylori* gene transcription is modulated in response to various stresses. In this section we summarize the results from several studies that profiled global gene expression changes to environmental conditions likely encountered by *H. pylori* during infection (Table 11.1).

Growth phase-dependent gene expression

Thompson *et al.* used an *in vitro* time course experiment to analyse the gene expression profile of *H. pylori* strain SS1 during transition from logarithmic to stationary phase of growth (Thompson *et al.*, 2003). Following stringent filtration of the data, this group found that most genes (80%) did not vary significantly over time, but the expression of a total of 325 genes was increased or decreased by at least two-fold. The repressed genes generally included ribosomal genes, genes involved in DNA synthesis, transcription and translation, genes encoding transport and binding proteins, and genes involved in energy metabolism. These transcriptional changes reflect lower growth rates as cells transition from log phase to stationary phase. The induced set of genes was quite heterogeneous but strikingly included many virulence genes such as the cytotoxin-associated gene *cagA*, the neutrophil-activating protein (*napA*), the major structural component of the flagella flaA, and genes involved in the stress response (*clpB* and *dnaK*). A complex array of stimuli encountered during stationary phase, including nutrient depletion, accumulation of toxic by-products and crowding probably account for many of these transcriptional changes. This study demonstrated the importance of examining several time points when examining the transcriptional response of *H. pylori* to environmental changes and of comparing bacterial cultures that have been isolated in identical growth phases.

Acid acclimation in H. pylori

Owing to the major fluctuations in pH that occur in the stomach, *H. pylori* most likely encounters a wide range of acidity during persistent infection of the gastric mucosa. Unlike enteric microorganisms that evolved mechanisms to survive acidic conditions when they transit through the stomach, *H. pylori* has developed unique acid-acclimation pathways that are compatible with growth in low pH, which have been the subject of excellent reviews (Sachs *et al.*, 2005). Several

Table 11.1 Genome-wide transcriptional profiling of stimulons in *H. pylori*

Perturbation	Conditions	No. of genes regulated	*H. pylori* strain used	Reference
Gastric cell line (AGS)	4 h cocultures	43	69a	(Kim et al., 2004)
Gerbil stomach	10 days post inoculation	N/A	69a	(Scott et al., 2007)
Growth phase	Time course: 6, 12, 18, 24, 30, 36, 42, 48 and 60 h (starting OD_{600} 0.05)	325	SS1	(Thompson et al., 2003)
Acid	pH 5; 30, 60, 90, 120 min	118	G27	(Merrell et al., 2003a)
	pH 7.4, 6.2, 5.5, 4.5; 30 min; +/– urea	~300	26695	(Wen et al., 2003)
	pH 7; 4 h vs. pH 5; 9 h	101	26695	(Bury-Mone et al., 2004)
Iron depletion/ repletion	Chelator (DPP 200 μmol/l) 2 h	183	SS1	(Merrell et al., 2003b)
	Exponential and stationary phase			
	Depletion with chelator (desferol); repletion with ferric chloride (100 μmol/l) 20 h	97	26695	(Ernst et al., 2005)

groups have examined the effect of short-term exposure (30–120 min) to acidic pH *in vitro*. Although the variety of approaches and procedures employed in these studies make it challenging to directly compare data sets, compilation of these results reveals some common themes.

Merrell *et al.* employed mildly acidic media (pH 5.0) to study gene expression over time (30, 60, 90 and 120 min) in *H. pylori* strain G27 (Merrell *et al.*, 2003a). A total of 118 genes (7% of the genome) showed at least a twofold expression change in at least one timepoint. As expected, genes of the urease operon were upregulated and several outer membrane protein genes were downregulated by acid. The expression of only one repressed gene, *cagA*, was validated using Western blot analysis. One caveat of this experiment is the decreased growth rate of *H. pylori* under acidic conditions. Although, the bacteria remained viable during the course of the experiment the inhibition of bacterial growth in acid could result in numerous changes in gene transcription that are indirect. This difference in growth rate also raises the risk of isolating RNA from bacteria that are in different growth phase, which may have major impacts on gene expression (Thompson *et al.*, 2003).

Wen *et al.* attempted to use more physiological conditions by carrying out their experiments in the presence or absence of 5 mmol/l urea (Wen *et al.*, 2003). Urea in gastric juice is maintained at the same level as in blood, between 1 and 3 mmol/l (Sachs *et al.*, 2005). Intrabacterial urease activity converts urea into ammonia and carbon dioxide, both of which enable acid acclimation in *H. pylori*. Wen *et al.* exposed *H. pylori* strain 26695 to decreasing levels of pH (7.4, 6.2, 5.5, 4.5) for 30 min with or without urea, after which the cytoplasmic pH was measured using a pH-sensitive fluorescent probe. As predicted, urea helped *H. pylori* to maintain a more neutral intracellular pH. Microarray analysis demonstrated a progressive increase in differentially regulated genes as the pH was lowered. At the lowest pH examined, pH 4.5, approximately 200 genes were upregulated and 100 downregulated. The addition of urea significantly decreased this number, presumably because it allowed *H. pylori* to raise the pH of its periplasm/cytoplasm. However, the transcriptional response to urea independent of pH was not rigorously examined, and we cannot exclude the possibility that urea affects the expression of *H. pylori* genes that do not belong to the acid acclimation stimulon. In this experiment, the growth rate of the bacteria was not measured, but the short exposure time (30 min) minimized changes in growth phase of the culture. The transcriptional changes for several genes were validated by real-time PCR analysis.

Bury-Moné *et al.* also used the *H. pylori* strain 26695 to compare the expression profile following incubations at pH 7.0 for 4 h and pH 5.0 for 9 h (Bury-Mone *et al.*, 2004). The authors chose distinct exposure times for the two conditions used in an effort to keep the bacteria in a similar growth phase (as measured by optical density). Following statistical analysis of the results, 56 and 45 genes were upregulated and downregulated under acidic conditions, respectively. To confirm these results, the authors carried out multiple functional assays such as real- time PCR, dots blots, and enzymatic activities for urease, amidase, and formamidase.

The direct comparison of the results of these three studies offers only limited overlap in the genes identified that may result from the use of different strains, growth phase of the culture, and the degree and nature of acid exposure. However, several important consistencies exist. Acidic pH upregulates several motility genes (*flgB* and *flaB*), genes involved in amino-acid metabolism (glutamate dehydrogenase *gdhA*, glutamine ABC transporter *glnH*), genes generating ammonia (ureI, ureF) and the gene for amidase. Consistently downregulated genes included *omp27*, *ptuA* and *frpB2*. The downregulation of several outer membrane proteins suggests *H. pylori* alters its membrane composition in low pH. Overall, these transcriptome analyses have greatly contributed to the demonstration of the existence of at least three regulons in the acid stimulon of *H. pylori*. The three major regulators in this network are the two-component system ArsRS (HP0165/0166), the metal-dependent regulator NikR (HP1338), and the ferric response regulator Fur (HP1027), which will be discussed in the next section.

Response to Iron availability

The bioavailability of metal ions such as iron and nickel are known to fluctuate considerably in the human stomach, as their solubility increases under acidic conditions. Although these metals are essential for the activity of a number of enzymes, their accumulation is highly toxic for the metabolism of bacteria (Touati, 2000). It is therefore not surprising that *H. pylori* has developed several strategies to tightly regulate the intracellular concentrations of these metal ions. Iron sources available in the gastric mucosa are lactoferrin, haem compounds, and iron derived from pepsin-degraded food. *H. pylori* can probably utilize haem compounds and ferric citrate as sole iron sources, but the use of human lactoferrin is still under debate (Belzer *et al.*, 2007). Studies investigating the global transcriptional response to iron have employed artificial conditions: iron depletion is obtained using chelators, whereas iron repletion results from the addition of ferric chloride to the medium.

Merell *et al.* analysed the response to iron starvation in the *H. pylori* strain SS1 by adding the chelator 2,2′-dipyridyl (DPP) to the media in the *H. pylori* strain SS1 (Merrell *et al.*, 2003b). Both the exponential and stationary phases of growth were examined and 138 and 75 genes were up and downregulated, corresponding to a total of 183 genes. It must be noted that the iron starvation conditions inhibited the growth of the bacteria, which was reflected in the microarray results with genes involved in protein translation being repressed. As expected, genes involved in iron sequestration, storage and transport were regulated by iron starvation: the ferritin *pfr*, was strongly repressed whereas the iron binding protein HP-NAP and the putative iron uptake receptors FrpB and FecA were induced. The expression of the transcription regulator Fur was induced by iron limitation and is likely responsible for many of the changes seen in the array. The extent of regulation directly and indirectly mediated by Fur will be discussed in the next section. A number of genes whose regulation by iron concentration were not previously appreciated were also increased, and included the amidases *amiE* and *amiF*, several motility genes, and the known virulence factors *cagA*, *vacA* and *napA*. To add credibility to these findings, the authors also added iron back to the media in the form of ferric chloride, and found that 28 genes overlapped with the previous data set. However, the total number and nature of genes that change under this condition was not described.

Ernst *et al.* examined the effect of iron limitation in *H. pylori* strain 26695 using deferoxamine mesylate (desferal) as a chelator and iron replenishment with ferric chloride (Ernst *et al.*, 2005). In contrast to the study by Merell, this group did not report any inhibition in bacterial growth under the conditions used, and bacterial RNA was isolated at a single time point in the late exponential phase (20 h). A total of 97 genes were regulated under changing iron concentrations. Only 10 of these genes were determined to be dependent on the regulator Fur (discussed below), indicating that other regulators are involved in the iron stimulon. It must be noted that the choice of chelator may have an important impact on the results. Unlike 2,2-dipyridyl (used by Merell *et al.*) that has high affinity for ferrous iron and is membrane permeable, desferal (used by Ernst *et al.*) is a siderophore-based iron chelator that removes ferric iron from the medium. Overall, these studies of iron stimulons highlight the complex transcriptional changes that result from this perturbation of the environment. These changes can be dissected out by elucidating the underlying role of individual regulators within the experimental condition chosen.

DNA microarrays to elucidate regulons and gene regulatory hierarchies

As compared to the genome content of other sequenced microorganisms, *H. pylori* uses a reduced repertoire of transcriptional regulators, perhaps reflecting its tight adaptation to the human stomach niche or suggesting that these regulators are multifunctional and have pleitropic effects on gene transcription. The studies of various stimulons discussed in the previous section just described illustrate this extensive regulation made of overlapping regulons and significant crosstalk. In this section we will summarize work from a number of groups that have defined several important regulons in *H. pylori* (Table 11.2).

The Fur regulon

To characterize the set of genes specifically controlled by the regulator Fur, Ernst *et al.* analysed the expression profile of a *fur* mutant in the presence and absence of iron in the cultured media (Ernst *et al.*, 2005). The experimental conditions used in this study were designed based on the finding that the Fur protein is bifunctional and has different DNA binding properties in its iron-bound form as compared to its iron-free form (Delany *et al.*, 2001). A total of 43 genes displayed Fur-dependent regulation. It is surprising that from the 97 iron-regulated genes identified only 10 displayed iron-dependent regulation in the *fur* mutant (with data lacking for 22 genes). This may reflect both the pleiotropic nature of the Fur regulator, and the fact that other regulators are involved in the *H. pylori* iron response. It is, however, troublesome that the microarray data obtained in this study generally lacked consistency within predicted operons.

Since its identification as a ferric responsive regulator, Fur has been implicated in the acid resistance of *H. pylori*. As discussed in the previous section, Bury-Moné *et al.* identified 101 genes that have altered expression during growth in acidic medium. When a double-mutant strain was generated in the regulators NikR and Fur, 36 out of the 101 genes were found to be dependent on the regulators NikR and Fur, implying that they play a major role in the acid acclimation pathways, but also arguing for the existence of other regulators (Bury-Mone *et al.*, 2004). Gancz, *et al.* sought to characterize the set of genes regulated by Fur under acidic conditions (Gancz *et al.*, 2006). Using similar conditions to a previous study by the same group (Merrell *et al.*, 2003b), a total of 95 genes were found to be differentially regulated in the *fur* null mutant under acidic conditions. When this list was compared with the 118 genes shown to be regulated by acid (Merrell *et al.*, 2003a), 12 were deduced to be strictly dependent on Fur. Among others, these include the aliphatic amidase *amiE*, the iron transport gene *frpB1*, and *pfr*. Of the 95 genes identified in the Fur regulon under acid conditions, 21 were also regulated by Fur under standard growth conditions for a total of 29 genes. In addition to traditional methods used to confirm the microarray results such as RNase protection assay, several Fur target genes predicted by bioinformatics analysis showed an electrophoretic mobility shift using recombinant *H. pylori* Fur protein. The genes giving evidence of Fur protein binding only represented a small fraction of the genes identified in the microarray, and thus many of the effects mediated by Fur appear indirect.

To discriminate between direct and indirect Fur regulation, Danielli *et al.* employed Fur chromatin immunoprecipitations (ChIP) to investigate sites at which the Fur protein binds to DNA *in vivo* by genome-wide location analyses (Danielli *et al.*, 2006). The experiment was performed using *H. pylori* strain G27 grown to exponential phase (OD 0.8–1.1) and incubated for 15 min in iron-supplemented and iron-chelated cultures. The immunoprecipitated DNA fragments were radioactively labelled and hybridized to a *H. pylori* macroarray. Iron-dependent enrichment for each target was calculated by dividing the IP values obtained after iron treatment by those obtained after chelation. As a control, an isogenic Δ*fur* mutant was also analysed and failed to reveal significant signal. A careful positional analysis of the distribution of Fur targets revealed a total of 200 loci targeted by Fur. These included genes known to be regulated by Fur such as *frpB1*, *pfr*, *fecAa*, *fecA2* and *fur*. A transcriptome analysis of the Fur regulon was then carried out by comparing wild type to a Δ*fur* mutant at two different time points (OD 0.6 and OD 1.1). Although the differentially regulated genes identified only partially overlapped with the studies discussed earlier, these two assays combined revealed a total of 59 genes under direct Fur regulatory control. These mainly comprise repressed genes but also include positively regulated genes such as *flaB*. It is intriguing that over two-thirds of Fur target genes that bind Fur do not appear to be under transcriptional regulation; the authors speculate that they may represent targets with yet undiscovered regulatory or structural functions of Fur. Although the resolution of this assay would be augmented with the use of high-density microarrays, this constitutes the first study to assess protein-DNA interactions for a regulatory protein *in vivo* at a genome-wide level in *H. pylori*.

NikR-mediated regulation

Similar to iron, nickel ions constitute a double-edged sword for bacteria: they are required for the synthesis and activity of a number of metalloenzymes but exhibit high toxicity at high intracellular concentration. Nickel metabolism has been of great interest because of its role as cofactor for the enzymes urease and hydrogenase, both of which are involved in acid resistance, mucosal colonization, and metabolism (Belzer *et al.*, 2007). A regulator of nickel uptake, NikR, was first identified in *E. coli* (De Pina *et al.*, 1999), and is encoded by the HP1338 gene in *H. pylori*. Two studies have characterized the NikR regulon by comparing the transcription profile of a *nikR* mutant to wild type exposed to different environments. As mentioned earlier, Bury-Moné *et al.* used a *fur/nicR* double-mutant under acidic condition to show that these two factors play a major role for adaptation to acid (Bury-Mone *et al.*, 2003). Contreras *et al.* generated a *nikR* null mutant in the *H. pylori* strain SS1 and grew cultures in excess nickel to show that NikR upregulates genes associated with nickel metabolism and nickel storage, and downregulates genes involved in ferric iron uptake and storage, stress responses, motility, and outer membrane proteins (Contreras *et al.*, 2003). The authors also sought to determined whether regulation by NikR was direct or indirect by using a gel retardation assay in conjunction with purified recombinant NikR protein. They found that NikR specifically binds to the hp1338–1339 intergenic sequence to repress the expression of two operons: *nikR-nadD* (hp1338–1337) and *exbB/exbD/tonB* (hp1339–1340–1341). The proteins in this second operon provides energy for the various ferric iron uptake systems, highlighting the interplay between the nickel and iron stimulons. In contrast, NikR failed to bind the upstream region of other nickel-activated genes. Since the expression of regulatory proteins such as the heat shock regulator HspR was also altered in the *nikR* mutant, it is reasonable to speculate that indirect effects explain these findings.

The two-component systems: HP0166-HP0165 (ArsRS), HP1364-HP1365 (CrdRS) and the orphan regulator HP1021

Genome sequencing revealed that *H. pylori* has few two-component regulatory systems as compared to other bacteria, with three histidine kinases and five response regulators identified. The ArsRS system has received the most attention due to its importance in the pH-responsive transcriptional control of the urease cluster and

Table 11.2 Microarray studies of regulons in *H. pylori*

Factor studied	Gene locus 26695 Annotation	Name	Function	Conditions	Strain used	Mutation	No. of genes regulated	Reference
Metal response	HP1027/HP1333	Fur/NikR	Ferric response regulator/nickel-responsive regulator	Acid	26695	aac(3)-IV (gentamicin) insertion for Fur aphA3 (kanamycin) insertion/deletion for NikR	N/A	(Bury-Mone *et al.*, 2004)
	HP1338	NikR	Nickel-responsive regulator	Excess nickel	SS1	aphA3 insertion/ deletion	43	(Contreras *et al.*, 2003)
	HP1027	Fur	Ferric response regulator	Iron	26695	aphA3 insertion	43	(Ernst *et al.*, 2005)
	HP1027	Fur	Ferric response regulator	Acid	G27	aphA3	95	(Gancz *et al.*, 2006)
Two-component systems	HP0165/0166	ArsRS	Acid-responsive regulator/signalling	Standard, grown for 24h	26695	CAT (chloramphenicol) insertion	7	(Forsyth *et al.*, 2002)
	HP0165/0166	ArsRS	Acid-responsive regulator/signalling	Acid pH5.0, 60min	G27	aphA3 insertion/ deletion	109	(Pflock *et al.*, 2006)
	HP0165	ArsS	Histidine kinase	Acid pH 5.0, 75min	J99	aphA3 insertion	68	(Loh and Cover, 2006)
	HP1364	CrdS	Histidine kinase	Acid pH 5.0, 75min	J99	aphA3 insertion	N/A	(Loh and Cover, 2006)
	HP1021		Orphan response regulator	Normal, exponential phase	26695	aphA3 insertion	79	(Pflock *et al.*, 2007)
Flagellar system/ motility	HP1122	FlgM	Anti-sigma factor	Normal, exponential phase	N6	aphA3 insertion/ deletion	N/A	(Josenhans and Suerbaum, 2002)
	HP1032	FliA	Alternative flagella sigma factor 28	Normal, exponential phase	N6	aphA3 insertion/ deletion	N/A	(Josenhans and Suerbaum, 2002)

	HP0714	RpoN	Alternative sigma factor 54	Normal, early and mid-log	N6, 88–3887 (motile 26695)	aphA3 insertion	N/A	(Niehus *et al.*, 2004)
	HP0703	FlgR	Response regulator (RpoN activator)	Normal, early and mid-log	N6, 88–3887 (motile 26695)	aphA3 insertion	N/A	(Niehus *et al.*, 2004)
	HP0244	FlgS	Histidine kinase specific for FlgR	Normal, early and mid-log	N6, 88–3887 (motile 26695)	aphA3 insertion	N/A	(Niehus *et al.*, 2004)
	HP1041	FlhA	Flagellar basal body protein involved in export	Normal, early and mid-log	N6, 88–3887 (motile 26695)	aphA3 insertion	N/A	(Niehus *et al.*, 2004)
	HP1035	FlhF	Flagellar biosynthesis protein	Normal, early and mid-log	N6, 88–3887 (motile 26695)	aphA3 insertion/ deletion	N/A	(Niehus *et al.*, 2004)
	HP1041/HP1122	FlhA/FlgM	Anti-sigma factor/ sigma 28	Normal, early and mid-log	N6	aphA3 insertion/CAT insertion/deletion	N/A	(Niehus *et al.*, 2004)
Heat-shock	HP1025	HspR	Heat shock response regulator	Normal, exponential phase	G27 derivative	aphA3 insertion/ deletion	N/A	(Roncarati *et al.*, 2007)
	HP0111	HrcA	Heat shock response regulator	Normal, exponential phase	G27 derivative	aphA3 insertion/ deletion	N/A	(Roncarati *et al.*, 2007)
	HP1025/HP0111	HspR/HrcA	Heat shock response regulator	Normal, exponential phase	G27 derivative	cat/aphA3	43	(Roncarati *et al.*, 2007)

other genes involved in acid adaptation. The response regulatory gene *arsR* is essential, so Forsyth *et al.* analysed the transcription profile of an *arsS*-deficient mutant of *H. pylori* strain 26695, and identified a total of seven genes that were differentially expressed in the mutant, most of them annotated as hypothetical and *H. pylori* specific (Forsyth *et al.*, 2002).

Since the ArsRS system was implicated in the *H. pylori* acid acclimation pathway, Pflock *et al.* isolated RNA prepared from an *arsS*-deficient mutant in strain G27, grown one hour at pH 5.0 in exponential phase, to show that the expression of 109 genes differed by at least twofold compared with the wild-type strain (Pflock *et al.*, 2006). Most of these genes (75/109) were activated in the mutant, suggesting that ArsR mainly acts as a repressor. Unfortunately, the mutation generated in arS in G27 affected the transcription of the whole operon ranging from HP0165 to HP0162, so changes in gene transcription may not solely be attributed to ArsS. The microarray data indicated that the ArsRS regulon includes the urease gene cluster, the amidase genes *amiE* and *amiF*, genes encoding outer membrane proteins, nickel storage proteins, and detoxifying enzymes involved in the oxidative stress response. Perhaps not surprisingly, several of these genes are also under the control of NikR and Fur. It is interesting that a small number of genes are regulated by ArsRS independently of the pH, perhaps indicating that this two-component system senses other environmental changes. Using DNase I footprinting analysis, ArsR was demonstrated to directly regulate *amiE*, *amiF*, and *rocF*, where the phosphorylated form of ArsR bound a region overlapping the −35 box of the respective promoters.

Loh *et al.* used a similar approach with an isogenic *ArsS* null mutant in *H. pylori* strain J99, which was exposed to acidic media (pH 5.0) for 75 min during exponential growth phase (Loh and Cover, 2006). This study found 68 genes differentially expressed. Despite a similar experimental design with Pflock *et al.*, only eight genes overlapped in both studies, including *ureAB*, *amiE*, *amiF*. The authors generated a complemented strain containing a wild type ArsS copy at the *rdxA* locus and demonstrated that the growth defect of the mutant in acid was rescued. A comparison of the transcriptional profile of this complemented mutant with the wild type strain would have strengthened the results. This group also demonstrated that the histidine kinase HP1364, originally identified as a regulator of the copper resistance determinant CrdAB–CzcAB, is required for pH-responsive gene regulation. Using the conditions described previously, a HP1364 null mutant in J99 revealed that transcription of 63 genes no longer responded to acidic pH. Although the complete set of data was not presented in the paper, it is intriguing that the pH-responsive HP1364-dependent regulon was largely similar to the ArsS-dependent regulon (Pflock *et al.*, 2007).

Pflock *et al.* sought to further analyse the role of the CrdRS system on acid responsive gene regulation (Pflock *et al.*, 2007). However, when they introduced null mutations of HP1364 in the *H. pylori* strains 26695 and G27 and used a variety of techniques such as immunoblot, primer extension analysis, and slot blot Northern hybridization to monitor gene expression, they did not find any changes in the set of genes known to be acid regulated. This major discrepancy with Loh *et al.* may only be reconciled as fundamental genetic differences between the *H. pylori* strains used.

Although the response regulator HP1021 does not have a known cognate sensor and it is unclear whether its activity is modulated in response to external or internal stimuli, its regulon was characterized by Pflock *et al.* (2007). A null mutation was constructed in the *H. pylori* strain 26695, and its expression profile analysed during logarithmic growth phase since the conditions for optimal activity of this regulator are unknown. A total of 79 differentially regulated genes were identified, including 17 genes encoding ribosomal proteins, reflecting the general growth defect associated with the loss of this regulator. Whereas most genes only showed modest changes, 10 genes revealed more than fourfold changes in expression levels in the mutant. Six of these belong to the operons HP0690 to HP0693 and HP0695 to HP0697. The authors speculated that these genes have a role in detoxifying acetone and/or allowing its use as a carbon source under nutrient-limiting conditions. Among other methods used to validate the microarray data, the authors used 2D-gel electrophoresis to identify 12 proteins consistently affected by HP1021 mutation. Of these, five were successfully identified by mass spectrometry and only 1 one did not appear on the microarray results. The authors also complemented the mutation by reintroducing the wild type allele at the same locus. Although the results were not shown in the paper, this complemented strain was analysed by microarray analysis and recapitulated the expression profile of a wild-type strain, ruling out second-site mutation effects.

The flagellar system of H. pylori

Flagellar motility is one of the most complex processes in *H. pylori*, involving more than 40 proteins. Motility is essential for colonization of the stomach but the transcriptional network that controls the expression of flagellar genes is not well understood. Since these genes are mostly scattered throughout the genome and are integrated within complex transcriptional circuitry, whole-genome microarrays have been a powerful tool to define this network. Based on studies from other bacteria, flagellar genes are temporally regulated and organized into a transcriptional hierarchy. At the top of the hierarchy, class 1 genes encode transcriptional regulators of class 2 promoters, which transcribe genes of the flagellar hook and basal body to form the type III secretion apparatus. Class 3 promoters generally transcribe late flagellar genes, encoding filament subunits and genes responsible for chemotaxis and flagellar rotation (Macnab, 1996)

The study by Josenhans *et al.* constituted one of the first studies to exploit DNA microarray technology for characterizing complex transcriptional networks in *H. pylori* (Josenhans and Suerbaum, 2002). This group generated null mutations in the newly identified anti-sigma factor gene FlgM, and in the flagella-specific sigma factor, FliA in *H. pylori* strain 26695 to define the genes that are under the control of these two regulators. In contrast to the commonly used PCR-based microarray, an oligonucleotide microarray was used in this study. As expected, FliA and FlgM had opposite effects on several class 3 genes, activating and repressing their transcription respectively. Of special interest, two novel operons were demonstrated to be under the control of FliA/FlgM, the first one comprising HP0472/OMP11, and the second one consisting of HP1052 and HP1051, which may function in the biogenesis or elongation of the flagellar sheath. However, caution must be taken when interpreting these results since the authors chose to study the non-motile 26695 strain, for which the flagellar network may be dysregulated.

In a more recent study by the same group, Niehus *et al.* sought to understand the integration of each regulon into the global regulatory circuitry of the *H. pylori* flagellar system (Niehus *et al.*, 2004). The authors used the motile strains N6 and 88–3887 to construct null mutations in RpoN, the FlgRS two-component system, and two other regulators FlhA and FlhF. Since FlgR is probably an activator of RpoN, the expression profile of RpoN, FlgR, and FlgS were very similar, controlling the transcription of class 2 flagellar genes encoding the flagellar hook protein (FlgE1), hook-filament adapter proteins (FlgK, FlgL), and the minor flagellin subunit FlaB. Other genes were only found to be regulated by FlgRS, including *hypA*, a nickel-binding protein involved in hydrogenase and urease maturation, *murG*, a glycosyltransferase, and six genes of unknown function (HP0114, HP0906, HP1076, HP1120, HP1233, and HP1154). FlhA and FlhF, both components of the flagellar basal body, were shown to be required for the transcription of class 2 and 3 flagellar genes. However, only FlhA cooperated with FlgM to generate feedback regulation of the RpoN-regulated genes. Since the transcription of several flagellar genes is known to be growth-phase dependent, special attention was taken to isolate RNA at two different time-points both in early and mid-logarithmic phase.

Heat shock protein regulation

Roncarati *et al.* sought to define the *H. pylori* transcriptional response to heat shock, important in the general stress response as well as in pathogenesis (Roncarati *et al.*, 2007). By mutating the two main regulators known, HspR and HrcA, either independently or together, the authors observed 43 genes in this regulon (showing 1.5-fold or more expression differences). Most of the genes previously shown to be under the control of these two regulators were detected such as the *groES-groEL* operon and the *grpE* and *dnaK* genes. The microarray data were confirmed with primer extension analysis, and DNA footprinting confirmed the direct binding of HrcA to the operator element of HspR and HrcA, which was increased in the presence of the GroESL chaperone. However, no direct action of HrcA or HspR could be detected for any other gene candidates and their mechanism of regulation currently remains unclear.

Conclusions from bacterial transcriptome studies

The *H. pylori* expression-profiling studies summarized here illustrate the heterogeneity in the data obtained from different groups, even though general themes can be drawn from the analysis of a stimulon/regulon. A direct comparison between these studies is often problematic for the reasons discussed earlier; one compelling strategy to facilitate such a comparison between experiments and research laboratories involves the use of a common reference, such as *H. pylori* genomic DNA. The expression patterns of various conditions established by different laboratories could therefore be easily compared using genomic DNA as the common denominator.

Beyond the characterization of the *H. pylori* Fur, NikR, ArsR/S, and flagellar regulons by microarray analysis, some of the studies determined whether the transcriptional regulation is direct or indirect. In addition to common procedures such as bioinformatic prediction of consensus operator sequences or DNAse I footprinting analysis, one group used a Chip on Chip experiment (chromatin immunoprecipitation followed by microarray hybridization) to study the genome wide binding of the regulatory protein Fur (Danielli *et al.*, 2006). This technology has the advantage of global analysis but requires effective antibodies. In the long term, this Chip on Chip technology will undoubtedly contribute to our understanding of the global DNA–protein interactions that occur as *H. pylori* faces changing environments.

One limitation of the microarrays used in most of the studies described here lies in their common use of a double-stranded PCR product probe for each ORF, which detect both sense and antisense gene expression without distinguishing between the two. Only a small number of studies use oligonucleotide arrays that only detect expression of the target strand. Therefore, upstream or downstream untranslated regions that may form small RNAs or antisense transcripts of ORFs are neglected in most analyses even though they are important regulators of gene expression in some organisms. Although no regulatory element of this type has so far been characterized in *H. pylori* the ferric regulator Fur is regulated at a posttranscriptional level by small regulatory RNA in both *E. coli* and *P. aeruginosa* (Masse and Gottesman, 2002; Wilderman *et al.*, 2004).

Microarray tracking of transposon mutants

While most microarray studies analyse gene content or gene expression, microarrays can also be used to map the position of introduced markers. One such application called MATT (Salama *et al.*, 2004) or TRASH (Sassetti *et al.*, 2001) uses microarrays to map the site of transposon induced mutations. These methods take advantage of the fact that transposon mutant libraries contain insertional mutations with the same sequence (that of the transposon) marking each mutation site. Using transcription from an outward facing promoter (Sassetti *et al.*, 2001), semi-random PCR (Manoil, 2000), or linker mediated PCR (Yamaoka *et al.*, 2000), the DNA next to the transposon insertion site can be amplified and labelled. Hybridization to a genomic microarray reveals the transposon insertion genomic location. The parallel nature of microarray hybridization experiments allows simultaneous tracking of a subset, or even an entire library of clones in a single hybridization.

Defining essential genes

Our group first used this method to map approximately 5400 transposon insertions from a transposon mutant library generated using *in vitro* transposition of a mini Tn-7 based vector in *H. pylori* strain G27 (Salama *et al.*, 2004). Each gene had between 0 and 19 transposon hits. Genes with no predicted transposon insertions were further analysed as candidate essential genes (not able to tolerate transposon insertion). The G27 genome contains 1552 genes represented on the microarray used. Of these genes, 344 (23%) had zero transposon hits and included many that are essential in other bacteria, e.g. subunits of DNA and RNA polymerase, a subset of ribosomal proteins, and several cell division proteins. Only 89 genes (6%) were expected to have zero hits if the distribution of transposon hits followed a Poisson distribution. Interestingly, some insertions mapped to genes predicted to be essential by another group (Chalker *et al.*, 2001). Follow-up sequencing of the precise transposon insertion sites showed that in two cases there were insertions within the gene suggesting they were not, in fact, essential in this strain and in one case the insertion mapped to an adjacent non-essential gene. This showed that the method in some cases only has resolution to map to a gene cluster and not a specific gene.

To validate the candidate list of essential genes further, several genes either with no annotation or annotated as non-essential in another bacterial species were tested further. To test for essentiality, independent mutant alleles were generated in strains bearing one or two copies of the candidate gene, by integrating a second copy at the *rdxA* locus. Of the five genes tested, three (60%) only gave transformants in a strain containing two copies of the gene indicating they were essential. This experiment was biased towards investigating genes with the highest likelihood of being false positives, suggesting that genes from this study bearing no transposon insertions were highly enriched for essential genes. Comparison of the essential gene list to those reported in other organisms showed significant overlap. Of 187 named genes, 102 were essential in *E. coli, H. influenzae or M. tuberculosis*. Pairwise comparison of the *H. pylori* essential gene set to these bacteria showed the most overlap with the Gram-positive *M. tuberculosis* (66 genes).

Screening for virulence genes

After establishing the MATT method for tracking transposon mutants using *in vitro* pools, this method was applied to screen for stomach colonization mutants (Baldwin *et al.*, 2007). A C57BL/6 mouse infection model was used to query a collection of 2400 transposon mutants in two different bacterial strain backgrounds for *H. pylori* genes required for establishing infection in the stomach. Of 758 different genes that could be detected by the hybridization protocol 223 (29%) had a predicted colonization defect. Many of the genes identified supported previous work on defining virulence factors in *H. pylori* and included urease genes, the nickel transporter *nikA*, the iron regulated transcription factor *fur*, the bacterial ferritin *pfr*, β1–4-galactosyl transferase, several chemotaxis genes, and the flagellar sheath localized adhesin *hpaA*. Several genes involved in responses to oxidative stress and damage repair were identified in this screen including a mix of some not previously tested and those identified in other work. Those with strong phenotypes included superoxide dismutase (*sodB*), a peroxyredoxin–thioredoxin like protein (HP0136), a heat shock protein (HSP90) homologue (HP0210). Two additional stress genes catalase (katA) and γ- glutamyltranspeptidase, an enzyme involved in glutathione metabolism, showed subtle or variable effects on colonization and persistence in this screen and the published literature. Similarly, two genes involved in DNA repair, *ruvC*, and endonuclease III, gave variable colonization phenotypes at early time points in other studies and gave inconsistent results in this screen. However, transposon mutants mapping to two DNA glycosylases (HP1347 and HP0602) showed attenuation, suggesting that the ability to repair DNA damage is likely very important for persistence *in vivo*.

A surprising finding of this study was loss of clones with insertions in the *cag* PAI. While many insertions within the PAI did not give a colonization phenotype, two genes were confirmed to have defects in competition with wild type in one strain background: HP0529, an *Agrobacterium virB8* homologue and HP0538 (*cagN*). Others have observed a competitive defect for PAI mutants at early, but not later time points in the mouse model (Marchetti and Rappuoli, 2002). Most studies, however, have showed no influence of the PAI on colonization and the PAI often becomes attenuated during adaptation of human isolates to the murine model (Philpott *et al.*, 2002). Genes involve in cell wall and cell surface structure synthesis and modification were identified. While such classes of genes have been found to be important for other pathogens, *H. pylori* has considerable redundancy in outer membrane proteins and adhesins, for example and finding phenotypes for such genes has been difficult. Other classes of genes that were unexpected included genes involved in DNA uptake and modification capabilities and a large number of genes encoding unannotated proteins. Elucidation of the functions of these genes in promoting stomach colonization may give new insights into *H. pylori*'s unique ability to persist in the human stomach.

Host transcriptome studies

In addition to probing bacterial genotypes and phenotypes, *H. pylori* researchers have taken advantage of microarrays designed to query human or mouse gene expression in order to learn about the host side of the host–pathogen interaction. Even before completion of the mouse and human genome sequencing projects, large collections of cDNA clones from tissue libraries and expressed sequence tag (EST) projects were compiled for printing microarrays. These collections have yielded extensive representation of the genome, allowing considerable insight into the transcriptional program induced in cells upon *H. pylori* exposure. With the completion of the mouse and human sequencing projects, newer generation arrays represent the total genes in the genome with fewer probes and in some cases researchers have chosen to focus on particular subsets of the genome with very dense sampling. To date studies have focused on host transcriptional responses to acute infection using tissue culture systems and mouse or primate models of infection. Microarrays have additionally been used to characterize the different cell types present in the stomach and their responses to *H. pylori* exposure in the mouse model. Using the mouse model, researchers have been able to devise partially successful vaccination protocols. Microarrays have been used to query differences in both the cell type composition and cellular phenotypes in stomach tissue from uninfected, chronically infected and protected animals. Finally progression of disease in the context of chronic infection has been probed in the mouse model with a MALT lymphoma outcome and from human and primate biopsies where disease progression is measured by loss of glandular architecture and development of precancerous lesions.

Cell culture studies

A number of studies have analysed the effects of *H. pylori* exposure on the transcriptional program of cultured cells in order to better understand the molecular pathways altered in host cells and to correlate these changes with some aspects of disease progression. Advantages of using cell culture include the ability to analyse a fairly homogeneous cell population, careful control of the bacterium to host cell ratio, synchronous growth of both cell types and fairly inexpensive performance of multiple biological replicates. These factors all contribute to increased detection of even subtle changes in expression with statistical significance. Cells in isolation and cells from transformed cancer lines, however, may not accurately reflect the cellular changes in the context of stomach tissue. Still, many genes highlighted in these studies could be validated based on previous observations or new experiments using RT-PCR, *in situ* hybridization or immunohistochemistry on human tissues from infected patients or infected animal models. The majority of studies have focused on epithelial cells, but there is increasing interest in how *H. pylori* interact with cells of the immune system and one study has profiled the transcriptional response of dendritic cells to *H. pylori* exposure.

Epithelial cells

H. pylori reside in the mucus layer overlying the gastric epithelium, with subsets found intimately attached to epithelial cells near the luminal surface and deep within gastric glands. Thus many transcriptional profiling studies have focused on the interaction of *H. pylori* with epithelial cell lines of gastric origin (KATO III (Cox *et al.*, 2001; Dossumbekova *et al.*, 2006), AGS (Chang *et al.*, 2006; Guillemin *et al.*, 2002; Sepulveda *et al.*, 2002), MKN45(Maeda *et al.*, 2001), SGC7901(Wang *et al.*, 2005; Yuan *et al.*, 2004)). Unfortunately, none of these lines form truly polarized monolayers or secrete mucus, features of epithelial cells thought to be modulated by *H. pylori* exposure. One study used the polarized human colon derived line T84 to specifically query whether transcriptional signatures of such processes could be detected (El-Etr *et al.*, 2004). *H. pylori* strains vary in their genetic make up (as described above) and no one strain has been adopted by the research community. When comparing study results, differences in the bacterium, host cells, multiplicity of infection and duration of bacterial exposure must be considered.

In spite of the diversity of cell lines, bacterial strains and infection protocols, comparison across the different studies profiling epithelial cell transcriptional responses to *H. pylori* exposure reveals remarkable similarities (Table 11.3). While the different array platforms and analysis schemes yielded different numbers and percentages of significantly regulated genes (8–1172; 0.3–6%), the functional categories and the genes with the highest fold change show remarkable consistency. The most predominant class of genes showing expression changes in almost all studies were those annotated as effectors or regulators of immune responses. Among the genes induced across studies are transcriptional regulators of proinflammatory programs such as NF-κB and the upstream regulator of NF-κB, AP-1 (composed of c-jun and c-fos heterodimers). A variety of chemokines and cytokines also showed changes including GRO-γ/MIP2-β/CXCL-3, IL-8, IL-2, IL-4, IL-7, IL-9, IL-13, and IL-23. Interestingly, a robust signature of tumour necrosis factor alpha (TNFα) and TNF targets and associated proteins such as A20/TNFα -induced protein 3 and ADAM10 (a transmembrane protease that may bind and process TNF-α precursor) appear in several studies. With respect to apoptosis, a very complex signature of both positive and negative regulators show changes in expression in response to *H. pylori*, including some of TNF targets, consistent with functional studies showing both induction and inhibition of apoptosis *in vivo*.

H. pylori induce epithelial cells in culture to undergo cell shape changes and display increased motility (Amieva *et al.*, 2003; Bagnoli *et al.*, 2005; Segal *et al.*, 1999). Reflecting this phenotype, structural genes as well as regulators of both the extracellular matrix and the cytoskeleton were found in several of the transcriptional profiling studies. Examples include matrix metalloproteases, keratins, and actin binding proteins such as cdc42 effector protein 2. Of particular note, the study utilizing polarized colon cells (El-Etr *et al.*, 2004) detected changes in a large number genes involved in regulation of cell junctions including β-catenin, cadherin family members, desmoglein, gap junction proteins and the claudin tight junction proteins, which had also been observed to change in studies using non-polarized cells (Guillemin *et al.*, 2002). β-Catenin was recently shown to be preferentially activated by a *H. pylori* strain selected for increased carcinogenesis in a gerbil infection model (Franco *et al.*, 2005). Another functional category affected by *H. pylori* exposure is genes involved in cell proliferation, which includes oncogenes and tumour suppressors such as cyclin D1. Whether the observation of alterations in cell turnover programs is a consequence of *H. pylori* disruption of the epithelium or of disease progression will require further studies.

Several studies have explored the contribution of various bacterial virulence genes to the host transcriptional signature including the secreted vacuolating cytotoxin (*vacA*) (El-Etr *et al.*, 2004; Guillemin *et al.*, 2002; Wang *et al.*, 2005; Yuan *et al.*, 2004), the proinflammatory outer membrane protein (*oipA/hopH*) (Dossumbekova *et al.*, 2006), and various components of the *cag* PAI (Dossumbekova *et al.*, 2006; El-Etr *et al.*, 2004; Guillemin *et al.*, 2002). Two studies found no significant transcriptional changes that depended on the presence of functional *vacA* in the AGS (Guillemin *et al.*, 2002) and T84 (El-Etr *et al.*, 2004) cell lines, while another group found a pronounced effect on the transcriptional response after mutation of *vacA* using either whole bacteria (Yuan *et al.*, 2004) or culture supernatants (Wang *et al.*, 2005) with SGC7901 cells and bacterial strain NCTC11638. Using whole bacteria, three hundred genes showed greater than 2.5 fold changes in gene expression with most showing decreased expression in wild type compared to the mutant. Many of the genes fell into the classes of cytoskeleton and cell proliferation and death. Interestingly, they observed collapse of the microtubule network after exposure of SGC7901 cells to wild-type, but not mutant, bacteria, while the actin network remained intact. In contrast to numerous other studies, this group observed decreased IL-8 production and increased apoptosis in the presence of *vacA* mutant bacteria. The SC7901 cell line is p53 minus (Jiang *et al.*, 2001), which may

Table 11.3 Cell types profiled by microarray with or without *H. pylori*

Cell type	Source of cells	No. of genes queried	No. of genes in signature	Predominant classes	Affects of Hp exposure	Reference
Parietal	Purified from mice	11,000	240	Energy production, secretion/ membrane expansion, acid, mitochondria	No change	(Mills *et al.*, 2001)
Parietal	Mouse microdissected	38,000	285	Energy production, acid, mitochondria, oxygen transport, angiogenesis	No change	(Mueller *et al.*, 2004)
Mucous/Pit	Mouse microdissected	38,000	701	Cytoskeleton, cell-junctions, signal transduction, innate and adaptive immunity	1000 genes differentially expressed	(Mueller *et al.*, 2004)
Zymogenic/chief	Mouse microdissected	38,000	95	Digestive and other enzymes, endoplasmic reticulum proteins	No change	(Mueller *et al.*, 2004)
Zymogenic/chief	Purified from mice	11,000	71	Digestive and other enzymes, secretory pathway, growth factors	No change	(Mills *et al.*, 2003)
Dendritic	Purified from mice	39,000	N/A	Cytokine/chemokine, antigen processing, DC-T-cell interactions, DC differentiation.	1546 differentially expressed, 70% MyD88 dependent	(Rad *et al.*, 2007)
Gastric epithelial cancer cell line	KATO III	57,800	N/A	Inflammation, tissue remodelling	232 genes differentially expressed	(Cox *et al.*, 2001)
Gastric epithelial cancer cell line	MKN45	2304	N/A	Inflammation, tissue remodelling	8 genes differentially expressed	(Maeda *et al.*, 2001)
Gastric epithelial cancer cell line	AGS	12,626	N/A	Cell proliferation, inflammation	718 genes differentially expressed	(Sepulveda *et al.*, 2002)
Gastric epithelial cancer cell line	AGS	22,500	N/A	Inflammation, tissue remodelling	106 PAI-dependent genes	(Guillemin *et al.*, 2002)
Colonic epithelial cancer cell line	T-84	43,000	N/A	Inflammation, cell junctions	607 genes differentially expressed	(El-Etr *et al.*, 2004)
Gastric epithelial cancer cell line	SGC7901	8464	N/A	cell proliferation, cytoskeleton	300 gene differentially expressed	(Yuan *et al.*, 2004)
Gastric epithelial cancer cell line	KATO III	22,277	N/A	N/A	1172 probes differentially expressed	(Dossumbekova *et al.*, 2006)
Gastric epithelial cancer cell line	AGS	12,812	N/A	Inflammation, tissue remodelling	522 genes differentially expressed	(Chang *et al.*, 2006)

explain its apparent extreme sensitivity to the presence of VacA.

While the effects of VacA on host gene expression appear to be cell line dependent, studies using three different cell lines have all shown a profound effect of the pathogenicity island on host gene expression (Dossumbekova *et al.*, 2006; El-Etr *et al.*, 2004; Guillemin *et al.*, 2002). As described in detail in Chapter 5 of this book, the *cag* pathogenicity island (PAI) encodes a type IV secretion system (T4SS) that translocates the protein effector CagA into host cells. There it becomes phosphorylated and activates host cell signal cascades that result in changes in cell shape and motility (Bourzac and Guillemin, 2005). A second effector of the T4SS appears to be peptidoglycan fragments that activate NFκB dependent secretion of IL-8 and other proinflammatory responses through the host protein Nod1 (Viala *et al.*, 2004). When Guillemin *et al.* compared mock infected AGS cells to those infected with bacteria in which the entire island was deleted (ΔPAI) the profiles were remarkably similar, indicating the majority of the host transcriptional response depends on the *cag* PAI (Guillemin *et al.*, 2002). They further dissected the contribution of individual components of the PAI including a T4SS component, *cagE*, which is required for both CagA delivery and IL-8 induction, the T4SS effector *cagA*, and a protein with unknown function, *cagN*, which affects neither CagA delivery or IL-8 induction. Interestingly, the profiles from ΔcagE and ΔPAI-infected cells were different. Since the ΔcagE strain does not form a functional T4SS, these results suggest that a partially functional complex or a T4SS subassembly on the surface of the bacterium allows a different interaction with cells than bacteria completely lacking the island. The PAI dependent transcriptional response included a *cagA* dependent response which was a subset of the *cagE*-dependent response. The active PAI response included 106 unique genes with the two most abundant classes of genes involving innate immune responses and regulation of cell shape and adhesion. Additional *cagA* dependent signatures (some of which showed lower induction, but were not absent in the Δ*cagA* infections) included repressed genes involved in DNA replication and cell cycle progression and induced genes involved in cytokine production. This partial phenotype suggests a combination of *cag* dependent and independent inputs into these cellular responses. In line with this observation, one group observed *cagA* dependent IL-8 induction at late time points (Brandt *et al.*, 2005). By grouping mutants with similar phenotypes and pooling time points with similar induction profiles as duplicates, statistical tests could be performed to elucidate classes of genes whose subtle expression differences could be revealed as dependent on the presence of *cagA*. Among these were genes involved in tight junctions (E-cadherin, α- and β-catenin). This led to subsequent work showing *H. pylori* interacts directly with tight junctions (Amieva *et al.*, 2003). Interestingly, when looking at all genes annotated as having tyrosine kinase or phosphatase activity, only Src-dependent tyrosine kinase lck and the tyrosine phosphatase SHP2 showed altered expression consistent with biochemical work implicating these genes as regulators of *cagA* dependent responses (Stein *et al.*, 2002). Another intriguing observation by this group was *cag* PAI-dependent induction of cholesterol biosynthesis genes and inhibition of cell cycle progression genes at late time points.

A second study, using the polarized colonic cell line T84, confirmed and extended the observation that the majority of the transcriptional response to *H. pylori* depends primarily on components of the *cag* PAI (El-Etr *et al.*, 2004). They attributed 92% of the transcriptional signature to the PAI, 71% to CagA and further determined, by profiling an EPIYA to EPISA phosphorylation site mutant, that only 32% of the *cagA* dependent genes required CagA tyrosine phosphorylation. The phosphorylation dependent genes included those annotated to be associated with the actin cytoskeleton, consistent with observations that cell scattering requires phosphorylation (Moese *et al.*, 2004). The phosphorylation-independent response included some genes involved in cell adhesion and junction formation and several signalling molecules. Comparison of the *cag* dependent transcriptional signature in T84 cells with previous experiments using the unpolarized gastric AGS cells revealed >50% overlap of genes that were adequately measured in both experiments.

Chang *et al.* provide some evidence that in addition to the PAI and *vacA* other bacterial genes may contribute to host gene expression changes (Chang *et al.*, 2006). They profiled AGS cells exposed to three different clinical isolates that all contained *cagA* and the *vacA* s1m2 allele. The three strains came from patients with three different clinical outcomes: duodenal ulcer (DU), gastric cancer (GC) and a MALT. ANOVA analysis of those genes whose expression differed from mock infected cells revealed 522 genes that significantly differed in the three groups. Comparing the signal from all genes the DU strain and GC strain profiles looked most similar ($r = 0.552$) (GC vs. MA $r = 0.393$, DU vs. MA $r = 0.355$, Pearson's correlation). A relatively small number of genes were found to show strain specific profiles (GC-4, DU-4 and MALT-12). When comparing the two more similar profiles (DU and GC) to the third (MALT), 32 genes showed more than two fold change and included pim-1, FEGF, junB, C/EBPβ, cyclin G2, fos, TNFα induced protein 3, CDC42 effector protein 2 and downregulation of HSP70. Interestingly, in the previous studies transcriptional changes for several of these genes depended on an active *cag* PAI. Thus while all three strains contained the *cagA* gene, the MALT strain may have contained a non-functional island due to deletion or mutation of another island component. The molecular pathways responsible for the small number of additional transcriptional changes and their significance for disease outcome remain to be determined.

Immune cells

While most *in vitro* studies of *H. pylori* pathogenesis have focused on the interaction of *H. pylori* with epithelial cells, work using animal infection models, in particular the mouse model, has revealed the importance of immune cells in driving *H. pylori* associated disease. In mice, as has been observed in humans, a T-cell helper type 1 (Th-1) response seems to drive pronounced cellular damage. Thus a major question is how a Th-1 response is generated by this extracellular bacterium and how immune cell function may be modulated by the bacteria. Consequently, interactions of *H. pylori* with antigen presenting cells and T-cells have become the focus of recent work.

The first transcriptional profiling paper of immune cells focused on the role of dendritic cells in *H. pylori* infection and in particular the role of toll like receptors (TLRs) (Rad *et al.*, 2007). Toll like receptors, present on many cells, including antigen presenting cells, recognize common constituents of microbial pathogens as well as host compounds produced in response to infection. Binding of the receptors follows a common signal transduction cascade that goes through the adapter molecule MyD88. Activation of TLR signalling can lead to dendritic cell maturation, causing productive interaction with other immune cells. Rad *et al.* first used *in vitro* experiments to show that *H. pylori* whole cells and cell extracts can induce dendritic cell maturation and secretion of IL-12, a Th-1 polarizing cytokine. Experiments with dendritic cells from a series of wild type and knockout mice showed this activation depends on MyD88 signalling from TLRs and not IL-1 or IL-18. Transcriptional profiling of wild type and MyD88−/− bone marrow-derived dendritic cells exposed to *H. pylori* revealed 1546 probe sets with significant expression differences after 6 h. Of these, 178 showed more than threefold difference in expression levels over un-stimulated dendritic cells. This signature included cytokines and chemokines as well as genes involved in antigen uptake and processing, dendritic cell-T-cell interactions and dendritic cell differentiation. Seventy percent of this response depended on MyD88. The exception was a large cluster of type-I IFN-induced genes. Only four genes were induced in MyD88−/− DCs and not wild type including *c-fos*, a gene whose expression has been shown to favour Th2 commitment (Agrawal *et al.*, 2003). These results are quite interesting in light of the fact that many *H. pylori* TLR agonists have been shown to have low activity due to sequence polymorphisms (flagellin-TLR5 agonist (Andersen-Nissen *et al.*, 2007; Andersen-Nissen *et al.*, 2005) and covalent modification (LPS-dephosphorylation; Tran *et al.*, 2005)). Elucidation of which TLRs mediate the MyD88 activation of dendritic cells by *H. pylori* could have important implications for understanding infection and vaccine research.

Mouse models

Expression associated with different cell subsets

As described previously, most of the bacteria in *H. pylori*-infected tissue are found in the mucus layer of the gastric epithelium, both in the upper portion of the gastric glands and deep within. Gastric glands are composed of three major cell types: mucus secreting pit cells in the upper third of the gland, acid producing parietal cells in the middle and the digestive enzyme producing zymogenic or chief cells at the bottom. All three cell types arise from gastric stem cells that reside in the middle of the gland. Two groups have invested considerable effort in an elegant series of papers to define the gene expression program of each of these major lineages in uninfected mice as well as during *H. pylori* exposure (Mills *et al.*, 2003; Mills *et al.*, 2001; Mueller *et al.*, 2004; Ramsey *et al.*, 2007). These two groups have used different mouse and *H. pylori* strains (FVB/N, Hp1 vs. Balb/C, SS1), as well as different array platforms (Affymetrix vs. cDNA spotted arrays) yet have obtained quite similar results.

Initial studies used cellular subsets purified from digested tissue based on size fractionation and binding to lectin coated beads specific for different cell types (e.g. *Dolicos biflorus* agglutinin (DBA) recognizes GalNacα(1–3)Gal containing glycans found exclusively on parietal cells (Mills *et al.*, 2001)). Subsequent studies used navigated laser capture dissection (nLCD) to purify individual cells from frozen sections viewed under a microscope. Both types of experiments are technically challenging and yield limited numbers of cells. Transcriptional profiling of such material has required pooling of material from multiple animals and amplification of the recovered RNA, pushing the limits of detection of the array platforms. In spite of these technical limitations, distinct molecular signatures of each cell type have been elucidated, largely by using programs to eliminate noise in the data and that measure enrichment of functional categories in the gene lists.

The first study used the lectin mediated fractionation approach to divide the stomach into 'parietal cell plus' (PC+) and 'everything else' fractions (PC–) (Mills *et al.*, 2001). Transcriptional profiling identified 240 genes in the PC+ signature most of which mapped to known or predicted functions including energy production, proton pumps, vesicular trafficking, membrane expansion, acid secretion, and free radical scavengers (presumably dealing with by products of oxidative phosphorylation). They also found enrichment of mitochondrial, calcium signalling and cell growth regulating genes. The gene list was validated by showing similar differential expression with qRT-PCR for a number of genes in cell preparations from tox176 vs. normal stomachs. Mice of the tox176 line lose their parietal cells due to expression of a modified diphtheria toxin in this lineage. They searched changes in gene expression that relate to infection by additional transcriptional profiling from germ-free mice infected with the Hp1 strain at 2 weeks and 8 weeks. They found no change in the PC+ cell transcriptional signature after infection. The PC– fractions showed significant transcriptional changes at 2 weeks which, excluding immunoglobulins, mapped to 99 genes. The functional annotations of these genes included INFγ-regulated genes, LPS regulated responses, actin related, extracellular matrix-cell motility and protein turnover. Up or downregulation of only 13 genes persisted after 8 weeks. This work concluded that while parietal cells have a unique transcriptional signature, it does not change dramatically during *H. pylori* infection. This result is perhaps surprising since parietal cells are thought to provide signals to the stem cell compartment to allow differentiation of the zymogenic lineage, which are often lost in chronic *H. pylori* infection. However, changes in gene expression in a small subset of the parietal cell population would not be measured in these experiments.

Subsequent studies focused specifically on the zymogenic cellular compartment. First using lectin purification (Mills *et al.*, 2003) and then using nLCD (Ramsey *et al.*, 2007) zymogenic cells were purified and gene expression was profiled relative to other cellular compartments of the stomach. The transcriptional signature of this compartment was smaller than the parietal cell compartment yielding 57 known genes and 14 expressed sequence tags (ESTs). The genes included a number of known genes involved in the production of digestive enzymes (e.g. cholecystokinin receptor A, α-amylase, trypsin 2/4, gastric intrinsic factor, alanyl aminopeptidase N (CD13)). There were also many genes associated with the secretory pathway including those involved on ribosome attachment/ERtranslocation, ER/Golgi function, general vesicular trafficking and vesicular contents. An unexpected gene identified was PdgfB (platelet-derived growth factor B) and was confirmed by qRT-PCR on nLCM dissected purified cell populations. The fact that Pdgf A and B, but not their receptor were expressed and parietal cells express another angiogenic factor – VEGFB – led to a hypothesis that these cell types together might contribute to angiogenesis in the gastric glands. Indeed in normal stomach, the distal glands lack both parietal cells and zymogenic cells consisting primarily of pit and neck cells and have 2 fold less capillary networks as measured by capillary bed imaging. Similarly, using transgenic mice lacking the parietal cell and zymogenic lineage (tox176), they observed two fold less microvascular density in the corpus region compared to normal mice while the density in the antrum (distal portion) was the same in transgenic mice and normal mice. In an attempt to identify factors involved in the differentiation of zymogenic cells, Ramsey *et al.* focused on the transcription factor signature. There were 41 genes in this class specific to the zymogenic cell type. To further narrow their search they looked for transcripts common to another tissue containing acinar secretory cells, the salivary glands, using publicly available expression profiles. They found a single basic helix–loop–helix containing protein, Mist1. Using available Mist1–/– mice they showed a defect for terminal differentiation of zymogenic cells in these animals. The identification of factors critical to zymogenic cell differentiation and angiogenesis in the stomach could provide tools to characterize early changes in differentiation that might be altered during *H. pylori* disease progression.

A final study used nLCD to profile transcriptional changes in pit cells, parietal cells and zymogenic cells from BalB/C mice 2, 7, 14 and 28 days post infection with *H. pylori* strain SS1(Mueller *et al.*, 2004). Unsupervised clustering of genes whose expression was fourfold higher than the reference sample (uninfected whole mouse stomach RNA) in at least three arrays (912 genes total) allowed segregation of each of the cell types. The pit cells had the most distinct profile encompassing 701 genes, the zymogenic cell signature included 95 genes and the parietal cell signature was composed of 285 genes. Two programs, GoMiner and DAVID, were used to identify functional categories enriched in the transcriptional signatures of each cell type. Categories enriched included cytoskeletal, cell-junctions, intracellular and cell-surface receptor-linked signal transduction. In addition, genes involved in both innate and adaptive immunity were induced including cytokines, chemokines, major histocompatibility class I mediated antigen processing and presentation. The chief cell signature was characterized by digestive and other enzymes, cyclin D2 and ER resident proteins (consistent with this cell type's production of heavily glycosylated secreted proteins). The parietal cell signature was composed of factors involved in energy generation, mitochondria, acid secretion, oxygen transport and angiogenesis. The results for the parietal and zymogenic profiles were largely consistent with the previous studies.

The only cell population for which there was a detectable signature for infection was the mucus producing pit cells. Clustering based on the pit cell signature showed complete segregation of infected and mock infected samples for the 28 day time point. Approximately 1000 genes were part of this signature. The largest functional class of genes induced were proinflammatory genes (IL-1β, TNFα dependent genes, a strong INF-γ response, chemokines (RANTES, GM-CSF), NF-κB, Jun-B and c-fos components of AP-1, down stream effectors of calmodulin, which can be part of a NF-κB activation pathway and a gene involved in biosynthesis of the proinflammatory lipid mediator leukotriene B4). This proinflammatory signature shares many similarities to the profiles observed using gastric epithelial cell lines. At later time points upregulation of genes known to be involved in gastric mucosal protection such as prostaglandin receptor E2, small proline rich protein 2A and trefoil factor-2 was observed. Again, upregulation of some genes, like trefoil factor-2, had been already observed in the cell culture studies. A number of gene associated with the cytoskeleton and actin were upregulated. Similar genes showed *cagA* dependent upregulation during co-culture experiments with tissue culture cells. Interestingly, the strain used in this mouse study, SS1, does not show measurable translocation of CagA during *in vitro* experiments suggesting that either the situation is different *in vivo*, or that other bacterial factors can also mediate these effects. Infection was associated with up regulation of the iron sequestering protein lactoferrin and the angiogenic factor VEGF. Though iron sequestration provides defence for many bacterial infections, some studies suggest *H. pylori* can use lactoferrin as an iron source (Husson *et al.*, 1993). A surprising finding of the GoMiner analysis was the downregulation of genes in the tumour-suppressor group including cell cycle progression and apoptosis regulators. There was also differential regulation of TGF-β receptor -1, Smad-2 and Smad-4. These results were confirmed for the late time point in 9 independent mice and real time analysis was used to confirm the induction of INF-γ inducible protein and small proline-rich protein 2A.

The fact that only the pit cells, which reside at the top of the gastric glands, showed a transcriptional response to *H. pylori* may reflect the presence of the majority of bacteria in this location. It will be interesting to determine whether *H. pylori* can indeed interact with parietal cell and zymogenic cell types, since bacteria have been observed deeper in the pits. Only a subset of parietal and zymogenic cells may be exposed to bacterial cells or their products and measurement of *H. pylori* effects on these cells may require single cell measurement methods rather than those using bulk cells.

Associated with vaccination/protection

Vaccine studies using the mouse model have revealed that while CD4+ T helper cells are required for protection, CD8+ cytotoxic T-cells and B-cells are not. The exact mechanisms of protection, however, remain unclear. Transcriptional profiling of whole stomach tissue after successful vaccination in comparison with tissue from infected or mock infected animals has begun to offer some insights. Three published studies have used different immunization protocols, mouse strains, and bacterial species. Additionally, the time points analysed (early only or late only) make direct comparisons between the studies difficult. Still some consistent trends have emerged (Table 11.4).

Mueller *et al.* (2003a) analysed Balb/C mice immunized with *H. felis* sonicate and cholera toxin adjuvant, then challenged with *H.* felis. In their protocol, mice were challenged 2 months after immunization and then sampled 22 months after challenge. Using this protocol, 14/19 challenged mice were protected. They analysed the transcriptome from whole stomach tissue, comparing protected animals vs. immunized but not protected vs. mock infected vs. infected. Unsupervised clustering of genes with a normalized ratio of more than two standard deviations the infection status (infected and unprotected clustered together). Of the samples from protected animals, perhaps not surprisingly 5/14 grouped with mock infected. The remaining 9/14 protected animals formed a third cluster. Genes upregulated during chronic (24 month) infection included lymphocyte-specific markers and signalling molecules previously linked to the development of MALT pathology (CD52 and calgranulin A) (see below). Downregulated during infection were many gastric enzymes and hormones- somatostatin, gastrin, phospholipase A2, and progastricsin reflecting massive loss of epithelial and endocrine cell types (atrophy).

They next searched for genes that specifically segregated control, immunized/protected, and immunized/unprotected samples. This yielded 776 genes whose transcription differed between two of the three groups with a 5% false-positive discovery rate. Most transcriptional changes distinguished infected from all other samples and included genes previously associated with infection. Two clusters correlated with protection. The first was a lymphocyte signature with many B-cell-specific genes such as the Ig genes encoding α, κ, λ, and joining chains, CD40, integrin β7 (a component of the homing receptor of B and T-cells for mucosal surfaces of the gastrointestinal tract). In other experiments, the same lymphocyte signature was found in animals from which bacteria had been eradicated by antibiotic therapy, which does not confer protection. Therefore this expression pattern is probably not sufficient for protection. The second cluster con-

Table 11.4 Investigation of transcriptional signatures of *H. pylori*-mediated disease progression and successful immunological protection

Disease model	No. of genes queried	No. of genes in signature	Predominant functional classes	Reference
Mouse/*H. felis* vaccination	38,000	776	Lymphocyte signature, adipocytes/adipokines	(Mueller *et al.*, 2003a)
Mouse/*H. pylori* vaccination	10,000	13	MHC class I, IFN-γ associated	(Rahn *et al.*, 2004)
Mouse/*Salmonella* expressing UreAB vaccination, *H. pylori* challenge	8187	453	IFN-γ associated, T-cell, MHCII, epithelial barrier maintenance/ defence, proliferation, adipokine	(Walduck *et al.*, 2004)
Mouse/*H. pylori* MALT development	12,000	532	T-cell activation, - cell activation, dendritic cell, NK cell, angiogenesis, cholesterol biosynthesis	(Syder *et al.*, 1999)
Mouse/*H. heilmanii* MALT lymphoma development	38,000	300 moderate to severe pathology, one gene (Calregulin) with malignancy	Extracellular matrix, DNA repair, lymphocyte, epithelial barrier maintenance	(Mueller *et al.*, 2003b)
Mouse/*H. felis* MALT development	38,000	125	MHC class I and II, B-cell malignancy associated signalling receptor and molecules, epithelial barrier maintenance/ defence	(Mueller *et al.*, 2005)
Primate/*H. pylori* disease progression	7000	206	Innate immune, inflammation, epithelial barrier maintenance, cell adhesion, extracellular matrix, Hsp70	(Huff *et al.*, 2004)
Human/*H. pylori* disease progression	847 inflammatory genes	232	Innate immune, cytokines, adhesion molecules, signalling, extracellular matrix, B- or T-cell activation	(Wen *et al.*, 2004)
Human/*H. pylori* disease progression	44,500	55, five genes associated with progression	Immune response, cell surface receptors, signal transduction, cell cycle, protein synthesis and modification, lipid transport and metabolism	(Tsai *et al.*, 2006)

tained genes expressed in fat cells/adipocytes – adipokines adipsin, resistin and adiponectin, fat cell specific fatty acid-binding protein 4, carbonic anhydrase 3 and CD36. Transcription of fat cell specific genes was up in protected, and down in unprotected, immunized stomachs relative to the controls. Also induced was histidine decarboxylase- an enzyme in the histamine synthesis pathway. Histamine modulates gastric acid secretion. Following up the fat cell signature, histological analysis showed that uninfected stomachs were surrounded by a layer of fat tissue. Infected individuals had diminished or disrupted fat layers and protected stomach exhibit a thicker fat layer than controls. Fat tissue of protected stomachs contained clusters of lymphocytes not found in controls. Staining of lymphocyte clusters in fat of protected animals or within the mucosa of infected animals showed that adipsin is present on fat cells and lymphocytes in protected animals and the B/T ratio is 1:4. In infected animals there were many more B-cells as well as occasional

T-cells found in the mucosa, but none of these cells showed adipsin staining.

A second study found evidence for a role of adipokines in protection (Walduck *et al.*, 2004). This study focused on the early immunization transcriptional signature (1, 3, 14 days) after challenge. Mice were immunized with *Salmonella* expressing the *H. pylori* urease A and B subunits and then challenged with *H. pylori* strain Hp76. In their immunization scheme day 1-infected and infected immunized had similar bacterial loads. At 3 and 14 days infected immunized animals had lower but detectable bacterial loads than infected animals. Whole stomach tissue transcriptional profiling allowed comparison of infected vs. mock at each time point and immunized, infected vs. infected at each time point. Using a criterion of threefold expression changes, 300 genes were associated with infection. They observed two major signatures that showed different temporal profiles. The early signature (24 h, 121 genes) showed downregulation of genes involved in innate immune responses (e.g. defensin-related cryptdins – Defcr), B-cell function (IfL, ly64, IgH-6, IL15ra) and stress response (Hsp70). The later signature (14 days) showed upregulation of gastrin and some immune genes along with downregulation of other immune genes (Irf7, Tnf) and genes controlling cell migration.

The immunization transcriptional signature was composed of 453 genes. At 24 hours 108 genes showed expression changes that were unique or additional to the infection signature. Prominent in the day 1 response was IFN-γ, antimicrobial peptide Defcr4 and several genes involved in proliferation. Again a distinct transcriptional signature was evident at 14 days and included IFN-γ dependent genes, T-cell-specific genes, MHCII, proteasome subunits. epithelial maintenance/defence (Dmbt1) and proliferation enhancing genes. Non-classical inflammatory genes such as the adipokine Acrp30 (adiponectin), fatty-acid binding protein Lipocalin 2 (Lcn2), and galanin receptor (a mediator of hunger perception and a modulator of leptin signalling). Leptin receptor was downregulated. Immunohistochemistry confirmed general increased expression of MHCII in immunized mice on day 1 and increased ICAM1 at day 14. CD4+ or CD8+ T-cells were not detected by immunohistochemistry, though were detectable by FACs as early as day 3. RT-PCR confirmed increased expression of Defcr4 and aquaphorin 5, which was up early then down late and may be involved in oedema associated with *H. pylori* infection. At day 14, Dmbt1, crp-ductin, surfactant protein D and Hsp70–3 expression changes were also confirmed by RT-PCR. Immunohistochemistry localized Aqp5 to both the crypt base and the luminal surface while Dmbt1 and Sftpd were predominantly localized to the base of the gastric units. The activation of proliferation genes was confirmed by BrdU incorporation studies showing higher cell proliferation of the antral neck cells of immunized mice compared with those from infected only mice at 14 days.

Another study profiled early transcriptional responses in whole stomach tissue after challenge of immunized animals and similar to the previous studies saw a profound IFN-γ response, though they did not report induction of adipokines seen in the other two studies (Rahn *et al.*, 2004). In this study mice were immunized with *H. pylori* strain SS1 sonicates and then challenged with the same strain. The immunized mice had more inflammation (though few or no detectable bacteria) than infected mice. Thus, stomach tissue from mice infected with *H. felis*, which showed as much or more inflammation than the protected mice, were also analysed. 314 genes showed transcriptional induction at least twofold over uninfected in the immunized protected mice. Of those, 13 were not induced in either the *H. pylori* infected mice (89 genes induced) or *H. felis* infected mice (327 genes upregulated). These included genes involved in MHC Class I expression, and the type I IFN receptor. The most dramatic difference was in a number of IFN-γ associated genes (IRF-7, GARG-16, ISGF3) and T-cell specific markers. There were several genes induced to a higher extent in the *H. felis* infected mice relative to *H. pylori* infected or immunized, protected mice including several MHC class II genes. Among the genes more induced during *H. felis* infection were markers of innate immunity and macrophage activation such as MRP8 and 14 calcium-binding proteins, serum amyloid A and lactotransferrin. Whether this signature represent a species specific induction of innate immune genes by *H. felis*, suppression of these responses by *H. pylori* or simply reflects the higher inflammatory cell infiltration seen in the *H. felis* infected animals remains to be determined. Phenotypic testing showed that splenic cells as well as CD4+ T-cells showed increased IFN-γ production when stimulated with *H. pylori* sonicates. Interestingly splenic cells, but not CD4+ T-cells, from immunized animals also responded more strongly to whole bacteria, indicating that natural killer cells may contribute to the IFN-γ response.

Several interesting themes emerge from these studies. Early work had suggested that shifting the immune response from a Th-1 response to a Th-2 response might enhance protection. Vaccine studies using different protocols and varying levels of efficacy showed a significant signature of IFN-γ regulated genes, suggesting that this Th-1 associated cytokine will be important in successful vaccination. Another even more surprising finding of the vaccine transcriptional profiling studies was the correlation between adipokine expression and protection. While Mueller *et al.* showed histological differences in the amount of fat present around the stomachs of protected vs. infected animals at very late time points, Walduck *et al.* detected an adipokine signature at very early time points. Since T-cells can also secrete these molecules, the source of adipokines during infection and upon immunization remains an open question. Finally, these studies have given some insight on the effectors of immunity after vaccination. This has been particularly perplexing since CD4+ T helper cells, but not B-cells or cytotoxic T-cell are required. The study by Rahn *et al.* suggests that natural killer cells may be important players, though in what capacity is not clear. Another possibility is that protection is mediated in large part by upregulation of epithelial defence factors such as crp-ductin and surfactant D proteins which bind bacteria and reduce their motility. This coupled with increased epithelial turnover and possibly altered mucin secretion may make the gastric niche unfavourable for *H. pylori* colonization. Interestingly, the cells making some of these factors are at the base of the gastric glands and distal to the site of the majority of bacteria and may require adaptive immune mediated 'help' to activate this expression program. Clearly more needs to be done to follow up when and in what cell types the components of the protection signature are expressed.

Disease progression towards MALT lymphoma

H. pylori associated disease correlates with the development of chronic inflammation. Both symptomatic and asymptomatic disease can show marked induction of inflammatory cell infiltrates and the development of MALT in the stomach. This apparently robust immune response, however, does not clear the infection and in rare cases can develop into MALT lymphoma. Transcriptional profiling has been used to characterize this tissue and changes that occur in the progression to lymphoma in the murine model (Mueller *et al.*, 2005; Mueller *et al.*, 2003b; Syder *et al.*, 2003).

H. pylori strain HP-1 shows tropism for cells bearing NeuAcα2,3Galβ1,4 glycans (Syder *et al.*, 1999). These glycans are normally expressed in a narrow strip at the forestomach/zymogenic region of glandular stomach junction and after two months infection lymphoid aggregates accumulate in this region while diffuse gastritis is seen in the rest of the zymogenic region (Syder *et al.*, 2003). Focusing on this region of the stomach Snyder *et al.* used nLCD followed by gene chip analysis to query the expression profile associated with this pathology. They found 361 genes and 171 EST's enriched in the bacterially associated lymphoid aggregates. These included genes associated with infiltration of T-cells, B-cells, macrophages, dendritic cells and natural killer cells. There was also evidence of increased expression of genes involved in regulating blood vessel formation (VEGF-A, and its receptor, fms-like tyrosine kinase-1). Genes involved in T and B cell activation (CD15, Zap70, T-cell factor 1, B lymphoid kinase, Wiscott–Aldrich syndrome homologue, Arp2/3, NFκB, Igs, complement component 3, INF-γ inducible protein 47) were also induced. There was reduced expression of some regulators of inflammation (tritetraprolin, PKC interacting cousin of thioredoxin2). Interestingly, genes involved in cholesterol biosynthesis were induced as had been seen in a previous study using epithelial cell culture (Guillemin *et al.*, 2002).

Two studies looked at the progression of lymphoid aggregates to MALT lymphoma. The first study (Mueller *et al.*, 2003b) tested the ability of *H. felis*, *H. pylori* and *H. heilmanii* to induce MALT. All showed MALT in some animals, but *H. heilmannii* to the greatest extent (by 18 months 60% of animals showed moderate to severe lymphoepithelial lesions (LELs), a precursor of MALT lymphoma). The second study took advantage of the observation that long-term infection with *H. felis* isolate CS1 (18 months) followed by eradication therapy, and then re-infection produced a higher frequency of progression to lymphoma than continual *H. felis* infection (Mueller *et al.*, 2005). Both studies initially extracted RNA from whole stomachs and used differentially expressed genes to determine whether samples segregated based on severity of pathology and looked for genes that statistically correlated with progression of disease. nLCD and histology were then used to determine the cell types expressing genes of interest. In the *H. heilmannii* experiment, clustering of the samples based on gene expression allowed segregation of control, mild, and moderate to severe pathology, but the progression from severe to malignant pathology could not be distinguished. In both studies stomach tissue from infected animals showed increased expression of lymphocyte signature genes including B-cell-specific genes (immunoglobulins, CD53, CD40, integrinβ7, B-cell markers required for interaction with activated T-cells via CD40 ligand and B-cell homing to the gastrointestinal tract). This signature was present even in the absence of histological evidence of bacteria or B-cells. A cluster of genes involved in fortification of epithelial barriers was induced in animals with current infection and included small proline-rich protein 2A, crp-ductin, trefoil peptides secretory granule proteoglycan, Ig heavy chain, CDC-linked kinase 3 and small inducible cytokines. A second cluster associated with the presence of bacteria at the time of harvest in the *H. felis* study included MHC genes (class I and II), cell surface receptors and signalling molecules highly expressed in premalignant or malignant but not normal B-cells (CD52, CD53, BclII-A1a,c and d).

The *H. heilmannii* study found that progression from mild to moderate/severe pathology had a large transcriptional signature (300 genes), some associated with human malignancies e.g. laminin receptor-1 (basement membrane degradation), MDR-1, nibrin (DNA double strand break repair), chitinase-3 like protein. This gene list as well as genes differentially regulated in all infected animals compared to controls were used to predict pathology of 29 additional animals profiled in the same way. This worked quite well with complete misclassification of only one animal (severe pathology predicted to be mild), suggesting that at least a portion of this transcriptional signature reflects the altered conditions in the stomach during disease progression. Transcriptional changes in only one gene could be significantly associated with progression to malignancy in the *H. heilmannii* study: calgranulin. Expression profiling of nLCD derived cellular subsets showed that the mucosal cell signature included calgranulin. In the study with *H. felis* eradication and re-infection, there was an additional transcriptional signature for re-infection, and thus more severe pathology, which included glutathione S-transferase, ferritin heavy chain, deoxycytidine kinase, SNF1-ki kinase, and proliferation-associated gene A. The presence of proliferation associated genes in this cluster prompted the authors to measure the extent of cell proliferation in the different groups. The lymphoid aggregates in the re-infected group had the highest proliferation rates – twofold higher than the continually infected group. Further histological analyses revealed that the proliferating cells have an unusual phenotype in that they show low expression of many B-cell markers ($B220^{-}$, $CD40^{low}$, $CD79\alpha^{low}$) but express high levels of IgM. Molecular profile of the infiltrating T-cells showed a CD4+ Th-2 phenotype. The predominant antigen presenting cell in the lymphomas were dendritic cells of the follicular phenotype (not myeloid or lymphoid). This dendritic cell subtype showed a high level expression of S100/calgranulin (the lone transcript associated with malignant progression in the *H. heilmannii* study). The presence of these follicular dendritic cells, and not T-cells correlated with progression to lymphoma in the *H. felis* study. Interestingly subsequent studies have shown that knock down of calgranulin (also called S100A8/9) alters the migratory phenotype of the SNU484 gastric cell line through loss of activation of matrix-metalloprotease 2 (Yong and Moon, 2007) and calgranulin A and B expression has been recently associated with high-grade dysplasia in Barrett's oesophagus (Bax *et al.*, 2007). Thus, calgranulin could be a marker for both epithelial and lymphoid cancers. Whether this gene is a biomarker for gastric cancer of epithelial origins remains to be tested.

Human (and primate) biopsies – disease progression

There have now been a few studies that have profiled expression patterns of human or rhesus macaque gastric biopsies. These studies show common themes with the experiments in cell culture studies and mouse models, though are less complete. Experimental observation of the early changes in the human stomach upon *H. pylori* infection is very difficult in the absence of experimental infection, but profiling of monkey infection is possible. An added advantage of the monkey model over rodent models is the ability to sequentially biopsy the same animal. Expression levels can thus be compared to baseline levels of the same animal. A major challenge associated with profiling biopsy material of either monkey or human origin is the limited amount of tissue. Improvements in RNA isolation methods and methods of amplification have made such studies possible, but to date the sensitivity is lower than in the mouse and cell culture studies and no studies on cell type specific profiles have been reported yet.

Using the rhesus macaque model, Huff *et al.* (2004) profiled transcription of whole antral gastric biopsy tissue after infection with the human *H. pylori* isolate J166 at 2, 8 and 24 weeks after infection using a human array platform. Approximately one-third of genes on the arrays showed baseline expression indicating there was sufficient homology between the human and monkey sequences. Comparison of post inoculation expression to that of baseline revealed 148 genes (84 up, 64 down) at 2 weeks; 129 genes (66 up, 63 down) at 8 weeks, 206 genes (105 up, 101 down) at 24 weeks that significantly changed. The largest group at all time points showed annotation of innate immune and inflammatory response. Other categories were chemokine-cytokine, heat shock, cell growth and differentiation, apoptosis, structural (many cytoskeletal), signal transduction, and transcription factor. Among the innate factors were human beta-defensin 2 (hBD-2), protease inhibitor 3, lipocalin-2 and surfactant. Mucin 1 was increased and may limit *H. pylori* cell–cell adhesion and binding to gastric epithelium. There was corresponding decreased expression of a transcription factor (ZEB1) which represses mucin1 in epithelial cells. Mucin 6 was also upregulated, which has been observed in other studies. Mucin5A, a receptor for BabA was not induced, but the infecting strains lose BabA expression (Solnick *et al.*, 2004). Another possibility is that the Mucin5A sequences between human and macaque have diverged. Several extracellular matrix remodelling enzymes showed changes (both induced and repressed). Several cell adhesion molecules were induced including CD44, ICAM-5, integrins α3 and αL, a TGF-binding protein, and galectins 3, 4, 8 which are involved in neutrophil recruitment and cell–cell and cell-matrix interaction. Galectins have also been shown to bind microbial components. A striking observation was downregulation of several Hsp70 family members. Hsp70 can be involved in amplification of inflammatory responses. The cytokine-chemokine signature was low and IL-8 induction was not observed. There seemed to be an immunosuppressive signature with downregulation of CXCL11, CXCL10, and glycodelin – a glycosylated protein that modulates T-cell function. Several cytoskeletal elements including actin binding proteins, elastin and villin were changed (many induced, some repressed). Among the growth factors, latent TGF-β binding protein was induced and platelet-derived growth factor receptor like protein was repressed. Myc showed repression at the 8 week time point. Both pro and anti-apoptotic gene expression was present similar to other studies suggesting a complex and reciprocal responses to *H. pylori*.

A pioneering study (Wen *et al.*, 2004) analysed the transcriptional profile of the chronic inflammatory response in human antral gastric biopsies from infected and uninfected patients (some after successful eradication) using an 847 gene inflammatory gene array.

The number of genes detected on the array was significantly higher in the infected group than the non infected group (83.9% vs. 71.3%). Clustering with all the genes allowed segregation of infected from uninfected samples and revealed three clusters of genes that were differentially expressed between the two groups. The genes upregulated during infection included IL-18R accessory protein, several TLRs (1, 4, 5, 6, but not 2), LPS binding protein, IL-2Rβ, TNF-α, IL-16, IFN-γR1, GM-CSFRβ, IL2Rγ, complement and complement receptors (CR1, CD21, C3, adhesion molecules (ICAM-1, VCAM-1, MadCAM-1), integrins ($\beta_{2,7},\alpha_{X,L,4}$) chemokine and receptors (BLC/BCA-1,MIP-3α, ENA-78, GRO-α,β,γ, CCR-7,6, CXCR-4), signalling molecules MAPK10, DRAK2, PKCB1,MMP-9, ADAM12, and B or T-cell activation (CD45, CD53, GL50/B7-H2, SLAM, B7-H1). Whether the upregulation of TLRs represents activation of these receptors or merely influx of cells bearing higher expression of the genes was not determined.

The authors went on to elucidate host and bacterial factors that correlate with the strength or extent of the inflammatory signature. Transcriptional analysis of corpus biopsies from a subset of patients showed an overlapping but muted response. The number of genes and intensity of the response also correlated with the degree of inflammation in the samples. The genotype of the bug appeared related to the clustering of the samples in that all the triple positive (*cagA*+, *vacAs1*, *babA2*+) fell in one of two nodes. CagA+ vs. CagA- samples showed differential expression in 35 genes while *babA2*+ vs. *babA2*– showed differential expression in 42 genes. Interestingly the infection signature was more prominent in the biopsies taken from patient infected with *babA2*- strains. Host factors including age and gender were analysed for their influence on the expression signature. While no signature could be correlated to either of these factors in the total sample set, there were genes that correlated with gender in both *H. pylori*+ and *H. pylori*– set. The robust signature of innate immune genes was rather surprising since *H. pylori* is thought to avoid TLR signalling. Included in the innate immune molecules was lactoferrin as has been seen in other studies. Upregulation of cytokine receptors for INF-γ, IL-2, IL-3, IL-12 and IL-18 supports the observation of a Th-1 driven response. The extensive chemokine signature presumably drives recruitment of other inflammatory cells including lymphocytes and dendritic cells. Interestingly, IL-8 expression was not detected in this experiment and this was confirmed by RT-PCR although this is the major cytokine induced in cell culture. Other CXC chemokines including GRO-α, β, γ were detected, suggesting they may play a greater role in neutrophil recruitment during human infection.

A final study attempted to identify genes during human infection by evaluating expression patterns of total RNA from gastric biopsies taken one year apart either after antibiotic eradication of infection ($n = 13$) or placebo control (and remained infected $n = 14$) on whole genome microarrays (Tsai *et al.*, 2006). The treatment group showed statistically significant regression of gastritis after 1 year while the placebo group did not. There was no increase in gastric atrophy or intestinal metaplasia in either group. The difference in expression between 1 year and baseline was computed for each gene. Genes were selected for cluster analysis based on having a net difference >3 standard deviations of the mean or the net differences in at least three patients which yielded 207 genes. Unsupervised clustering using these genes largely segregated the patients from the treatment and placebo group and further analysis revealed 30 genes whose expression differed between the baseline and 1 year biopsies. All 30 genes showed lower expression at 1 year. The majority (5) were associated with immune response and inflammation. Other functional categories represented by multiple genes included cell–cell adhesion (3) and lipid transport and metabolism (4). While changes of expression for genes in these general classes were found in the cell culture and mouse studies, no specific genes were common with this study. In the placebo group 32 genes were significantly up regulated and 23 genes downregulated.

In order to identify genes associated with disease progression, baseline gene expression was compared between subjects with gastritis and intestinal metaplasia, revealing upregulation of 11 genes and downregulation of one gene.

Intestinal metaplasia associated upregulated genes were involved in defence/immune response (TFF3, DAF, DPP4, NT5E), surface antigen/receptor activity (GPA33), fatty acid metabolism (FABP1), protein biosynthesis and metabolism (PEPD, PRSS and ASS), cytoskeleton (VIL1) and metabolism (AGXT1). The down regulated gene was ABCC5- associated with cellular transport. They then queried genes down regulated in the treatment group and upregulated in the placebo group. Five genes met this criterion: FABP1, MTP, MUC13, S100A10 and TACSTD1. The differential expression of FABP1 was confirmed by immunohistochemistry. Perhaps the most interesting finding of this study was that even in the context of chronic infection, eradication of the bacteria could reverse the transcriptional signature of the stomach tissue. While it has been difficult to document reversal of pathology in eradication trials due to the time required to see effects (Mera *et al.*, 2005), the cellular transcription signature provides an alternate metric to measure at earlier time points.

Conclusions from transcriptional profiling studies

While the human and primate biopsy studies are only beginning and have either used suboptimal (primate RNA profiled on human arrays) or incomplete (human tissue profiled on immunologic sub-array), early indications suggest common signatures with the cell culture and mouse model studies. One interesting principle that has emerged from these studies is a strong signature of innate mucosal defence genes. While not as pronounced in the cell culture studies, all the mouse infection studies and the primate study show expression of iron sequestering genes, such as lactoferrin, alteration of mucus profiles and the crp-ductin genes. The role of these gene products in shifting the balance from a controlled infection to one that causes severe pathology will be an exciting area of future study. Another collective insight is that T- and B-cells play clear roles during infection but may have novel phenotypes (for example unusual IgM producing B-cell found in fat tissue associated with protection). Other immune cells such as dendritic cells and natural killer cells have emerged as possible contributors to disease specific profiles and protection phenotypes. An unexpected role for cholesterol, lipid production and fat cell products has emerged from both cell line and whole animal studies. The challenges that lie ahead will be to extend the cell line and animal studies to more human studies, to elucidate the cell types responsible for some of these transcripts and the regulatory networks that lead to these expression changes.

Overview and outlook

Microarrays have proved to be a powerful tool for the study of *H. pylori* pathogenesis. The experiments to date have demonstrated that *H. pylori* have a surprisingly plastic genome when comparing isolates both within and between patients suggesting a high rate of genetic changes. Paradoxically, it has been difficult to observe *de novo* genomic changes experimentally which may mean the rates of change are lower than initially predicted. Advances in this area will likely require genome wide experiments of more single colony isolates and may benefit from newer technologies that allow detection of single nucleotide polymorphisms including tiling arrays and comparative whole genome sequencing. We now have a basic framework of the major transcriptional regulatory networks *H. pylori* uses to combat a variety of environmental stresses likely experienced during colonization of the stomach through *in vitro* work. A major mystery is the mechanism by which a number of genes in the acid and metal dependant stimulons, which are not engaged by the known transcription factors, are regulated. This could mean there are many un-annotated transcription factors hiding in the genome or that other mechanisms are at play, which may include methylation, regulatory RNAs or epigenetic mechanisms. A clear next challenge is to measure the global transcriptional state *in vivo* to elucidate which and under what conditions regulons become active in the stomach. Finally transcriptional profiling of the murine host has provided a highly sophisticated view of specific cell types in the stomach that do and do not respond to *H. pylori* infection, disease progression and even successful vaccination. Again the major challenge is to extend these studies more fully to the human host.

References

Achtman, M., Azuma, T., Berg, D.E., Ito, Y., Morelli, G., Pan, Z.J., Suerbaum, S., Thompson, S.A., van der Ende, A., and van Doorn, L.J. (1999). Recombination and clonal groupings within *Helicobacter pylori* from different geographical regions. Mol. Microbiol. *32*, 459–470.

Agrawal, S., Agrawal, A., Doughty, B., Gerwitz, A., Blenis, J., Van Dyke, T., and Pulendran, B. (2003). Cutting edge: different Toll-like receptor agonists instruct dendritic cells to induce distinct Th responses via differential modulation of extracellular signal-regulated kinase-mitogen-activated protein kinase and c-Fos. J. Immunol. *171*, 4984–4989.

Albert, T.J., Dailidiene, D., Dailide, G., Norton, J.E., Kalia, A., Richmond, T.A., Molla, M., Singh, J., Green, R.D., and Berg, D.E. (2005). Mutational discover in bacterial genomes: metronidazole resistance in *Helicobacter pylori*. Nature Methods *2*, in press.

Alm, R.A., Ling, L.S., Moir, D.T., King, B.L., Brown, E.D., Doig, P.C., Smith, D.R., Noonan, B., Guild, B.C., deJonge, B.L., *et al.* (1999). Genomic-sequence comparison of two unrelated isolates of the human gastric pathogen *Helicobacter pylori*. Nature *397*, 176–180.

Amieva, M.R., Vogelmann, R., Covacci, A., Tompkins, L.S., Nelson, W.J., and Falkow, S. (2003). Disruption of the epithelial apical-junctional complex by *Helicobacter pylori* CagA. Science *300*, 1430–1434.

Andersen-Nissen, E., Smith, K.D., Bonneau, R., Strong, R.K., and Aderem, A. (2007). A conserved surface on Toll-like receptor 5 recognizes bacterial flagellin. J. Exp. Med. *204*, 393–403.

Andersen-Nissen, E., Smith, K.D., Strobe, K.L., Barrett, S.L., Cookson, B.T., Logan, S.M., and Aderem, A. (2005). Evasion of Toll-like receptor 5 by flagellated bacteria. Proc. Natl. Acad. Sci. USA *102*, 9247–9252.

Bagnoli, F., Buti, L., Tompkins, L., Covacci, A., and Amieva, M.R. (2005). *Helicobacter pylori* CagA induces a transition from polarized to invasive phenotypes in MDCK cells. Proc. Natl. Acad. Sci. USA *102*, 16339–16344.

Baldwin, D.N., Shepherd, B., Kraemer, P., Hall, M.K., Sycuro, L.K., Pinto-Santini, D.M., and Salama, N.R. (2007). Identification of *Helicobacter pylori* genes that contribute to stomach colonization. Infect Immun *75*, 1005–1016.

Bax, D.A., Siersema, P.D., Haringsma, J., Kuipers, E.J., Vos, A.J., Van Dekken, H., Van Vliet, A.H., and Kusters, J.G. (2007). High-grade dysplasia in Barrett's esophagus is associated with increased expression of calgranulin A and B. Scand. J. Gastroenterol. *42*, 902–910.

Belzer, C., Stoof, J., and van Vliet, A.H. (2007). Metal-responsive gene regulation and metal transport in *Helicobacter* species. Biometals *20*, 417–429.

Bjorkholm, B., Lundin, A., Sillen, A., Guillemin, K., Salama, N., Rubio, C., Gordon, J.I., Falk, P., and Engstrand, L. (2001). Comparison of genetic divergence and fitness between two subclones of *Helicobacter pylori*. Infect. Immun. *69*, 7832–7838.

Bjorkholm, B.M., Guruge, J.L., Oh, J.D., Syder, A.J., Salama, N., Guillemin, K., Falkow, S., Nilsson, C., Falk, P.G., Engstrand, L., and Gordon, J.I. (2002). Colonization of germ-free transgenic mice with genotyped *Helicobacter pylori* strains from a case–control study of gastric cancer reveals a correlation between host responses and HsdS components of type I restriction-modification systems. J. Biol. Chem. *277*, 34191–34197.

Bourzac, K.M., and Guillemin, K. (2005). *Helicobacter pylori*-host cell interactions mediated by type IV secretion. Cell Microbiol. *7*, 911–919.

Brandt, S., Kwok, T., Hartig, R., Konig, W., and Backert, S. (2005). NF-kappaB activation and potentiation of proinflammatory responses by the *Helicobacter pylori* CagA protein. Proc. Natl. Acad. Sci. USA *102*, 9300–9305.

Bury-Mone, S., Skouloubris, S., Dauga, C., Thiberge, J.M., Dailidiene, D., Berg, D.E., Labigne, A., and De Reuse, H. (2003). Presence of active aliphatic amidases in *Helicobacter* species able to colonize the stomach. Infect. Immun. *71*, 5613–5622.

Bury-Mone, S., Thiberge, J.M., Contreras, M., Maitournam, A., Labigne, A., and De Reuse, H. (2004). Responsiveness to acidity via metal ion regulators mediates virulence in the gastric pathogen *Helicobacter pylori*. Mol. Microbio. *53*, 623–638.

Chalker, A.F., Minehart, H.W., Hughes, N.J., Koretke, K.K., Lonetto, M.A., Brinkman, K.K., Warren, P.V., Lupas, A., Stanhope, M.J., Brown, J.R., and Hoffman, P.S. (2001). Systematic identification of selective essential genes in *Helicobacter pylori* by genome prioritization and allelic replacement mutagenesis. J. Bacterio.l *183*, 1259–1268.

Chang, Y.T., Wu, M.S., Chang, Y.J., Chen, C.C., Lin, Y.S., Hsieh, T., Yang, P.C., and Lin, J.T. (2006). Distinct gene expression profiles in gastric epithelial cells induced by different clinical isolates of *Helicobacter pylori*--implication of bacteria and host interaction in gastric carcinogenesis. Hepatogastroenterology *53*, 484–490.

Contreras, M., Thiberge, J.M., Mandrand-Berthelot, M.A., and Labigne, A. (2003). Characterization of the roles of NikR, a nickel-responsive pleiotropic autoregulator of *Helicobacter pylori*. Mol. Microbiol. *49*, 947–963.

Cox, J.M., Clayton, C.L., Tomita, T., Wallace, D.M., Robinson, P.A., and Crabtree, J.E. (2001). cDNA array analysis of cag pathogenicity island-associated *Helicobacter pylori* epithelial cell response genes. Infect. Immun. *69*, 6970–6980.

Dalma-Weiszhausz, D.D., Warrington, J., Tanimoto, E.Y., and Miyada, C.G. (2006). The affymetrix GeneChip platform: an overview. Methods Enzymol. *410*, 3–28.

Danielli, A., Roncarati, D., Delany, I., Chiarini, V., Rappuoli, R., and Scarlato, V. (2006). In vivo dissection of the *Helicobacter pylori* Fur regulatory circuit by genome-wide location analysis. J.Bacteriol. *188*, 4654–4662.

De Pina, K., Desjardin, V., Mandrand-Berthelot, M.A., Giordano, G., and Wu, L.F. (1999). Isolation and characterization of the nikR gene encoding a nickel-responsive regulator in *Escherichia coli*. J. Bacteriol. *181*, 670–674.

Delany, I., [other authors] (2001). The Fur repressor controls transcription of iron-activated and -repressed genes in *Helicobacter pylori*. Mol. Microbiol. *42*, 1297–1309.

Doherty, J.M., Carmichael, L.K., and Mills, J.C. (2006). GOurmet: a tool for quantitative comparison and visualization of gene expression profiles based on gene ontology (GO) distributions. BMC Bioinformatics *7*, 151.

Dossumbekova, A., Prinz, C., Mages, J., Lang, R., Kusters, J.G., Van Vliet, A.H., Reindl, W., Backert, S., Saur, D., Schmid, R.M., and Rad, R. (2006). *Helicobacter pylori* HopH (OipA) and bacterial pathogenicity: genetic and functional genomic analysis of hopH gene polymorphisms. J. Infect. Dis. *194*, 1346–1355.

Dudoit, S., Gentleman, R.C., and Quackenbush, J. (2003). Open source software for the analysis of microarray data. Biotechniques *Suppl*, 45–51.

Eisen, M.B., Spellman, P.T., Brown, P.O., and Botstein, D. (1998). Cluster analysis and display of genome-wide expression patterns. Proc. Natl. Acad. Sci. USA *95*, 14863–14868.

El-Etr, S.H., Mueller, A., Tompkins, L.S., Falkow, S., and Merrell, D.S. (2004). Phosphorylation-independent effects of CagA during interaction between *Helicobacter pylori* and T84 polarized monolayers. J. Infect. Dis. *190*, 1516–1523.

Ernst, F.D., Bereswill, S., Waidner, B., Stoof, J., Mader, U., Kusters, J.G., Kuipers, E.J., Kist, M., van Vliet, A.H., and Homuth, G. (2005). Transcriptional profiling of *Helicobacter pylori* Fur- and iron-regulated gene expression. Microbiology *151*, 533–546.

Falush, D., Wirth, T., Linz, B., Pritchard, J.K., Stephens, M., Kidd, M., Blaser, M.J., Graham, D.Y., Vacher, S., Perez-Perez, G.I., *et al.* (2003). Traces of human migrations in *Helicobacter pylori* populations. Science *299*, 1582–1585.

Forsyth, M.H., Cao, P., Garcia, P.P., Hall, J.D., and Cover, T.L. (2002). Genome-wide transcriptional profiling in a histidine kinase mutant of *Helicobacter pylori* identifies members of a regulon. J. Bacteriol. *184*, 4630–4635.

Franco, A.T., Israel, D.A., Washington, M.K., Krishna, U., Fox, J.G., Rogers, A.B., Neish, A.S., Collier-Hyams, L., Perez-Perez, G.I., Hatakeyama, M., *et al.* (2005). Activation of beta-catenin by carcinogenic *Helicobacter pylori*. Proc. Natl. Acad. Sci. USA *102*, 10646 10651.

Gancz, H., Censini, S., and Merrell, D.S. (2006). Iron and pH homeostasis intersect at the level of Fur regulation in the gastric pathogen *Helicobacter pylori*. Infect. Immun. *74*, 602–614.

Gressmann, H., Linz, B., Ghai, R., Pleissner, K.P., Schlapbach, R., Yamaoka, Y., Kraft, C., Suerbaum, S., Meyer, T.F., and Achtman, M. (2005). Gain and loss of multiple genes during the evolution of *Helicobacter pylori*. PLoS Genet. *1*, e43.

Guillemin, K., Salama, N.R., Tompkins, L.S., and Falkow, S. (2002). Cag pathogenicity island-specific responses of gastric epithelial cells to *Helicobacter pylori* infection. Proc. Nat. Acad. Sci. USA *99*, 15136–15141.

Heller, M.J. (2002). DNA microarray technology: devices, systems, and applications. Annu. Rev. Biomed. Eng. *4*, 129–153.

Huang, D.W., Sherman, B.T., Tan, Q., Collins, J.R., Alvord, W.G., Roayaei, J., Stephens, R., Baseler, M.W., Lane, H.C., and Lempicki, R.A. (2007). The DAVID Gene Functional Classification Tool: a novel biological module-centric algorithm to functionally analyze large gene lists. Genome Biol. *8*, R183.

Huff, J.L., Hansen, L.M., and Solnick, J.V. (2004). Gastric transcription profile of *Helicobacter pylori* infection in the rhesus macaque. Infect. Immun. *72*, 5216–5226.

Husson, M.O., Legrand, D., Spik, G., and Leclerc, H. (1993). Iron acquisition by *Helicobacter pylori*: importance of human lactoferrin. Infect. Immun. *61*, 2694–2697.

Israel, D.A., Salama, N., Krishna, U., Rieger, U.M., Atherton, J.C., Falkow, S., and Peek, R.M., Jr. (2001). *Helicobacter pylori* genetic diversity within the gastric niche of a single human host. Proc. Natl. Acad. Sci. USA *98*, 14625–14630.

Jiang, X.H., Wong, B.C., Lin, M.C., Zhu, G.H., Kung, H.F., Jiang, S.H., Yang, D., and Lam, S.K. (2001). Functional p53 is required for triptolide-induced apoptosis and AP-1 and nuclear factor-kappaB activation in gastric cancer cells. Oncogene *20*, 8009–8018.

Josenhans, C., and Suerbaum, S. (2002). The role of motility as a virulence factor in bacteria. Int. J. Med. Microbiol. *291*, 605–614.

Joyce, E.A., Chan, K., Salama, N.R., and Falkow, S. (2002). Redefining bacterial populations: a post-genomic reformation. Nat. Rev. Gene.t *3*, 462–473.

Kim, C.C., Joyce, E.A., Chan, K., and Falkow, S. (2002). Improved analytical methods for microarray-based genome-composition analysis. Genome Biol. *3*, RESEARCH0065.

Kim, N., Marcus, E.A., Wen, Y., Weeks, D.L., Scott, D.R., Jung, H.C., Song, I.S., and Sachs, G. (2004). Genes of *Helicobacter pylori* regulated by attachment to AGS cells. Infect. Immu.n *72*, 2358–2368.

Kivi, M., Rodin, S., Kupershmidt, I., Lundin, A., Tindberg, Y., Granstrom, M., and Engstrand, L. (2007). *Helicobacter pylori* genome variability in a framework of familial transmission. BMC Microbiol. *7*, 54.

Kraft, C., Stack, A., Josenhans, C., Niehus, E., Dietrich, G., Correa, P., Fox, J.G., Falush, D., and Suerbaum, S. (2006). Genomic changes during chronic *Helicobacter pylori* infection. J. Bacteriol. *188*, 249–254.

Linz, B., Balloux, F., Moodley, Y., Manica, A., Liu, H., Roumagnac, P., Falush, D., Stamer, C., Prugnolle, F., van der Merwe, S.W., *et al.* (2007). An African origin for the intimate association between humans and *Helicobacter pylori*. Nature *445*, 915–918.

Loh, J.T., and Cover, T.L. (2006). Requirement of histidine kinases HP0165 and HP1364 for acid resistance in *Helicobacter pylori*. Infect. Immun. *74*, 3052–3059.

Macnab, R.M. (1996). Flagella and motility. In *Escherichia coli* and *Salmonella* Cellular and Molecular Biology, F. C. Neidhardt, ed. (Washington, D.C., ASM Press), pp. 123–145.

Maeda, S., Otsuka, M., Hirata, Y., Mitsuno, Y., Yoshida, H., Shiratori, Y., Masuho, Y., Muramatsu, M., Seki,

N., and Omata, M. (2001). cDNA microarray analysis of *Helicobacter pylori*-mediated alteration of gene expression in gastric cancer cells. Biochem. Biophys. Res. Commun. *284*, 443–449.

Manoil, C. (2000). Tagging exported proteins using *Escherichia coli* alkaline phosphatase gene fusions. Methods Enzymol. *326*, 35–47.

Marchetti, M., and Rappuoli, R. (2002). Isogenic mutants of the cag pathogenicity island of *Helicobacter pylori* in the mouse model of infection: effects on colonization efficiency. Microbiology *148*, 1447–1456.

Masse, E., and Gottesman, S. (2002). A small RNA regulates the expression of genes involved in iron metabolism in *Escherichia coli*. Proc. Natl. Acad. Sc.i USA *99*, 4620–4625.

Mera, R., Fontham, E.T., Bravo, L.E., Bravo, J.C., Piazuelo, M.B., Camargo, M.C., and Correa, P. (2005). Long term follow up of patients treated for *Helicobacter pylori* infection. Gut *54*, 1536–1540.

Merrell, D.S., Goodrich, M.L., Otto, G., Tompkins, L.S., and Falkow, S. (2003a). pH-regulated gene expression of the gastric pathogen *Helicobacter pylori*. Infect Immun *71*, 3529–3539.

Merrell, D.S., Thompson, L.J., Kim, C.C., Mitchell, H., Tompkins, L.S., Lee, A., and Falkow, S. (2003b). Growth phase-dependent response of *Helicobacter pylori* to iron starvation. Infect. Immun. *71*, 6510–6525.

Mills, J.C., Andersson, N., Stappenbeck, T.S., Chen, C.C., and Gordon, J.I. (2003). Molecular characterization of mouse gastric zymogenic cells. J.Biol. Chem. *278*, 46138–46145.

Mills, J.C., Syder, A.J., Hong, C.V., Guruge, J.L., Raaii, F., and Gordon, J.I. (2001). A molecular profile of the mouse gastric parietal cell with and without exposure to *Helicobacter pylori*. Proc. Natl. Acad. Sci. USA *98*, 13687–13692.

Moese, S., Selbach, M., Kwok, T., Brinkmann, V., Konig, W., Meyer, T.F., and Backert, S. (2004). *Helicobacter pylori* induces AGS cell motility and elongation via independent signaling pathways. Infect. Immun. *72*, 3646–3649.

Mueller, A., Merrell, D.S., Grimm, J., and Falkow, S. (2004). Profiling of microdissected gastric epithelial cells reveals a cell type-specific response to *Helicobacter pylori* infection. Gastroenterology *127*, 1446–1462.

Mueller, A., O'Rourke, J., Chu, P., Chu, A., Dixon, M.F., Bouley, D.M., Lee, A., and Falkow, S. (2005). The role of antigenic drive and tumor-infiltrating accessory cells in the pathogenesis of *Helicobacter*-induced mucosa-associated lymphoid tissue lymphoma. Am. J. Pathol. *167*, 797–812.

Mueller, A., O'Rourke, J., Chu, P., Kim, C.C., Sutton, P., Lee, A., and Falkow, S. (2003a). Protective immunity against *Helicobacter* is characterized by a unique transcriptional signature. Proc. Natl. Acad. Sci. USA *100*, 12289–12294.

Mueller, A., O'Rourke, J., Grimm, J., Guillemin, K., Dixon, M.F., Lee, A., and Falkow, S. (2003b). Distinct gene expression profiles characterize the histopathological stages of disease in *Helicobacter*-induced mucosa-associated lymphoid tissue lymphoma. Proc. Natl. Acad. Sci. USA *100*, 1292–1297.

Niehus, E., Gressmann, H., Ye, F., Schlapbach, R., Dehio, M., Dehio, C., Stack, A., Meyer, T.F., Suerbaum, S., and Josenhans, C. (2004). Genome-wide analysis of transcriptional hierarchy and feedback regulation in the flagellar system of *Helicobacter pylori*. Mol. Microbiol. *52*, 947–961.

Oh, J.D., Kling-Backhed, H., Giannakis, M., Xu, J., Fulton, R.S., Fulton, L.A., Cordum, H.S., Wang, C., Elliott, G., Edwards, J., *et al.* (2006). The complete genome sequence of a chronic atrophic gastritis *Helicobacter pylori* strain: evolution during disease progression. Proc. Natl. Acad. Sci. USA *103*, 9999–10004.

Pflock, M., Bathon, M., Schar, J., Muller, S., Mollenkopf, H., Meyer, T.F., and Beier, D. (2007). The orphan response regulator HP1021 of *Helicobacter pylori* regulates transcription of a gene cluster presumably involved in acetone metabolism. J. Bacteriol. *189*, 2339–2349.

Pflock, M., Finsterer, N., Joseph, B., Mollenkopf, H., Meyer, T.F., and Beier, D. (2006). Characterization of the ArsRS regulon of *Helicobacter pylori*, involved in acid adaptation. J. Bacteriol. *188*, 3449–3462.

Philpott, D.J., Belaid, D., Troubadour, P., Thiberge, J.M., Tankovic, J., Labigne, A., and Ferrero, R.L. (2002). Reduced activation of inflammatory responses in host cells by mouse-adapted *Helicobacter pylori* isolates. Cell Microbiol. *4*, 285–296.

Rad, R., Brenner, L., Krug, A., Voland, P., Mages, J., Lang, R., Schwendy, S., Reindl, W., Dossumbekova, A., Ballhorn, W., *et al.* (2007). Toll-like receptor-dependent activation of antigen-presenting cells affects adaptive immunity to *Helicobacter pylori*. Gastroenterology *133*, 150–163 e153.

Rahn, W., Redline, R.W., and Blanchard, T.G. (2004). Molecular analysis of *Helicobacter pylori*-associated gastric inflammation in naive versus previously immunized mice. Vaccine *23*, 807–818.

Ramsey, V.G., Doherty, J.M., Chen, C.C., Stappenbeck, T.S., Konieczny, S.F., and Mills, J.C. (2007). The maturation of mucus-secreting gastric epithelial progenitors into digestive-enzyme secreting zymogenic cells requires Mist1. Development *134*, 211–222.

Reimers, M., and Carey, V.J. (2006). Bioconductor: an open source framework for bioinformatics and computational biology. Methods Enzymol. *411*, 119–134.

Roncarati, D., Danielli, A., Spohn, G., Delany, I., and Scarlato, V. (2007). Transcriptional regulation of stress response and motility functions in *Helicobacter pylori* is mediated by HspR and HrcA. J. Bacteriol. *189*, 7234–7243.

Sachs, G., Weeks, D.L., Wen, Y., Marcus, E.A., Scott, D.R., and Melchers, K. (2005). Acid acclimation by *Helicobacter pylori*. Physiology (Bethesda) *20*, 429–438.

Salama, N., Guillemin, K., McDaniel, T.K., Sherlock, G., Tompkins, L., and Falkow, S. (2000). A whole-genome microarray reveals genetic diversity among *Helicobacter pylori* strains. Proc. Natl. Acad. Sci. USA *97*, 14668–14673.

Salama, N.R., Gonzalez-Valencia, G., Deatherage, B., Aviles-Jimenez, F., Atherton, J.C., Graham, D.Y., and Torres, J. (2007). Genetic analysis of *Helicobacter pylori* strain populations colonizing the stomach at different times post-infection. J. Bacteriol. [Details?]

Salama, N.R., Shepherd, B., and Falkow, S. (2004). Global transposon mutagenesis and essential gene analysis of *Helicobacter pylori*. J Bacteriol *186*, 7926–7935.

Sassetti, C.M., Boyd, D.H., and Rubin, E.J. (2001). Comprehensive identification of conditionally essential genes in mycobacteria. Proc. Natl. Acad. Sci. USA *98*, 12712–12717.

Schena, M., Shalon, D., Davis, R.W., and Brown, P.O. (1995). Quantitative monitoring of gene expression patterns with a complementary DNA microarray. Science *270*, 467–470.

Scott, D.R., Marcus, E.A., Wen, Y., Oh, J., and Sachs, G. (2007). Gene expression in vivo shows that *Helicobacter pylori* colonizes an acidic niche on the gastric surface. Proc. Natl. Acad. Sci. USA *104*, 7235–7240.

Segal, E.D., Cha, J., Lo, J., Falkow, S., and Tompkins, L.S. (1999). Altered states: involvement of phosphorylated CagA in the induction of host cellular growth changes by *Helicobacter pylori*. Proc. Natl. Acad. Sci. USA *96*, 14559–14564.

Sepulveda, A.R., Tao, H., Carloni, E., Sepulveda, J., Graham, D.Y., and Peterson, L.E. (2002). Screening of gene expression profiles in gastric epithelial cells induced by *Helicobacter pylori* using microarray analysis. Aliment. Pharmacol Ther. *16* Suppl. 2, 145–157.

Solnick, J.V., Hansen, L.M., Salama, N.R., Boonjakuakul, J.K., and Syvanen, M. (2004). Modification of *Helicobacter pylori* outer membrane protein expression during experimental infection of rhesus macaques. Proc .Natl. Acad. Sci. USA *101*, 2106–2111.

Stein, M., Bagnoli, F., Halenbeck, R., Rappuoli, R., Fantl, W.J., and Covacci, A. (2002). c-Src/Lyn kinases activate *Helicobacter pylori* CagA through tyrosine phosphorylation of the EPIYA motifs. Mol. Microbiol. *43*, 971–980.

Suerbaum, S., Smith, J.M., Bapumia, K., Morelli, G., Smith, N.H., Kunstmann, E., Dyrek, I., and Achtman, M. (1998). Free recombination within *Helicobacter pylori*. Proc. Natl. Acad. Sci. USA 95, 12619–12624.

Syder, A.J., Guruge, J.L., Li, Q., Hu, Y., Oleksiewicz, C.M., Lorenz, R.G., Karam, S.M., Falk, P.G., and Gordon, J.I. (1999). *Helicobacter pylori* attaches to NeuAc alpha 2, 3Gal beta 1, 4 glycoconjugates produced in the stomach of transgenic mice lacking parietal cells. Mol. Cell *3*, 263–274.

Syder, A.J., Oh, J.D., Guruge, J.L., O'Donnell, D., Karlsson, M., Mills, J.C., Bjorkholm, B.M., and Gordon, J.I. (2003). The impact of parietal cells on *Helicobacter pylori* tropism and host pathology: an analysis using gnotobiotic normal and transgenic mice. Pro.c Natl. Acad. Sci. USA *100*, 3467–3472.

Thompson, L.J., Danon, S.J., Wilson, J.E., O'Rourke, J.L., Salama, N.R., Falkow, S., Mitchell, H., and Lee, A. (2004). Chronic *Helicobacter pylori* infection with Sydney strain 1 and a newly identified mouse-adapted strain (Sydney strain 2000) in C57BL/6 and BALB/c mice. Infect. Immun. *72*, 4668–4679.

Thompson, L.J., Merrell, D.S., Neilan, B.A., Mitchell, H., Lee, A., and Falkow, S. (2003). Gene expression profiling of *Helicobacter pylori* reveals a growth-phase-dependent switch in virulence gene expression. Infect Immun *71*, 2643–2655.

Tomb, J.F., White, O., Kerlavage, A.R., Clayton, R.A., Sutton, G.G., Fleischmann, R.D., Ketchum, K.A., Klenk, H.P., Gill, S., Dougherty, B.A., *et al.* (1997). The complete genome sequence of the gastric pathogen *Helicobacter pylori*. Nature *388*, 539–547.

Touati, D. (2000). Iron and oxidative stress in bacteria. Arch Biochem Biophys *373*, 1–6.

Tran, A.X., Stead, C.M., and Trent, M.S. (2005). Remodeling of *Helicobacter pylori* lipopolysaccharide. J Endotoxin Res *11*, 161–166.

Tsai, C.J., Herrera-Goepfert, R., Tibshirani, R.J., Yang, S., Mohar, A., Guarner, J., and Parsonnet, J. (2006). Changes of gene expression in gastric preneoplasia following *Helicobacter pylori* eradication therapy. Cancer Epidemiol Biomarkers Prev *15*, 272–280.

Tusher, V.G., Tibshirani, R., and Chu, G. (2001). Significance analysis of microarrays applied to the ionizing radiation response. Proc Natl Acad Sci USA *98*, 5116–5121.

Viala, J., Chaput, C., Boneca, I.G., Cardona, A., Girardin, S.E., Moran, A.P., Athman, R., Memet, S., Huerre, M.R., Coyle, A.J., *et al.* (2004). Nod1 responds to peptidoglycan delivered by the *Helicobacter pylori* cag pathogenicity island. Nat. Immunol. 5, 1166–1174.

Walduck, A., Schmitt, A., Lucas, B., Aebischer, T., and Meyer, T.F. (2004). Transcription profiling analysis of the mechanisms of vaccine-induced protection against *H. pylori*. Faseb J. *18*, 1955–1957.

Wang, H.T., Li, Z.H., Yuan, J.P., Zhao, W., Shi, X.D., Tong, S.Q., and Guo, X.K. (2005). Effect of *Helicobacter pylori* VacA on gene expression of gastric cancer cells. World J. Gastroenterol. *11*, 109–113.

Wen, S., Felley, C.P., Bouzourene, H., Reimers, M., Michetti, P., and Pan-Hammarstrom, Q. (2004). Inflammatory gene profiles in gastric mucosa during *Helicobacter pylori* infection in humans. J. Immunol. *172*, 2595–2606.

Wen, Y., Marcus, E.A., Matrubutham, U., Gleeson, M.A., Scott, D.R., and Sachs, G. (2003). Acid-adaptive genes of *Helicobacter pylori*. Infect Immun *71*, 5921–5939.

Wilderman, P.J., Sowa, N.A., FitzGerald, D.J., FitzGerald, P.C., Gottesman, S., Ochsner, U.A., and Vasil, M.L. (2004). Identification of tandem duplicate regulatory small RNAs in Pseudomonas aeruginosa involved in iron homeostasis. Proc. Natl. Acad. Sci. USA *101*, 9792–9797.

Wolber, P.K., Collins, P.J., Lucas, A.B., De Witte, A., and Shannon, K.W. (2006). The Agilent in situ-synthesized microarray platform. Methods Enzymol *410*, 28–57.

Yamaoka, Y., Kwon, D.H., and Graham, D.Y. (2000). A M(r) 34,000 proinflammatory outer membrane protein (*oipA*) of *Helicobacter pylori*. Proc. Natl. Acad. Sci. USA 97, 7533–7538.

Yong, H.Y., and Moon, A. (2007). Roles of calcium-binding proteins, S100A8 and S100A9, in invasive phenotype of human gastric cancer cells. Arch. Pharm. Res. *30*, 75–81.

Yuan, J.P., Li, T., Chen, H.B., Li, Z.H., Yang, G.Z., Hu, B.Y., Shi, X.D., Tong, S.Q., Li, Y.X., and Guo, X.K. (2004). Analysis of gene expression profile in gastric cancer cells stimulated with *Helicobacter pylori* isogenic strains. J. Med. Microbiol. 53, 965–974.

Zeeberg, B.R., Qin, H., Narasimhan, S., Sunshine, M., Cao, H., Kane, D.W., Reimers, M., Stephens, R.M., Bryant, D., Burt, S.K., *et al.* (2005). High-throughput GoMiner, an 'industrial-strength' integrative gene ontology tool for interpretation of multiple-microarray experiments, with application to studies of common variable immune deficiency (CVID). BMC Bioinformatics 6, 168.

The Application of Proteomics Technology to *Helicobacter pylori*-associated Gastroduodenal Disease: State-of-the-Art and Future Clinical Potentials

12

Ming-Shiang Wu, Lu-Ping Chow, Jaw-Town Lin and Shyh-Horng Chiou

Abstract

The current disease paradigm suggests that host genetics and *Helicobacter pylori* virulence play an important role in modulation of the final outcome. Elucidation of the bidirectional relationship between the host and bacteria is thus essential to clarify pathogenesis and development of new prevention and treatment strategies. Proteomics technology has provided unprecedented opportunity to comprehensively survey a cell's translational landscape and may allow in-depth analyses of host and pathogen interaction, either separately or interactively. Using this high-throughput platform and taking advantage of complete sequences for both the *H. pylori* and human genome in the database, some promising results and important information have been reported. Among the recent contributions relating to *H. pylori*, this review focuses on proteomics-based characterization of intrinsic and extrinsic perturbations and strain differences of *H. pylori*, epithelial cell response to *H. pylori* infection, identification of diagnostic/prognostic biomarkers, and uncovering of immunogenic proteins for vaccine targets. Future combination of evolving new proteomics technologies with clinical phenotypes and genotypes information will enhance the understanding of disease pathogenesis, lead to a more precise predication of variable outcomes, and facilitate the development of effective biomarkers for diagnosis, treatment and prevention of *H. pylori* infection.

Epidemiology and unmet clinical needs of *H. pylori* infection

The discovery and successful culture of *H. pylori* has revolutionized the diagnosis and treatment of gastroduodenal disease. *H. pylori* is a microaerophilic spiral-shaped Gram-negative bacterium that efficiently colonizes the human stomach. Epidemiological studies have suggested that *H. pylori* infects at least half of the world's population, and in developing countries, the infectious process occurs early and reaches 60–80% in adulthood (Suerbaum *et al.*, 2002). The annual incidence of *H. pylori* acquisition by adults in much lower in developed countries. To survive in the highly acidic environment, *H. pylori* can release ammonia from urea by urease, thereby neutralizing the gastric acid in stomach. However, ammonia is also a toxic substance to epithelial alls, so that its production plus other factors may often induce inflammation on epithelial cells. The prolonged infection and inflammation due to persistent *H. pylori* colonization ultimately lead to chronic gastritis, peptic ulcer, mucosa-associated lymphoid tissue lymphoma and gastric adenocarcinoma (Normark *et al.*, 2003). Currently, *H. pylori* is defined as class I carcinogen by WHO and eradication of *H. pylori* is considered as the first line treatment of peptic ulcer disease.

Although *H. pylori* has been identified as a major pathogen leading to the development of a wide range of gastroduodenal disease, only a small portion of infected patients have more severe gastric pathology such as peptic ulcer (10–15%), and gastric malignancies (1–2%) (Bodger *et al.*,1998). Why the majority (80–90%) of *H. pylori*-infected individuals remain asymptomatic, with only a small portion developing disease, is unknown. Furthermore, what are the factors which determine whether an infected individual will develop severe diseases remain ill-defined. Recently, *H. pylori* infection has been reported to be negatively associated with the development of obesity and gastroesophageal reflux disease (Chow *et al.*, 1998; Wu *et al.*, 2005). Taken together, indiscriminate eradication of *H. pylori* in general population is not advisable and identification of high-risk population is crucial in targeted prevention and treatment of *H. pylori*-related gastroduodenal disease.

The current disease paradigm suggests that gastric immune and inflammatory responses due to interaction of host genetic and bacterial virulence factors determine the final outcomes of *H. pylori* infection. *H. pylori* contain a set of sophisticated virulence factors and mechanisms for their delivery and enable it to survive, multiply and eventually cause disease. The repertoire consists of surface-exposed adhesions or enzymes and secreted or translocated effectors molecules. A hallmark is its *cag* type IV secretion system which as been characterized to be related to pathogenesis (Blaser *et al.*, 2001). Both host factors and the characteristics of infecting strains have been postulated to contribute to the variable outcomes and have been the focus of intensive investigations (Blaser *et al.*, 2002). Among the strain-specific virulence factors for *H. pylori*-related gastroduodenal disease, the cytotoxin-associated gene pathogenicity island, outer membrane inflammatory protein, duodenal ulcer-promoting gene and blood group antigen-binding adhesion are the only factors that to date qualify as virulence factors (Lu *et al.*, 2005). On the other hand, single nucleotide polymorphisms of cytokine genes in IL-1, TNF-α and IL-10 etc have been documented to be host-related factors (El-Omar, 2006). Collectively, the elucidation of host–pathogen interaction and discovery of new highly sensitive and specific discovery of biomarkers for early prognostic stratification holds the key to future prevention and treatment of *H. pylori*-related gastroduodenal diseases.

The evolving proteomics technology sheds new light on investigation of *H. pylori*-related diseases

Technical advances coupled with complete deciphering of the human and *H. pylori* genome sequences herald a new era in the identification of high-risk population and elucidation of pathogenesis (Tomb *et al.*, 1997; Alm *et al.*, 1999). Among important contributions relating to *H. pylori*, global gene expression has been shown in *H. pylori*, whether interacting with host cells *in vitro*, in animal models or in infected tissues (Mills *et al.*, 2001; Guillemin *et al.*, 2002). Additional high throughput methods to investigate the bidirectional relationship include proteomics. These studies could delineate the cross-talk between the host and *H. pylori* and enhance the understanding of disease pathogenesis.

Proteomics is the study of the proteome (the collection of all the proteins expressed from the genome in all isoforms, polymorphisms and post-translational modifications). The core component of proteomics technology is two-dimensional gel electrophoresis (2D-GE) and mass spectrometry (MS). 2D-GE separation of complex protein mixtures and the subsequent analysis of isolated proteins spots by MS allow fast and accurate identification of proteins. The comparison of spots from different sample separated on customized 2D gels allows the detection of punctual differences in their mobility and facilitates tracing back differences in protein expression, presence of isoforms, splice variants and posttranslational modifications by mass spectrometry. Application of such technologies in the investigation often creates new possibilities in the elucidation of pathogenesis and the discovery of novel treatment target and early disease markers. In the field of *H. pylori* related gastroduodenal disease, this review focuses on proteomics-based characterization of intrinsic and extrinsic perturbations and strain differences of *H. pylori*, epithelial cell responses to *H. pylori*, identification of diagnostic/prognostic biomarkers, and uncovering of immunogenic proteins for vaccine targets.

Proteomics-based characterization of intrinsic and extrinsic perturbations and strain differences of *H. pylori*

Stomach is not a friendly niche for *H. pylori* to survive. It was presumed that *H. pylori* must have evolved a number of strategies to circumvent the hostile environment. Recent advances in proteomics technologies permit the evaluation of systematic changes in protein expression when *H. pylori* encountered intrinsic or extrinsic perturbations, such as pH and oxidative stress. Slonczewski and colleagues first identified pH-dependent stress responses in *H. pylori* using 2D-GE (Slonczewski *et al.*, 2002). Another study compared protein differences in *H. pylori* grown either on solid agar-based media or broth culture media and found alkaline pH in broth media might induce UreB and GroEL expression to 140-fold (Uwins *et al.*, 2006). The oxidative stress has been assumed to be one of the key determinants leading to *H. pylori*-related pathogenesis. Such oxidative stress may also exert bidirectional effects on host and pathogens. The *H. pylori* induced oxidative stress can cause lesions in host cells, whereas *H. pylori* will also face the attack of oxygen-related free radicals by the host. To address the influence of oxidative stress and its underlying mechanisms, we have compared protein expression profiles of *H. pylori* incubated under normal microaerophilic (5% O_2) and aerobic stress (20% O_2) condition (Chuang *et al.*, 2005). The results revealed that more than ten proteins were differentially expressed. Most notably, the protein expression level of UreE (an essential metallochaperone for urease activity) and alkyl hydroperoxide reductase (AhpC) with antioxidant potential were greatly reduced under oxygen tension. We have further characterized AhpC, which can display both antioxidative activities and molecular chaperone properties under different conditions (Chuang *et al.*, 2006). Another study using mutant strains with deficiencies in oxidative stress combating proteins has demonstrated overexpression of iron-sequestering protein NapA, catalase or NADPH quinine reductase MdaB could protect bacterial DNA of mutant strains from oxidative stress damage (Olczak *et al.*, 2005). These studies indicate that *H. pylori* possesses a number of defensive mechanisms to protect against oxidative stress (Chuang *et al.*, 2005, 2006; Olczak *et al.*, 2005).

Several studies have utilized proteomics to investigate the role of mutant proteins of *H. pylori* (Lee *et al.* 2004; Carlsohn *et al.*, 2006; Terry *et al.*, 2006). Lee *et al.* showed that the ferric uptake regulator (Fur) protein of *H. pylori* functions as a classical transcriptional repressor (Lee *et al.* 2004). Carlsohn and colleagues, by combining proteomic analysis and *in vivo* studies, suggested that a surface-localized protein HpaA is essential for the colonization of *H. pylori* in mice (Carlsohn *et al.*, 2006a). Terry *et al.* used proteomic mapping of a suppressor of non-chemotactic cheW mutants and showed that *H. pylori* contains a new chemotaxis protein (Terry *et al.*, 2006).

The protein composition in different fraction of *H. pylori*, i.e. subproteome, may be investigated by proteomics which can facilitate identification of bacterial proteins important in pathogen-host interactions. Bae and colleagues described strategies for the enhancement of basic proteins in proteome projects of *H. pylori* (Bae *et al.*, 2003). Baik *et al.* performed a proteomic analysis of outer membrane proteins (OMPs) of *H. pylori* strain 26695. A sarcosine-insoluble outer membrane fraction was resolved by 2DGE with immobilized pH gradient strips. 16 kinds of OMP, including nine immunoreactive proteins by immunoblot analysis, were characterized (Baik *et al.*, 2004). Using a proteomic approach, Backert and colleagues investigated subproteomes of soluble and structure-bound *H. pylori* proteins. They establish a dynamic 2-DE-PAGE-database with multiple subproteomes of *H. pylori*. Four *cag* PAI protein, numerous OMPs, VacA, other potential virulence factors, and few ribosomal proteins were detected in the structure-bound fraction. In contrast, catalase, gamma-glutamyl-transpeptidase, and NapA were found in the soluble fraction (Backert *et al.*, 2005a). Another study performed by Pyndiah *et al.* investigated cytoplasmic and membrane complex of the *H. pylori* strain J99. Beside identification of complexes involved in *H. pylori* physiology, this study also described protein interaction involving known pathogenic factors such as urease with GroEL, or cagA with DNA gyrase Gyr A (Pyndiah *et al.*,

2007). Recently, the protein-protein interaction map of *H. pylori* was established by Rain and colleagues (Rain *et al.*, 2001). In addition, proteomic profiles of *H. pylori* can be used to type clinical isolates and distinguish different strains. Carlsohn and colleagues used a mass spectrometry-based method to characterize OMP profile of clinical isolates either with duodenal ulcer, gastric cancer or asymptomatic gastritis. Over 60 unique membrane or membrane-associated proteins, including 30 of the 33 theoretically predicted OMPs, were identified from the strains. Among them, Omp26 was expressed by all disease-related strains but was present in one out of the five strains from asymptomatic carriers (Carlsohn *et al.*, 2006b). Also by a similar approach, researchers can separate strains causing iron deficiency anaemia (IDA) from non-IDA *H. pylori* (Park *et al.*, 2006). Such typing method by proteomic profiles correlated with the results of genome scanning (e.g. *cagA*, *vacA*, *iceA* and *babA* status) (Govorun *et al.*, 2004).

Proteomics analysis of epithelial cell responses to *H. pylori* infection

H. pylori can induce expression of proinflammatory genes and changes in the actin cytoskeleton, resulting in phenotypic changes in gastric epithelial cells. By profiling the alterations in the protein pool of gastric epithelial cells, either using *in vitro* co-culture system, animal models or human gastric specimens, it is possible to have a greater understanding of the cellular response and unravel the greater complexity in previously unexplored signalling network. Such studies may provide valuable information for the elucidation of the pathogenesis and may be useful as prognostic indices of *H. pylori*-related gastric disorders (Lim *et al.*, 2004).

Lim *et al.* treated AGS cells with *H. pylori* at a bacterium–cell ratio of 300:1 for 12 h. Four differentially expressed proteins, whose expression levels were increased by more than two-fold in *H. pylori*-infected cells, were identified. These included 14-3-3 protein α/β, cullin homologue 3, α-enolase, and ezrin, which were known to be related to cell proliferation, cell adhesion, and carcinogenesis (Lim *et al.*, 2004). We used the similar *in vitro* model of *H. pylori*-infected gastric epithelial AGS cells with a multiplicity of infection of 100 and evaluated the effectiveness of *H. pylori* infection by functional analyses (Chan *et al.*, 2006). Changes in protein patterns were identified using a proteomics approach consisting of 2D fluorescence difference gel electrophoresis (2D-DIGE) and mass spectrometry. Our results revealed 28 proteins were found to be altered. The expression profiles of eight of these proteins, lamin r-1 chain precursor, valosin-containing protein, heat shock 70 kDa protein, mitochondrial matrix protein P1, FK506-binding protein 4, T-complex protein 1, enolase α, and 14-3-3 β were further examined in cancerous and paired surrounding normal tissues by immunoblot assay and immunohistochemical staining to identify molecular targets that may be involved in the pathogenesis of *H. pylori*-induced gastric disease. On the basis of our results, valosin-containing protein, mitochondrial matrix protein P1, T-complex protein 1, enolase α, and 14-3-3β may play a crucial role in *H. pylori*-induced gastric carcinogenesis by mediating antiapoptotic and proliferative responses (Chan *et al.*, 2006). Also by infected culture AGS models, Backert and colleagues combined two global approaches, i.e. cDNA microarray and proteomics, to investigate the host cell responses to wild type *H. pylori* strain or isogenic T4SS (type IV secretion system) mutant. Their results suggested that host proteins such as serine/threonine and tyrosine kinases, transcription factors, cell cycle related components and actin cytoskeletal signalling molecules as potential targets of individual *H. pylori* virulence determinants (Backert *et al.*, 2005b).

There were several studies reporting proteomic analysis of the human gastric mucosa from *H. pylori* infected patients (Lee *et al.*, 2003; Baek *et al.*, 2004; Nardone *et al.*, 2007). Compared with non-infected tissues, Baek *et al.* found four increased proteins (78 kDa glucose-regulated protein precursor, endoplasmin precursor, aldehyde dehydrogenase 2 and L-lactate dehydrogenase B chain) and four decreased proteins (intracellular chloride channel protein 1, glutathione S- transferase, heat shock protein 60 and cytokeratin 8) in infected tissues (Baek *et al.*, 2004). Another study by Nardone and colleagues reported increased expression of Rho guanosine diphosphatase dissociation inhibitor α and heat shock protein 27, and decreased expression of glutathione transferase and gastrokine 1 in gastric mucosa of *H. pylori*-positive dyspeptic patients (Nardone *et al.*, 2007). Coupled with leaser capture microdissection, Lee *et al.* use 2D-DIGE to analysed protein constituents of spasmolytic polypeptide expression metaplasia (SPEM), a gastric mucosa metaplasia that develops in the atrophic fundus mucosa infected with *H. pylori*. SPEM has been identified in the atrophic mucosa surrounding a high percentage of gastric adenocarcinoma and may represent a precursor lineage of malignancy. They found 28 differentially expressed proteins between SPEM and surface cells. These proteins would have the potential for development of new immunohistochemical antibodies to further study this important metaplasia (Lee *et al.*, 2003). Proteomics technology was also applied in clarification of drug effect on *H. pylori* infection or substrate related to *H. pylori*-induced signalling changes (Oh *et al.*, 2005; McCaig *et al.*, 2006). Oh *et al.* in an animal model, showed α-tocopherol pretreatment significantly prevented the gastric mucosa damage, caused by water immersion restraint stress in the presence of *H. pylori* infection through induction of HSP27 expression (Oh *et al.*, 2005). *H. pylori* infection may induce matrix metalloproteinase (MMP)-7 productions, which act on growth factor and matrix proteins. McCaig and colleagues used proteomic methods to identify the substrate of MMP-7. Their results showed insulin-like growth factor binding protein (IGFBP)-5 is a substrate of MMP-7 (McCaig *et al.*, 2006). Because insulin-like growth factor acts on both stromal and epithelial cells, their findings suggest that MMP-7 expression contributes to redefining the niche occupied by dividing cells and leading to hyperproliferation in *H. pylori* infection (McCaig *et al.*, 2006).

Proteomics-based identification of diagnostic/prognostic biomarkers

Traditional biomarkers for gastroduodenal diseases include measurement of serum pepsinogen and gastrin level for gastric inflammation and CEA level for gastric cancer. However, the risk profiles based on assays of these tests have limited sensitivity and specificity in identification of susceptibility to complications and in prognosis tailored to individuals. Proteomics is a potential tool for the discovery and application of novel biomarkers in increased precision in diagnosis and prognosis of *H. pylori*-related gastroduodenal diseases. Since individual biomarkers will have limited sensitivity and specificity, proteomics may further provide the advantage to seek for a panel of complementary biomarkers that will have more robust operational characteristics (Arab *et al.*, 2006).

There were several investigations that used 2D-GE to identify biomarkers for gastric caner. In an effort to identify tumour-associated proteins from plasma-bearing mice that may be used as diagnostic biomarkers, Juan *et al.* developed a strategy that combined a tumour xenotransplantation model in nude mice with comparative proteomic technology. They found several mouse acute phase proteins were overexpressed. Among them, serum amyloid A was observed to be upregulated and confirmed on patients with gastric cancer (Juan *et al.*, 2004). This group of researchers also used quantitative proteomic techniques together with microarray chips to establish comprehensive proteome and transcriptome profiles of the metastatic gastric cancer cells and non-invasive gastric cancer cells. Notably, a large subset of differentially expressed proteins was associated with tumour metastasis, including proteins functioning in cell–cell and cell–extracellular matrix adhesion, cell motility, proliferation, and tumour immunity (Juan *et al.*, 2006). In contrast, three studies utilized proteomic approach to globally analyse the protein profiles of paired surgical specimens of primary gastric cancer and non-tumour mucosa aiming at identifying specific disease-associated proteins as potential clinical biomarkers and for carcinogenetic study (He *et al.*, 2004: Jang *et al.*, 2004; Nishigaki *et al.*, 2005). With this approach, He *et al.* noticed that a 18 kDa antrum mucosa protein was dramatically under-expressed in cancer tissues, implicating a possible special pathological role for this protein in gastric carcinogenesis (He *et al.*, 2004). Jang *et al.* noted that eight proteins (14-3-3 zeta, calcyclin, keratin, apoprotein A-1 precursor, proteasome activator complex subunit, nucleoside diphosphate kinase, nicotinamide

N-methyltransferase, and pyridoxal kinase) were overexpressed, whereas five proteins (CA11, prohibin, peroxiredoxin4, serum amyloid P component, and NADH-ubiquinone oxireductase 23 kDa subunit) were downregulated (Jang *et al.*, 2004). Another study by Nishigaki *et al.* identified nine proteins with increased expression and 13 proteins with decreased expression in gastric cancer. The two most notable groups included proteins involved in mitotic checkpoint (MAD1L1 and EB1) and mitochondrial functions (CLPP, COX5A, and ECH1). Immunohistochemical analyses confirmed that the level of expression of MAD1L1, HSP27, and CYR61 were altered in gastric cancer (Nishigaki *et al.*, 2005). Kim *et al.* used 2D-GE and lectin blot followed by electrospray ionization-MS to analyse human gastric cancer tissues. This glycomic approach permitted identification of glycol- or putative glycosylated proteins which may be associated with tumorigenesis (Kim *et al.*, 2004). Proteomics technology may also be used for identification of metastatic biomarkers and mechanisms of drug resistance to gastric cancer. Takikawa and colleagues used 2D-DIGE to identify specific proteins differentially expressed between a highly metastatic gastric cancer cell line and its parental cell line. The 13 differentially expressed proteins were mainly involved in protein synthesis (transfer RNA synthesis), metabolism (flavoprotein subunit, pyruvate kinase, adenylate kinase), receptor and signal transduction (annexin I and A2), the cytoskeleton (keratin 5, cytokeratin 8) and cell cycling (ts11). These protein changes might serve as novel markers for the prediction of postoperative recurrence (Takikawa *et al.*, 2006). Yang *et al.* investigated the mechanisms of mutidrug resistance (MDR) of vincristine-resistant human gastric cancer cell line and found 24 differential proteins might be responsible for mechanisms of MDR in human gastric cancer (Yang *et al.*, 2006). Recently, Ellmark and colleagues used antibody microarray based technology to identify protein expression signature associated with *H. pylori* infection and gastric cancer. They constructed large-scale antibody microarrays, composed of 127 different antibodies against mostly immunoregulatory antigens and applied this recombinant microarray to malignant and normal stomach tissue samples from *H. pylori*-positive and -negative patients. Their results demonstrated that distinct tumour- as well as infection-associated protein expression signature could be identified from these complex tissue proteomes, as well as biomarkers such as IL-9, IL-11, and MCP-4, previously not found in these diseases (Ellmark *et al.*, 2006).

As gastric cancer and duodenal ulcer are considered clinically divergent, we compared two-dimensional immunoblots of an acid-glycine extract of *H. pylori* probed with serum samples from 15 patients with gastric cancer and 15 with duodenal ulcer to find gastric cancer-related antigens, which were subsequently identified by mass spectrometry. Many protein spots were recognized by more than one serum, and 24 of these were better recognized by gastric cancer sera. The proteins showing higher frequency of recognition in gastric cancer group are threonine synthase, rod shape-determining protein, S-adenosylmethionine synthetase, peptide chain release factor 1, DNA-directed RNA polymerase alpha subunit, co-chaperonin GroES (monomeric and dimeric forms), response regulator OmpR, and membrane fusion protein. Of these proteins, GroES was identified as a dominant gastric cancer-related antigen with a much higher seropositivity of gastric cancer samples (64.2%, $n = 95$) compared with 30.9% for gastritis ($n = 94$) and 35.5% for duodenal ulcer ($n = 124$). GroES seropositivity was more commonly associated with antral gastric cancer than with non-antral ones (odds ratio = 2.7; 95% confidence interval, 1.1–6.7). In peripheral blood mononuclear cells, GroES stimulated production of IL-8, IL-6, granulocyte macrophage colony-stimulating factor, IL-1$\beta\alpha$, tumour necrosis factor-α, cyclooxygenase-2, and prostaglandin E_2. Moreover when incubated with gastric epithelial cells, GroES induced expression of IL-8, cell proliferation, and upregulation of c-jun, c-fos, and cyclin D1 but caused downregulation of p27 (Kip1). Our results indicated that GroES of *H. pylori* is a novel gastric cancer-associated virulence factor and may contribute to gastric carcinogenesis via induction of inflammation and promotion of cell proliferation (Lin *et al.*, 2006). In addition, we have found 11 proteins were strongly recognized by serum IgG from duodenal ulcer (DU) patients. Among them, translation elongation factor EF-G (FusA), catalase (KatA), and urease alpha subunit (UreA) were identified as DU-related antigens, showing a higher seropositivity in DU samples ($n = 124$) than in gastric cancer samples ($n = 95$) (FusA: 70.2% vs. 45.3%, KatA: 50.8% vs. 41.1%, UreA: 44.45 vs.27.4%). A protein array containing the three DU-related antigens was further developed to provide a more rapid and more sensitive method for detecting serum antibody patterns of DU patients (Lin *et al.*, 2007).

Mini *et al.* also applied a comparative proteomic and immunoproteomic analysis to identify the antigen patterns of three different *H. pylori* strains which were probed against single sera from *H. pylori*-positive patients affected by gastric cancer or DU. The different antigenic repertories introduced the importance of the strain to be used for immunoblotting as a diagnostic test (Mini *et al.*, 2006).

Surface enhanced laser desorption/ionization-time of flight- mass spectrometry (SELDI-TOF-MS) is a recently developed innovative approach for high-throughput proteomic studies. In brief, this technology consists of ProteinChip array, the reader and the software. Proteins of interest from biological samples bind selectively to various chip chemistries (hydrophobic, ionic, cationic, and metal-binding). After washing away nonspecifically bound proteins, an energy absorbing matrix in added to the retained sample. Then laser energy is applied and the sample produces spectra of mass: charge ratio (m/z), which is analysed by TOF-MS. It thus provides a proteomic platform in which patterns of proteomic signatures from high dimensional MS data are used as a diagnostic classifier. This approach offers the advantages of speed, simplicity and sensitivity. Hynes *et al.* first applied this technology to analyse surface extracts from *H. pylori* and established that different patterns could be generated in response to bile stress among various pathogenic *Helicobacter* spp. which may give insights into the ability of these strains to colonize different niches (Hynes *et al.*, 2003). Using different chip surfaces, Das *et al.* have shown differential protein expression profile of gastric epithelial cells following *H. pylori* infection (Das *et al.* 2005).The protein pattern identified by SELD-TOF-MS can be used to diagnose gastric cancer. Ebert and colleagues first applied this technology coupled with a pattern-matching algorithm and serum samples to establish the diagnosis of gastric cancer. Their results revealed gastric cancer and controls can be correctly classified and concluded that ProteiChip technology in conjunction with bioinformatics allowed the highly sensitive and specific recognition of gastric cancer patients (Ebert *et al.*, 2004). Based on this finding, they further identified thrombin light chins as the single best mass for differentiation of gastric cancer patients from individuals with dyspepsia with a sensitivity of 89.9 % and specificity of 90% (Ebert *et al.*, 2005). The same group researchers recently used magnet bead-assisted MALDI-TOF MS to search serum markers of gastric cancer and found elevation of fibrinopeptide A was observed in gastric cancer patients (Ebert *et al.*, 2006). Several groups of researchers also confirmed that a uniform serum proteomic fingerprint identified by SELDI-TOF-MS could have the potential as the noninvasive diagnosis of gastric cancer (Liang *et al.*, 2006; Poon *et al.*, 2006; Su *et al.*, 2007).

Proteomics-based uncovering immunogenic proteins for vaccine targets

Until recently, the development of vaccines for use in humans relied on the response to attenuated or whole-cell preparations, or empirically selected antigens (Nilsson, 2002). Effective vaccines could provide solutions to control *H. pylori* infection; however, vaccine development has been hampered by the slow identification of protective antigens. Proteomics provides global information about relevant antigen properties and thus might be ideally suited for identifying promising antigen subsets (Bumann *et al.*, 2004). Using a detergent-solubized *H. pylori* preparation and 2D-PAGE, Nilsson *et al.* were able to identify 40 proteins which was expected to speed up the identification of expressed virulence proteins and vaccine targets (Nilsson *et al.*, 2000). Chakravarti and colleagues described two different approaches: the first involved rapid identification of a series of monoclonal antibody reactive proteins from N-terminal sequence tags. The other approach involved identification of proteins in outer membrane preparations by 2D-GE followed by

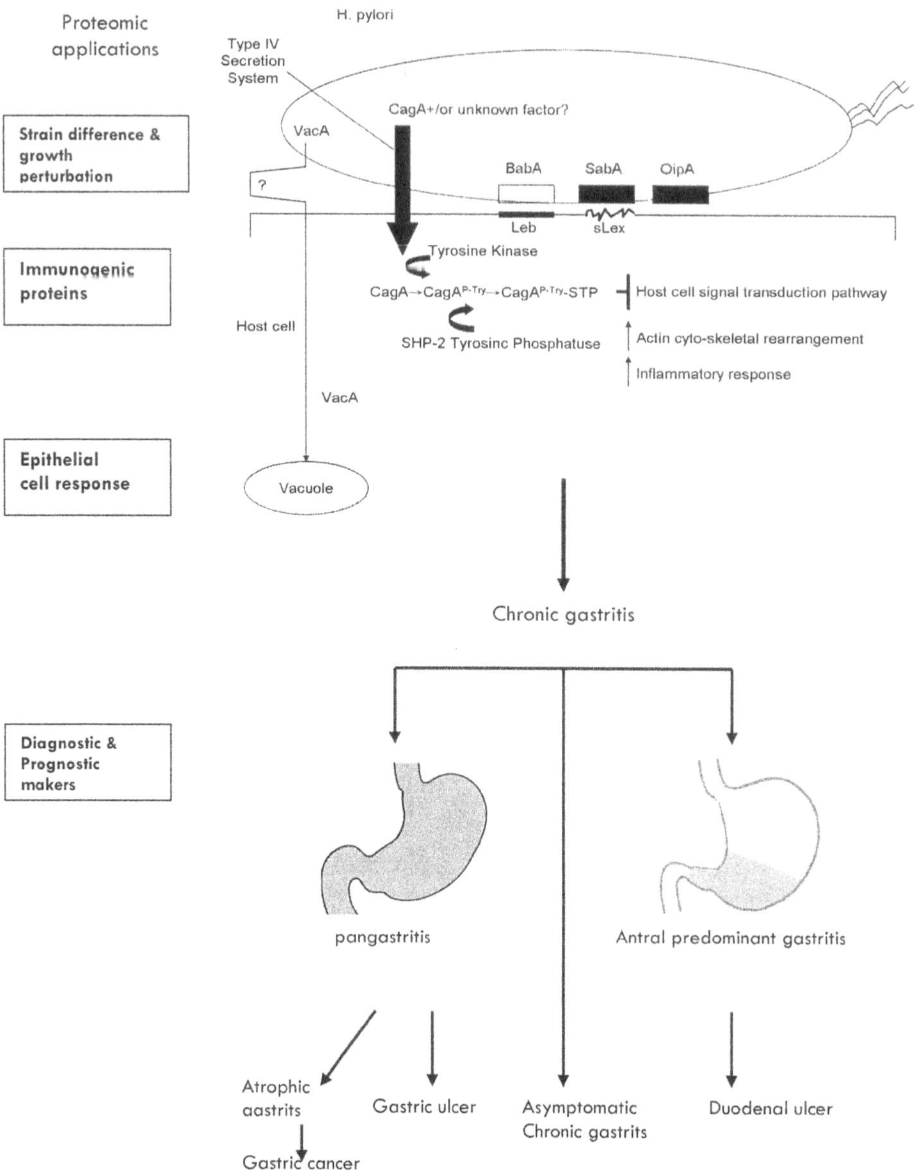

Figure 12.1 Overview of *H. pylori*-host interaction and the various outcomes after infection. Potential proteomic applications include (1) proteomics-based characterization of intrinsic and extrinsic perturbations and strain differences of *H. pylori*, (2) elucidation of epithelial cell response to *H. pylori* infection, (3) identification of diagnostic/prognostic biomarkers and (4) uncovering of immunogenic proteins for vaccine targets. Adapted from Mol. Cell Biochem. 2003; 253: 209–215.

trypsin digestion and peptide mass map analysis (Chakravarti *et al.*, 2000). In a recent review, Bumann *et al.* described how a combined *Helicobacter* subproteome analysis resulted in the rapid identification of novel, highly protective antigen and illustrated the great potential of pathogen proteomics for vaccine development (Bumann *et al.*, 2004). A web-accessible proteome database for microbial research, including *H. pylori* proteomics data, was also stored in a proteomics database (http://www.mplib-berlin.mpg.de/2D-PAGE/) (Pleissner *et al.*, 2004).

Conclusions

Elucidation of the bidirectional relationship between the host and *H. pylori* is essential to clarify pathogenesis and development of new prevention and treatment strategies. Proteomics technology

State-of-the-Art and Future Clinical Potentials

is a tremendously powerful tool for simultaneously monitor proteins and protein variation on a large scale in various biological samples. This new technology has provided unprecedented opportunity to comprehensively survey a cell's translational landscape and may allow in-depth analyses of host and pathogen interaction, either separately or interactively. Using this high-throughput platform and taking advantage of complete sequences for both the *H. pylori* and human genome in the database, some promising results and important information have been reported. Fig. 12.1 summarizes the current status of proteomic applications in investigations of the *H. pylori*–host interaction and the various outcomes after infection. Future combination of evolving new proteomics technologies with clinical phenotypes and genotypes information will enhance the understanding of disease pathogenesis, lead to a more precise prediction of variable outcomes, and facilitate the development of effective biomarkers for diagnosis, treatment and prevention of *H. pylori* infection.

References

Alm, R.A., Ling, L.S., Moir, D.T., King, B.L., Brown, E.D., Doig, P.C., Smith, D.R., Noonan, B., Guild, B.C., and de Jong, B.L., *et al.* (1999). Genomic-sequence comparison of two unrelated isolates of human pathogen *Helicobacter pylori*. Nature 397, 176–180.

Arab, S., Gramolini, A.O., Ping, P., Kislinger, T., Stanley, B., van Eyk, J., Ouzounian, M., Maclennan, D.H., Emil, A., and Liu, P.P. (2006). Cardiovascular proteomics: tools to develop novel biomarkers and potential applications. J. Am. Coll. Cardiol. 48, 1733–1741.

Backert, S., Kwok, T., Schmid, M., Selbach, M., Moese, S., Peek, R.M. Jr., Konig, W., and Meyer, T.F. (2005a). Subproteomes of soluble and structure-bound *Helicobacter pylori* proteins analyzed by two-dimensional gel electrophoresis and mass spectrometry. Proteomics 5, 1331–1345.

Backert, S., Gressmann, H., Kwok, T., Zimny-Armdt, U., Konig, W., Jungblut, P.R., and Meyer, T.F. (2005b). Gene expression and protein profiling of AGS gastric epithelial cells upon infection with *Helicobacter pylori*. Proteomics 5, 3902–3918.

Bae, S.H., Harris, A.G., Hains, P.G., Chen, H., Garfin, D.E., Hazell, S.C., Paik, Y.K., Walsh, B.J., and Cordwell, S.J. (2003). Strategies for the enrichment and identification of basic proteins in proteome projects. Proteomics 3, 569–579.

Baek, H.Y., Lim, J.W., Kim, H., Kim, J.M., Kim, J.S., Jung, H.C., and Kim, K.H. (2004). Oxidative-stress-related proteome changes in *Helicobacter pylori*-infected human gastric mucosa. Biochem. J. 15, 291–299.

Baik, S.C., Kim, K.M., Song, S.M., Kim, D.S., Jun, J.S., Lee, S.G., Song, J.Y., Park, J.U., Kang, H.C., Lee, W.K., Cho, M.J., Youn, H.S., Ko, G.H., and Rhee, K.H. (2004). Proteomic analysis of the sarcosine-insoluable outer membrane fraction of *Helicobacter pylori* strain 26695. J. Bacteriol. 186, 949–955.

Blaser, M.J., and Berg, D.E. (2001). *Helicobacter pylori* genetic diversity and risk of human disease. J. Clin. Invest. 107, 767–773.

Blaser, M.J. (2002). Polymorphic bacteria persisting in polymorphic hosts: assessing *Helicobacter pylori*-related risk for gastric cancer. J. Natl. Cancer Inst. 94, 1662–1663.

Bodger, K., and Crabtree, J.E. (1998). *Helicobacter pylori* and gastric inflammation. Br. Med. Bull. 54, 139–150.

Bumann, D., and Jungbult, P.R. (2004). *Helicobacter pylori* vaccine development based on combined subproteome analysis. Proteomics 4, 2843–2848.

Carlsohn, E., Nystrom, J., Bolin, J., Nilsson, C.L., and Svennerholm, A.M. (2006a). HpaA is essential for *Helicobacter pylori* colonization in mice. Infect. Immun. 74, 920–926.

Carsohn, E., Nystrom, J., Karlsson, H., Svennerholm, A.M., and Nilsson, C.L. (2006b). Characterization of the outer membrane protein profile from disease-related *Helicobacter pylori* isolates by subcellular fractionation and nano-LC FT-ICR MS analysis. J. Proteome Res. 5, 3197–3204.

Chakravarti, D. N., Fiske, M.J., Fletcher, L.D., and Zagursky, R.J. (2000). Application of genomics and proteomics for identification of bacterial gene products as potential vaccine candidates. Vaccine 19, 601–612.

Chan, C.H., Ko, C.C., Chang, J.G., Chen. S.F., Wu, M.S., Lin, J.T., and Chow, L.P. (2006). Subcellular and functional proteomic analysis of the cellular responses induced by *Helicobacter pylori*. Mol. Cell. Proteomics 5, 702–713.

Chen, Y.R., Juan, H.F., Huang, H.C., Huang, H.H., Lee, Y.J., Liao, M.Y., Tseng, C.W., Lin, L.L., Chen, J.Y., and Wang, M.J., *et al.* (2006). Quantitative proteomic and genomic profiling reveals metastasis-related protein expression patterns in gastric cancer cells. J. Proteome Res. 5, 2727–2742.

Chow, W.H., Blaser, M.J., Blot, W.J., Gammon, M.D., Vaughan, T.L., Risch, H.A., Perez-Perez, G.I., Schoenberg, J.B., Stanford, J.L., Rotterdam, H., West, A.B., and Fraumeni, J.F., Jr. (1998). An inverse relation between *cagA*+ strains of *Helicobacter pylori* infection and risk of esophageal and gastric adenocarcinoma. Cancer Res. 15, 588–590.

Chuang, M.H., Wu, M.S., Lin J.T., and Chiou, S.H. (2005). Proteomic analysis of proteome expressed by *Helicobacter pylori* under oxidative stress. Proteomics 5:3895–3901.

Chuang, M.H., Wu, M.S., Lo, W.L., Lin, J.T., Wong, C.H., and Chiou, S.H. (2006). The antioxidant protein alkylhydroperoxide reductase of *Helicobacter pylori* switches from a peroxide reductase to molecular chaperone function. Proc. Natl. Acad. Sci. USA. 103, 2552–2557.

Das, S., Sierra, J.C., Soman, K.V., Suarez, G., Mohammad, A.A., Dang, T.A., Luxon, B.A., and Reyes, V.E. (2005). Differential protein expression profiling of gastric epithelial cells following *Helicobacter pylori* infection using ProteinChip. J. Proteome Res. 4, 920–930.

Ebert, M.P., Meuer, J., Wiemer, J.C., Schulz, H. U., Reymond, M.A., Traugott, U., and Malfertheiner, P. (2004) Identification of gastric cancer patients by serum protein profiling. J. Proteom Res. 3, 1261–1266.

Ebert, M.P., Lamer, S., Meuer, J., Malfertheiner, P., Remond, M., Buschmann, T., Rocken, C., and Seibert, V. (2005). Identification of the thrombin light chain a as the single best mass for differentiation of gastric cancer patients from individuals with dyspepsia by proteome analysis. J. Proteome Res. 4, 586–590.

Ebert, M.P., Niemeyer, D., Deininger, S.O., Wex, T., Kinppig, C., Hoffmann, J., Sauer, J., Albrecht, W., Malferteiner, P., and Rocken, C. (2006). Identification and confirmation of increased fibrinopeptide a serum protein levels in gastric cancer sera by magnet bead assisted MALDI-TOF mass spectrometry. J. Proteome Res. 5, 2152–2158.

Ellmark, P., Ingvarsson, J., Carlsson, A., Lundin, B.S., Wingren, C., and Borrebaeck, C.A. (2006). Idenitification of protein expression signature of *Helicobacter pylori* infection and gastric adenocarcinoma using recombinant antibody microarray. Mol. Cell. Proteomics 5, 1638–1646.

El-Omar, E.M. (2006). Role of host genes in sporadic gastric cancer. Best Pract. Res. Clin. Gastroenterol. 20, 675–686.

Govorun, V.M., Lokhov, P.G., Moshkovskii, S.A., Momynaliev, K.T., Selesnyova, O.V., Kudryavtseva, L.V., Serebryakova, M.V., Tikhonova, O.V., Goufman, E.I., and Archakov, A.L. (2004). Comparative analysis of different typing methods for *Helicobacter pylori* clinical isolates. Biochemistry (Mosc) 69, 536–541.

Guillemin, K., Salama, N.R., Tompkins, L.S., and Falkow, S. (2002). Cag pathogenicity island-specific responses of gastric epithelial cells to *Helicobacter pylori* infection. Proc. Natl. Acad. Sci. USA. 99:15136–15141.

He, Q.Y., Cheung, Y.H., Leung, S.Y., Yuen, S.T., Chu, K.M., and Chiu, J.F. (2004). Diverse proteomic alterations in gastric adenocarcinoma. Proteomics 4, 3276–3287.

Hynes, S.O., McGuire, J., Falt, T, and Walstrom, T. (2003). The rapid detection of low molecular mass proteins differentially expressed under biological stress for four *Helicobacter* spp. using ProteinChip technology. Proteomics, 3:263–278.

Jang, J.S., Cho, H.Y., Lee, Y.J., Ha, W.S., and Kim, H.W. (2004). The differential proteome profile of stomach cancer: identification of the biomarkers candidates. Oncol. Res. 14, 491–499.

Juan, H.F., Chen, J.H., Hsu, W.T., Huang, S.C., Chen, S.T., Yi-Chung Lin, J., Chang, Y.W., Chiang, Y., Wen, L.L., and Chan, D.C., *et al.* (2004). Identification of tumor-associated plasma biomarkers using proteomic techniques: from mouse to human. Proteomics 4, 2766–2775.

Kim, Y.S., Hwang, S.Y., Oh, S., Sohn, H., Kang, H.Y., Lee, J.H., Cho, E.W., Kim, H.Y., Yoo, J.S., and Kim, N.S., *et al.* (2004). Identification of target proteins of N-acetylglucosaminyl-transferase V and fucosyltransferase 8 in human gastric tissue by glycomic approach. Proteomics 4, 3353–3358.

Lee, H.W., Choe, Y.H., Kim, D.K., Jung, S.Y., and Lee, N.G. (2004). Proteomic analysis of a ferric uptake regulator mutant of *Helicobacter pylori* regulation of *Helicobacter pylori* gene expression by ferric uptake regulator and iron. Proteomics 4, 2014–2027.

Lee, J.R., Baxter, T.M., Yamaguchi, H., Wang, T.C., Goldenring, J.R., and Anderson, M.G. (2003). Differential protein analysis of spasmolytic polypeptide expressing metaplasia using laser capture microdissection and two-dimensional difference gel electrophoresis. Appl. Immunohistochem. Mol. Morphol. 11, 188–193.

Liang, Y., Fang, M., Li, J., Liu, C.B., Rudd, JA., Kang, H.F., and Tew, D.T. (2006). Serum proteomic patterns for gastric lesions as revealed by SELDI mass spectrometry. Exp. Mol. Pathol. 2006, 81, 176–180.

Lim, J.W., Kim, H., Kim, J.M., Kim, J.S., Jung, H.C., and Kim, K.H. (2004). Cellular stress-related protein expression in *Helicobacter pylori*-infected gastric epithelial AGS cells. Int. J. Biochem. Cell Biol. 36, 1624–1634.

Lin, Y.F., Wu, M.S., Chang, C.C., Chang, Y.C., Lin, J.T., Chen, D.S., and Chow, L.P. (2006). Gro ES, a novel virulence factor of *Helicobacter pylori* in inflammation and gastric carcinogenesis. Mol. Cell. Proteomics 5, 1484–1496.

Lin, Y. F., Chen, C. Y., Tsai, M. H., Wu, M.S., Wang, Y.C., Chuang, E.Y., Lin, J.T., Yang, P.C., and Chow, L.P. (2007). Duodenal ulcer-related antigen of *Helicobacter pylori*: immunoproteome and protein microarray approaches. Mol. Cell. Proteomics 6, 1018–1026.

Lu, H., Yamaoka, Y., and Graham, D.Y. (2005). *Helicobacter pylori* virulence factors: facts and fantasies. Curr. Opin. Gastroenterol. 21, 653–659.

McCaig, C., Duval, C., Hemers, E., Steele, I., Pritchard, D.M., Przemeck, S., Dimaline, R., Ahmed, S., Bodger, K., and Kerrigan, D.D., *et al.* (2006). The role of matrix metalloproteinase-7 in refining the gastric microenvironment in response to *Helicobacter pylori*. Gastroenterology 130, 1754–1763.

Mills, J.C., Syder, A.J., Hong, C.V., Gurge, J.L., Raaii, F., and Gordon, J.I. (2001). A molecular profile of the mouse gastric parietal cell with and without exposure to *Helicobacter pylori*. Proc. Natl. Acad. Sci. USA. 98, 13687–13692.

Mini, R., Bernardini, G., Salzano, A.M., Renzone, G., Scaloni, A., Figura, N., and Santucci, A. (2006). Comparative proteomics and immunoproteomics of *Helicobacter pylori* related to different gastric pathologies. J. Chromatogr. B Analyt. Technol. Biomed. Life Sci. 833, 63–79.

Nardone, G., Rippa, E., Martin, G., Rocco, A., Siciliano, R.A., Fiengo, A., Cacace, G., and Malorni, A. (2007). Gastrokine 1 expression in patients with and without *Helicobacter pylori* infection. Dig. Liver Dis. 39, 122–129.

Nilsson, C.L., Larsson, T., Gustafsson, E., Karlsson, K.A., and Davidsson, P. (2000). Identification of protein vaccine candidates from *Helicobacter pylori* using a preparative two-dimensional electrophoretic procedures and mass spectrometry. Anal. Chem. 72, 2148–2153.

Nilsson, C.L. (2002). Bacterial proteomics and vaccine development. Am. J. Pharmacogenomics 2, 59–65.

Nishigaki, R., Osaki, M., Hiratsuka, M., Toda, T., Murakami, K., Jeang, K.T., Ito, H., Inoue, T., and Oshimura, M. (2005). Proteomic identification of differentially-expressed genes in human gastric carcinomas. Proteomics 5, 3205–3213.

Normark, S., Nilsson, C., Normark, B.H., and Hornef, M.W. (2003). Persistent infection with *Helicobacter pylori* and the development of gastric cancer. Adv. Cancer Res. 90, 63–89.

Oh, T.Y., Yeo, M., Han, S.U. Cho, Y.K., Kim, Y.B., Chung, M.H., Kim, Y.S., Cho, S.W., and Hahm, K.B. (2005). Synergism of *Helicobacter pylori* infection and stress on the augmentation of gastric mucosal damage and its prevention with alpha-tocopherol. Free Radic. Biol. Med. 38, 1447–1457.

Olczak, A.A., Wang, G., and Maier, R.J. (2005). Up-expression of NapA and other oxidative stress proteins in a compensatory response to loss of major *Helicobacter pylori* stress resistance factors. Free Radic. Res. 39, 1173–1182.

Park, S.A., Lee, H.W., Hong, M.H., Choi, Y.W., Choe, Y.H., Ahn, B.Y., Cho, Y.J., Kim, D.S., and Lee, N.G. (2006). Comparative proteomic analysis of *Helicobacter pylori* strains associated with iron deficiency anemia. Proteomics 6, 1319–1328.

Pleissner, K.P., Eifert, T., Buettner, S., Schmidt, F., Boehme, M., Meyer, T.F., Kaufmann, S.H, and Jungblut, P.R. (2004). Web-accessible proteome databases for microbial research. Proteomics 4, 1305–1313.

Poon, T.C., Sung, J.J., Chow, S.M., Ng, E.K., Yu, A.C., Chu, E.S., Hui, A.M., and Leung, W.K. (2006). Diagnosis of gastric cancer by serum protein fingerprinting. Gastroenterology 130, 1858–1864.

Pyndiah, S., Lasserre, J.P., Menard, A., Claverol, S., Prouzet-Mauleon, V., Megraud, F., Zerbib, F., and Bonneu, M. (2007). Two-dimensional blue native/SDS gel electrophoresis of multiprotein complexes from *Helicobacter pylori*. Mol. Cell. Proteomics 6, 193–206.

Rain, J.C., Selig, L, De Reuse, H., Battaglia, V., Reverdy, C., Simon, S., Lenzen, G., Petel, F., Wojcik, J., Schachter, V., Chemama, Y., Labigne, A., and Legrain, P. (2001). The protein-protein interaction map of *Helicobacter pylori*. Nature 409, 211–215.

Slonczewski, J.L., and Kirkpatrick, C. (2002). Proteomic analysis of pH-dependent stress responses in *Escherichia coli* and *Helicobacter pylori* using two-dimensional gel electrophoresis. Methods Enzymol. 2002, 358, 228–242.

Su, Y., Shen, J., Qian, H., Ma, H., Ma, L., Zhang, W., Meng, L., Li, Z., Wu, J., and Jin, G., *et al.* (2007). Diagnosis of gastric cancer using decision tree classification of mass spectra data. Cancer Sci. 98, 37–43.

Suerbaum, S., and Michetti, P. (2002). *Helicobacter pylori* infection. N. Engl. J. Med. 347, 1175–1186.

Takikawa, M., Akiyama, Y., Maruyama, K., Suzuki, A., Liu, F., Tai, S., Ohshita, C., Kawaguchi, Y., Bandow, E., and Yonemura, Y., *et al.* (2006). Proteomic analysis of a highly metastatic gastric cancer cell line using two-dimensional differential gel electrophoresis. Oncol. Rep. 16, 705–711.

Terry, K., Go, A.C., and Ottemann, K.M. (2006). Proteomic mapping of a suppressor of non-chemotactic chew mutants reveals that *Helicobacter pylori* contains a new chemotaxis protein. Mol. Microbiol. 61, 871–882.

Tomb, J. F., White, O., Kerlavage, A. R., Clayton, R.A., Sutton, G.G., Fleischmann, R.D., Ketchum, K.A., Klenk, H.P., Gill, S., and Dougherty, B.A., *et al.* (1997). The complete genome sequence of the gastric pathogen *Helicobacter pylori*. Nature 388, 539–547.

Uwins, C., Deitrich, C., Argo, E., Stewart, E., Davidson, I., and Cash, P. (2006). Growth-induced changes in the proteome of *Helicobacter pylori*. Electrophoresis 27, 1136–1146.

Wu, M.S., Lee, W.J., Wang, H.S., Huang, S.P., and Lin, J.T. (2005). A case–control study of association of *Helicobacter pylori* infection with morbid obesity in Taiwan. Arch. Intern. Med. 165, 1552–1555.

Yang, Y.X., Xiao, Z.Q., Chen, Z.C., Zhang, G.Y., Yi, H., Zhang, P.F., Li, J.L., and Zhu, G. (2006). Analysis of multidrug resistance in vincristine-resistant human gastric can cer cell line SGC7901/VCR. Proteomics 6, 2009–2021.

Index

D

E

F

G

H

T

U

V

W

X

Z

www.ingramcontent.com/pod-product-compliance
Lightning Source LLC
LaVergne TN
LVHW062121200926
848484LV00007B/1733
* 9 7 8 1 9 0 4 4 5 5 3 1 8 *